Anatoly Ivanovich Kiselev
Alexander Alexeevich Medvedev
Valery Alexandrovich Menshikov

# Astronautics

## Summary and Prospects

Springer-Verlag Wien GmbH

Anatoly Ivanovich Kiselev
Alexander Alexeevich Medvedev
Valery Alexandrovich Menshikov

Translated from Russia by *Valery Sherbakov, Nikolai Novichkov, Alexander Nechaev*

Originally published as *Cosmonautics on the Frontier of Millenniums. Summary and Prospects*
© Mashinostroenie-Polyot, Moscow, 2001

© 2003  Springer-Verlag Wien
Originally published by Springer-Verlag Wien in 2003

Printed on acid-free and chlorine-free bleached paper
SPIN: 10896075
CIP data applied for

With  245 partly coloured Figures

ISBN 978-3-211-83890-7      ISBN 978-3-7091-0648-8 (eBook)
DOI 10.1007/978-3-7091-0648-8

# Preface to the English language edition

The monograph by A.I.Kiselev, A.A.Medvedev, V.A.Menshikov, distinguished organizers and noted experts in the development and production of space rocketry, is a continuation of the constructive dialogue and cooperation between Russia and the West in such an important and promising field of human activity as astronautics. The authors have made an outstanding personal contribution to the international effort in space research, in the development of piloted flight programs and international space stations.

The book has been published twice in Russia and is popular not only with experts in astronautics but with all those interested in Russian cosmonautics and her role in the world's space research efforts.

At the turn of millennium, the authors have gathered together a huge amount of facts and drawn conclusions concerning the creation and operation of space equipment over more than half a century, and in doing, so indicate how astronautics will conceivably develop in the 21st century.

Today's achievements in applied astronautics are utilized in many fields of human activity. Satellite communication and satellite television are now common place as the telephone became fifty years ago. Information from satellites is not only used by heads of states but also for farming, forestry and fishing purposes and ecological monitoring. Road, maritime and air traffic control seems inconceivable nowadays without satellite systems.

Not only today's astronautics, but also its prospects must be clearly understood and thoroughly investigated. What will the future multi-function systems look like? All these issues are considered here. The section devoted to ecological problems makes the book especially significant and highly topical. It is probably the first time that Russia's space researchers have discussed the subject so openly. This, indisputably, gives joy and inspires hope. Inspiring still more hope, however, is the constructive approach displayed by the authors. Not confining themselves to a mere description of space activity's negative impact on the environment, they describe in concrete terms the measures being taken by some aerospace agencies to reduce this impact and propose ways and means for further work in this area.

The book contains a large amount of facts that can be understood without specialist knowledge and that are accompanied by a many photographs, charts and diagrams. Foreign readers will find in it answers to many of their questions, and the book will thus be interesting both to experts pursuing astronautics professionally and to lay readers at large.

NASA administrator
(1992–2001)
D.Goldin

## Preface to the Russian edition

The monograph by A.I.Kiselev, A.A. Medvedev and V.A.Menshikov, Astronautics: Summary and Prospects, aroused enthusiasm both among experts and the public at large. This is due to the felicitous choice of presentation that combines a simple description of complex space matters with scientific substantiation of the subject matter described. The wealth of color photos makes the book still more attractive, and it was nominated for an award at the 14th International Moscow Book Fair, being singled out as the "best publication of the book fair".

The book's popularity led to a second edition, substantially revised and enlarged. Since the first edition did not sufficiently cover the issues of space impact on ecology and the prospective development of space systems, the authors revised the entire volume, including in it the chapter "Space activity and ecology" and the section "Multi-function space systems".

Using the federal monitoring system, now in the phase of system engineering, as an example, the authors consider the basic principles of building and the likely principal features of the future multi-role space systems capable of fulfilling a wide range of missions.

Nowadays, space activity increasingly criticized by dedicated ecologists. This is not unfair. The development and utilization of space, like any other human activity, is invariably accompanied by such an untoward effect as environmental pollution. This criticism, however, is often prejudiced and opinionated. The authors attempt to look into the matter impartially and without bias. It is particularly important that the section "Space activity and ecology" does not merely state the facts involved in the negative impact of space activity on the environment, but also supplies a detailed description as to how this impact can be cushioned in the near and in the more distant future.

The graphic material has been updated and enlarged with due regard for readers' comments.

Finally, it should be noted that foreign readers, too, will find the book highly interesting.

General Director of the Russian Aviation and Space Agency
Yu.N.Koptev

# The contents

Introduction 1

**Part 1. Trends in global space exploration** 17
**Global astronautics, a general survey** 19
**1.1 Orbital facilities as they are** 26
    1.1.1 The main trends in development of orbital facilities 30
    1.1.2 Communication facilities 32
    1.1.3 Civil use facilities 37
    1.1.4 Commercial facilities of ERS 38
**1.2 Foreign launch facilities** 40
    1.2.1 Heavy lift launchers 47
        American heavy lift launchers 47
        Heavy lift launchers of European Space Agency 50
        The Japanese heavy lift launchers 52
    1.2.2 Medium lift launchers 53
        Medium lift launchers of the USA 53
        Medium lift launchers of European Space Agency 65
        Medium lift launchers of China 67
        Medium lift launchers of Japan 69
        Medium lift launchers of India 70
    1.2.3 Small lift launch vehicles 70
        Small lift launch vehicles of the USA 70
        Europe's small lift launchers 79
        Small lift launchers of other countries 83
    1.2.4 Reusable space transport systems 87
**1.3 Spaceports of the world** 99
    1.3.1 Spaceports and test sites of the USA 100
    1.3.2 Spaceports of other countries 104
**1.4 Spacecraft control systems abroad. The present state of the art and**
**the prospect for the future** 113
    1.4.1 General description of foreign control systems for spacecraft 113
        SC control systems in USA 114
        SC control systems of European Space Agency (ESA) 130
        SC control systems of France 132
        SC control systems of UK 133
        SC control systems of China 134
        SC control systems of Japan 135
        SC control systems of international consortiums and commercial firms 135
    1.4.2 Evolution and trends in development of SC control complexes abroad 137
**1.5 Trends in restructuring Aerospace Industry** 147
    1.5.1 Why restructure? 148
        Military-political causes 148
        Economic factors 150
        Resources related reasons 152
    1.5.2 Analysis of restructuring of aerospace companies 153
    1.5.3 The aims of restructuring 162
**1.6 Reliability as the basis of efficient functioning of space systemsin the future** 170
    1.6.1 Insuring reliability of advanced launch systems 170
    1.6.2 Supporting active operation life of durable spacecraft (DSC) 177
        Longevity level attained by foreign spacecraft 178
        The longevity reached on domestic-made spacecraft 179
    1.6.3 Insuring the reliability and safety of long life space stations 180
        The main principles of insuring the reliability and safety of long life orbital stations 182
        Design and engineering principles of insuring the reliability and safety of long live
        orbital stations (LLOS) 183

Principles of maintenance and repair of long live orbital stations (LLOS)         184
Safeguarding the required reliability and safety of the international space station  185
1.6.4 Insuring the quality and reliability of the Russian segment in international
space programs                                                                    188
1.6.5 Optimization of strategies used to deploy and replenish
multi-satellite space systems based on reliability and cost criteria              190
1.6.6 Insuring failure-free operation of advanced durable spacecraft and carrier
rockets                                                                           193
1.6.7 Development of space systems and components
quality control methods and their application for evaluation
and maintenance of product quality                                                196
1.6.8 Space insurance: summing up the past and looking to the future              197
**1.7 Space and national security**                                               208


**Part 2. The Main Trends in Development of Astronautics in Russia**              241
**2.1 A leap in improvement of orbital facilities**                               248
2.1.1 Space monitoring systems                                                    252
2.1.2 Navigation systems                                                          255
2.1.3 Space energy, production and medicine                                       255
2.1.4 Fundamental research                                                        256
2.1.5 Space manufacturing technologies and materials study                        258
Space biotechnologies and genetic engineering                                     259
Space biology and medicine                                                        260
Safeguarding the asteroid safety                                                  262
2.1.6 Creating prerequisites in science and technology                            262
2.1.7 Multi-functional constructions                                              267
2.1.8 Advanced space materials                                                    271
2.1.9 Multi-functional space systems                                              277
**2.2 Russia's launch vehicles**                                                  284
2.2.1 Carrier rockets                                                             290
2.2.2 Carrier rocket booster units                                                304
2.2.3 Carrier rockets built around the ICBM withdrawn from service                310
2.2.4 Advanced carrier rockets                                                    317
2.2.5 New generation launch vehicles based on multi-purpose rocket booster        322
2.2.6 Reusable space systems                                                      326
**2.3 Manned astronautics as the trend line**                                     333
**2.4 Power plants and propulsion units of space rocketry**                       347
2.4.1 Prospective trends in improvement of power plants and propulsion units      355
2.4.2 Remote power supply system                                                  360
2.4.3 Onboard solar power installations                                           364
2.4.4 Nuclear power plants and electrical propulsion units                        367
2.4.5 Prospective trends in improvement of chemical rocket engine                 375
2.4.6 Electric rocket engines                                                     382
2.4.7 Unconventional rocket engines and techniques of space travel                383
**2.5 The necessity of dual use technologies in space**                           386


**Part 3. The Ground-Based Infrastructure**                                       395
**3.1 Russias's spaceports, state and prospects**                                 397
3.1.1 The history of spaceports' construction                                     397
3.1.2 Russia's spaceports' ground infrastructure today                            405
3.1.3 The main trends in development of Russia's spaceports                        413
3.1.4 Comparative estimation of variants of launch complexes                      419

3.1.5 Problems to be addressed in building spaceports. Trends in
        development of spaceports                                                                      426
**3.2 The state and prospects of development of spacecraft ground-based
      facilities and control systems**                                                                 430
    3.2.1 General principles of building a ground-based control complex                                430
        Aim and application of a ground-based control complex                                          430
        Demands made of the main characteristics of GBCC                                               431
        The main systems of GBCC                                                                        432
        Command and instrumentation posts of GBCC                                                      434
        Information relay space system                                                                 437
        Structural peculiarities of far-out spacecraft GBCC                                            439
        Choice of design characteristics of long distance space communication radio lines             440
        Antenna systems for long range space communication                                            444
    3.2.2 The state and the main trends in development of GBCC                                         449
        The experience of creating and operating Russia's ground-based SC control complexes            449
        The existing structure of SC ground-based complexes and control systems                       452
        The state and prospective development of complexes and systems of the single
        Federal SC GBACC                                                                               457
        The principles of building multi-purpose information relay space system                        465
        The state and prospects of development of automation systems                                   470
**3.3 Operation systems. The state and prospective development**                                       476
**3.4 The prospects of development of software for space**                                             501

**Part 4. Space exploration and ecology**                                                              507
**4.1 Space contribution to ecology**                                                                  509
**4.2 The impact of space and missile technology on the environment**                                  517
**4.3 Pollution of the Earth's surface**                                                               526
    4.3.1 Launching sites and ecology                                                                  526
    4.3.2 Pollution from operating launching sites                                                     533
    4.3.3 Pollution by spillage of rocket fuel components                                              539
    4.3.4 The fall of rocket parts                                                                     543
    4.3.5 Impact of spacecraft and their fragments                                                     548
**4.4 Near-earth space and space exploration**                                                         553
    4.4.1 Launchers' engine emission                                                                   553
    4.4.2 Space junk                                                                                   556
    4.4.3 Radioactive contamination of NES                                                             561
**4.5 Environment-friendly space technology has no alternatives**                                      562
**4.6 International law aspects of space ecology**                                                      571

**Conclusion**                                                                                         575
**Bibliography**                                                                                       577
**Abbreviations**                                                                                      581
**Index**                                                                                              587
**About the authors**                                                                                  592

# Introduction

Astronautics as a science, and then as an applied industry, took shape in the mid-20th century. This, however, was preceded by the exciting history of the birth and development of the idea of space flights, first generated by fantasy and then supported by theoretical investigations and practical experiments. Originally, the human fantasy flew into space using fabulous means or the forces of nature (tornadoes and hurricanes). On the eve of the 20th century, fiction authors resorted in their descriptions to technical devices such as balloons and super-powerful cannons and, finally, to rockets as such. Generations of young romantics grew up enchanted by the works of J.Verne, H.Wells, A.Tolstoy and A.Kazantsev, whose subject matter was space travel.

The fantasts' fantasies spurred on the scientists, leading to K.E.Tsiolkovsky's words: "First came the thoughts, fantasies and fairytales, then the precise cal-culations."

Publication in the early 20th century of theoretical works by space pioneers K.E.Tsiolkovsky, F.A.Tsander, Yu.V.Kondratyuk, R.H.Goddard, H.Hanswint, R.Esnault-Pelterie, H.Oberth and V.Homan reined the flight of fantasy, but gave rise to new trends in science. Thus attempts were made to find out how astronautics could benefit mankind and what impact it would make on people.

It should be noted that the idea of combining human activity on Earth with that in space was first propounded by the founder of theoretical space research, K.E.Tsiolkovsky. His axiom, "The Earth is the cradle of reason, but one cannot live in the cradle for ever" was in fact an alternative – either Earth or space. Tsiolkovsky never believed that man's getaway into space would be caused by a depletion of life on Earth. On the contrary, he propounded the rational trans-formation of our planet's nature by virtue of human intellect. People, he asserted, would change "the Earth's surface, its oceans, atmosphere, plants and themselves. They will control the climate and reign supreme in the solar system as well as on Earth, which for eons to come will remain the human habitat."

In the USSR, the beginning of practical work on space programs is associated with the names of S.P.Korolev and M.K.Tikhonravov.

In early 1945, M.K.Tikhonravov organized a team of experts at the Research Institute of Propulsion Systems, setting before them the task of engineering a piloted high-altitude rocket vehicle (a cabin with two astronauts) intended for studying the upper atmosphere. The team included N.G.Chernyshev, P.I.Ivanov, V.N.Galkovsky, G.M.Moskalenko and others. It was decided to use as a basis a single-stage rocket powered by a liquid fuel motor and designed for vertical take-off to an altitude of up to 200 km.

This project (dubbed BP-190) envisioned fulfillment of the following tasks:

- the study of weightlessness during man's brief free flight in a hermetically closed cabin
- the study of the cabin's mass center movement and its movement around the mass center after separation from the carrier rocket
- obtaining data concerning the upper atmosphere; checking the operability of the systems (separation, descent, stabilization, landing, etc.) that make up the high-altitude cabin

K.E.Tsiolkovsky and M.K. Tikhonravov,
the designer of the first Soviet liquid propellant rocket GIRD-09

The BP-190 project was the first to offer the following solutions that found use in modern spacecraft (SC):

- parachute-aided descent system; soft-landing brake rocket motor; separation system using explosive bolts
- electric-contact boom for pre-ignition of the soft-landing motor; catapult-free sealed cabin with life support systems
- cabin stabilization system for use beyond dense atmospheric layers with the aid of small thrust nozzles

Overall, the BP-190 project was a package of new technological solutions and concepts well-proven now by the course of development of space rocketry in Russia and other countries. In 1946, M.K.Tikhonravov presented the ideas of the BP-190 project to I.V.Stalin. In 1947, Tikhonravov and his team started working on the rocket complex, and in the late 1940s – early 1950s demonstrated the possibility of reaching orbital velocity and launching an artificial Earth satellite (AES) using rocket facilities then available in the country.

In 1950 – 1953, the efforts of the Tikhonravov team focused on the study of problems pertaining to the creation of carrier rocket and artificial satellite com-ponents.

The team of organizers of GIRD headed by S.P.Korolev and F.A.Tsander, designer of a number of rocket pilot engines

In his report to the government in 1954 on the possibility of developing AES, S.P.Korolev wrote: "As per your direction here is presented for your consideration the memorandum by comrade M.K.Tikhonravov 'On the artificial Earth satellite'…". In the statement of work done over 1954, S.P.Korolev noted: "We would deem it reasonable to carry out the draft project of the ASE itself with regard for work now under way (worthy here of special attention is the work of M.K.Tikhonravov…)."

So work began on preparing the launch of PS-1, the first artificial Earth satel-lite. The first council of chief designers was formed, headed by S.P.Korolev, who subsequently directed space programs in the USSR that later became the world's leading country in space exploration. The OKB-1 – TsKBM – NPO Energia corporation established under S.P Korolev became in the early 1950s the center of space science and industry in the USSR.

R.H.Goddard with the first liquid propellant fuel rocket (1925)

Astronautics is unique in that many things originally predicted by fantasts and later confirmed by scientists came true within an incredibly short space of time. The mere forty-odd years that have passed since the launch of the Earth's first artificial satellite on 4 October 1957 have provided space history with a series of remarkable achievements, attained first by the Soviet Union and the US, and then by other space powers.

Orbiting the Earth are now thousands of satellites. Spacecraft have reached the surface of the moon, Venus and Mars. Research equipment has been sent to Jupiter, Mercury and Saturn to obtain knowledge about those distant planets of the solar system.

The launch into space on 12 April 1961 of the first man, Yu.A.Gagarin was one of the great triumphs of astronautics. It was followed by a formation flight, man's walk in space, the creation of orbital stations Salyut and Mir... The USSR was for a long period the world's leader in piloted programs.

Very telling is the trend of transition from launching solitary SC, primarily for military purposes, to building large-scale space systems for a variety of tasks (including those in economic and scientific research) and to integrating the space industries of various countries.

The team of organizers of GIRD headed by S.P.Korolev and F.A.Tsander, designer of a number of rocket pilot engines

Chief Designers' Council, comprising M.S.Ryazansky, N.A.Pilyugin, S.P.Korolev, V.P.Glushko, V.P.Barmin and V.I.Kuznetsov

What are the achievements of space science in the 20th century? Powerful liq-uid-propellant rocket motors have been built to give the required cosmic speeds to carrier rockets. The contribution made here by V.P.Glushko is hard to over-estimate. The creation of such motors became possible thanks to the realization of new scientific concepts and solutions that reduced to a minimum the losses sustained on turbo-pump units' drives. The development of carrier rockets and liquid-propellant rocket motors promoted the evolution of thermal, hydro- and gas dynamics, the theory of heat transmission and strength, instrumentation equipment, vacuum and plasma technology. Impetus was given to the further development of solid propellants and other types of rocket motors.

In the early 1950s, the Soviet scientists M.V.Keldysh, V.A.Kotelnikov, A.Yu.Ishlinsky, L.I.Sedov, B.V.Rauschenbach and others developed mathe-matical principles and navigation-and-ballistics support for space flights.

The problems that emerged during the preparation and realization of space flights also forced the intensive development of such general sciences as celestial and theoretical mechanics. The wide use of new mathematical methods and the creation of sophisticated computing machines made it possible to tackle even the most complicated tasks of designing spacecrafts' orbits and controlling them in flight. This gave rise to a new science, the dynamics of space flight.

Design bureaus headed by N.A.Pilyugin and V.I.Kuznetsov created unique systems used to control high-reliability space rockets.

At the same time, V.P.Glushko and A.M.Isayev created the world's most advanced school of practical rocket motor building. The theoretical foundation of the school had been laid as far back as the 1930s at the dawn of Russia's rocket building. Russia's leadership in this area has been maintained to this day.

Thanks to the hard and constructive efforts of design bureaus headed by V.M.Myasishchev, V.N.Chelomei and D.A.Polukhin, large-size, extra-strong capsules have been manufactured. This has served as a basis for the powerful intercontinental rockets, UR-200, UR-500, UR-700, for constructing at a later date piloted stations such as Salyut, Almaz and Mir, 20-ton class modules such as Kvant, Kristal, Priroda and Spektr, modern modules, Zarya and Zvezda, for the International Space Station (ISS) and the Proton family of carrier rockets. The constructive cooperation between those design bureaus and the Khrunichev Machine Building plant brought forth by the early 21st century the Angara family of carriers, smaller spacecraft complexes and ISS modules. The amalgamation of design bureaus and the plant, accompanied by the restructuring of those entities gave birth to Russia's largest corporation – the Khrunichev State Research and Production Space Center.

In creating carrier rockets based on ballistic missiles, much work had been done by KB Yuzhnoye (design bureau) headed by M.K.Yangel. In terms of reliability, those carrier rockets outperform all others known to the world's astronautics. That same KB created under the supervision of V.F.Utkin, the Zenit medium class carrier, representative of the second generation of carrier rockets.

Over four decades the capabilities of carriers' and spacecrafts' control systems have grown considerably. While in 1957-58, the error tolerated in the ejection of artificial satellites into circular Earth orbit varied within several tens of kilometers, in the mid-1960s the system accuracy was so high that the craft

S.P.Korolev with the first team of cosmonauts

launched to the Moon could land on its surface with a deviation from the target point within a mere 5 km. The control systems designed by N.A.Pilyugin had been for a long time among the best in the world.

The great achievements of astronautics in the area of space communication, TV broadcasting, relay services and navigation, and the transition to high rate lines made it possible as early as 1965 to transfer to the Earth a photographic image of Mars from a distance above 200 million km. In 1980, an image of Saturn was transmitted to Earth from a distance of around 1.5 billion km. The applied mechanics' research and production association, originally established as a branch of the Korolev OKB (experimental-design bureau), and headed for many years by M.F.Reshetnev, is one of the world's leaders in the development of this type of spacecraft.

Commercial satellite communications systems are being created that embrace practically all the countries of the world, and ensure a two-way communication with all subscribers. This kind of communication has proved most reliable and is becoming increasingly profitable. Relay systems enable monitoring and control of space constellations from a single post on Earth. Created and in operation now are satellite navigation systems. Without them the use of today's transportation vehicles such as commercial ships, civil aircraft, military hardware etc. is unthinkable.

Qualitative changes also took place in piloted flights. The feasibility of working outside a spaceship was first proved by the Soviet cosmonauts in the 1960 – 1970s. In 1980-90s it was demonstrated that a man can live and work in weightlessness for more than a year. Also during flights, a great number of experiments were conducted in the fields of engineering, geophysics and astronomy.

Extremely important are investigations in space medicine and life support systems. In order to establish what assignments can be expected to be performed by a man in space, especially in the event of prolonged flights, a thorough study of man and his life-support systems is necessary.

One of the first space experiments was photographing Earth. It showed what a wealth of information observation from space can supply for the discovery and rational use of natural resources. The development of photographic and optic-electronic complexes for Earth-sounding, cartography, natural resources' explo-ration, ecology monitoring as well as the creation of medium class carriers based on R-7A rockets is now the domain of the former branch No3 of the OKB. The branch was first converted into TsSKB (central special purpose design bureau), which was later changed into today's GRNPTs TsSKB-Progress (state center for rocket research and production) headed by D.I.Kozlov.

In 1967, during the automatic docking of two unmanned artificial Earth satel-lites, Kosmos-186 and Kosmos-188, a major engineering and scientific problem, namely the encounter and docking of SC in space, was resolved. This made it possible to create within a fairly short period the first orbital station (USSR), and to choose a more rational pattern of spacecrafts' flights to the moon with astronauts landing on its surface (USA). In 1981, the first flight of a reusable space transport system, Space Shuttle, was performed, followed in 1991 by the launch of Russia's Energia-Buran system.

On the whole, the solution of various tasks involved in space exploration – from the launching of artificial Earth satellites to that of interplanetary spacecraft and piloted spaceships and stations – supplied a great deal of valuable information regarding the universe and the planets of the solar system and promoted the technological progress. The Earth's satellites together with sounding rockets provided detailed data concerning Earth's adjacent space, helping to detect radiation zones. In the course of their investigation, better knowledge was obtained as to how Earth interacts with the charged particles issued by the sun. The interplanetary space flights have helped us to understand more thoroughly the nature of many natural phenomena, such as the wind, solar storms, meteorite showers, etc.

The spacecraft launched to the moon sent back pictures of its surface, photo-graphed, among other things, its side invisible from Earth with a resolution that exceeded by far the capabilities of ground-based equipment.

Samples of lunar soil were taken and automatic self-propelled vehicles, Lunokhod-1 and Lunok-hod-2 were delivered to the moon's surface.

Automatic spacecraft made it possible to obtain extra information about Earth's shape and gravitation field, to refine the knowledge of Earth's profile and its magnetic field. Artificial satellites have taught us more about the mass, shape and orbit of the moon. The observations, too, resulting from a space-craft's trajectories have supplied more precise data concerning the mass of Venus and Mars.

The design, manufacture and operation of very complex space systems essen-tially contributed to the development of advanced technology. Automatic spacecraft sent to planets are actually robots controlled from Earth by radio sig-nals. The need to develop reliable systems for tackling such kinds of tasks has led to a better understanding of the problems of analysis and syn-thesis of various complicated engineering systems. Such systems find use both in space research  and other fields of human activity. The demands of astro-nautics have necessitated the designing of integrated automatic devices oper-ating under rigid constraints of the carriers' lifting capacity and the inherent conditions of space. It was an extra stimulus that provoked a rapid improve-ment of automatic machinery and microelectronics.

These programs have been implemented largely due to the efforts of design offices headed by G.N.Babakin, G.Ya.Guskov, V.M.Kovtunenko, D.I.Kozlov, N.N.Sheremet'yevsky and others.

Astronautics has brought about a new trend in technology and construction in the form of the spaceport construction industry. The pioneers of the new trend were research and development groups led by the prominent scientists V.P.Barmin and V.N.Solov'yov. Nowadays, there are more than a dozen space-ports with unique ground-based automated complexes, test stations and other sophisticated assets for the preparation of spacecraft and carrier rockets for launches. Russia performs a great number of launches from its world-renowned spaceports Baikonur and Plesetsk, and executes experimental launches from its spaceport Svobodnyy, now being built in the eastern part of the country.

Today's needs for communication and remote control over long ranges have led to the development of high quality monitoring and control systems that have promoted the development of technical means for tracking spacecraft and measuring the parameters of their movement over interplanetary ranges, thus opening up new areas for application of satellites. In modern astronautics this is one of the priority trends. A ground-based automated control complex developed by M.R.Ryazansky and L.I.Gusev supports to this day the func-tioning of Russia's orbital constellation.

The progress in space technology has created systems of space meteoro-logical support, which receive with the prescribed periodicity the pictures of

Earth's cloud canopy and conduct observations in various spectrum ranges. The data supplied by meteorological satellites serve as a basis for the production of effi-cient weather forecasts, primarily for large regions. Currently, practically all countries of the world use meteorological data.

The results obtained in satellite-aided geodesy are particularly important for military assignments, for mapping natural resources, improving the precision of trajectory measurements and the study of Earth. The use of space systems pro-vides a unique opportunity to maintain ecological monitoring of Earth and global control of natural resources. The results of space survey have proved an effective means of monitoring the growth of farm crops, detection of vegetation diseases, measurement of some soil-related phenomena and the water environment etc. The combination of various methods of space survey provides practically authentic, complete and detailed information about natural resources and the state of the environment.

In addition to the trends that have already emerged, there will appear new ones in the use of space technology, e.g. the organization of production processes that are impractical on Earth. Weightlessness can be used to produce crystals for semiconductor connections. In the conditions of zero gravity, freely-flowing liquid metal and other materials are easy to mold by means of weak magnetic fields. This will pave the way for the fabrication of ingots of prescribed shape without their crystallization into moulds being necessary as on Earth. The specific feature of such ingots is an almost absolute absence of internal tensions and high purity.

The use of space assets is critical in the creation of Russia's single information network, in giving a global capability to telecommunications, especially during the large-scale introduction into the country of the Internet. The future of the Internet depends on the wide use of high-rate, wide-band space communication channels since the possession and exchange of information in the 21st century will be just as significant as the possession of a nuclear weapon.

Russia's piloted astronautic policy aims at the further development of science, the rational use of Earth's natural resources, the solution of problems involved in the ecological monitoring of land and oceans. This necessitates the creation of piloted space systems both for flights in near-Earth orbits and for the realization of mankind's cherished ambition – flights to other planets.

The implementation of such aims is inseparably tied to the creation of new engines, for instance ionic and photonic, capable of flying in space without large supplies of propellant, or using natural forces, e.g. gravitation, the torsion field and other means.

The creation of new unique types of space rocketry as well as methods of space exploration, the conduct of space experiments on automatic and piloted spaceships, and stations in near-Earth space as well as in the orbits of the solar systems' planets provide fertile ground for researchers and designers of various countries to pool their efforts.

Another field of space activity that calls for cooperation is the solution to ecological problems coming in the wake of space exploration. The increasing scale of space exploration has brought ever-growing pressure on the environment, and is polluting the land, oceans and the lower atmosphere. The main source of pollution are regrettably imperfect launch vehicles. Man must concentrate his efforts on reducing such pollution. Near-Earth space is over-crowded with tens of thousands of man-made objects, among them spacecraft and their debris (last stages of carrier rockets, fairings, adapters and discarded components). In addition to the problem of a polluted Earth, mankind will soon face yet another challenge, the littering of near-Earth space.

S.A.Afanasyev, A.I.Kiselev. V.N.Chelomei and L.A.Borisov

The issues of space exploration have been and are being tackled in the former USSR and today's Russia by a number of organizations and companies headed by a galaxy of successors to the first Council of Chief Designers, Yu.P.Semenov, N.A.Anfimov, I.V.Barmin. G.P.Biryukov, B.I.Gubanov, G.A.Yefremov, A.G.Kozlov, B.I.Katorgin, G.Ye. Lozino-Lozinsky and others.

Research and development in the USSR have been accompanied by the serial production of space hardware. The creation of the Energia – Buran complex called for the cooperation of more than 1000 companies. At short notice, the directors of manufacturing plants, A.A.Chizhov, V.D.Vachnadze,

A.I.Kiselev, A.A.Makarov, L.D.Kuchma, I.I.Klebanov, S.S.Bovkun and many others, re-worked the facilities in their charge and organized the production of space equipment. Worthy of special mention is the contribution made by a number of outstanding organizers of the space industry: D.F.Ustinov, K.N.Rudnev, V.M.Ryabikov, L.V.Smirnov, S.A.Afanasyev, O.D.Baklanov, V.Kh.Doguzhiyev, O.N.Shishkin, Yu.N.Koptev, A.G.Karas, A.A.Maksimov and V.L.Ivanov.

The successful launch in 1962 of the Kosmos-4 ushered in the use of space for national defense. Initially, that mission was assigned to the Defense Ministry's Research Institute NII-4. Later on, an agency, dubbed TsNII-50, was isolated from it and specifically entrusted with that function. It was where space systems of military and dual application were devised and developed. The key role in the process was played by such prominent military scientists as T.I.Levin, G.P.Melnikov, I.V.Meshcherikov, Yu.A.Mozzhorin, P.Ye.El'yasberg, I.I.Yatsunsky and others.

It is generally recognized that the use of space hardware increases by 1.5- 2 times the efficiency of armed forces' operations. The specific nature of wars

Discussing the prospects of Russia's astronautics during President V.V.Putin's visit to the Khrunichev Research and Production Space Center

and armed conflicts of the late 20th century shows that the role of space in ful-filling tasks during military confrontations is on the increase. Only space recon-naissance, navigation and communication permit the whole length of the enemy's defense line to be seen, provide global communication and the precise pin-pointing of any object. All this allows military operations to be started "with a rush", i.e. on unprepared terrain and in remote regions. Only the use of space hardware assures protection of territories against nuclear missile threats originating from whatever aggressor. Thus space is becoming the basis of na-tional military capability, a pronounced trend as the new millennium progresses.

In such conditions, new approaches are needed for the development of advanced types of space hardware radically different from the current generation of space equipment.

The existing generation of orbital hardware is mainly purpose-built equipment based on sealed constructions designed to operate on specific types of launch vehicles. In the new millennium, it will be necessary to create multi-function spacecraft based on unsealed modular platforms. Essential, too, is the develop-ment of a series of standardized launch vehicles that ensure high efficiency at low operational cost. Only under such circumstances, based on the potential of the space rocket industry, will Russia boost its economy in the 21st

Wernher von Braun in his office

century, improve the quality of its scientific research, international coopera-tion, social and economic life and strengthen its defense capability. This even-tually will con-solidate the country's position on the international scene.

The key role in the creation of Russia's space rocket science and technolo-gy has been and is being played by leading enterprises within the space rock-et in-dustry such as GKNPTs (Khrunichev State Research and Production Space Center), RKK Energia (Rocket Space Corporation), TsSKB (Central Special Design Bureau), KBOM (General Engineering Design Bureau), KBTM (Trans-port Engineering Design Bureau), Energomash (All-Union Design and Tech-nological Institute of Power Engineering), KBKhA (Design Bureau of Auto-mated Chemical Process) and others. Work proceeds under the guidance of Ro-saviakosmos (Russian Aviation and Space Agency).

Russian astronautics is going through a difficult period. The funding of space programs has plummeted, and a number of space companies are in a near-to-desperate state. But Russian science is not standing still. Even under such adverse conditions space systems for the 21st century continue to be designed.

The exploration of space abroad began with the launch on 1 February 1958 of the US spacecraft Explorer. The program was headed by Wernher von Braun, until 1945 a leading expert on rocketry in Germany and later working in the US. Based on the Redstone ballistic missile he created the Jupiter-S carrier rock-et which was used to launch the Explorer-1. On 20 February, 1962, the Atlas car-rier, developed under the guidance of K.Bossart, launched into orbit the Mercury spaceship piloted by the first US astronaut J.Glenn.

However, none of those achievements blazed the trail since they were repeti-tions of what Russian astronautics had done previously. That is why the US government strove to gain leadership in the space race. In some fields of space activity, on some stretches of the marathon space race, the effort bore fruit.

In 1964, the US was the first to put SC in geostationary orbit. The greatest success, however, was the landing of the American astronauts on the moon in the Apollo-11 spaceship and the walk of the first people, N.Armstrong and E.Aldrin, on the moon's surface. This achievement was made possible by the development under von Braun's direction of the Saturn carriers built in 1964-67 within the framework of the Apollo program.

The Saturn carriers were a family of two- and three-stage carriers of heavy and super-heavy class using standardized assembly units. The two-stage vari-ant, Saturn-1, provided for the launching of a 10.2-ton payload into low Earth orbit. The three-stage variant, Saturn-5, provided for the launching of a 139-ton payload (47 tons to the trajectory of the moon flight).

One of US space technology's great successes was the creation of the Space Shuttle, a reusable space system with an orbital stage featuring aerodynamic properties. Its first launch took place in April 1981. Even though not all the pos-sibilities offered by reusability were fully utilized, and not all designed economic parameters were achieved, it was, undoubtedly, a giant (no matter how costly) step in space exploration.

The first successes of the USSR and the US galvanized some other countries into more energetic activity in space. The US carriers were used to launch the first British spacecraft Ariel-1 (1962), the first Canadian SC Alouette-1 (1962) and the first Italian SC San-Marco (1964). However, launches by means of other countries' carriers made the owners of SC dependent here on the USA. That is why work began on the creation of domestically manufactured carriers. The biggest success here was achieved by France, who launched as early as 1965 the A-1 spacecraft using its own carrier, Diamant-A. Later exploiting this success, France developed the Ariane family of carriers, which proved to be one of the most cost-effective.

An indisputable success of international astronautics was the implementation of the experimental flight Soyuz – Apollo, whose final phase – the launching and docking in orbit of the Soyuz and Apollo spaceships – took place in July 1975. That flight ushered in international programs that had been successfully developing in the last quarter of the 20th century. Their unequivocal success was the manufacture, launch and assembly in orbit of an international space station. Especially important became the international cooperation in space services, whereby the key role in Russia is played by the Khrunichev State Research and Production Space Center that has established jointly with Lockheed Martin the joint venture ILS. With its Proton and Atlas rockets, the venture is one of the leaders in the market of space launches.

Relying on their many years of experience in the design and fabrication of space rocket systems and using both native and foreign analyses and works on astronautics known to them, the authors give here their own viewpoint regarding the development of astronautics in the 21st century. Time will soon show whether they are right.

The authors feel indebted to academicians of the Russian Academy of Sciences, N.A.Anfimov and A.A.Galyeyev, the General Director of Rosaviakosmos, Yu.N.Koptev, as well as general designers of rocketry and space systems from all over Russia for reviewing the book and supplying valuable advice concerning its contents.

The authors will appreciate all comments, suggestions and criticisms that they feel sure will follow the book's publication, confirming their belief that the issues of space are vital and demand the close attention of scientists, practical engineers and all those to whom the future is of concern.

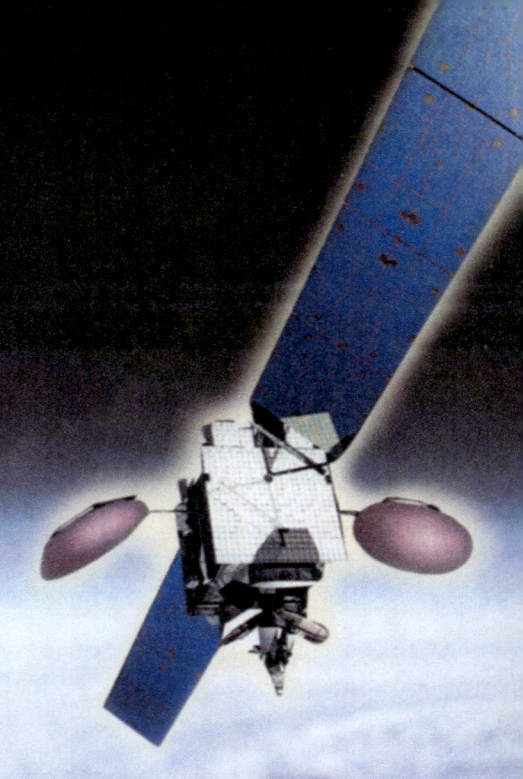

# PART 1

# Trends in global space exploration

# 1 Global astronautics, a general survey

On October 4, 1957, the Soviet Union placed the Earth's first artificial satellite in orbit, an event that paved the way to the space race which to date has reached an unprecedented scale. At its initial stage that proceeded in the era of the "cold" war, the driving forces of this marathon race were political and military. The prestige and security of the nation (in a broad sense of the terms) were the principal considerations that drove the leaders in space research and development to spend money generously and without hesitation. By the mid 1960s the intensity of launches reached a record figure. For example, in 1966 foreign countries launched 101 spacecraft. The transition from sporadically launched separate spacecraft to the spaced-based permanently operating systems occurred very quickly and in an avalanche fashion (Fig. 1).

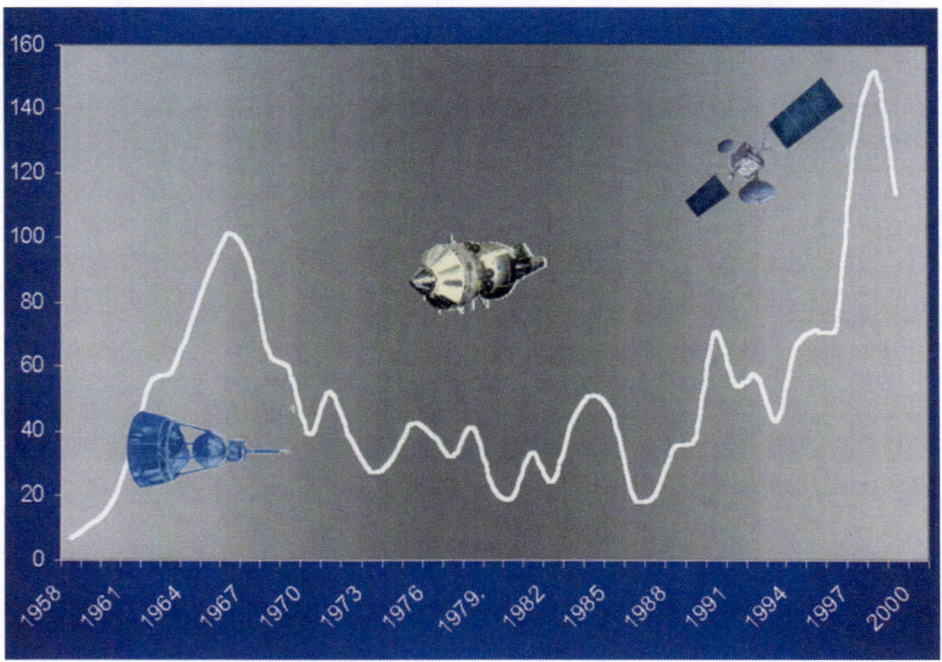

**Fig. 1.** Change in the number of SC launched by foreign countries

The decline in intensity of spacecraft launches in the early 1970s is explained by the fact that the baseline space systems (reconnaissance and surveillance, communications, meteorology, missile assault warning) had been by that time deployed and began to be used as standard systems on a permanent basis, receiving evolutionary improvements. The reduction in numbers of spacecraft put in orbit was also due to achievements in science and technology which considerably extended their operating service life.

The activity in space exploration never flagged for a day. New countries kept joining the space research effort which trend was particularly pronounced in the most promising field of space exploration, the satellite communications.

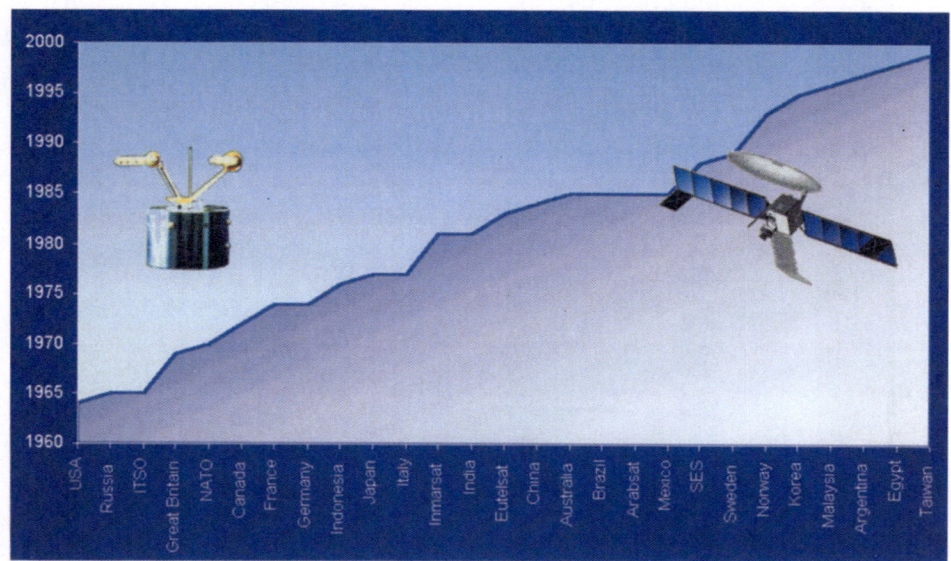

**Fig. 2.** The beginning of utilization of own communication SC by various countries

Nowadays, all the world's leading powers and many developing nations are somehow or other involved in space research. It should be noted however that the involvement varies greatly from the use of separate communications channels to a full-scale and comprehensive use of space facilities for fulfilling a wide range of civilian and military tasks. Interestingly, the individual tasks (e.g. communications) are tackled by quite a number of countries (up to 170–180) whereas versatile space exploration that covers military, civilian, economic, scientific, social and other aspects of human life are the affordable domain of just a handful of states.

Table 1 shows the capabilities in space research of the states that are engaged in state-supported space programs or possess at least their own assets for space research, albeit of foreign manufacture.

Table 2 shows that around 30 countries have state-supported space programs and are implementing them.

Nineteen countries have production capacities and scientific potential enabling them to develop and produce their own spacecraft. It should be noted however that most of them are capable of creating only small spacecraft for experimental use. For instance, out of 23 countries having space communication facilities, 17 use spacecraft developed by foreign companies.

**Table 1.** Involvement of states in various types of space exploration

| Item No | Country | Types of work in space | | | | | | | | | | | |
|---|---|---|---|---|---|---|---|---|---|---|---|---|---|
| | | 1 | 2 | 3 | 4 | 5 | 6 | 7 | 8 | 9 | 10 | 11 | 12 |
| 1. | Argentina | ** | | | | | ** | | | | ** | ** | |
| 2. | Australia | ** | ** | | | | | | | | | | |
| 3. | Austria | | | | | | | | | | | ** | |
| 4. | Belgium | ** | | | | | | | | | | | |
| 5. | Brazil | ** | ** | ** | * | | | * | ** | | | ** | |
| 6. | Canada | ** | ** | | | | ** | ** | | | | | |
| 7. | Chile | ** | ** | | | | | | | | | ** | |
| 8. | China | ** | ** | ** | ** | ** | ** | ** | ** | | ** | ** | * |
| 9. | Czech Republic | ** | ** | | | | | | | | | ** | |
| 10. | Denmark | ** | | | | | | | | | | ** | |
| 11. | Egypt | | | | | | ** | | | | | | |
| 12. | Finland | ** | | | | | | | | | | ** | |
| 13. | France | ** | ** | ** | ** | ** | ** | ** | | | ** | ** | * |
| 14. | Germany | ** | ** | | | * | ** | | | | ** | ** | |
| 15. | Great Britain | ** | ** | | | ** | ** | | | | ** | ** | |
| 16. | Hong Kong | | | | | | ** | | | | | | |
| 17. | India | ** | ** | ** | ** | * | ** | ** | ** | | ** | ** | |
| 18. | Indonesia | ** | | | | | ** | | | | | | |
| 19. | Iran | | | * | | | | | | | | | |
| 20. | Israel | ** | ** | ** | ** | * | | * | | | | ** | |
| 21. | Italy | ** | ** | * | * | | ** | | | | ** | ** | |
| 22. | Japan | ** | ** | ** | ** | * | ** | ** | ** | | ** | ** | * |
| 23. | Luxemburg | | | | | | ** | | | | | | |
| 24. | Mexico | ** | ** | | | | ** | | | | | | |
| 25 | Netherlands | ** | | | | | | | | | | ** | ** |

| Item No | Country | Types of work in space | | | | | | | | | | | |
|---|---|---|---|---|---|---|---|---|---|---|---|---|---|
| | | 1 | 2 | 3 | 4 | 5 | 6 | 7 | 8 | 9 | 10 | 11 | 12 |
| 26. | North Korea | | | * | | | | | | | | | |
| 27. | Norway | ** | ** | | | | | | | | | ** | |
| 28. | Pakistan | ** | ** | | | | | | | | | ** | |
| 29. | Philippines | | | | | | ** | | | | | | |
| 30. | Portugal | | ** | | | | | | | | | ** | |
| 31. | Russia | ** | ** | ** | ** | ** | ** | ** | ** | ** | ** | ** | ** |
| 32. | Singapore | | * | | | | | | | | | ** | |
| 33. | South Africa | ** | * | | | | | * | | | | | |
| 34. | South Korea | ** | * | * | | | ** | * | | | | ** | |
| 35. | Spain | ** | ** | * | | | ** | | | | | ** | |
| 36. | Sweden | ** | ** | | | | ** | | | | | ** | |
| 37. | Taiwan | ** | * | | | | | | | | | ** | |
| 38. | Thailand | ** | | | | | ** | * | | | | | |
| 39. | Turkey | ** | | | | | ** | | | | | | |
| 40. | USA | ** | ** | ** | ** | ** | ** | ** | ** | ** | ** | ** | ** |

Symbols
1 – state-supported program
2 – manufacture of SC
3 – manufacture of boosters
4 – launching of SC
5 – military tasks
6 – communications
7 – ERS
8 – meteorology
9 – navigation
10 – science
11 – experiments
12 – piloted flights
** – in operation
* – possible

Most of those countries use their space equipment to tackle civilian tasks. Many of them have ways and means of using their space facilities to satisfy their defense related needs. This is true in particular of the data obtained from spacecraft employed to research the Earth's natural resources. Scores of countries have now recourse to such information and the level of their information support keeps growing. Space communication facilities and meteorological observation systems are also easily available to scores of countries and can be put to military uses.

However, as noted above, it is only the USA, France, China, Japan and India that have a developed infrastructure enabling them to tackle the complex tasks of exploration and utilization of space. Therefore, any talk about the wide use of space both in practical and potential terms makes sense only with reference to those countries. In addition, Great Britain and Germany look to the military uses of space. Israel, too, which is now in the possession of homemade orbital injection facilities and spacecraft, can be nominally placed among such countries. The study of the processes that are underway in different countries engaged in such a new effort as space research and development revealed certain trends and regularities specific to those processes and interesting in terms of space marketing.

1. The number of states involved in research and applied programs with the use of space facilities keeps growing. Many industrialized and industrializing countries strive to become "space powers" for reasons either of prestige or economy. This urges them to develop and maintain the level of technologies required by the space market.

2. The independent development of space by the countries proceeds as a rule gradually, from the low near-earth orbits up to the higher and then interplanetary ones to the extent permitted by their space rocket technologies.

3. It is becoming ever more common practice to join the efforts of countries, organizations and individual firms for doing major research and implementing applied programs. Such cooperation assumes the form of international associations and consortiums. The instance of the European Space Agency (ESA) comprised of 14 countries is in this respect very telling.

4. The transition from research and experiments to practical use of the unique capabilities afforded by space for satisfaction of human demands is the ultimate goal of all nations involved in space exploration. Under such circumstances it is characteristic of the industrializing countries to usually deploy applied space facilities, for example communication equipment designed and manufactured to their orders by industrialized countries.

5. The practical use of space is becoming ever more complementary in nature. One the one hand, the civilian space equipment (including commercial types) is being used on ever growing scale for military purposes. On the other, military space hardware is receiving ever-wider acceptance as civil use commodities.

6. At the initial stages of space exploration and utilization, all space programs in all countries are financed by national budgets. The pattern and amount of spending of the world's leading space powers on space programs in 1998 is shown in Table 2.

7. The commercial use of space and provision of services in its exploration and utilization on a commercial basis are continuously expanding. The share of private capital in funding space exploration is also growing.

Nowadays, business in space is becoming more and more of a private enterprise. Cuts of federal spending on space exploration are offset by investments in commercial projects, especially in the creation of the satellite communications network. The space related revenues increase annually by 20%.

**Table 2.** States' funding of space exploration in the leading space powers, US$ million

| Items of expenditure | USA | Japan | France | Germany | Italy | China | Belgium | Great Britain | India | Canada | Total |
|---|---|---|---|---|---|---|---|---|---|---|---|
| Advanced projects for creating space rocketry | 3120 | 694 | 490 | 142 | 83 | 126 | 9 | 3 | 17 | 26 | 4710 |
| Fundamental research | 1977 | 400 | 315 | 156 | 114 | | | 13 | 6 | 23 | 3004 |
| Expendable boosters | 1021 | 400 | 760 | 233 | 115 | 163 | 79 | 8 | 109 | | 2888 |
| ERS | 1341 | 18 | 420 | 164 | 176 | | 117 | 112 | 60 | 75 | 2483 |
| Telecommunications | 781 | 520 | 126 | 190 | 105 | 127 | 60 | 32 | 26 | 38 | 2005 |
| Navigation | 139 | 50 | 68 | 148 | 90 | 43 | 61 | 81 | 40 | 19 | 739 |
| Space-based production | 504 | 182 | | | | | | | | | 686 |
| Orbital maintenance equipment | 165 | 100 | | | | | | | | | 265 |
| Piloted flights | 5195 | 36 | 21 | 107 | 57 | 21 | | 48 | | 67 | 5552 |
| Total | 14243 | 2400 | 2200 | 1140 | 740 | 480 | 326 | 297 | 258 | 248 | 22332 |

8. The sphere and scale of commercial services in space are growing. This is especially noticeable in the area of satellite communications where initially Intelsat system's services were used by just a few states that had large land-based stations. As time went by, the commercial satellite communications system came to serve hundreds and thousands of private firms, banks, communication and telecasting networks, thousands of mobile transport means, millions of individual satellite TV and radio receivers.

9. The steady advancement of personal satellite communication systems provides for high quality transmission of voice, high speed data flow, and multimedia. Also, it enables conferencing, interactive communication, and access to the Internet.

10. The space commercial infrastructure is now in the making. Designed to meet the demand for services and materials manufacture, it is progressing steadily. Most of commercial systems of satellite communication deployed to date have their own ground-based spacecraft control stations. A variety of commercial launch vehicles have been created. In production now are commercial multi-mission all-purpose space platforms.

11. A space industry has been created that specializes in development and manufacture of spacecraft, orbital launch and ground-based space complexes.

12. Competition is stiffening on the world's market of space products and services both between countries and between private commercial organizations and firms. This is giving impetus to the restructuring (toward consolidation) of the space industry. An enlargement of space enterprises proceeds on an unprecedented scale.

13. The space systems are becoming ever more sophisticated, which, in turn, complicates technological and ecological problems associated with their manufacture. More stringent demands are being made of space technologies.

14. A new branch of jurisprudence, space law, has emerged and is steadily developing. This regulates relations in space exploration both at home and abroad.

Thus, space exploration as a new type of economically very promising human activity generates problems similar to those which mankind faced during the emergence and evolution of various new sciences and technologies (e.g. aeronautics, radio communication, and others). The trends peculiar to exploration and utilization of space are in many ways similar to those encountered in many fields of human research work. The main difference of space exploration from other activities apparently lies in the unmatched rate at which countries and their organizations intensify their efforts in space research and development.

## 1.1 Orbital facilities as they are

By convention, the orbital facilities can be divided, depending on their assigned missions, into several large groups: civil spacecraft (SC), commercial spacecraft and military spacecraft. These groups, in their turn, can be subdivided into subgroups as per their target function: communications SC, Earth remote sensing (ERS), navigation support SC, meteorological support SC, research and experiment SC, piloted SC, reconnaissance and surveillance SC.

The conventional division of SC according to assigned missions and target functions is presented in Fig. 3. Note that division of communication SC into low altitude and geosynchronous types is critically important since those belong to vehicles of different classes.

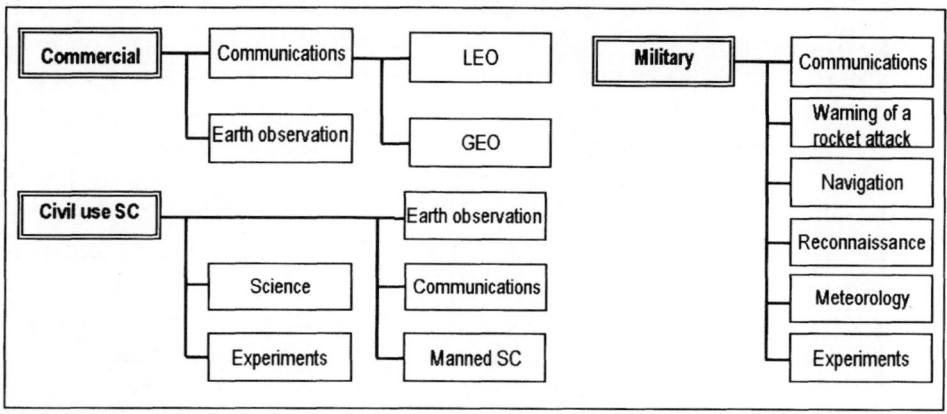

**Fig. 3.** Conventional categorization of SC as per target function

The number of SC launched all over the world tends to grow steadily, though during some years it may decrease. During the decade of 1991 – 2000, about 1100 SC had been launched into space. The projected number of SC to be launched between 2001 and 2010 is in the region of 2000.

Fig. 4 shows changes in numbers of launches of SC in the world over the years since 1990 to this day and the anticipated numbers till 2010.

As can be seen from Fig. 4, the first decade of the 21st century is characterized by a considerable growth of SC launches. This is accounted for by the deployment of Orbcomm and Globalstar low orbit mobile communication systems and Teledesic, Spaceway, Cyber-Star/SkyBridge and other wide-band multimedia communication systems.

The expected number of commercial communications SC in 2001-2010 will be around 1,500 (Fig. 5), which is above 70% of the total number of launched SC (Fig. 6) or up to 50% of the total volume of the market of SC (Fig. 7).

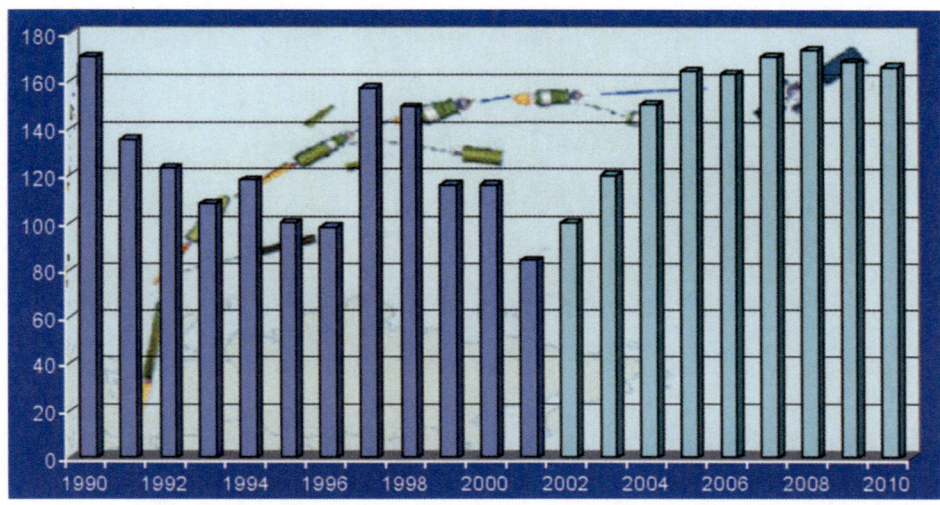

Fig. 4. Changes in number of launches of SC in the world on yearly basis

Such a big number of communications SC is indicative of the rapid progress in all types of communications. Generated by the need for information support of man's various activities, this progress has become possible due to improvement of electronics technologies and methods of signals transmissions and processing. The effectiveness of investments made at various times determined the rate of improvement of a particular type of communication.

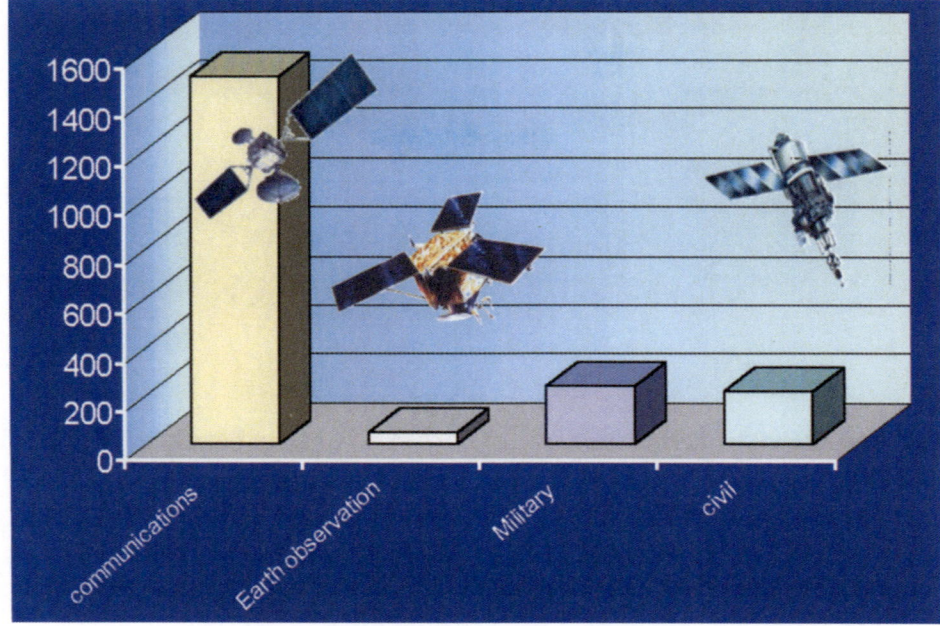

Fig. 5. Number of various types of SC launched in 2001-2010

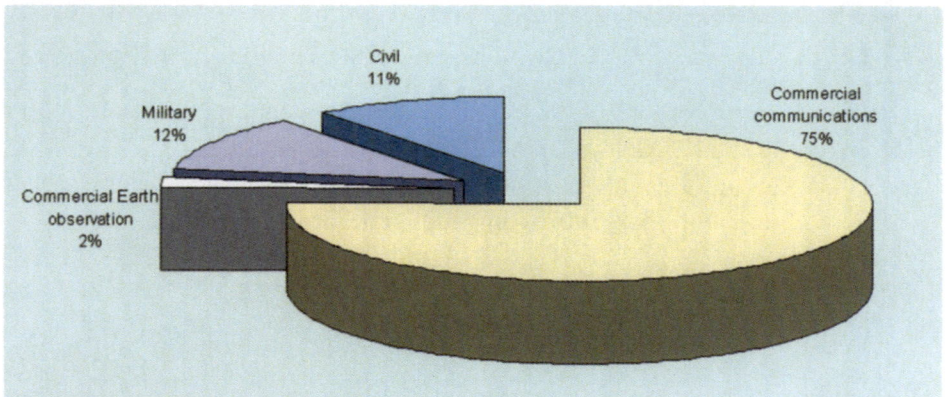

**Fig. 6.** Share of various types of SC in the number of launched craft in 2001-2010

The development of satellite communication has peculiarities of its own which are most closely associated with the use of space rocketry, enhanced risks for investors, complexity of infrastructure of satellite control systems, necessity to periodically replace them because of physical restrictions of their service life, and insufficient information at the initial stages of satellite manufacture about what and how will affect the equipment's operation in space.

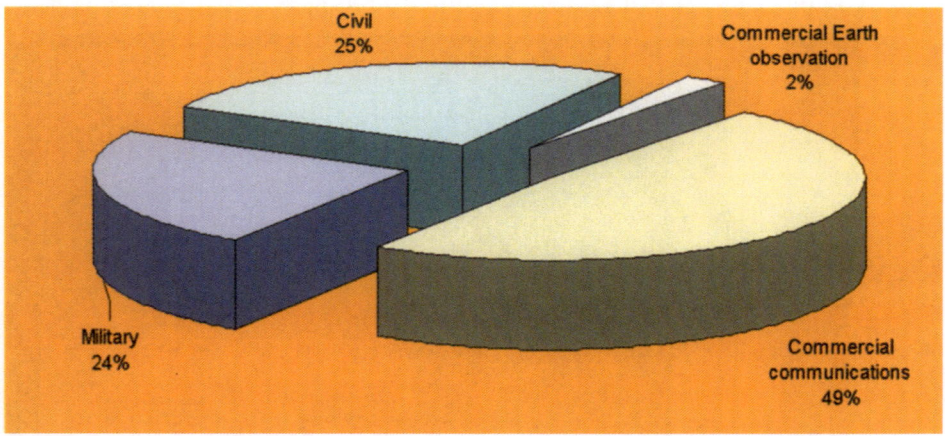

**Fig. 7.** Share of SC of various application in the total volume of SC market during 2001-2010

Initially, satellites were used to organize extended main communication lines. Later on, as technological capabilities improved, regional and corporate networks began to appear on the basis of terminals like VSAT (very small aperture antenna). Communication satellites began to be used by the mass consumer to receive TV transmission channels.

The market immediately responded to this by an increase in number of TV satellites and by improving the standards of digital television, both which trends not only heightened the signals' quality but also made it possible to receive from one satellite hundreds of TV programs via one reasonably small antenna.

Up to 70% of the communication satellites are now doing jobs for TV. In the coming years this ratio will still persist.

The satellite communication today tackles a range of tasks and provides a wide variety of services. The real restrictions come about only as a result of a manufacturer's inadequate production capabilities or because of insufficient funding.

Apart from the communication, commercial use of space facilities is projected for earth remote sensing. This use however will not be so extensive as for communications.

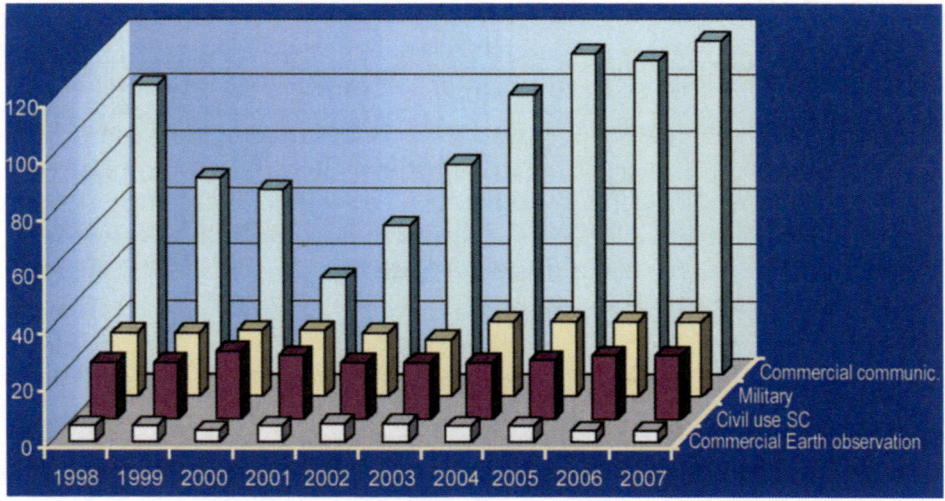

**Fig. 8.** Number of SC of various application in the total of craft launched during 1998-2007

Fig. 8 shows the distribution by years of various types of spacecraft during 1998 – 2007. Fig. 9 presents the volume of market for various SC during the same period.

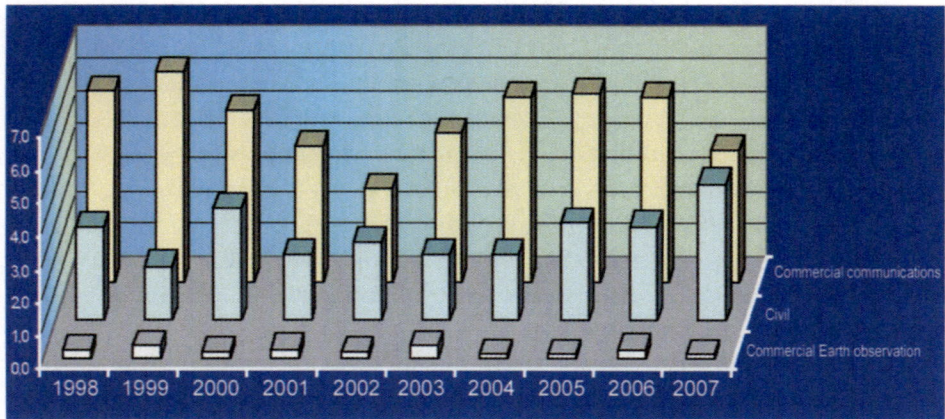

**Fig. 9.** Changing of the market volume for various SC during 1998-2007

## 1.1.1 The main trends in development of orbital facilities

The new revolutionary technologies have radically changed the image, performance and cost of orbital facilities in the 21st century.

Experts single out the following main trends in development of advanced SC:

**1. On-board processing, power supply, communication equipment.** The new principles of using SC according to which the user of services pays only if he actually uses them, necessitated the creation of on-board facilities for processing signals. This is required for filtration and routing of data as well as for their recovery on board the SC. In the coming 10 to 15 years a major breakthrough is expected in the area of data processing on board SC. This stems from achievements in computer science.

The power supply of SC is to be increased by creating an improved flywheels system which could be incorporated in high-performance power accumulation and storage units. Studies are underway as to how to increase the efficiency of solar cells to their theoretical limit (37-40%).

In the area of communications, the development of phased antenna arrays (PAA) with electronic control of the polar pattern and variable configuration is considered a particularly promising trend. A reduction in cost of PAAs is expected thanks to combining the imposed requirements with the achievements in antenna design, in materials manufacture and in production of semiconductor-based instruments.

**2. Power plants.** Electric rocket power plants (ERPP) provide for the use of a much smaller power supply compared to the traditional chemical packages. However, their use for purposes other than orientation and stabilization is restricted by their low thrust performance. The increase in the thrust of ERPP expected in the very near future will make it possible to use them for interorbital transitions. Thus, with the power of on-board ERPP in the region of 50 kW, the transition of a communication satellite by means of the ERPP from a low orbit to a geosynchronous one will occur in less than a month.

**3. Miniaturization.** The achievements in miniaturization of component packages of space hardware will essentially cut down the weight of both prime equipment and supporting systems. For example, the weight of the orientation unit will decrease over the period of 1980 – 2020 by practically ten times. Fig. 10 shows the change in the weight of SC structural elements relative to the gross weight of SC.

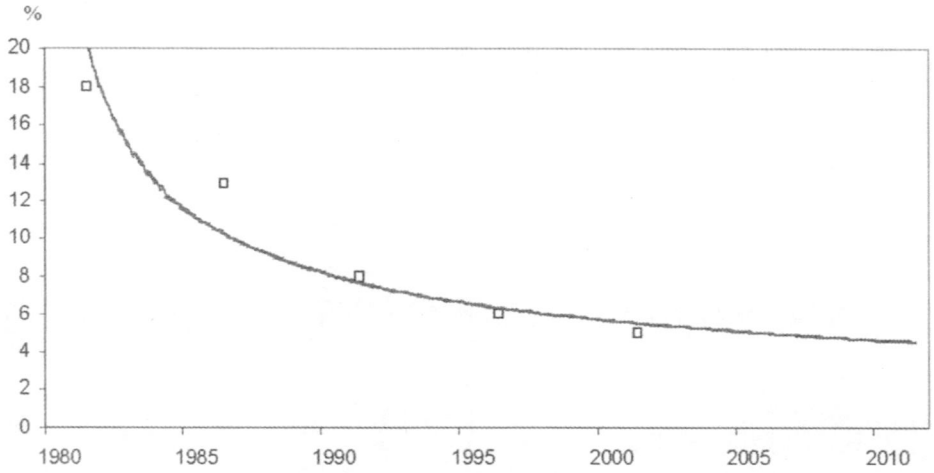

**Fig. 10.** Changes in weight of structure relative to the total weight of SC, %

The attempts to lessen the weight and size of equipment will in the foreseeable future be a characteristic trait in the development of orbital facilities placed in low to medium orbits. Also considered is the possibility of creating systems with distributed parameters based on the use of a large number of microsatellites (weighing from 10 to 100 kg) and nanosatellites (weighing less than 10 kg). Envisaged in the longer term (2010-2020) is the use of satellites weighing around 1 kg.

**4. Inflatable structures** can contribute to miniaturization (by placing SC in small containers in order to reduce the launching costs) and simultaneously provide for putting in space huge structures measuring up to 300 m.

**5. Optic systems.** The use of optic systems or laser-based inter-satellite communication equipment will increase the speed of data transmission to the level of 0.5 gigabit /sec. Except for the beam pointing accuracy, there are no serious obstacles to using laser-based communication between SC.

**6. Robotics.** Development is underway of modules of robot operated SC. Such SC will be able to establish communication between themselves, gather information, perform various operations, including repair, without human interference.

### 1.1.2 Communication facilities

Currently, the market of telecommunications is developing at a fast rate indeed. Its volume in 1998 exceeded US$ 1 trillion. In it, the segment of space communications accounted for about US$ 27 billion or 2.3% of the world's total market. However, within 10 years it's expected to grow to 6%. This means that the annual growth rate of this segment of the market will amount to 16.1% and its volume will increase in 2008 to US$ 182 billion. On various segments of the communication equipment market, the share of space communication will vary between 5 and 15%.

The universal introduction of personal computers and the associated expansion of the Internet and its capabilities have a profound impact on the progress of satellite communication. Under such circumstances, a high speed information flow exchange has become critically important. The companies operating communication SC redouble their efforts trying to meet demands of the expanding market.

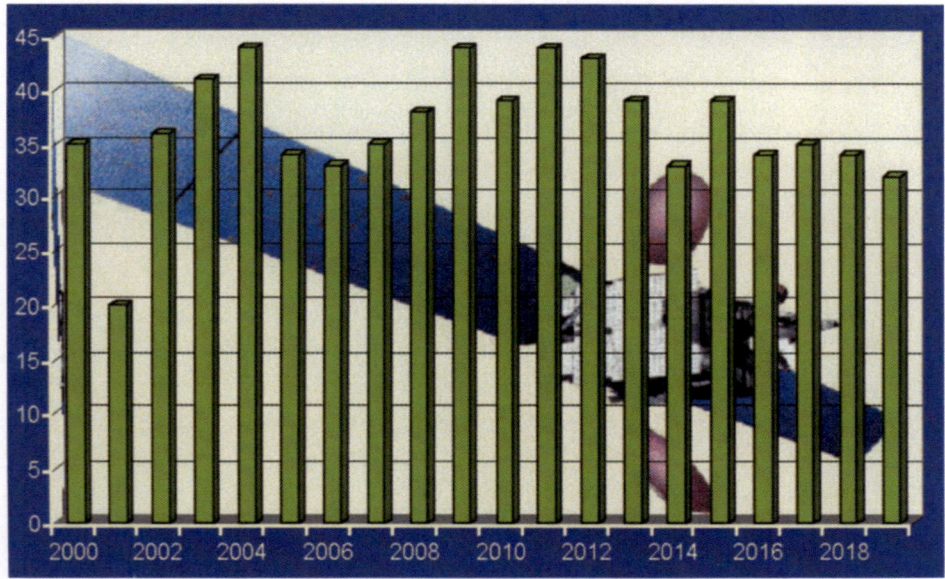

**Fig. 11.** Distribution by years of communication SC launched into geostationary orbit during 2000-2019

The market of space communication is now divided into the following segments:
- fixed communication (communication between stationary objects)
- mobile communication (that is communication between moving objects)

The fixed satellite communication provided by space systems on the geostationary orbit is to date the governing factor on the market of space telecommunication. It accounts for up to 90% of revenues received. Today, the satellite industry of fixed communication is having its boom time. The number of active relay-stations in orbit during the period since 1980 has increased practically ten-fold. This tendency will persist till at least 2010.

Until 1998, virtually all foreign communication SC used to be placed in a geostationary orbit. By 1997, more than 140 satellite communication systems had been registered and around 200 were either claimed for or at the stage of coordination. The number of commercial types of communication satellites put in geostationary orbit during 2000 – 2019 is shown in Fig. 11.

However, in the early 1990s a lot of communication system projects came along which intended to use a large number of low-orbit SC. In the late 1990s the implementation of those projects began. The low-orbit space systems of mobile voice communication featuring a global coverage constitute today a new and quite an important segment of the market. It is expected that the total number of low-orbit communications SC launched into orbit during 2000 – 2019 will be in excess of 1,800. Fig. 12. shows the number of communications SC launched into low orbits during that period on a yearly basis.

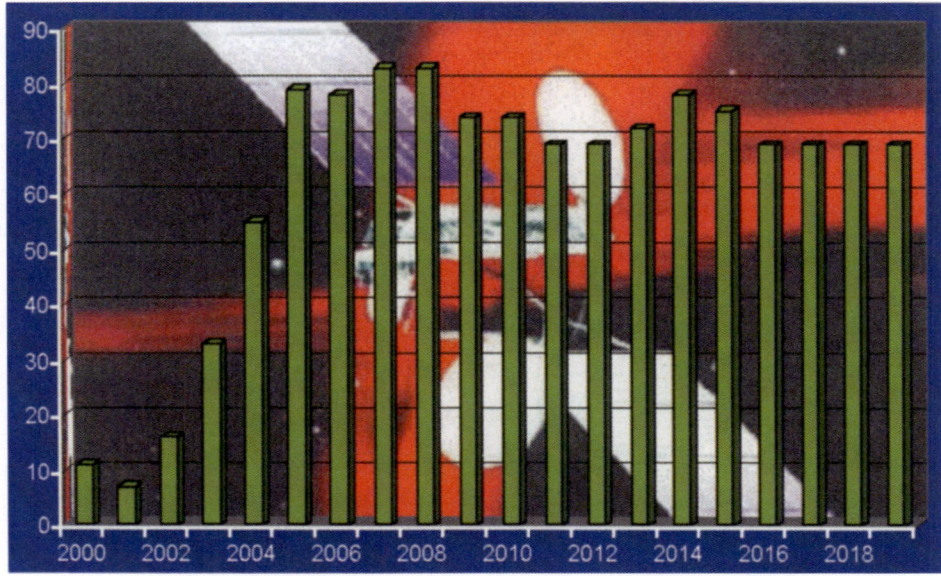

**Fig. 12.** Distribution by years of communication SC launched into near-earth orbit during 2000 – 2019

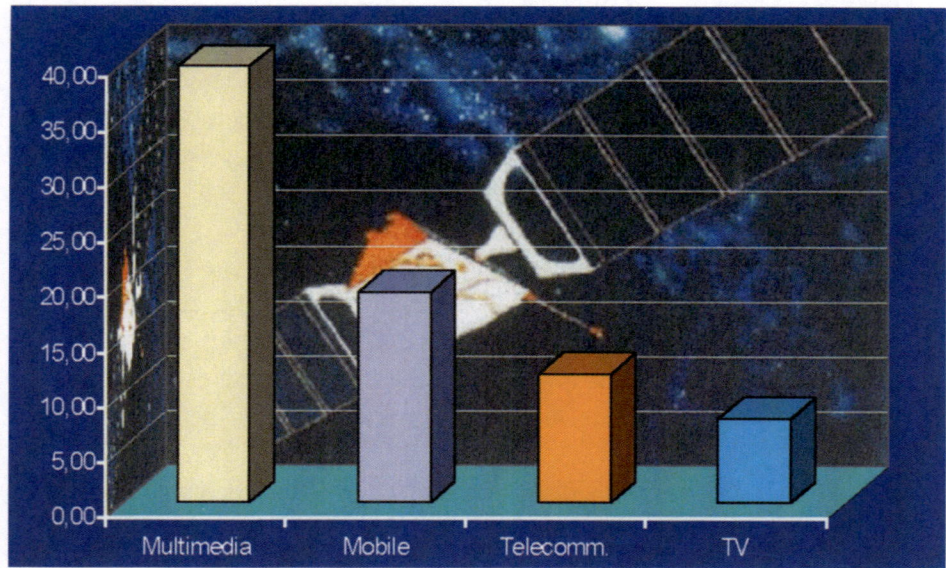

**Fig. 13.** Volume of market of communication SC of various application launched
during 2001-2010, US$ billion

The volume of the world's market for commercial communications SC
during 1998 – 2007 will reach about US$ 60 billion. Categorization of this
volume based on types of communications space vehicles and their share
in 1998 – 2007 are shown in Figures 13. and 14.

The USA is now indisputably the world's leader in production and sales
of communications SC. The Americans pioneered the way for practically all
possible types of satellite communication. This became possible due to the
large and diverse fleet of carrier rockets and the advanced electronics industry.

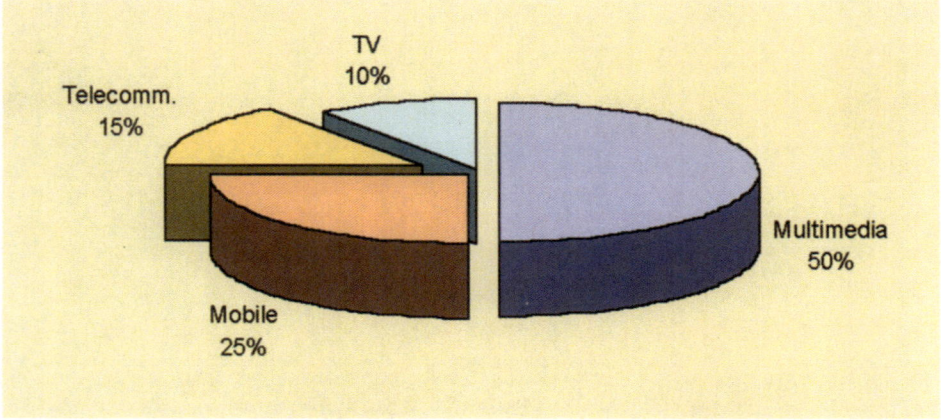

**Fig. 14.** Share of SC of various application in the total volume of the market of communication
SC launched during 2001 – 2010

The American major aerospace companies Boeing (previously Hughes Space and Communications International), Lockheed Martin Astro Space, and Space Systems/Loral have won practically all the world's market of communications SC. All the companies manufacturing communications satellites have two to three basic models. Depending on the customers' requirements, one or another basic model is taken, which is then outfitted with the required number of relay-stations operating within the range ordered by the customer.

The projected share of the leading companies in the total volume of communications SC manufactured in 2000 – 2019 is shown in Fig. 15.

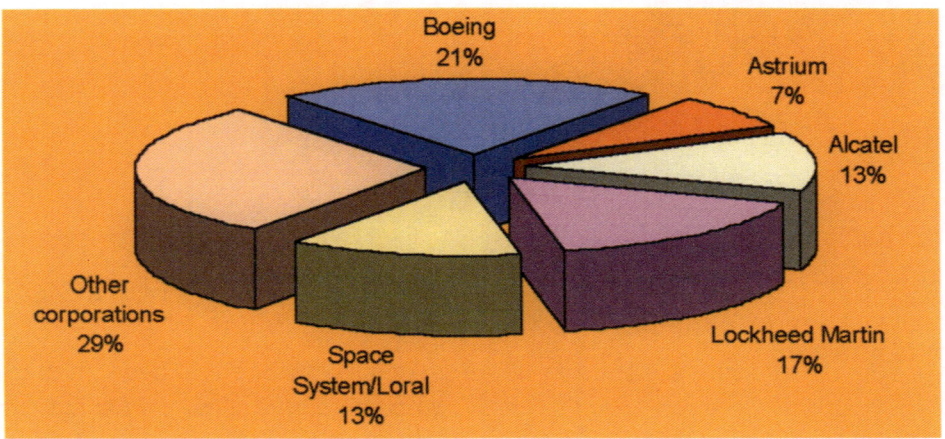

**Fig. 15.** Distribution of communication SC by principal manufacturers, as created during 2000 – 2019

The leading role among the American manufacturers of communication SC and communication systems has been played by Hughes (now a division of Boeing). This situation is not likely to change in the foreseeable future. Most private companies and organizations of different countries prefer SC of this company to all others. Its basic variant HS-601 has become over recent years the most wanted product on the world's satellite market. Fig. 1.1.14 shows the exterior of SC as installed on the HS-601.

Even the US Navy has ordered a batch of UHF-Follow-On (UFO) military communications satellites based on this model.

The runner-up among the American manufacturers of communication satellites is Lockheed Martin Astro Space. Among the customers of this company is the International Maritime Satellite Telecommunications Organization (INMARSAT). Lockheed Martin Astro Space also handles orders for the USAF (DSCS-3 and Milstar satellites).

The Intelsat organization has preferred the Space Systems/Loral concern. The Intelsat-7 satellites are manufactured on the basis of the standard model FS-1300. SC have 40 relay-stations operating in two ranges. They provide

22,500 two-way telephone channels (up to 112,500 in a digitally compressed mode) and 3 TV channels. Some of the TV channels of those craft are used to beam a direct television broadcast to America, Europe, Asia, Northern Africa, and the Middle East.

Fig. 16.
Communication SC based on HS-601

The emergence of communication systems based on a large number of low-orbit SC enabled quite a few companies, until then minor players in satellite communication business, make their presence known. Among them are Motorola (Iridium and Teledesic systems), Orbital Sciences Corp. (Orbcomm system), Alenia Spazio (Globalstar system) and others.

In the last decade, Western Europe opened up to competition its own companies capable of rivaling the USA in the manufacture of communication satellites. Alcatel Espace of France and the Astrium international company now actively promote their satellites on the world's market.

Other countries involved in space race have also started to create spacecraft. During last few years, Israel, India and Italy came up with their first communication craft for the geostationary orbit. In the late 1980s China created its first Dunfanhun communication satellite for a stationary orbit. It is equipped with 24 relay-stations and is capable of transmitting simultaneously six full-color TV programs and servicing 8,000 telephone lines. The satellite has been developed by the Chinese Academy of Space Technologies in Beijing jointly with the German company Deutsche Aerospace. In terms of communication equipment it is close to vehicles manufactured by Hughes, US, but is inferior to western models in terms of navigation system and stabilization technology. The satellite weighs 2,232 kg. The designed service life is 8 years. China plans to export this type of satellite.

However, having launched three Dunfanhun vehicles, China, too, turned to Americans. Hughes manufactured for China the Chinasat satellite which was created by Lockheed Martin on the basis of its newest model, AS-7000. Forty-eight relay-stations are installed on the satellite. It is designed for servicing all China's territory and a big portion of Asia.

### 1.1.3 Civil use facilities

The category of the so-called civil use SC includes satellites developed and launched under state supported programs, excepting military SC. Placed into the same category can be exploration and experimental SC, communication SC, land survey SC, including meteorological SC and ERS SC (Fig. 3).

Foreign experts predict that the total number of civil use SC launched during 1998 – 2007 will be 200, that is about 12%.

The number of various types of civil use SC launched during 1998 – 2007 is shown in Fig. 17.

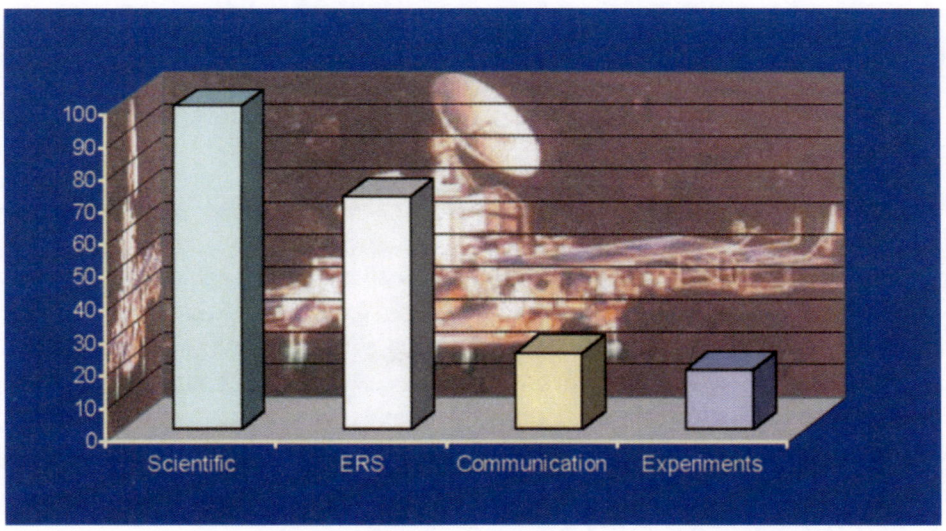

**Fig. 17.** Number of civil use SC of various application launched during  1998 – 2007

The volume of the world's market of civil use SC during 1998 – 2007 will reach around US$ 26 billion, that is about 25%. The distribution of this volume by years during 1998 – 2007 is shown in Fig. 18.

The supplied diagrams show that during the specified period the civil space programs will focus mostly on research done with the help of SC. Coming next are land survey SC. Communication facilities receive little attention in civil use programs since practically all of them are used commercially.

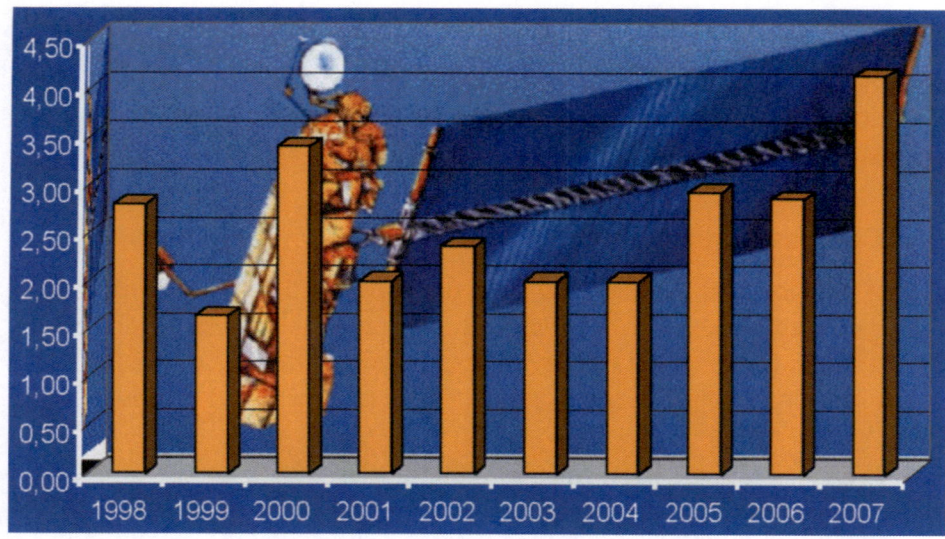

**Fig. 18.** Distribution by years of the total market of civil SC, US$ billion

## 1.1.4 Commercial facilities of ERS

Commercial facilities of Earth remote sensing (ERS) from space are only making their first steps. The data supplied by ERS are now increasingly used in farming, regional development, construction, and extractive industries. The existing space facilities of ERS, such as SPOT, Landsat and others are not purely commercial in spite of the market-regulated mechanism of distributing the information obtained. Those systems are subsidized by state agencies since operating them at the present time does not pay off.

Nonetheless, attempts are being made to create commercial SC for ERS. Referred to such type are the currently operating OrbView and Ikonos as well as systems now in the making and with the slated commissioning date within the next few years, e.g. EarlyBird and others.

Shown in Fig. 19 are currently operating ERS SC.

The number of commercial ERS SC projected by foreign experts for the period between 1998 and 2007 is around 50 units, that is about 3%.

Fig. 20 shows the distribution by years of commercial ERS SC launched during 1998 – 2007. The volume of the world's market of commercial ERS SC during 1998 – 2007 will make up less than US$ 2.5 billion, that is a mere 2% of the total volume.

**Fig. 19.** American Landsat-7 ERS SC

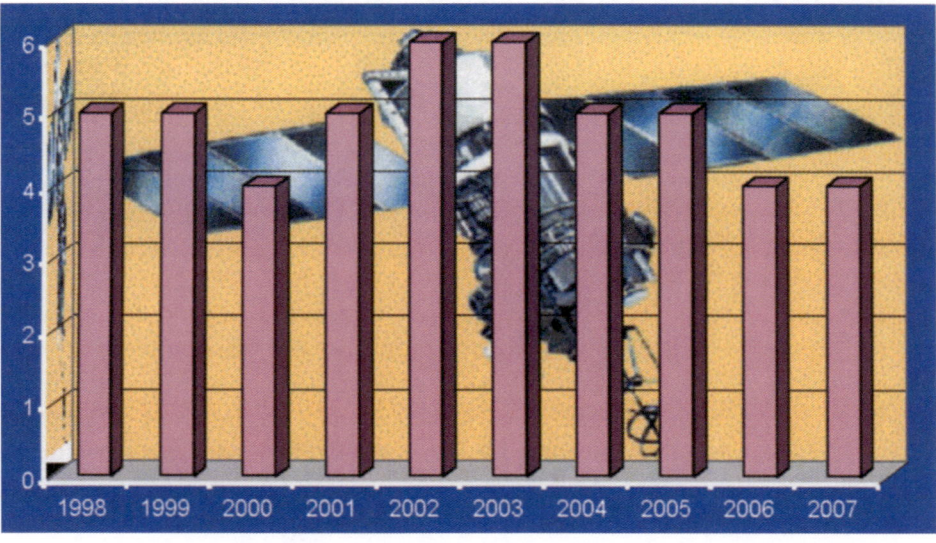

**Fig. 20.** Distribution by years of the number of commercial ERS SC launched during 1998-2007

## 1.2 Foreign launch Facilities

Space launch facilities are sophisticated technical transport systems designed to deliver payloads to a target orbit in space.

All existing space launch systems as well as facilities whose operation is envisioned in the foreseeable future (25 to 30 years) are essentially jet-propelled vehicles.

The first evidence of using devices that employ the jet propulsion principle appeared in the Chinese chronicles relating the siege of Beijing by Mongols in 1232. The Chinese used both signal and combat powder rockets whose flight range at the time was out to 400 m.

Starting in the 14 century rockets also began to be used in Europe to deliver explosives to enemy positions (Italy and later France). However, the very poor accuracy of those devices prohibited their use on a wider scale. Centuries went by, but no remarkable progress was registered in the development of rocketry.

Early in the 19th century the first theoretical works appeared that were devoted to building and using powder rockets. By the end of the 19th century there were already projects that envisioned the use of rocket engines for flights in airless space (N.Kibalchich, G.Hanswidt).

At the beginning of the 20th century there appeared serious theoretical investigations into rocketry. Considered among others were the issues related to the use of rockets for delivery of payloads into space (K.Tsiolkovsky, Yu.Kondratyuk, F.Tsander). In the 1930s the jet propulsion theory came to be used in many experimental endeavors which resulted in creation and widespread use of rocket weapons during the Second World War. After the War, powerful combat complexes of intercontinental ballistic missiles were created and deployed. Such complexes ensured delivery of nuclear munitions to any point on Earth.

Beginning October 4, 1957, the rockets began to be used for delivery of special purpose equipment to target orbits in space. Such equipment was intended for performance of a variety of tasks. Thus the space race started in which launch facilities were to play the key role. In the shortest possible time the USA and the USSR had built launchers capable of putting multi-ton payloads into space.

Understanding that pursuing independent programs in space research requires independent capabilities of launching payloads into space, many countries strove to create their own such technologies. Few, however, succeeded. The launch facilities prove to be a costly luxury.

The expendable rocket launchers that are in current use, had been normally developed on the basis of intercontinental ballistic missiles (ICBM), the cost effectiveness being the governing factor in building them. The cost reduction issue of launching payloads became particularly important due to the spreading commercialization of space. The state agencies of the leading countries involved in space exploration are also concerned about the cost reduction of launches on account of decreasing budgetary allocations to space programs.

The analysis of space related work done abroad shows that the royal road leading to reduction of the cost of launching payloads into space by means of a jet propulsion engine is the multiple use of component elements. At the initial stage this is the multiple reproduction of the carrier components, that is, the use of modular structures. The next step is the building of modular launch facilities and the creation of a family of carriers based on standard modules. The final phase is the multiple use of the launch facilities proper (Fig. 21).

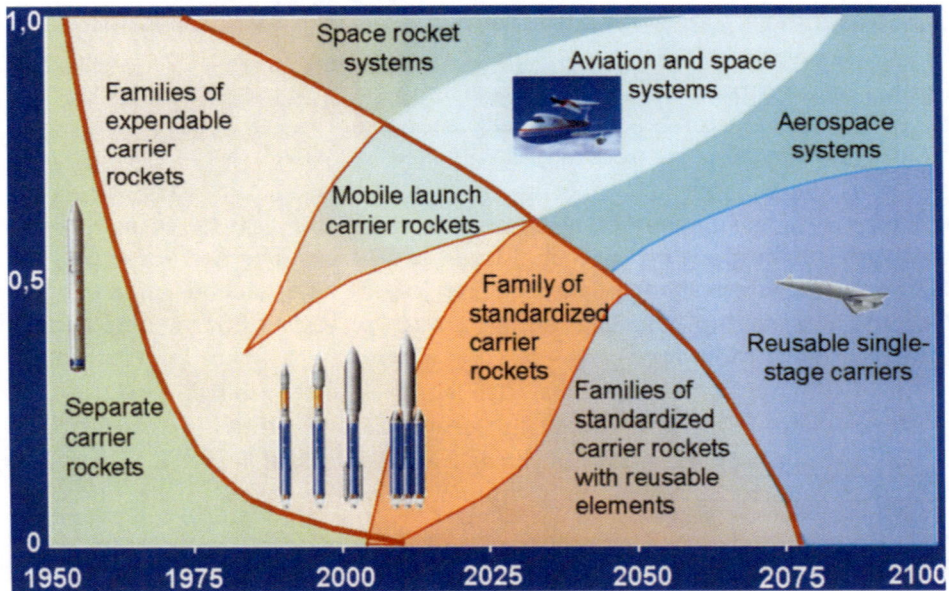

**Fig. 21.** Change of share of various types of launchers

*Launch vehicles of modular construction* consisting of a large number of identical elements have long attracted considerable attention of researchers space technologies. Combined with the use of elements created and optimized for other programs, such an approach held much promise in terms of economy. Interestingly, the term "multiple module construction" is in this case used with reference to the layout diagram not only of separate stages but the launch vehicle as a whole.

As far back as in the mid 1970s, OTRAG consortium was formed in Germany that aimed to develop and put in operation a cheap commercial launch vehicle comprised of a large number of separate units. Depending on the component package, the OTRAG type of launchers was to launch various payloads, e.g. a 200 kg payload to an orbit at an altitude of 100 km (a standard module of four separate units), an up to 10 ton payload to an orbit at altitude of 300 km or 1.5 ton of payload to a geostationary orbit or GEO (a variant comprised of 600 separate units).

The cheapness of the launchers was to be achieved due to the low cost of separate units in the event of their series production, which uses elements and assemblies already manufactured by various industries.

The original plan was to slim down the cost of delivering payloads similar to those launched by Titan-3 type of launcher (up to 1.5 ton off to a GEO) to US$ 20 – 25 million. The first launch of the experimental model of the OTRAG launcher was carried out at a proving ground in Zaire in 1977. Operational launches were supposed to start in 1981. The plans, however, had not met with success. The project was scrapped for financial and political reasons.

In the early 1980s, the American company Space Services Inc. (EER Systems since 1990) developed a family of Conestoga launch vehicles based on the Castor-4 solid propellant booster rocket. The experimental single stage model of the rocket was launched in 1982.

Plans were made in 1994 to launch a three-stage variant of the Conestoga-1229 launch vehicle capable of delivering 220 kg of payload to the solar synchronous orbit at an altitude of 700 km. The first stage was the Castor-4 solid propellant rocket engine (SPRE), the second one was the Casor-4 shortened in half, and the third was the Star-48V SPRE. The launch, however, never took place.

In 1995 a four-stage Conestoga-1620 launch vehicle was launched from the Wallops Island Test Center (WITC). It provided for putting into space 600 kg of payload. Used here as the first stage were 4 Castor-4 SPREs, 2 Castor-4 SPREs as the second, one Castor-4 SPRE was used twice as the third, the Star-48V SPRE as the fourth. The launch site created specifically for that purpose was the first in the USA privately owned launching complex. The cost of manufacturing the Conestoga-1620 rocket is estimated at US$ 18 million.

The launch failed. No more reports appeared about the Conestoga launches.

Yet another company, the American Rocket, developed in the early 1980s a four-stage ILV-1 launch vehicle using 22 hybrid rocket engines. The rocket stages were bundles of engines mounted around a common tank with liquid oxygen. The idea was to use 12 engines in the first stage, 4 in the second and third respectively, and 2 in the fourth. The payload delivered by such a launch vehicle to the low polar orbit could reach 700 kg.

All the cases described represent essentially "piecing together" of identical units of both separate stages and the vehicle as a whole. None of the projects succeeded because of financial difficulties of the companies engaged in development and the slack interest in them on behalf of potential customers.

Nowadays, the principle of the multiple module construction gained acceptance in building modular types of launch vehicles.

*The modular principle of building the launch facilities* became particularly manifest in the EELV (Evolved Expendable Launch Vehicle) research projects carried out to order of the US DoD.

The program of creating new generation expendable EELV launch vehicles has been underway in the USA since 1995. Its object was to replace the now obsolescent and costly carrier rockets.

The implementation of the program of creating new generation expendable launch vehicles (ELVs) will considerably cut down the launch cost (according to some data, by 25 to 50%).

The study of concepts presented on a competitive basis singled out the projects of Boeing and Lockheed Martin. Wishing to secure the permanent competition as long as the EELV program was in force, the DoD employed the strategy of procuring products of two suppliers at a time.

Both products use the modular principle of building launch vehicles.

**The EELV project as offered by Boeing.**

The Delta-4 family of carriers developed by Boeing includes five variants of carriers, i.e. light, medium, heavy and two intermediate ones.

All the variants have a common booster core (CBC) fitted with an RS-68 liquid oxygen and hydrogen engine. Used as upper stages are the PH Delta second stages of the previous generations:

- for lightweight launch vehicle – the Delta-2 derivatives using storable propellant;
- for medium weight launch vehicle – the Delta-3 derivatives using cryogenic propellant;
- for heavy weight launch vehicle – a modified Delta-3 derivative using cryogenic fuel.

The heavy launch vehicles employ three CBC units. The intermediate class of launch vehicles is supposed to use strap-on solid propellant boosters derived from the Delta-3. Boeing proposed to use liquid propellant rocket engines (LPRE) as part of rescue capsules, which were to descend onto a sea surface by means of a parachute. Tests had been conducted that confirmed the feasibility of the concept. The customer, however, declined the proposals.

**The EELV project as offered by Lockheed Martin.**

The EELV family of launchers (also referred to as Atlas-4) developed by Lockheed Martin is based on a single CBC equipped with the Russian-made RD-180 oxygen-kerosene engine. Two variants of upper stages are going to be used:

- Agena-2000 that uses storable propellant;
- Centaur that uses oxygen-hydrogen propellant.

The medium lift vehicle (MLV) employs the Agena-2000 stage. The use of the Centaur stage enables the carrier to deliver payloads to geostationary orbit.

The heavy lift vehicle (HLV) is comprised of three one-piece units and one of upper stages. In either design the first stage of heavy launch vehicles employs three modules. This tallies with the results of research done by experts of the Khrunichev Research Center. Those results showed in the first place that building launch vehicles is economically feasible on the basis of standardized modules and, second, that the optimum effect is obtained with the use of 3 to 4 modules.

In 2001 the first launch of the EELV medium lift vehicle is to take place.

Another project offered by Lockheed Martin and aimed to use standardized modules is the development of launch vehicles for civil customers. The principal goal of creating the new family was to cut down costs while enhancing the reliability. The capabilities of those launch vehicles were determined on the basis of existing and projected demands of the market, which unlike in the case of EELV were oriented to the payloads as asked for by the US DoD.

The backbone of the family will be, as in the case of the EELV family, the common central rocket unit fitted with the RD-180 engine. Used here as the second stage is the Centaur-3 booster with one or two engines. Plans are made to use solid rocket boosters, SRB (from one to five).

The idea is to put together four series of launch vehicles having the following capabilities to deliver payloads to low near-earth orbits (H=185 km, i=28.5°): 300th – 10 tons and 12.7 tons (for one and two power units on the second stage respectively); 400th – 9.85 tons and 12.5 tons; 500th – without boosters – 8.3 tons and 10.3 tons; 500th – with five boosters – 16.35 tons and 20.05 tons; HLV – 19.05 tons or 6.35 tons on GEO. It is planned to use essentially new manufacturing equipment that makes it possible to reduce the time of preparing launch vehicles (down to 10 days for 300th and 400th series) and reach the number of 19 launches per year.

The modular types of expendable launch vehicles were regarded as intermediate space transport facilities in transition to the future commercially viable reusable launchers.

*The use of reusable injection facilities* is the principal means of cost reduction in delivery of payloads to space.

The capability to reuse launchers holds much promise in cutting down the specific cost of orbital injection as compared to the existing ELVs.

The analysis shows that all reusable launchers now in development can be categorized, depending on how they launch and land, into several types.

As per type of launch: vertically, horizontally and aerially started.

As per type of landing: vertically and horizontally landed.

Predominant among the projects of reusable means of injection are single-stage launchers (SSLs), which however does not testify to the final choice in their favor. This, in a way, is "wishful thinking". The work on defining the optimum means of building the required facilities for the purpose is not yet complete. Such work is being carried out, in particular, by Boeing and Lockheed Martin, the companies that set the standard for launchers best suited to meet the USA requirements for the access to space after 2005.

The USA is holding the leading position among the countries that are in the possession of launch facilities.

The USA operates several types of ELVs of various lifting capability (Atlas, Delta, Titan, Pegasus, Taurus and others) plus the reusable Space Shuttle transportation system (SSTS). The space objects are launched at the rate of 30 to 40 per year.

In 1994 the White House issued the directive "The national space transport systems politics" elaborated by the National Council for Science and Technology. It defined the USA politics in building launch facilities.

The directive pointed out two strategic lines of activity:

1. Maintaining at the existing level and improvement of performance of the existing expendable launch vehicles.

2. Investments in research and development aimed to create and operate such reusable space transportation systems (RSTS) of the new generation that essentially reduce the cost of space flights (within 10 years the cost of delivery of a 1 kg payload to a near-earth orbit (NEO) was to come down from US$ 20,000 to 2,000).

Four tasks had been set:

1. To map out the national policy in space flights related expenses which policy would be in line with restrictions of the current budget and capabilities provided by new technologies. According to the new policy, the Department of Defense (DoD) will bear the brunt of responsibility for modernization of existing expendable launch vehicles. NASA will supervise the scientific research, design and development aimed to create reusable systems.

2. To delineate the policy in the use by federal entities of foreign-made launch facilities and their components. The end of the "cold war" has made it possible for the USA to use foreign achievements in science and technology, including those of Russia, and to resort to other states' technologies and launch facilities provided that it does not impair the national security of the USA, its foreign policy and principles of the commercial market of launches.

3. To formulate the policy of using by federal entities of the redundant ballistic missiles for space launches.
The document obliges the government to be mindful of the need for commercial launches and sets certain criteria concerning the use of ballistic missiles.

4. To ensure the increasing role of the private sector in taking decisions at the federal level which pertain to scientific research, design and development in the area of space flights.

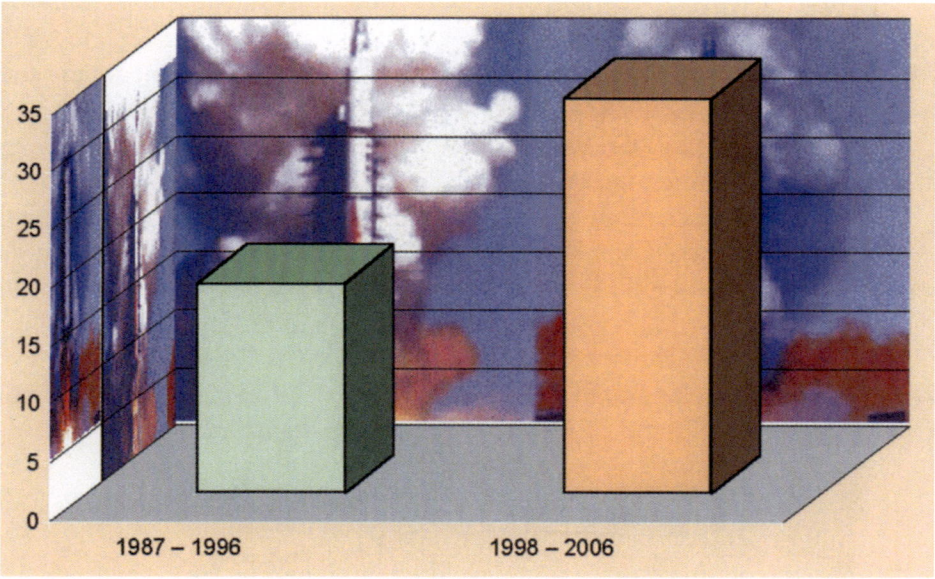

**Fig. 22.** Volume of the world's launch services market, US$ billion.

Compared to the previous national policy in the area of space flights, this document urges the Ministry of Trade and Transportation to seek for chances of cooperation between the government and industry and to give them due consideration in implementing the plans of NASA and the Department of Defense.

The Arianespace consortium that operates the Ariane family of launch vehicles is now carrying out up to 15 launches per year. The types most widely used for launching SC are Ariane-44L and Ariane-44LP. The coming years will witness the commissioning of a more powerful Ariane-5 launch vehicle. It is expected that the share of commercial launches performed with the help of the Ariane-5 launcher will grow while the Ariane-4 will be gradually phased out.

Since 1970 China has been launching the CZ series of vehicles (the Long March) which are used to put spacecraft into orbits, up to stationary ones. The operation is characterized by a high degree of reliability (0.85). The frequency of launching the Chinese vehicles can be as high as 5 to 7 per year.

Japan modernizes now its fleet of launch vehicles. In addition to the modified heavy launcher H-2A, it will put in operation medium lift launchers, J-2 and M-5. The Japanese vehicles will be launched twice to four times a year. The volume of market for launch services in the previous period and in the future till the year 2006, as well as the structure of this market, are presented in Figures 22 and 23.

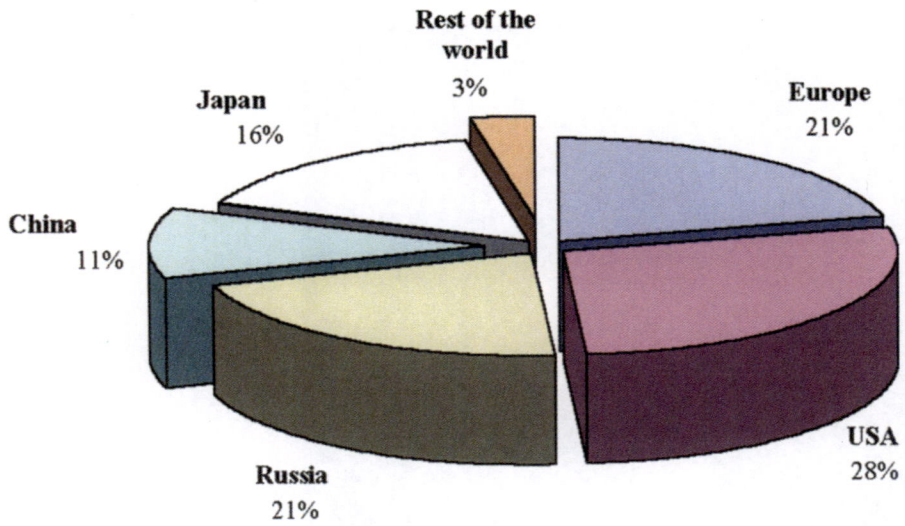

**Fig. 23.** Structure of the world's launching services market in 1990 – 2006

## 1.2.1 Heavy lift launchers

### American heavy lift launchers

The heavy lift launchers are available to the USA, the countries of the European Space Agency (ESA) and Japan. In the possession of the medium lift launchers (MLVs) are the USA, ESA, China, Japan and the Ukraine. India is likely to join soon those countries.

The first heavy lift launchers had been created by the Americans in 1964 – 1967 to service the Apollo lunar program. The most powerful of them, Saturn-5 (Fig. 24), provided for sending into NEO at an altitude of 500 km a roughly 120 ton payload. With the completion of the Apollo and Skylab program such launchers became unnecessary.

The USA now uses Titan-4 heavy lift launchers of the Titan family manufactured by Lockheed Martin (Fig. 25). The first launch of this vehicle took place in 1989.

**Fig. 24.**
Saturn-5 launch vehicle

The Titan-4 launch vehicle is different from others in that it employs powerful solid propellant boosters consisting of 7 fuel sections. In addition, the Titan-4 component package provides for use of the IUS solid propellant two-stage booster and the Centaur oxygen-hydrogen unit (a more powerful variant) as the last stages. The launchers' booster shells are made by graphite epoxy filament winding.

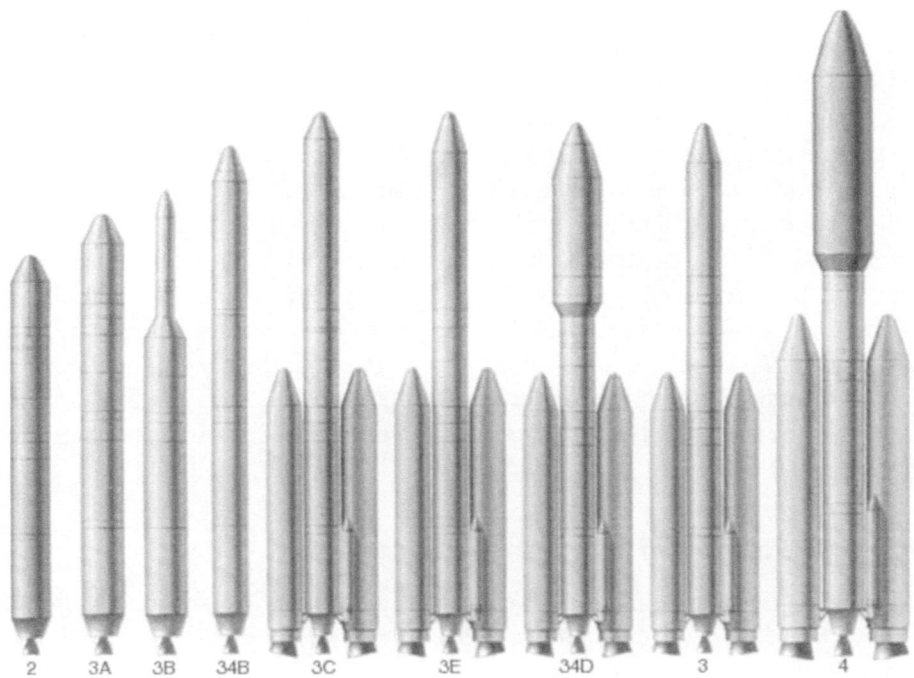

2       3A      3B      34B       3C           3E            34D           3              4

**Fig. 25.** The Titan family of launchers

**Table 3.** Characteristics of Titan-4 launcher

| Characteristics | Titan-4A | Titan-4B |
|---|---|---|
| Lifting capacity, t | | |
| – low near-earth orbit (28.3°) | 17,77 | 21,9 |
| – geostationary orbit | 4,536 | 5,76 |
| – solar synchronous orbit | | 14,09 |
| Number of stages | 2+2 strap-on boosters | 2+2 strap-on boosters |
| Size, m | | |
| – length 62,1 | 63,14 | |
| – max. diameter | 3,05 | 3,05 |
| Take-off weight, t | 868,0 | 939,3 |

The Titan–4 vehicles are launched from the Air Force's Base at Cape Canaveral Space Center and from Vandenberg Air Force Base. The payloads are large-sized military SC or NASA's SC for deep space research. There are no plans for commercial use of Titan-4 vehicles. The most powerful variant of launch vehicles since the expedition of Saturn–5 to the

Moon is the Titan–4B version which has been in operation since 1997. The basic difference of this version is its more powerful solid propellant strap-on boosters manufactured by Alliant Techsystems. Those use a new type of fuel, polybutadiene having a terminal hydroxyl group. The fuel is contained in tanks made of composite material (graphite epoxy). The shells of such boosters are manufactured by means of winding, whereas the shells of the Titan–4 vehicles' boosters are made from D6AC grade of steel. The cost of launching each version of the Titan–4 launcher is estimated at US$ 350 million. On October 12, 1998, Lockheed Martin Astronautics (LMA) received an order from the Air Force to the tune of US$ 1.327 billion for completion of production of 40 Titan–4 launchers and launching 39 of them before 2002. Starting in 1989, 25 Titan–4 rockets have already been launched, 22 of them being the Titan–4A version. Under the new contract, the more powerful Titan–4B model that had been launched three times previously, would be used in another 14 launches.

## Heavy lift launchers of European Space Agency

The step-by-step upgrade of the existing systems, the traditional principle in aircraft building, had been adopted for creating European launchers. This is represented by various models of launchers, including the Ariane-4 vehicle. Contrary to them, the heavy lift Ariane-5 is a new step forward in all respects. That is why the West European experts believe that this launcher will become the new series' first model.

The Ariane-5 launcher (Fig. 26) is expected to be used for sending spacecraft into an orbit which is intermediate relative to a geostationary one or for sending it into a low near-earth orbit.

The tank of the first stage is manufactured from aluminum alloy 3 mm thick. It has a heat-resistant coating made of polyurethane 2 cm thick. Involved in its manufacture are such French companies as Aerospatiale (integrator), SEP (engine) and CRYOSPACE (tank); Holland's Fokker (engine mounting frame); Belgium's SABKA (engine start-up system). Mounted on the stages is the Vulcan-HM60 cryogenic liquid propellant rocket engine (LPRE).

The second stage of the Ariane-5 launcher is supposed to keep the vehicle in the ready-to-launch condition for 60 days after it has been charged with rocket propellant. An L-9 LPRE manufactured by DASA, Germany, is installed on the stages. It provides for four activations. It is used to obtain pitch-and-yaw control.

The Ariane-5 launcher has 2 P-230 solid propellant boosters. The boosters are lined with heat-protective coating. The nozzle orifice is made from carbon reinforced with carbon fiber. The elastically supported bearing provides for the nozzle deflection by 6.0 to 6.6°. The shell of the booster is made from thermally treated D6AC grade of steel. Used as a heat insulator here is the ethylene propylene monomer and silicon dioxide or Kevlar.

**Fig. 26.** Ariane-5 launcher

Spelda and Speltra special detachable conic-cylindrical payload modules are installed in the cargo-bay shroud to enable a simultaneous launch of two or three SC.

In terms of lifting capacity the Ariane-5 launcher is to date one of the world's most powerful launch vehicles. By using them, ESA anticipates to essentially consolidate its position on the international market of launch vehicles where the competition may stiffen all the more with the emergence of new players to the game. ESA hopes that this launcher will come in useful in connection with the expected growth in demand for such facilities on the international market. Those are required for creating international orbital space stations.

The first launch of the Ariane-5 attempted in June 1996 failed. An error in the software resulted in an explosion a mere 40 seconds after the launch.

Late in October 1997 a second, now successful, launch of this vehicle had been undertaken. A third launch within the framework of qualification tests took place in October 1998. The first operational flight of the Ariane–5 launch vehicle took place in December 1999 with the XXM astronomic exploration spacecraft. In 2002 the Ariane-5 was used to send into a polar orbit at an altitude of about 800 km the European Envisat ERS spacecraft weighing 8.5 tons. The modernization plans (Evolution program) provide for building a series of the Ariane-5 vehicles of various lifting capacity: 5E/S (7.1 tons for the transfer orbit ), 5E/SV (8 tons), 5E/CA (10 tons), 5E/CB (12 tons).

According to estimates, the reliability ratio of the Ariane-5 launcher must be 0.985.

**Table 4.** Characteristics of the Ariane-5 basic model

| Characteristics | Basic model |
|---|---|
| Lifting capacity, t | |
| – low near-earth orbit (28.5°) | 18.0 |
| – geostationary transfer orbit | 6.8 |
| – solar synchronous orbit (98.6°) | 10.0 |
| – lunar transfer orbit | 4.45 |
| Number of stages | 2+2 strap-on boosters |
| Overall size, m | |
| – length | 51.37 |
| – max. diameter | 5.4 |
| Take-off weight, t | 746 |

The cost of launching the Ariane-5 vehicle is now estimated at US$ 120 – 130 million. The cost is expected to be reduced by 10% as compared to the Ariane–4 vehicle.

## The Japanese heavy lift launchers

NASDA works on improving the H-2 launch vehicle under new space programs, setting its sights on creation of the H-2A, a more powerful version of the preceding vehicles. The first stage of this vehicle is to receive the LE-7A cryogenic LPRE, with the LE-5B cryogenic LPRE going to the second. The self-same 2 strap-on boosters, as used on the H-2 vehicle, are to be installed as strap-on boosters.

The H-2A vehicle is to deliver payloads weighing up to 20 tons to low NEOs and those weighing up to 4.15 tons to a transfer orbit.

The H-2A vehicle dubbed H-2A-202 is to become the basic model for a whole family of launch vehicles (Fig. 27).

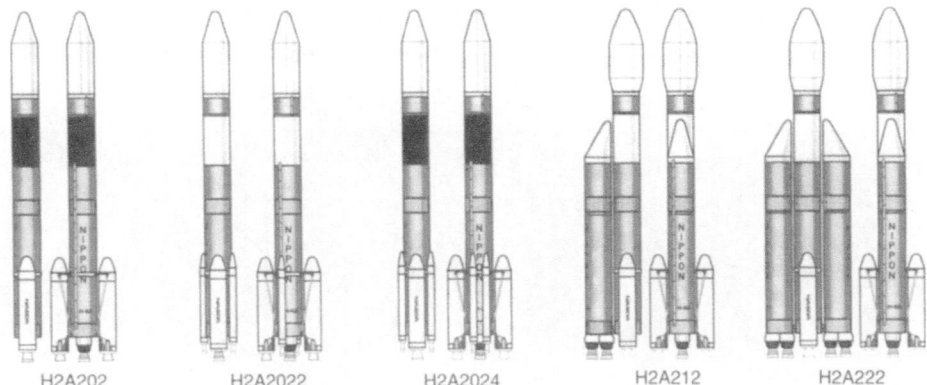

**Fig. 27.** The H-2A based family of Japanese launchers

The H-2A-2022 vehicle fitted with extra two Castor-4 boosters is supposed to put into a transfer orbit payloads weighing up to 4.5 tons. The H-2A-2024 version with four Castor-4 boosters is to deliver up to 5 tons of payload.

The launching of heavy lift communications spacecraft or flying the HTV transport spaceships for visiting international space stations (ISSs) is expected to be accomplished by using the H-2A-212 vehicle capable of putting up to 7.5 tons of payload into a transfer orbit. This is an combination of the cryogenic central unit, two solid propellant power units and one liquid propellant launching booster. In using two liquid propellant boosters (H-2A-222 launcher) the weight of payload increases to 9.5 tons.

The first successful launch of the H-2A vehicle from the Tanegashima Space Research Center took place in August 2001.

It is expected that the cost of the H-2 launcher will be cut down to US$ 80 million (currently it varies between US$ 120 and 160 million).

### 1.2.2 Medium lift launchers

### Medium lift launchers of the USA

Lockheed Martin uses for commercial launches 3 two-stage launch vehicles of the Atlas-Centaur family: Atlas-2, Atlas-2A, Atlas-2AS. Their predecessor, the Atlas-1 launch vehicle, has been out of operation since 1997. These launch vehicles are capable of sending SC to a geostationary transfer orbit (GTO) and differ mainly in their lifting capacity.

The Atlas-2 launch vehicle, in operation since 1991, has been produced since 1995 in Denver only (Colorado) because the San-Diego-based facility (California) is being shut down. Compared to the Atlas-1, the Atlas-2 launch vehicle (Fig. 28) has elongated fuel tanks for liquid oxygen and hydrogen, which provides for a larger rocket propellant loading. Four insulation panels of the Atlas-1 vehicles are replaced with panels made of foam and connected with tanks.

**Fig. 28**. Atlas-2 launch vehicle at the SLC-3E launching site of the Western proving ground (Vandenberg Air Force Base)

The Atlas-2A vehicle has been in operation since 1992. This is an elongated variant of the Atlas-2 with augmented engines on the Centaur elongated last stage. The welded skirt is made of aluminum alloy.

The Atlas-2AS launch vehicle variant that has been in operation since 1993 has an increased thrust as compared to other models of the Atlas-2. Two out of 4 strap-on boosters of the first stage activate at the launching site and are jettisoned in flight 54 seconds after they have begun to operate. The second couple of boosters activates 57.7 seconds after the start of the operation and is jettisoned after the 114th second. In terms of performance the Atlas-2 AS launch vehicle is a close replica of the West European Ariane-4 vehicle.

The Atlas-2 launch vehicles are launched both from the launching sites at Cape Canaveral and the Vandenberg Air Force Base. The cost of launching the Atlas-2 vehicle is estimated at roughly US$ 76 million, that of the Atlas-2AS at US$ 93-98 million.

**Table 5**. Characteristics of the Atlas-2 family of launch vehicles

| Characteristics | Atlas-2 | Atlas-2A | Atlas-2AS |
|---|---|---|---|
| Lifting capacity, t | | | |
| – low near-earth orbit | 6.78 (27.0°) | 7.316 (28.5°) | 8.61 (28.5°) |
| – geostationary transfer orbit | 2.95 (28.5°) | 3.066 (27.0°) | 3.8 (27.0°) |
| Number of stages | 21/2 | 21/2 | 21/2 |
| Size, m | | | |
| – length | 46.8 | (47.5) | (47.5) |
| – max. diameter | 3.05 | 3.05 | 3.05 |
| Take-off weight, t | 187.56 | 187.70 | 237.497 |

The Atlas-3 model is the further development of the Atlas type of vehicles. It differs mainly in that it uses at the first stage the RD-180 engine manufactured by NPO Energomash. Plans are made to operate two models of launch vehicles, Atlas-3A and Atlas-3B which will provide for putting into orbit 4 and 4.5 tons of payload respectively. The first launch of the Atlas-3 vehicle took place on May 25, 2000.

The most successful in terms of reliability were the models offered by McDonnell Douglas (now a division of Boeing). Of those, operated now on the commercial basis, are the Delta-2-7920 two-stage vehicle (for putting into orbits with a low power consumption) and the Delta-2-7925 three-stage vehicle first launched early in 1989. Launches are carried out from the sites at Cape Canaveral and Vandenberg Air Force Base.

The adapter between the first and second stages is of isogradic construction (with evenly arranged elements of rigidity). The staging occurs after 8 seconds as the first stage ceases to operate following the detonation of explosive bolts. The second stage engine starts to operate 5 seconds after staging.

**Table 6**. Characteristics of the Delta-2 launch vehicle

| Characteristics | Delta-2-7925 | Delta-2-7325 | Delta-2-7325 |
|---|---|---|---|
| Lifting capacity, t | | | |
| – low near-earth orbit | 5.039 | <5.0 | <5.0 |
| – geostationary transfer orbit | 1.842 | <1.8 | <1.8 |
| – solar synchronous orbit | 3.175 | <3.0 | <3.0 |
| Number of stages | 3+9 strap-on boosters | 3+4 strap-on boosters | 3+3 strap-on boosters |
| Size, m | | | |
| – length | 38.41 | 38.41 | 38.41 |
| – max. diameter | 2.44 | 2.44 | 2.44 |
| Take-off weight, t | 231.87 | <230.0 | <230,0 |

Cargo-bay shroud is made of aluminum alloy. Being developed now is a shroud made of composite material.

The cost of launching the Delta-2 vehicle is around US$ 46 million.

Fig. 29.
Delta-3 launch vehicle

McDonnell Douglas has offered to the foreign market of space carriers a new medium lift launch vehicle (MLV), the Delta-3 (Fig. 29) which is categorized in the USA as an intermediate class vehicle (i.e. between medium and heavy classes). According to its developers, this more powerful vehicle is required for launching larger communications satellites. The distinctive feature of the new launch vehicle is the employment, on the second stage, of the modernized RL-10 LPRE that uses cryogenic propellant components. The stage is supplied by Pratt & Whitney. This LPRE is similar to the one on the Delta-2 vehicle except that it has a larger fuel tank located at the second enlarged stage. The Delta-3 vehicle comprises 9 strap-on boosters for the first enlarged stage manufactured by Alliant Techsystems which enables the vehicle to deliver twice as much payload as in case of the Delta-2. Another difference is that the Delta-3 vehicle has a controllable nozzle for regulating the thrust vector and is 1.1 to 1.2 m larger than the model of the Delta-2 being in current operation. One of the principal goals in creating it was the reduction of the dry weight of the Delta-3 through the wide use of composite materials and the reduction of the total number of constituent parts in its component package. According to manufacturers' calculations, this launch vehicle will not be inferior to the Delta-2 in terms of reliability. The two oxidizer tanks for the Delta-3 are manufactured by Boeing. Mitsubishi, Japan, manufactures from aluminum the fuel tanks of the first stage and liquid hydrogen tanks of the new last stage of isogradic construction.

Excepting some minor changes, the same launching site at Cape Canaveral is expected to be used for launching the Delta-3 vehicle, as for the Delta-2.

The authors of the launch vehicle hoped it would rival the Atlas and the West European Ariane-4 launchers.

With the take-off weight of 230 tons the Delta-3 vehicle provides for launching a 8.35 ton payload to an altitude of 180 km with an orbital inclination of 28°. In the event of stationary transfer orbit the payload weight is 3.81 tons.

Boeing suggests to also use the Delta-3 vehicle with six strap-on boosters. This is cheaper than using the standard vehicle with nine boosters. At the same time, the performance of such version is somewhat higher than that of the Delta-2 launcher.

The first launch of the Delta-3 that took place in August 1998 failed. The second launch of the Delta-3 with the Orion-3 communications spacecraft took place in 1999 (the SC was delivered to the off-nominal orbit).

The approximate cost of launching the vehicle is US$ 75 million.

Much has been done in the USA to improve the Titan family of launch vehicles through the use of large strap-on solid propellant boosters measuring 3.05 m in diameter. Due to the effort the vehicles of this family came to be used for launching military and civil SC varying widely in weight.

The Titan-2 two-stage space launch vehicle is a carrier converted from an intercontinental ballistic missile (ICBM) of the same designation. Its first launch with space payload took place in 1964. The Titan-2 rockets are developed by Lockheed Martin under the supervision of the department of space systems, the USA Air Force Systems Command. The vehicles are assembled at the Lockheed Martin facility in the town of Middle-River (near Baltimore, Maryland). The Titan-2 vehicle uses a self-igniting propellant.

The first and the second stages use integral fuel tanks made from machined sheets of aluminum alloy with a large content of copper. The thickness of the wall in the lower part of the first stage fuel tank reaches 4.5 mm. The tank bottoms are made of chemically milled sheets of aluminum alloy. The fuel tanks of either stage have a factor of safety equal to 1.25.

To pressurize the tanks, vaporized propellant is used. To achieve this, part of propellants bled after exit from the pumps is passed via pipelines heated by the gas generator of the turbopump unit. As this occurs, propellants become vaporized whereupon they are fed through openings in the bottom to respective fuel tanks. The body skin is put together from separate panels of aluminum alloy 9.5 mm thick. The panels are joined by electric arc welding with the use of tungsten electrodes. The staging proceeds in the "hot" mode, i.e. the second stage engine activates before separation. "Ports" are provided on the first stage adapter for gas exit.

The radioinertial arrangement is used here as a control system. Operating as an actuator device of the control system are the hinged engines of the first and second stages. The pitch, roll, and yaw control of the first stage is accomplished by means of hinged engines. The pitch-and-yaw control of the second stage is accomplished as in the first stage while the roll control is obtained by means of special swiveling nozzles with the use of turbine exhaust gas.

**Table 7**. Characteristics of Titan-2 types of launchers

| Characteristics | Titan-2B | Titan-2G | Titan-2S | Titan-2L |
|---|---|---|---|---|
| Lifting capacity, t | | | | |
| – low near-earth orbit | 3.175 (28.6°) | 3.175 (28.6°) | 3.70 | 8.165 (28.5°) |
| – geostationary transfer orbit | 1.043 | | | |
| – solar synchronous orbit | 3.028 | | | |
| Number of stages | | 2 strap-on boosters | 2+2…8 strap-on boosters | 2+2 powerful |
| Size, m | | | | |
| – length | 31.4 | 31.4 | | |
| – max. diameter | 3.05 | 3.05 | 3.05 | 3.05 |
| Take-off weight, t | | 153.7 | | |

The Titan-2 vehicle has several upgraded modifications. They are the Titan-2B (basic as per one source), the Titan-2G (basic as per other source), the Titan-2S and the Titan-2L (the latter, now in development, is the most powerful).

Compared to the ICBM the Titan-2 modification is equipped with improved adapters, orientation system and low-thrust engine. The Titan-2B modification has an improved electric wiring, onboard electronics and cargo-bay shroud. The Titan-2S modification has additional strap-on boosters (from 2 to 8) with the Castor-4A solid propellant rocket engines (SPRE) and an elongated first stage. Two more powerful strap-on boosters are expected to be added in the Titan-2L modification.

Lockheed Martin Astronautics has converted into rocket launchers 14 Titan-2 ICBMs withdrawn from combat duty.

Since 1995, on the initiative of the Department of Defense in the USA, expendable launch vehicles (EELV) of the new generation have been developed. They are supposed to replace by 2006 the obsolescent and costly Delta-2, Atlas-2, Titan-2 medium lift and Titan-4 heavy lift launchers that are being used now.

The implementation of the program of creating a family of EELV at the turn of the 21st century will boost the USA potential for military use of space (space navigation, early detection of launches of combat missiles, communications equipment). The key goal of the program is to cut down the cost of launches by 25 to 50%.

In 1998, the Air Force designated Boeing and Lockheed Martin to be prime developers of the launch vehicles to be built under the EELV program. Each company was contracted for US$ 500 million to complete design and development work on new transport systems. Also, separate agreements were concluded for using such systems to deploy military satellites in 2002–2006 fiscal years. According to the reached agreements, Boeing is to launch 19 of its rockets to the amount of US$ 1.38 billion, with 9 launches worth US$ 650 million being the share of Lockheed Martin.

The original EELV project of Boeing and Lockheed Martin designed rocket families composed of three types and being similar in their component packages. Those were small lift type (designated as S), medium (M) and heavy (H). Both families are designed on the basis of liquid-fuel stages referred to as common central blocks (CCB). The small and medium lift rockets differed from one another in their upper stages. The heavy lift rocket was additionally fitted with two extra launching boosters made on the basis of CCB.

However, right before the principal contracts had been concluded, both companies decided to abandon the S class of vehicles in order to reduce the cost of developing the rockets. The delivery of the required type of spacecraft (from 4 to 4.5 tons on the polar orbit or about 2 tons on the transfer orbit) was to be accomplished through medium lift launchers and the currently available transport systems. But since the small lift launcher projects prepared under the EELV program can gain wider acceptance in the future, the specifications of such launchers have been included in the proposed description of new means of delivery.

*Launch vehicles of Boeing*

The key element of the Delta-4 family of launchers is the first cryogenic stage measuring 38 m in height and 5 m in diameter. A new oxygen-hydrogen engine, RS-68, has been developed specifically for this unit. It boasts the thrust of 294 tons and is simple in design, which makes it cheap to manufacture. It is thanks to the low cost of operation of the Delta-4 launchers providing the specific cost of cargo delivery in the region of US$ 13,200/kg that in distribution of orders Boeing had been favored over others. The Delta-4 family of launchers is portrayed in Fig. 30.

The Delta-4S small lift launcher is fitted with a CCB, with a second stage borrowed from the Delta-2 launcher and, if necessary, with a Star-48B solid propellant booster unit. The power characteristics of this transport system enable it to send into a polar orbit up to 4.47 tons of cargo and up to 2.2 tons into a transfer orbit.

The difference of the Delta-4M medium lift launcher from the previous model lies in a second stage with one RL-10B-2 oxygen-hydrogen engine and a nose cap 4 m in diameter.

Such a component package of the launcher provides for sending 7.2 tons of cargo into a polar orbit and 4.54 tons into a transfer orbit. Boeing is considering the possibility of fitting the Delta-4M launcher with two or four solid propellant boosters of Alliant Techsystems in order to enhance the launching capability of the medium lift launchers. Three new models designated Delta-4M+4.2, Delta-4M+5.2 and Delta-4M+5.4 make it possible to deliver to a transfer orbit satellites weighing 5.7, 4.8 and 6.6 tons, respectively.

Fig. 30.
Family of Delta-4 launchers

Apart from the improved second stage (fuel tanks diameter increased to 5 m) and the nose cap 4 m in diameter, the Delta-4H heavy lift launcher will also use two launch boosters created on the basis of the first stage. This will increase its lifting capacity to 22.5 and 15 tons in sending payloads respectively to polar and transfer orbits. The Delta-4M and 4H will be the first space transport systems using in their component packages only cryognic stages.

The launches of the Delta-4 vehicles are slated to take place at Vandenberg Air Force Base from the SLC-6 launching pad and from the LC-37 pad at Cape Canaveral. Boeing has allocated US$ 250 million for modernization of the Delta-4 ground-based launch infrastructure at Cape Canaveral. The selected contractor, Raytheon Engineers and Constructors, has improved the launching site, built a new horizontal integration facility (HIF) and done other required jobs.

What makes the Delta-4 launcher different from its predecessors is the horizontal assembly of the product which dramatically reduces the cost and duration of work on the assembly site. For example, the assembly of launchers inside the HIF is supposed to start 14/21 days prior to launch, T time, (for medium/heavy lift vehicles respectively). The pressurization of cargo is made 10/12 days prior to T, erection of the launcher on site is accomplished 8/9 days, stacking of payload 5 days prior to T. Overall, the duration of the preflight preparation of the Delta-4 is to decrease from 24 to 6–8 days as compared to the currently operated Delta-2.

Boeing is deploying the main facility for assembly of the Delta-4's first stages at Decatur, Alabama. The required rigging and manufacturing equipment will be ready for installation at the facility in the first quarter of 2000. The nose caps are planned to be manufactured at the Pueblo plant, Colorado, which is now assembling caps for the Titan-4 launchers. The cap of the Delta-4H is being created on the basis of that launcher's cap.

Most of the Delta-4 structural elements made of composite materials are to be supplied by Alliant Techsystems Space and Strategic Systems Group, which is now deploying the required facilities at Iuka, Mississippi. The amount of work to be done by that company under the Delta-4 program, should it succeed, may be worth US$ 1 billion.

The first launch of the Delta-4 was scheduled for 2002.

*Launch vehicles of Lockheed Martin*

The yet unnamed launchers of Lockheed Martin will be designated as LM EELV (in some sources the Atlas-4 designation may occur). The family of the LM EELV is shown in Fig. 31.

The small lift LM EELV vehicle was supposed to be equipped with a hydrogen-kerosene stage fitted with the Russian-made RD-180 engine developing a 390 ton thrust and with the Agena-2000 stage running on nitrogen tetroxide and monomethyl hydrazine. With such a component package the launcher would be capable of delivering 3.9 tons of payload to a low polar orbit and 1.84 ton to a transfer orbit.

The Agena-2000 stage has been designed by Atlantic Research on the basis of a booster made in the late 1950s. The construction of the stage, which had

proved itself over decades, was going to be improved through the use of
the newest technologies and some optimized subassemblies borrowed from
other LPREs. In particular, a possibility was considered to use on the stage
a gimbal suspension from the Delta vehicles, pipeline valves from the
Ariane-5, titanium nozzles of the LPRE from Apollo spaceship, etc. However,
the cost of modifying the stages appeared so high it would thwart the attempts
to reduce the cost of delivering payloads to space, a requirement imposed
by the EELV program. Therefore, Lockheed Martin decided to abandon the
small lift launch vehicles.

Fig. 31. The LM EELV family

The medium lift LM EELV is a combination of the first oxygen-kerosene stage and the Centaur modernized cryogenic booster unit with one RL-10-A4 oxygen-hydrogen engine. Thus in terms of configuration it repeats the Atlas-3 launcher. The lifting capacity of the new transport system will be 7.3 tons in delivery to a low polar orbit and 3.855 tons in delivery to a transfer orbit. Same as Boeing, Lockheed Martin studies the feasibility of fitting the medium lift launcher with several solid propellant boosters. In doing so, representatives of both companies declare that the modernized medium lift launchers will be used mainly for launching commercial communications satellites whose weight is expected to grow in the future.

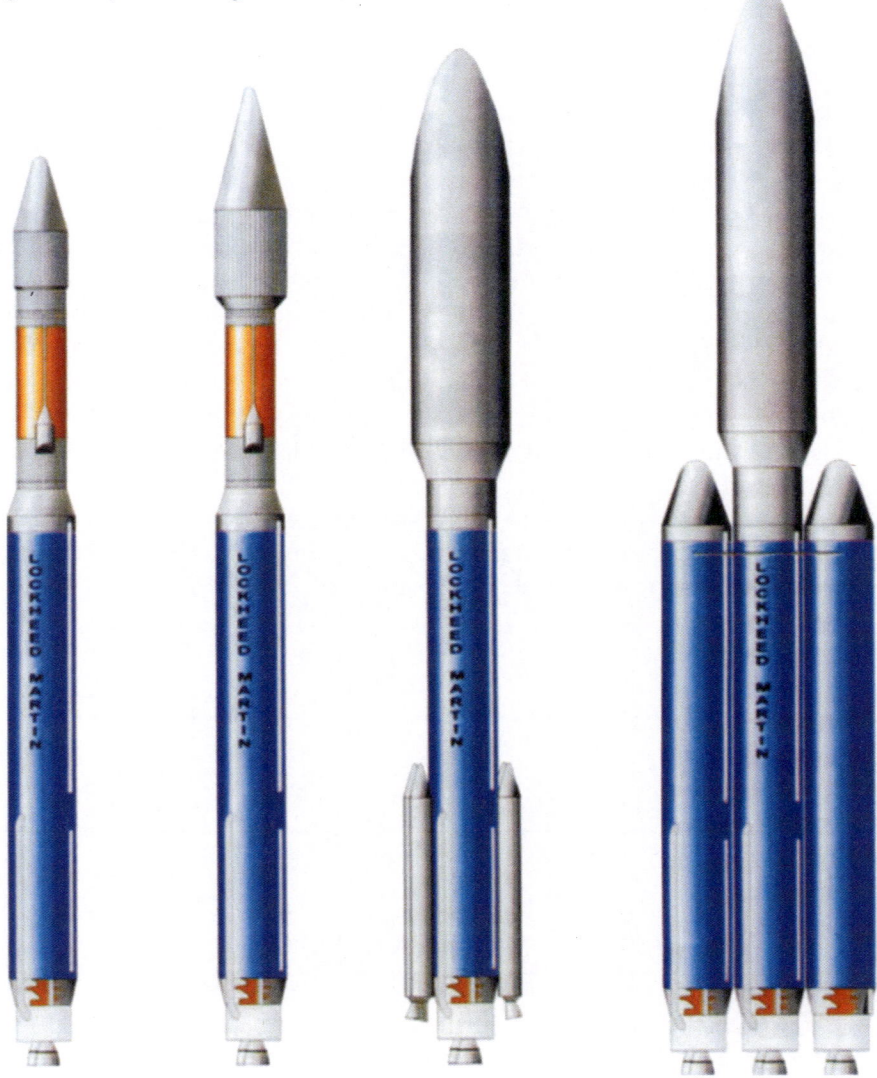

**Fig. 32**. Atlas-5 family of launchers

The heavy lift LM EELV with a lifting capacity of 18.6 and 6.1 tons during insertion into the polar and transfer orbits will differ respectively from the previous model in having two launching boosters made on the basis of the first stage.

The launches of LM EELV whose assembly must proceed vertically will be carried out from LC-41 launching site at Cape Canaveral and SLC-3W site of Vandenberg Air Force Base.

Lockheed Martin also undertook to develop one more family of launchers that would reduce the cost of launches by virtue of their modular construction. This is the Atlas-5 commercial transport system (Fig. 32).

The reasons for those two very similar efforts are explained by the fact that the launchers developed to order of state agencies are allegedly not allowed to be used commercially. It is also possible that the LM EELV family (Atlas-4) developed under the EELV program was not the developers' best job. Based on the amount of contracts concluded by the US DoD with Boeing and Lockheed Martin equal respectively to US$ 1.38 billion and US$ 615 million as well as the number of launches envisioned by those contracts (19 and 9 respectively) it may be deduced that Boeing enjoys certain preferences in the EELV program.

Taken as a basis for creating the Atlas-5 family of launchers is the common core booster (CCB). Used as the second stage on the Atlas-5 vehicles is the Centaur-3 booster unit with one or two RL10A-4-2 oxygen-hydrogen engines. The Atlas-5 is going to use solid rocket boosters. Three types of nose caps are used for the Atlas-5 series: 3 meter medium variant; 4 meter long and extended variants; 5 meter short, medium and long variants.

Plans are made to use four series of the Atlas-5 launchers: 300th, 400th and 500th (as per diameter of the cap) and heavy launch vehicle (HLV).

The weight of payloads delivered to various orbits by basic types of launchers of the Atlas-5 family (with one engine comprised by the rocket unit and without launching boosters) is shown in Table 8.

**Table 8.** Weight of payload delivered by Atlas-5 family of launchers to various orbits, ton

| Type of orbit | Series | | | |
|---|---|---|---|---|
| | 300 | 400 | 500 | HLV |
| GTO | 5.1 | 5.0 | 4.1 | 13.2 |
| GEO | | | 1.5 | 6.4 |
| Polar (H=185 km) | 8.2 | 8.0 | 6.7 | 19.0 |
| Low (H=185 km, i=28,5°) | 10.0 | 9.9 | 8.3 | 19.0 |

The first launch of the Atlas-5 launcher of series 400 was scheduled for mid-2002. The use of EELV launchers of the new generation will enable the USA to control up to 50% of the market of commercial launches instead of today's 30%, thus ousting somewhat ESA with its Ariane launchers.

The cost of delivery of payloads with the help of EELV launch vehicle based on the Delta-4 must not exceed US$ 35 million; based on LM EELV must not exceed US$ 60 million. According to other sources, the cost of delivery of payload with the help of one EELV launcher can be as high as US$ 85 million.

## Medium lift launchers of European Space Agency

The West European space consortium, Arianespace, has been operating for over 10 years the Ariane-4 medium lift launchers (the first launch of the Ariane-4 vehicle took place on June 15 1988). Its safety factor is 0.97. Six modifications of this launcher have been built which put around 1,400 satellites into space. Given below is the description of modifications now in use.

In the course of implementation of the program for creating the Ariane-4 three-stage launcher six modifications of this family had been built which differed from one another mainly in the number and type of strap-on boosters and, hence, in their lifting capacity: Ariane-40, Ariane-42P, Ariane-44P, Ariane-42L, Ariane-44LP, Ariane-44L (Fig. 33).

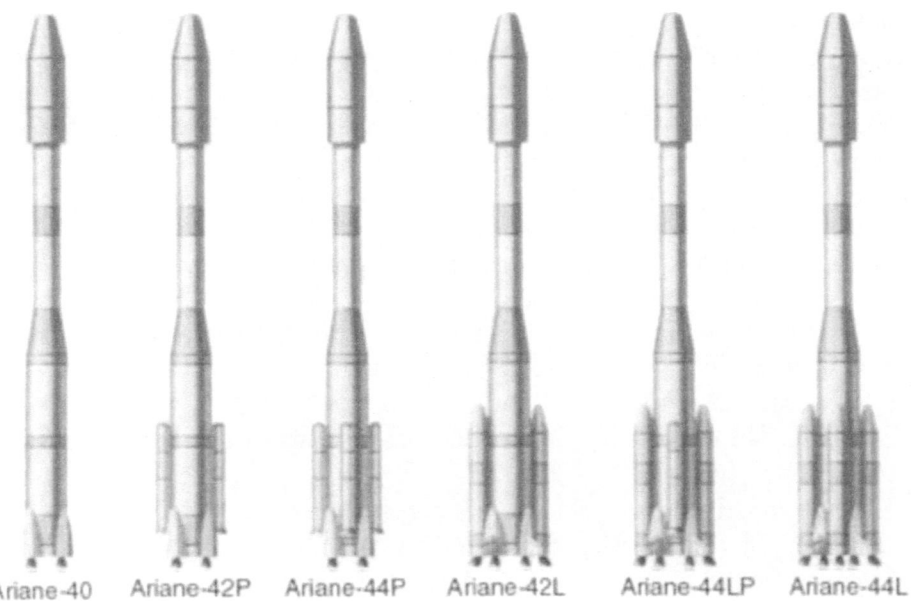

Ariane-40    Ariane-42P    Ariane-44P    Ariane-42L    Ariane-44LP    Ariane-44L

**Fig. 33.** The Ariane-4 family of launchers

Four Viking-5 sustainer LPRE are used on the first stage of the Ariane-4 launchers. Two identical tanks made of 15CDV6 steel 3.8 m in diameter and 10.09 mm in thickness are separated by an adapter 2.688 m in height and a water tank 0.73 m in height with the forefront hemispherical bottom reinforced with glass fiber. Water-cooling is also provided for strap-on boosters (running on liquid fuel). The front skirt of the stage with the height of 1.5 m is supported by 8 brake assemblies manufactured by BPD, Italy.

A Viking-4B LPRE is installed on the second stage of the Ariane-4 launcher. The stage has stiffening members. The fuel tanks of aluminum alloy form a cylinder 6.515 m high with hemispherical partitions that divide them into two vessels. The feeding pipeline for nitrogen tetra oxide (oxidizer) is passing through the lower tank. The engine fastening frame comprises a cylindrical section 188 mm high and a cone 1.35 m high that carries a fastening flange with a gimbal suspension. The frame is connected with a conical adapter between the first and the second stages of the launcher by means of a rear conical skirt 1.57 m high. The skirt incorporates a toroidal tank for cooling water of the Viking-4 engine (the total diameter 2.24 m, pipeline diameter 340 mm). The front skirt 1.245 m high is connected with an adapter between the second and the third stages made of carbonized fiber and incorporates three spherical containers with helium for tank pressurization.

Because the third stage of the Ariane-4 launcher is intended for sending spacecraft into GTO, its structure has also been reinforced with stiffening members. The rear-end adapter between the stages is of carbonized fiber manufactured by Fokker. This lightens it by 53 kg compared to the aluminum adapter of the predecessor, the Ariane-3 launcher, and provides for an increase of the SC weight by 12.3 kg. The cryogenic fuel is in the tanks made of aluminum alloy 7020. The tanks are separated by a phenolic polymer partition. Abutting on the tanks is a 450 mm high front skirt joined with the instruments section. A HM-7B cryogenic LPRE manufactured by West European SEP company is used on this stage (since 1994).

Solid propellant strap-on boosters are only used on the Ariane-42P, Ariane-44P and Ariane-44LP modifications. The solid propellant rocket engines (SPREs) of the boosters activate during the vehicle's launch 4.2 seconds subsequent to the primary engines and separate 89 seconds after activation on the Ariane-2LP, 78 seconds on the Ariane-44P and 67 seconds on the Ariane-44LP launchers. The shell of such booster is made from AISI-4130 grade of steel 5 mm thick.

Liquid propellant boosters manufactured by DASA, Germany, and other West European companies are comparable in size and performance to the second stage and are used alongside solid propellant boosters on the Ariane-42L, Ariane-44LP and Ariane-44L launchers. The tanks are stainless steel 2.1 mm thick. Water-cooling is used here which is accomplished via inter-tank adapter-skirt of the first stage. The separation of boosters occurs pyrotechnically after 149.1 seconds at an altitude of 37.5 km after the charge of SPRE burns out (in 143.6 seconds). Specifications of the Ariane-4 launchers are shown in Table 9.

In spite of the widening scope of work under the Ariane-5 program, the management of the Arianespace consortium expects to continue operating the Ariane-4 launchers well after 2003. Plans are made to develop a new Ariane-4-Lite launcher which would reduce the cost of sending payloads into transfer or polar orbits by 35%.

**Table 9.** Characteristics of the Ariane-4 family of launchers

| Characteristics | Modification of Ariane-4 launcher | | | | | |
|---|---|---|---|---|---|---|
| | -40 | -42P | -44P | 42L | -44LP | -44L |
| Lifting capacity, t | | | | | | |
| – low near-earth orbit | 4.6 | 6.0 | 6.5 | 7.0 | 7.0 | 7.0 |
| – geostationary transfer orbit | 1.9 | 2.6 | 3.0 | 3.2 | 3.7 | 4.2 |
| – solar synchronous orbit | 2.7 | 3.4 | 4.1 | 4.5 | 5.0 | 6.0 |
| Number of stages | 3 | 3+2 solid propellant boosters | 3+4 solid propellant boosters | 3+2 liquid propellant boosters | 3+4 solid propellant and 2 liquid propellant boosters | 3+4 liquid propellant boosters |
| Size, m | | | | | | |
| – length | >56.35 | >56.0 | >56.0 | <60.0 | <60.0 | 60.13 |
| – max. diameter | 3.8 | 3.8 | 3.8 | 3.8 | 3.8 | 3.8 |
| Take-off weight, t | 245 | 324 | 356 | 363 | 421 | 484 |

## Medium lift launchers of China

Two modifications of the Chanzhan-3 (CZ-3) basic three-stage launcher – Chanzhan-3A (CZ-3A) and Chanzhan-3B (CZ-3B), operated now by China – can be categorized as intermediate class (according to American classification). The latter two do not fundamentally differ from the basic type. Chanzhan-2E (CZ-2E) can also be referred to that group. Foreign experts consider these launchers identical to the American Delta-2 medium lift launchers. For the first time in their space effort the Chinese used a cryogenic power unit on their launcher. The outside view of the CZ-3B launcher is shown in Fig. 34.

The CZ-3 vehicle was first lifted off in 1984 and has performed by now over 10 launches. Its safety factor is 77.8%. The CZ-3A modification first started in early 1994, the CZ-3B in early 1996. The CZ-3 series are intended for sending SC into GTO from Sichan space base. Characteristics of the CZ type of launchers are shown in Table 10.

Four YF-20B or YF-21B LPREs are installed on the first stage of the CZ-3A launcher. Installed on the second stage is the YF-22B LPRE plus four YF-23B small thrust control LPREs suspended on hinges. The third stage is the first

Chinese-made stage with 2 YF-75 restartable LPREs. The tanks for liquid oxygen and hydrogen with a common partition are heat-insulated by means of a heat protective coating, which is sprayed polyurethane foam. The first activation provides direction to the orbit, the second one enables insertion into the geostationary transfer orbit.

Fig. 34.
CZ-3B launcher

A choice of 2 glass fiber fairings is provided, 8.887 m long, 3.35 m in diameter. The fastening assembly of the payload honeycomb arrangement with aluminum filler is secured by bolts to the instruments compartment and affixed with its front to the payload section adapter by means of bolts. Two explosive bolts accomplish the separation of payload.

Installed on the three stages of the CZ-3B launcher are the same LPREs as on CZ-3A. This launcher, however, has 2–4 strap-on boosters with a YF-20 LPRE.

In 1990, the first CZ-2E two-stage vehicle of the same class had been launched from the spaceport of Sichan. By now, some 10 launches have been executed.

Installed on the first stage of the launcher are 4 YF-20 LPREs, YF-22 or YF-23 LPRE on the second, and one YF-20 LPRE on each of the four strap-on boosters.

After China's consent to bring the cost of launching SC with the help of its CZ-3 launchers closer to the prices of the world's market the cost of launching by means of such vehicles amounts to around US$ 70 million.

One more medium lift series of launchers is being developed in China.

**Table 10.** Characteristics of CZ-3A, CZ-3B and CZ-2E launch vehicles

| Characteristics | CZ-3A | CZ-3B | CZ-2E |
|---|---|---|---|
| Lifting capacity, t | | | |
| – low near-earth orbit | 8.5 | 12.0 | 8.8 |
| – geostationary transfer orbit | 2.3 | 4.85 | 3.46 (with commercial perigee stage EPKM) |
| – solar synchronous orbit | | 5.7 | - |
| – interplanetary flights orbits | | 5.5 | - |
| – Number of stages | 3 | 3+4 strap-on boosters | 2+4 strap-on boosters |
| Size, m | | | |
| – length | 52.52 | 55.55 | 49.7 |
| – max. diameter | 3.35 | 3.35 | 3.35 |
| Take-off weight, t | 240 | 425 | 462 |

## Medium lift launchers of Japan

The H-2 launch vehicle developed by Mitsubishi under the supervision of NASDA essentially enhances Japan's capability to launch large-sized SC into space. According to the American classification it could be categorized as intermediate (between medium and heavy) class. There is some similarity between the construction of the H-2 vehicle and that of Western Europe's new heavy lift Ariane-5. They differ mainly in size and the total thrust.

**Table 11.** Characteristics of H-2 launch vehicle

| Characteristics | |
|---|---|
| Lifting capacity, t | |
| – low near-earth orbit | 10.0 |
| – geostationary transfer orbit | 4.0 |
| – geostationary orbit | 2.0 |
| – solar-synchronous orbit | 4.3 |
| Number of stages | 2+2 strap-on boosters |
| Size, m | |
| – length | 51.1 |
| – max. diameter | 4.0 |
| Take-off weight, t | 278.0 |

The component package of the H-2 has been selected so as to reduce the total cost of development and to simultaneously enhance the reliability compared to the existing launch vehicles. The manufacture of the H-2 launcher anticipated the use of exclusively Japanese technologies.

The first stage of the launcher is fitted with an LE-7 LPRE fueled with liquid oxygen and hydrogen manufactured by Mitsubishi on principles similar to those employed in the engine of the American Space Shuttle. Installed on the second stage of the vehicle is the LE-5A LPRE of Mitsubishi, also running on liquid oxygen and hydrogen. In terms of size and performance this LPRE is comparable to the Vulcan LPRE of the Ariane-5 vehicle. The launch vehicle has two solid propellant strap-on boosters manufactured by Nissan. The vehicle is equipped with a new inertial control system using three ring-laser gyroscopes.

The cargo-bay shroud has two modifications, both being aluminum honeycomb structures.

The spaceport for launching H-2 vehicles is Tanegashima.

According to estimates, the cost of launching H-2 vehicles exceeds that of commercial MLVs of the USA, ESA, China and Russia and is in the region of US$ 175 million.

## Medium lift launchers of India

Since 1986 the Hindustan Aeronautics, India, has been developing under the supervision of ISRO a three-stage medium lift geosynchronous space launch vehicle (GSLV) for placing satellites in GEO. Installed on the first stage of the GSLV is an S-125 solid propellant engine of Indian manufacture, and Vicas LPRE, also Indian-made, on the second one. Used here as the third stage is the 12 KRB cryogenic booster unit developed by the Khrunichev State Space Research and Production Center. Vicas LPREs are also to be installed on 4 strap-on boosters.

The GSLV has the take-off weight of 402 tons and is 50.9 m long. It can place a 2.5 ton payload in a GTO.

According to ISRO, the cost of commercial launch by means of GSLV must be in the region of US$ 80 million. The first launch of the GSLV from the Indian launching site on the island of Shrikharikota took place in April, 2001.

The Indian experts believe that the use of GSLV will make India fully independent in carrying out its national space program and will enable it to abandon the practice of sending its own satellites into GEO with the help of foreign-made launch vehicles.

## 1.2.3 Small lift launch vehicles

## Small lift launch vehicles of the USA

In 1993, Lockheed Martin started working on the program aimed to create a family of small and medium lift vehicles (Lockheed Launch Vehicle, LLV). The first launch of a vehicle of that family, a small lift two-stage solid propellant LLV-1, that had been repeatedly postponed due to various malfunctions, did take place in August 1995, but failed.

After the merger of Lockheed and Martin Marietta in 1995, the program was dubbed Lockheed Martin Launch Vehicle (LMLV) while the launchers got the names of LMLV-1 (Athena-1), LMLV-2 (Athena-2) and LMLV-3 (Athena-3). The outward view of the Athena-1 vehicle is shown in Fig. 35.

The component package of the LMLV is as follows:
- LMLV-1 with lifting capacity around 1.0 ton and above – two-stage, 14.0 m long, 66.3 ton take-off weight;
- LMLV-2 with lifting capacity around 2.0 tons and above – three-stage, 23.0 m long, 121.5 ton take-off weight;
- LMLV-3 with lifting capacity up to 4.0 tons – three-stage, with strap-on boosters.

These launchers are expected to be used for placing payloads in low near earth orbit. At the same time, there are plans to develop additional stages enabling the payloads to reach GEO and other planets.

Installed on the first stage of the LMLV is the Castor-120 SPRE manufactured by Thiokol. The same SPRE is installed on the second stage of the LMLV-2 and LMLV-3. Installed on the last stage (second for LMLV-1 and third for LMLV-2 and LMLV-3) is the Orbus-21D SPRE manufactured by Pratt & Whitney (division of the United Technology Corporation). Installed on strap-on boosters of the LMLV-3 is the Castor-4A SPRE manufactured by Thiokol.

**Fig. 35.**
Athena-1 launcher

The Athena-2 launcher was first used to launch the AMC Lunar Prospector on January 7, 1998. The cost of the launcher in that case was US$ 26 million. The first launch of the Athena-1 took place in January 1999. The expected cost of launches is US$ 14 million for Athena-1 and US$ 18 million for Athena-2.

The LMLV family of launchers is characterized by short launch preparation time with a small number of personnel involved in maintenance which ensures low costs. According to the Mplan, the launch team will comprise 25 to 30 people, with only eight work places in the launch control van.

The installation of payload in LMLV should proceed, according to the program, three days prior to the launch. A team of operators, working one shift, must prepare LMLV-1 for launch in 15 days, LMLV-2 in 18 days, LMLV-3 in 25 days.

Foreign experts believe that LMLV-1 and LMLV-2 will rival Pegasus and Taurus launchers manufactured by Orbital Sciences company which currently dominates the market of small lift launchers.

EER Systems took the initiative to develop the Conestoga carrier rocket. The launch of one of Conestoga modifications took place in 1982. The second launch of the Conestoga-1620 solid propellant rocket carrier took place in 1995 at the launching site of Wallops Island.

The payload was the METEOR-1 orbital platform designed to carry out a variety of industrial technology experiments in flight. The launch, however, was unsuccessful and at the 40th second of the flight the rocket was detonated.

The Conestoga-1620 carrier rocket measures 15.4 m in height and is a combination of seven Castor-4 solid propellant engines. Four SPREs form the first stage, two the second, one central forms the third. The Star-63F SPRE is used as the upper (fourth) stage.

With its take-off weight of 87.3 tons, the Conestoga-1620 rocket allows to launch a 1.2 ton payload to an altitude of 180 km with an inclination of 28° and objects weighing around 1.0 ton to polar orbits at the same altitude. In the first launch the weight of payload was 944 kg. It was supposed to be delivered to an orbit at an altitude of 450 km and inclination of 40°.

The planned cost of launch was in the region of US$ 10–20 million.

Since 1983 Microcosm and Space Machine Engineering companies (Torrance, California) have been developing to the order and with financial backing of the US Air Force Philips Laboratory a family of Scorpios multi-role carrier rockets of a new generation whose distinctive features are the simple design, including that of the power unit, and the use of the newest composite materials and advanced manufacturing technologies with massive employment of computers, all of which is accomplished at a relatively low cost. No gantry crane will be required to prepare the rockets for launches or to service them. Nor will it be necessary to use a service tower for servicing the payload. For example, the Liberty launcher of this family measures a mere 9 m in height, thus providing an easy access to the payload section.

The Scorpios basic launcher has a five-stage construction. The first 4 stages have 7 bundles of engines, 7 in each, i.e. a total of 49 engines.

It is planned to build launchers of the following classes: SR-S – sounding rocket (ultra-small lift), SR-1 – sounding rocket, SR-3 – small lift launcher (for micro spacecraft), Liberty – small lift launcher, Exodus – medium lift launcher.

As regards the characteristics of the launchers, only the weight of payload sent into orbit is known. They are shown in Table 12.

The cost of putting payloads into orbit with the help of the Scorpios MLV must not exceed US$ 8 million.

Platforms International Inc. company (Redland, California), the developer of unmanned systems recently established a space division and is negotiating for supplies of equipment required to implement the first project of this division – the development of the SpaceRay launcher for launching commercial satellites.

**Table 12.** Weight of payload delivered by Scorpios

| Modification of Scorpios launcher | Altitude of orbit, km | Payload weight, t |
|---|---|---|
| SR-S | 200 | 0.1 |
| SR-1 | 200 | 0.408 |
| SR-3 | low NEO | 0.077 |
| Liberty | low NEO | 1.0 |
| Exodus | low NEO | 6.804 |

In the developers' opinion, the SpaceRay is the only system that meets all the requirements imposed by the Commercial Space Transportation Study (CSTS) on the ideal commercial launcher in the way of lifting capacity, accessibility, reliability, cost effectiveness and user-friendliness.

According to representatives of the Platforms International, the SpaceRay system stands a good chance of winning the leading position specifically in this segment of the market since it uses already available and flight proven components and doesn't have to either develop new technologies or create new prototypes. The leadership is expected to be won by placing orders for development of the SpaceRay system with organizations that are well experienced in designing space equipment.

The developers of the SpaceRay expect to reduce by half the cost of placing satellites in orbit as compared to the rival systems at the sacrifice of the extremely costly ground-based infrastructure normally used in such cases. A fully reusable rocket with horizontal launch and landing capability will make it possible to "launch on demand" and "pay upon the accomplished fact of launching" which is different from the currently accepted practice of three years' contracts with step-by-step prepayments. In view of the wide variety of payloads and their various configurations it will become possible to reduce the payback period from several months to several hours.

The US Air Force has begun to realize its plan to use the decommissioned Minuteman ICBM as a carrier rocket. The required optimization is going to be done under the Multi-Service Launch System (MSLS) program now being implemented by the Air Force. The program calls for the retrofitting of 43 Minuteman missiles, to make them suitable to launch sub-orbital and orbital objects.

The difference between the sub-orbital and orbital Minuteman rockets lies in the number of stages. Two stages of the Minuteman-2 ICBM are required for sub-orbital injection of objects. To reach an orbit, a third stage of the Minuteman-2 or Minuteman-3 is required, or any other upper stage.

The MSLS program started in 1992 when the Air Force concluded a roughly US$ 40 million contract for optimization of the Minuteman ICBM for sub-orbital injections and for one demonstration launch in early 1996. Lockheed Martin Astronauts (Denver, Colorado) has been acting as an executor of the program since 1995.

The legal basis of the program is the directive of USA President "The Policy in National Space Transportation Systems" which allows to place in orbit the satellites of the state's agencies with the use of decommissioned militry missiles. In this case, the following demands must be satisfied:
• the use of rockets for handling the tasks of the customer organization;
• compliance with provisions of agreements on arms control;
• lesser cost as compared to the currently available commercial carrier rockets;
• approval by the secretary of defense of each decision relating to orbital launches of retrofitted rockets.

The pivot of the MSLS program is the demonstration of cost effectiveness of using the decommissioned rockets. According to experts, the cost of launching the Pegasus, the chief rival of the Minuteman, is between US\$ 10 and 12 million as compared to the estimated cost in the region of US\$ 5–8 million for the MSLS program.

According to the Air Force data, the cost of converting one Minuteman ICBM into a sub-orbital rocket is US\$ 3.5 million, that of converting it into an orbital objects launcher is US\$ 4 million. The cost of preparing the carrier rocket for launch and of using the launching facilities is estimated at US\$ 2–4 million. In addition, unofficial estimates show the Pegasus to be a rather costly means of launching small size objects, whereas launchers created under the MSLS program can launch experimental satellites weighing from 225 to 360 kg that cannot be delivered to orbit by any other means.

The Air Force decided to launch in 1998 a MightySat-2 satellite, equipped with classified experimental instruments, by using the Minuteman ICBM as a carrier rocket. Previously, its launch was planned to be executed by using the Space Shuttle reusable space transport system.

The new three-stage Minuteman launcher will use as a sustainer stage two lower stages of the Minuteman-2 ICBM and the third stage of the Minuteman-3 ICBM. This launcher will ensure the delivery to orbit of objects weighing up to 150 kg. To implement the project, the Air Force has contracted with the Spectrum Astro company (Phoenix, Arizona), via the Phillips Laboratory, for manufacture of 5 satellite platforms along with all supporting equipment, for assembly of the launchers and for performance of purpose-oriented operations. The cost of the deal is estimated at US\$ 23.5 million.

Orbital Sciences Corporation (OSC) took the initiative to develop the Pegasus aerodynamically supported missile with an air start. The first launch of the Pegasus from the B-52 aircraft was executed in 1990.

The Pegasus vehicle is a solid propellant three-stage rocket with a takeoff weight of 18.6 tons. It is 14.9 m long. The payload delivered to a circular orbit at an altitude of 460 km with an inclination of 90° was 0.27 ton.

The launch cost is US\$ 11.5 million.

In 1994 a modified variant of the launcher, dubbed Pegasus-XL (Fig. 37), began to be used. Its characteristics are similar to those of the basic model of the Pegasus. However, the elongated first and second stages contribute an 18% increase to the lifting capacity. With its take-off weight of 23.6 tons the Pegasus-XL provides for injection of a 0.45 ton payload into a low near-earth orbit at an altitude of 256 km and inclination of 28°.

The launch cost is US\$ 6.5–13.5 million.

As of January 1, 2002, 29 launches of the Pegasus vehicles have been executed, with 61 SC sent into space.

In addition, there are reports that OSC has been planning to develop a

Pegasus-Turbo, a new and more powerful variant of the aerodynamically supported missile.

To increase the weight of payload sent into low near-earth orbit to 1.02 ton, the company planned to install two extra turbojet engines on the first stage. According to experts' assessments, the new modification of the launcher must be 25–30% more expensive than the existing one. However, due to the enhanced lifting capacity, the specific cost of injection of 1 kg of payload into space must be 40% lower. The take-off weight of the Pegasus-Turbo launcher will amount to 20.8 tons.

In terms of lifting capacity, the Pegasus-Turbo is expected to be an intermediate link between the Pegasus and the Taurus ground-based launcher, the product of the same company.

**Fig. 36.** Pegasus-XL launcher under the fuselage of the L-1011 carrier aircraft

The Taurus launcher has been developed by OSC as a reliable and cost effective means of launching SC weighing up to 1,360 kg into low near-earth orbits and those weighing up to 360 kg into geosynchronous orbit (Fig. 37).

The Taurus is a four-stage rocket derived from the air-based variant of the Pegasus. The Taurus launcher stages are officially numbered from zero to third. The total length of the launcher is about 27 m, the take-off weight is 68 tons.

**Fig. 37.** Taurus launcher

The project employed advanced technologies in the structure and control system. The carrier had been originally created as a transportable vehicle whose launch could be executed in a short time and from whichever place the customer desired.

The first launch of the Taurus vehicle took place on March 13, 1994, and was a success. The rocket sent into orbit two military research spacecraft, STEP-MO and DARPASAT. In the time subsequent to the event, OSC had developed an improved variant capable of carrying a greater payload and having adequate room for it. During the launch on 10 February 1998 the Taurus 2210 vehicle had a more powerful first stage. Used in the first launch as the first stage was the redesigned and optimized first stage of the Peacekeeper ICBM (MX). In the second launch a similar, though specially manufactured rocket unit was used, equipped with the Castor-120 engine made by Thiokol Corporation. In addition, a wider nose cap was used (2,337 mm in diameter instead of 1,575 mm). To accommodate two extra SC together with the prime payload, a dual payload attach fitting (DPAF) was used. It was jettisoned in orbit. Together with the DPAF the third stage is 2.21 m long, its diameter is 1.6 m and it weighs 511 kg.

In all, as of January 1, 2002, six launches of the Taurus vehicles have been executed. The last one took place in September 2001 and failed.

OSC plans to develop a modern medium lift launcher of this family, Taurus-2. In terms of payload this is the same class of vehicle as the American Delta-2 and Titan-2 carriers. It is to be equipped with the last stage having a LPRE of German manufacture fuelled with a two-component propellant originally developed for the Ariane-5 launcher. The Taurus-2 is based on the newest concept that resulted from the Americans' effort to replace the obsolete expendable launchers by those superior in cost-efficiency ratio.

The US Administration and Congress give more support to those developments than to the costly programs aimed to build costly reusable carriers.

According to CEO of Orbital Sciences, the company is prepared to make available US$ 20 to 25 million for development of the Taurus-2 launcher provided that sufficient number of orders is obtained for launching payloads with the help of this carrier. The company had spent roughly similar finances to create the basic version of the Taurus vehicle and a modification of the Pegasus-XL vehicle with an air start.

The Taurus-2 vehicle has several rivals. Among them are the LMLV family of carriers.

The draft of the Taurus-2 vehicle had to be prepared as far back as 1995 with the first demonstration flight slated for the late 1996 – early 1997. However, as of January 1, 2002, no launches of the Taurus-2 had been executed.

The Taurus-2 launcher is a two- or three-stage modification of the solid propellant Taurus vehicle with an inertial control system. Installed on the first stage of the Taurus-2 vehicle is the Castor-120 SPRE manufactured by Thiokol. This is a modified SPRE of the first stage of the MX ICBM.

The second stage also uses the Castor-120 SPRE. The launcher's last stage LPREs called Estus run on monomethylhydrazine (fuel) and nitrogen tetraoxide (oxidizer). The Estus LPRE has been developed, as mentioned above, as the last stage engine of the now being created Ariane-5 launcher. The Taurus-2 launcher can deliver a 2.3 ton payload to a circular near-earth orbit at an altitude of 185 km and with inclination of 28,5°. Its lifting capacity can be increased by using on the first stage strap-on launch boosters with Castor-4 SPREs. In case of using eight above said SPREs the lifting capacity can reach 5.0 tons. In sending a SC into a GEO with the help of the Taurus-2 having eight strap-on boosters with Star-48 SPREs on the stage of final ascent the weight of payload can be as high as 1.8 tons.

The designers tried to create a launch vehicle that would use a fairing and an adapter similar to those on the Delta launcher. The designers believe those are standard for spacecraft of this weight category. Such a launch vehicle with a fairing (similar to the one on the Delta vehicle) 3.05 m in diameter is exposed to the same dynamic loads. The SC is joined with the Taurus-2 launcher by means of similar electrical and mechanical plug-in connections as in the Delta vehicle.

The available information regarding the specific cost of launching the Taurus-2 vehicle is controversial (varying between US$ 2.2 and 11.0 per kg of payload). In any case, according to experts, this is enough for competition against the existing and proposed launch vehicles of similar lifting capacity.

An American organization, California Commercial Spaceport (CCS), is working on creation of a more advanced launching facility for carrier rockets using the Castor-120 SPRE. The Taurus-2 will also be able to start from that launching facility. In addition, Orbital Sciences is incorporated by the Spaceport Florida Authority (SFA) group that received a US$ 2.15 million grant from the US DoD for building a launching facility for small and medium lift launch vehicles on the territory of the Air Force Eastern Test Range.

Martin Marietta has created a Titan-2SLV, a two-stage carrier rocket using liquid propellant engine. This is a modified Titan-2 ICBM withdrawn from service. The first launch of one of the Titan-2 rockets modified for use as a carrier took place in 1988. Such carrier rockets were designated Titan-2SLV. By early 1995, 5 out of 13 rockets had been utilized. The launches of the Titan-2SLV can be executed from Vandenberg Air Force Military Base (SLC-4W launch facility) and from Cape Canaveral (launch site 19).

The modifications attempted by Martin Marietta involve the following jobs:
- replacement of the standard front compartment by a structure designed to accommodate a payload;
- replacement of the nose cap by a cargo-bay shroud;
- installation of telemetric equipment and rocket emergency detonation device;
- alteration of the on-board cable network.

The Titan-2 ICBM converted into the Titan-2SLV launch vehicle should have its engines overhauled and test-fired after a repeated assembly at the facility of the Aerojet General company.

Also, preventive inspections of rocket fuel lines are to be made and hydrostatic tests carried out.

Rocket engines will be installed at the Vandenberg Air Force Base.

The take-off weight of the Titan-2SLV is 150 tons. This carrier is capable of delivering the following types of payload:

- up to 3 tons to a low circular orbit at an altitude of 185 km and with inclination of 28.5°;
- 2.36 tons to a low circular orbit at an altitude of 185 km and with inclination of 63.5°;
- 2.17 tons to a low polar orbit at an altitude of 185 km.

The cost of launching the Titan-2SLV is estimated at US$ 30–40 million.

## Europe's small lift launchers

**In Spain** under the supervision of the Spanish National Institute of Aerospace Technologies (INTA) funded by the Ministry of Defense a project is underway aimed to create a Capricornio three-stage solid propellant small lift carrier vehicle for launching small spacecraft. The first stage of the carrier is the American-made Castor-4B SPRE, the upper stages are Spain's products.

The take-off weight of the carrier measuring 18.25 m in length is 15 tons. The carrier is capable of delivering up to 140 kg of payload to low near-earth orbits.

The first launch of the Capricornio with a satellite designed to gather and transmit data from Spain's stations in Antarctica was scheduled for 1998 from the launching site of Isla de EL Hierro on Canary Islands. The satellite had been created by the Barcelona-based Microelectronic Center. The complete cost of building the carrier was estimated at US$ 32 million, that of the launch – at US$ 9 million.

**The Italian Space Agency** (ASI) announced as far back as 1997 that a Vega family of launch vehicles had been developed in the country. Since 1988 the BPD Difesa e Spazio company had been working on that family of vehicles. The original idea was to increase the lifting capacity of the American-made Scout-G1 carrier by equipping it with two solid propellant strap-on PAP boosters borrowed from the Ariane-4 carrier. A round-trip capsule was planned to be installed for carrying out micro-gravitation experiments. Launches were planned to be executed from the San Marco floating launch pad located near the equator in the Formosa Bay off the coast of Kenya. The cost of work was estimated at US$ 200 million. The first launch was scheduled for 1995.

This project, known as Scout-2 or San Marco Scout, being a purely Italian job, had been altered to suit the Zefiro SPRE created on the basis of PAP boosters but different in having oscillating nozzles. In 1991 four static firing tests had been carried out. The first flight test of a bundle of three Zefiro engines conducted on 18 March 1992 at the military proving ground in Sardinia was partially successful. Non-flying tests were due to be completed by mid-1995. Plans were made to carry out two qualification flights of the carrier at the end of the same year in order to start operation in mid-1996.

As it happened, in 1994, the schedule was changed and the rocket got its current name, Vega (Fig. 38). In addition to the basic variant ,Vega-K0, other versions had been proposed, K2 and K4 with two or four Zefiro SPREs as strap-on boosters. The latter, however, were eventually discarded. The Vega-K0 four-stage variant became the basic one. The replacement of the Zefiro first stage engine by the Castor-120 type of engine (installed on the modern modification of the American Taurus launcher) and the abandonment of the fourth stage resulted in a heavier and more powerful variant, the Vega-K3. The work put through by the BPD Difesa e Spazio company at its own expense aimed to fully optimize the Zefiro engine and determine the component package of the Vega's subsystems. The first launch of the Vega carrier was due to be executed in 2005.

**Fig. 38.** Vega launcher

Since the recovery of the San Marco platform requires substantial financial funding (the last launch had been carried out from there back in 1988) the launches of the Vega carriers will be executed from the Kourou spaceport where the ELA-1 launching site will be re-equipped for them.

In February 1998, the Italian company BPD Difesa e Spazio announced its collaboration with Aerospatiale of France aimed to create a carrier that would improve on the Ariane carrier (ACLV-1 – Ariane Complementary Launch Vehicle). The small solid propellant launcher was named Lance-Proteus. Its first stage is made on the basis of the lower segment of the EAP strap-on solid propellant booster of the Ariane-5 launcher. Installed on the second stage is the Zefiro SPRE and a new P-7 engine on the third one (it is possible to use stages of the existing French-made maritime ballistic missile or a commercially available engine). The liquid module of re-acceleration must precisely place the satellite in orbit. Such a carrier can be developed within four years should roughly US$ 330 million funding is made available.

Experts of Aerospatiale are sure about the market for such a rocket. This applies to launches of the Proteus class small research spacecraft, Earth remote sensing spacecraft, and to spacecraft for Italy's Skymed/Cosmo data transmission network. The Proteus modular platform for accommodation of various instruments used to monitor the Earth's surface, provide communication and carry out research, has been developed by CNES and by Aerospatiale's division of satellites. The development of the Lance-Proteus carrier will be of special importance in creating a truly independent system of Europe's surveillance or "a galaxy" of small spacecraft. This carrier can be launched up to six times a year.

The carrier can be launched from the ELA-3 launch complex in Kourou (French Guiana). In this case, used as a firing table is a base to which one of the two EAP boosters of the Ariane-5 vehicle is currently affixed. The cost of launch is not yet known, but the preliminary inquest suggests the figure in the region of US$ 20 million.

By replacing the first stage engine on the Zefiro SPRE with a shortened nozzle, the BPD company can create a rocket comparable in terms of performance with the Vega-K0 carrier whose launch will cost around US$ 12 million.

The BPD has been planning ever since April 1997 to turn out a Lance/Proteus carrier variant in which the P7 SPRE and the final ascent unit would be replaced by a combination of two stages with LPRE fueled with NT+UDMH (nitrogen tetraoxide + unsymmetrical dimethylhydrazine) developed jointly with the NPO Yuzhnoye (research and production association) at Dnepropetrovsk, Ukraine.

**In France**, work on the small lift launchers began in 1990 with development of the DLA system. The program worth around 2 billion French francs implied the creation by 1998 of two variants of a three-stage solid propellant rocket. Installed on the first stage of the scaled-down DLA-P variant was the P92 engine built on the basis of a segment of the EAP booster of the Ariane-5 rocket. Installed on the second stage was the P30 engine while the third stage was the liquid type L5 (a scaled-down version of the L9.7 stage of the Ariane launcher). The DLA-S first stage of the scaled-up version was supposed to be built on the basis of a EAP full-sized booster. Both versions' fairing was borrowed from the Ariane-4.

The DLA concept received no financial support and was replaced in 1993 by Europe's ESL small lift launcher, which is a fully solid propellant three-stage rocket having identical first and second stages, P50, plus a third stage, P7. The final ascent module with single component LPE is integrated into the carrier control system. The development was due to be completed in 1998. The cost of launching from the ELA-3 launch base will amount to roughly US$ 20 million. Launching a scaled-down version with P50 – P7 – P7 stages costs 13% less.

In spite of its own work on the San Marco Scout program, the Italian agency ASI took part in research under the ESL project. The development had not been completed for a number of reasons, specifically for insufficient funding and difficulties in finding the market for the prospective carrier. As a result, Italy proposed to use in the project the Zefiro SPRE. Delays in the first launch of the Ariane-5 and the succeeding accident in June 1996 forced CNES to refuse any further involvement in the initiative "for saving the budget". In 1996 a ACLV-1 project emerged that had to improve on the Ariane. In charge of the project were SEP, Aerospatiale and FiatAvio. The idea was to use the P85 engine on the first stage, Zefiro on the second and a ten ton class SPRE on the third. In 1997 SEP gave up the development work estimated at US$ 250 million, deeming that the prospective market would not justify the expenses incurred.

**The Israeli corporation** Israel Aircraft Industries (IAI) has developed on the basis of the Jericho-2 medium range missile the Shavit (comet) three-stage carrier whose first launch was executed in 1988 from an airbase in Palmahim.

In 1998, while launching the Ofeq-4 satellite an accident occurred on the Shavit carrier in the second stage area. Officially, that was the first failure in the Shavit program. However, according to unconfirmed information, there had been one more accident in the early 1990s. In the first two launches of 1988 and 1990 the Shavit carrier had been used that was the first variant of this family of rockets. Installed on the first two stages of the carrier were similar SPREs while the third stage had been fitted with the AUS-51e-acceleration engine manufactured by Rafael (Haifa).

The first stage engine of the succeeding variant, Shavit-1, proposed in 1995 was manufactured by the Israel Military Industries company (IMI) and was elongated in shape. The Shavit-2 variant, dubbed "Next", was proposed for the first time in 1992 for commercial launches. It also had an elongated second stage and a new final ascent module driven by single component fuel.

The IAI corporation plans to promote the carrier on the international market of space launches, for which purpose it proposes to create new variants of the Shavit with participation of American and European partners. There are reports that IAI's main partner in the Shavit-2 (Next) project was to become Coleman Aerospace (Orlando, Florida) now engaged in supplies of the Hera ballistic targets and involved in launches, for the US Air Force, of spacecraft to sub-orbital trajectories. The above-mentioned project would focus on the most advanced technologies known to date in the relevant areas. Another partner was to be Atlantic Research Corporation (ARC) (Gainesville, Virginia) holding an exclusive license of IMI and Rafael companies, Israel, for marketing in the USA the Shavit-1 rocket's stages. It is also reported that Matra Marconi Space (MMS) of France might also participate in the joint venture. Currently, however, some evidence has been available from the company that no progress has been made under the Shavit project.

The IAI also planned to use the Shavit family of launchers for international commercials launches. Apart from the US-based launch complexes, a possibility was considered of using those in the Kourou spaceport and in Brazil's Alcantara Space Research Center.

## Small lift launchers of other countries

Since the late 1980s **Brazil** has been developing with participation of the Institute of Aeronautics and Space (IAEB) its first carrier rocket, VLS.

Originally, the commissioning of the VLS three-stage solid propellant rocket was slated for 1992. The first launch, however, had to be postponed to 1997 because of a number of serious technical problems. The first attempts to launch the carrier in 1997 and the second one in 1999 failed. The cost of developing the VLS carrier is estimated at US$ 450 million.

With its take-off weight of 50 tons, the VLS carrier fitted with four solid propellant boosters can send payloads weighing up to 200 kg into a near-earth orbit at an altitude of 750 km and with inclination of 25°. The cost of the launch is US$ 6.5 million.

The convenient location of the launching complex from which the VLS carrier had been launched (Alcantara spaceport lying at a latitude of 2°7' north, at a longitude of 44°23' west) inspires hope in Brazil for getting orders to launch commercial cargoes.

Two more modifications of the carrier are being developed now, VLS-2 with lifting capacity of 0.6 ton and VLS-3 with lifting capacity of 0.1 ton.

**India** began to use its own carriers for practical purposes in the 1980s. In 1980–1983, with the help of the first SLV launcher it sent into low orbits three Rohini space probes weighing 40 kg each.

On the basis of that launcher a more powerful four-stage ASLV carrier had been developed which was capable of sending a 150 kg payload into an equatorial orbit to an altitude of 400 km. The take-off weight of the carrier was 39 tons. The first and second launches of the carrier that took place respectively in 1987 and 1988 ended in accidents. However, the successful launch of the ASLV-D3 in May 1992 injected into orbit a SROSS space research probe.

In January 1994 one more Indian carrier had been launched. The launch was successful. All in all, since the start of operation four launches of the ASLV carrier had been executed. The ASLV-D4 may be the last rocket from this series.

Work is going on now aimed at creating a new carrier rocket, PSLV (Fig. 39). The second test launch of the new PSLV carrier executed on 15 October 1994 from the spaceport of Shrikharikota passed off without a hitch: an Earth remote sensing, IRS-P2, weighing 804 kg had been placed in a circular solar-synchronous orbit at an altitude of 820 km and with inclination of 99°. During the first test flight in September 1993, whose cost is estimated at US$ 144 million, a failure of the on-board digital computer caused an off-design staging which made the rocket deviate from the designed path and deliver the payload – a multi-purpose IRS-IE satellite weighing around 900 kg – to the off-design low orbit.

**Fig. 39.**
India's PSLV carrier rocket

As of January 1, 2002, another six successful launches of the PSLV have been executed.

The carrier rocket consists of four stages (the first and third stages use solid propellant, the second and fourth use liquid) and a solid propellant power unit. It is fitted with a nose cap 3.2 m in diameter, 8.3 m in length.

With the take-off weight of 275 tons, the PSLV launcher provides for launching payloads weighing up to 3 tons to a low near-earth orbit at an altitude of 400 km, 1 ton payloads to a solar-synchronous orbit at an altitude of 900 km and with inclination of 99°, and 0,45 ton payloads to a transfer orbit.

The commissioning of the PSLV launchers became an important milestone in implementation of India's national space program. This transportation system enables India to essentially lessen its dependence on other countries for carrying out its own space research.

Using PSLV and GSLV systems, India expects to obtain certain results on the market of commercial launches of spacecraft.

**Japan** operates a whole series of small lift launch vehicles.

Japan's Mu-3S three-stage solid propellant launcher belonging to the ISAS organization measures 20 m in length and has a take-off weight of 54 tons. The payload delivered to a low NEO (altitude at perigee = 450 km, altitude at apogee = 570 km) amounts to 220 – 300 kg.

There is a modification of the launcher, Mu-3H, that uses on the first stage a more powerful SPRE with a thrust of 114 tons.

The first launch of the rocket took place in 1974. This is a co-product of several companies headed by Nissan Motors.

The Mu-3S2 is also a three-stage solid propellant rocket. It is longer in size (28.2 m) and has a larger take-off weight (62 tons). The payload injected by such a launcher into a low NEO at an altitude of 250 km reaches 770 kg.

It is jointly manufactured by a group of companies headed by Nissan Motors to the order of the Institute of Space and Aeronautical Science (ISAS).

The first launch of that rocket was executed in 1985. The operation of the Mu-3S2 has now ended.

A new three-stage rocket, M-5, whose development began in 1990, is to become one of the world's largest solid propellant carrier rockets. With the height of 31 m and diameter of 2.5 m, its take-off weight will amount to 139 tons. The M-5 must provide for injection of a 2.2 ton payload into a low NEO at an altitude of 200 km and with inclination of 31°, of a 2.0 ton payload into an orbit at an altitude of 500 km with inclination of 31°, and of a 0.8 ton payload into a GTO.

The total cost of building the M-5 carrier is estimated at US\$ 133 million, with the launch cost not exceeding US\$ 36 million. Should the latter characteristic be obtained in practice, the specific expenditure sustained in injection of payloads by means of this space transport system, will diminish by nearly four times as compared against the M-3S-2 carrier, i.e. from US$ 70,000 to 18,000 per kilo of cargo.

The choice of SPREs for the M-5 carrier is determined by their cheapness, simplicity, high reliability and the experience gained by the ISAS in operating them. Among other merits of the SPREs are their high thrust and, hence, fast speed pick-up.

All three stages of the M-5 carrier are manufactured by Nissan Motors.

The first launch of the M-5 vehicle took place in 1997. This carrier will be mainly used for delivery of scientific research SC.

In 1990 the ISAS began jointly with Nissan Motors to study the possibility of creating a three-stage solid propellant rocket of ALV (air-launched vehicle) type that is launched from an aircraft or some other platform.

˙The new space transport system will be a combination of the second and third stages of the M-5 rocket. Proposed for use as the upper stage is the third stage of the M-3 S-2 carrier. The take-off weight of such a system will be around 52 tons, with the height of 17 m. Equipped with supporting planes for flight in the upper atmospheric layers, the rocket is supposed to be launched at an altitude of 10 km as the aircraft's moves with a speed of 200 m/sec.

According to preliminary estimates, the rocket will be powerful enough to inject spacecraft weighing up to 1.27 ton into a NEO at an altitude of 250 km. The developers also expect to attain a specific cost of delivery of payloads no more than US$ 2,556/kg, that is, the expenses incurred by the launch must not exceed US$ 3.2 million.

Sanctioned by the Space Activities Commission (SAC), the NASDA began in 1991 to design a space transportation system aimed to deliver small and medium SC. In order to reduce to the minimum the expenses involved, the new three-stage solid propellant rocket, designated J-1, was created on the basis of technologies available to Japan at that time. The ISAS, specializing in development and operation of solid propellant rocket components, was to play an important role in the project.

The first stage of the J-1 rocket had been developed on the basis of the launch booster of the H-2 carrier. Used as other stages were the assemblies of the second and third stages of the M-3S-2 rocket.

The take-off weight of the carrier is 87 tons, the height is 33 m. The weight of payload delivered to a low NEO is around 1 ton. The cost of development of the J-1 rocket is estimated at US$ 90 million.

The first launch of the J-1 took place in 1996. The cost of the launch was US$ 43 million. The use of new SPREs and simplification of the pre-flight preparations were to have reduced the cost of launch to US$ 20 million.

This, however, had not been achieved. The cost of the J-1 was 2 – 3 times that of foreign carriers of the similar class. The second launch of the J-1 was scheduled for the first half of 2002.

Studied as an alternative project is the possibility of creating a J-2 carrier a two-stage liquid propellant rocket using the Russian-made NK-33 engines.

**China,** too, is in the possession of several types of small lift carriers. A combined three-stage CZ-1D carrier has the take-off weight of 81 tons and measures 28.2 m in length. The payload delivered by it to a low NEO at an altitude of 300 km with inclination of 57° is 0.75 ton.

The carrier has been in operation since 1991. The launch cost is around US$ 10 million.

The CZ-2C has been in operation since 1975. This is a two-stage rocket using exclusively a liquid propellant. Its take-off weight is 191 tons, it measures 38.4 m in length. It provides for launching a 2 ton payload to a circular orbit at an altitude of 400 x 185 km.

The modified CZ-2C carrier launches commercial cargoes. In 1997–1998 it had launched six Iridium SC. Plans were made to use those carriers later on for maintaining orbital groups of the Iridium system.

The estimated cost of a single launch of the CZ-2C carrier is in the region of US$ 25–35 million.

As of March 1, 2002, 21 launches of this carrier have been executed.

The **South Korean** company Hyundai is developing under a program sponsored by the Ministry of Science and Technology a three-stage liquid propellant carrier, HD-1L, fueled with liquid oxygen and kerosene.
The rockett measures 30 in length and has a take-off weight of 100 tons.

The HD-1L must deliver about 1 ton of payload to low near-earth orbits.

The HD-1L carrier is due to go into service in 2015.

### 1.2.4 Reusable space transport systems

Currently, there is only one operating RSTS, the American Space Shuttle Transportion System (Fig. 40).

Regular operational launches of the SSTS began in November 1982. By January 1, 1999, 93 flights had been carried out, one of which (twenty fifth) ended in a RSTS failure with a loss of the Challenger orbital stage (OS).

**Fig. 40.**
Space Shuttle Transportation System

The SSTS is a two-stage rocket system with two solid propellant boosters (SPB). Sustainer engines of the first stage are located on the second stage (also serving as the OS) which makes reusability possible. The orbital stage has an aerodynamic capability and accomplishes a horizontal landing on an airdrome upon return from space.

In designing the system, it was presumed that practically all its components would be used many times over. The first stage was to save by means of parachutes the SPB for their subsequent recovery and reuse. It was planned in the longer term to preserve the main fuel tank. The plan, however, did not work out. The only fully reusable element of the system is the orbital stage. Now in operation are four OS (Atlantis, Discovery, Columbia, Endeavor).

The Space Shuttle provides for delivery of a 24.9 ton payload to a circular orbit at an altitude of 200 km and inclination of 28.5° with return of a 15 ton payload from the orbit.

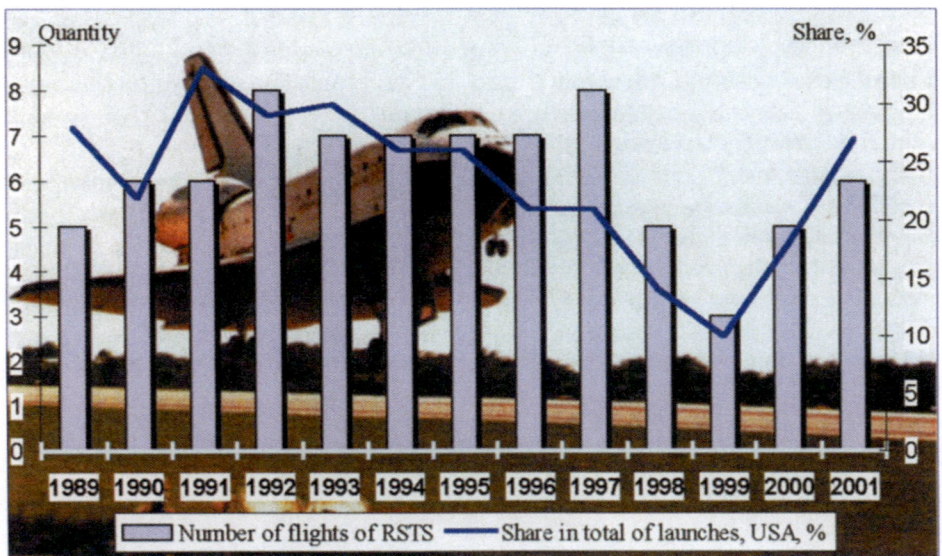

**Fig. 41.** Share of SSTS flights in total launches, in USA

It should be noted that the design objectives were never achieved in terms of reusability. More so, the underlying concept that anticipated the return of large payloads from an orbit for their subsequent reuse found no practical application. The US DoD refused to use SSTS for delivery of military heavy SC, and this despite the fact that the size and lifting capacity of the system had been elaborated to meet the demands of the military. Up to eight flights in a year undertaken for NASA is the ultimate benefit reaped from the SSTS (Fig. 41). In a situation like this any talk about cost-effectiveness or rate of return with reference to the RSTS makes no sense. The cost of delivery of 1 kg of payload by using the Space Shuttle is the highest in its class and nearly double the similar characteristics of expendable launchers. Nonetheless, the operation of the Space Shuttle is planned to continue at least till 2012.

For all its shortfalls, the SSTS is, indisputably, a step forward. The somewhat premature emergence of this system caused rather by political factors than technological or economic reasons, does not belittle in any way the advantages offered by the capability to reuse launching facilities. Therefore, the efforts to create reusable space transportation systems do not only slacken, but gather momentum with every year both in the USA and other countries.

So what projects are being tackled now with the aim of creating advanced RSTSs that would be used in the 21st century?

The RLV program (reusable launch vehicle) is being implemented by NASA in compliance with the directive of NSTP. The program seeks to create an automatic reusable space transport vehicle that would cut dramatically the cost of delivery of cargoes to space.

The estimated cost of the entire program is uncertain and varies between US$ 6 and 36 billion. Up to 80% of the funding is expected to be provided by private firms with payment made only for development of new technologies associated with a high degree of technological risks.

Fig. 42.
X-34 Experimental vehicle

If reliability of a new RSTS created under the reusable launch vehicle (RLV) program reaches 0.98, NASA and the US DoD will gradually abandon the Space Shuttle and ELVs for launching their SC.

The building and testing of technologies for the advanced RSTS had been accomplished under the X-34 and X-33 programs.

The X-34 program originally sought to create a prototype of a reusable air-based two-stage STS (space transportation system) for launching small lift spacecraft (0.54–1.13 ton). The contract for development of the system had been concluded in 1995.

As work went on, the project had been repeatedly amended. In 1997, NASA jointly with the OSC company approved a preliminary design of a supersonic vehicle intended for optimization of technologies used in advanced RSTSs. A vehicle weighing 20.3 tons and measuring 17.78 m in length with a wingspan of 8.45 m will be discharged from the L-1011 aircraft. After discharge, the LPRE activates and the vehicle completes within 15 minutes a loop-like flight with a speed of M=8 at an altitude of 76 km out to the range of 820 km with a subsequent landing on a runway. The exterior view of the X-34 vehicle is shown in Fig. 42.

Demonstration flights of the X-34 were due to start in 2000.

The X-33 is a scale demonstrator of technologies used in the VentureStar advanced reusable carrier whose operation was scheduled at the start of the millennium. It had been developed by Lockheed Martin to the order of NASA.

In designing the new vehicles Lockheed Martin relied heavily on its experience in science and technology gained while working under the transatmospheric vehicle (TAV) military project as well as on its own developments of the Aero-Ballistic Rocket single-stage transportation system.

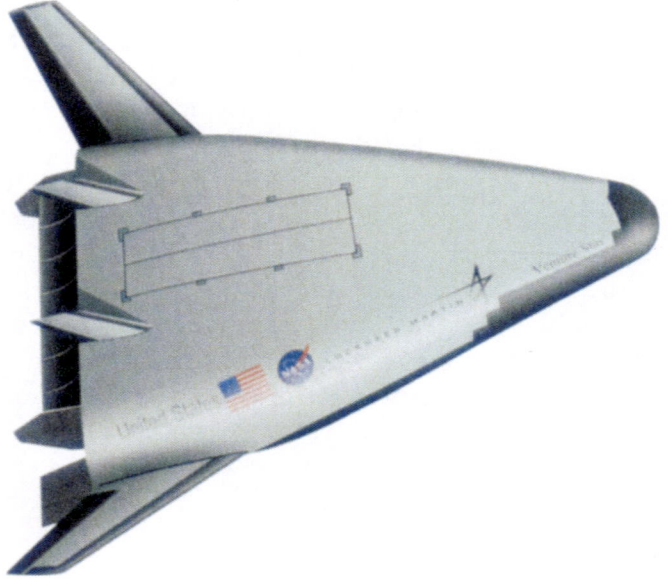

**Fig. 43.**
VentureStar reusable space transportation system

Plans were made to carry out 15 flights of the X-33 vehicle from the Edwards airbase (California) where a launch site is outfitted. The X-33 was to carry out suborbital flights at a high altitude with a speed of up to M=15, thus imitating the condition of orbital insertion.

The vehicle uses a metallic thermal protection except on the nose, fore edges of the fins and steers where carbon-carbon composite (CCC) material is fitted.

Among other innovating technologies are the Linear Aerospike linear motor and a liquid oxygen tank made of composite materials.

In all, NASA has spent around US$ 1 billion on the X-34 and X-33 programs. However, in March 2002 the financing of the program was suspended.

Within the framework of the program for creating an advanced RSTS, Lockheed Martin proposed the VentureStar vertical take-off/horizontal landing vehicle. Here, the "lifting body" with two vertical stabilizers had been chosen as the aerodynamic configuration. For accomplishing flight control during descent, rotary flaps were provided. The exterior view of the RSTS is shown in Fig. 43.

The vehicle is to be 37.2 m long, 37 m wide, 8 m in height. With the take-off weight of 725 tons the system can place payloads of up to 18 tons in low near-earth orbit. The cargo compartment is 13.5 m long, 4.5 m wide.

It was planned to fly the VentureStar vehicle up to 50 times per year. The low specific cost of cargo delivery equal to US$ 2,200/kg could enable Lockheed Martin to win a considerable segment of the market of commercial launches. However, once the X-33 and X-34 programs are scrapped, the creation of the VentureStar vehicle will become impossible.

Within the framework of the RLV program a possibility is being studied of creating a launch complex outside the USA. The main criterion in choosing the location is the proximity to the equator and a high altitude above sea level.

Another program, Future-X, involves design and development work on component packages for advanced space transport systems.

This program will cover the complete infrastructure that provides for delivery of space vehicles to various work orbits, including geostationary ones.

The Future-X program envisages work in the following three major sectors:
- development and ground tests of the main components of the transport systems (fuel tanks, power units, heat resistant coatings, electronics, etc.) as well as some structural elements of the space vehicles themselves;
- optimization of the newest high-performance technologies whose development is associated with a high degree of risk (Pathfinder project);
- execution of test flights of experimental prototypes of transport systems with the purpose of refining the techniques of operational ground servicing (Trailblazer project).

The experimental vehicles being created under the Future-X program for demonstration trials are to become an intermediate link between the experimental research prototypes of small and heavy types. The items of the former type are rather quick to build and are intended for optimization of a particular element; they are tested once a year or so. The HLV are used to carry out a comprehensive test program intended for several years; in course of its implementation efforts are made to minimize duration of the post-flight servicing and of pre-launch preparation of the vehicles.

The technologies being developed under this program can reduce the specific cost of delivery of cargoes to space down to US$ 888/kg. Hypersonic vehicles created with the use of state-of-the-art technologies will be able to fly over oceans in a matter of few hours and land on standard aviation airdromes. This is what arouses interest on the part of private companies.

Within the framework of the Future program, Boeing plans to develop an advanced technology vehicle (ATV). The reusable unmanned ATV (Fig. 44) will serve as a flying laboratory for optimization of the newest technologies, including the advanced design of the fuselage, power unit and operating techniques useful for a variety of flying machines. The ATV must demonstrate the advantages of using the so-called "green components" of the fuel based, for instance, on the environment-friendly hydrogen peroxide and kerosene.

Fig. 44.
Advanced technology vehicle

The ATV system being created on the basis of the X-48 space maneuver craft is 8.34 m long, has a wingspan of 4.32 m and a take-off weight of 5,580 kg. As distinct from the X-33 and X-34 vehicles, the ATV will fly at a speed corresponding to M=25, that is, it will reach an orbit and safely return from it. The ATV can make over 40 orbital flights.

The ATV is planned to be sent into orbit in a cargo compartment of the Space Shuttle. It will become the first experimental flying vehicle capable of making a fully controllable flight both in orbit and in atmosphere.

The flight tests under the Future-X program are expected to be completed in 36 to 48 months after which time NASA expects to use the flying vehicles to demonstrate other new technologies.

The total cost of the project is estimated at roughly US$ 150 million. NASA and Boeing contribute equal shares.

The Kistler Aerospace company (Kirkland, Washington) is developing an unmanned two-stage fully reusable STS, K-1.

The K-1 launches vertically, the soft landing of its stages is accomplished by means of parachutes and air bags. With such an arrangement the cost of delivering 1 kg of payload must be half of that obtained on existing expendable launch vehicles.

The K-1 is being developed and manufactured by the Northrop Grumman company.

The take-off weight of the K-1 will be 220 tons. The weight of payload delivered to the NEO at an altitude of 720 km and inclination of 37° reaches 3.2 tons. The system is designed for 100 missions.

Each stage of the K-1 system will be fueled with liquid oxygen and kerosene. It is expected that three Russian-built engines, NK-33, will be used on the first stage and their improved variant, the NK-43, on the second. The engines had been manufactured in the 1970s by the NPO Trud (research and production association, now the NK Engines Research and Engineering Complex) for the N-1 space launcher.

The first stage provides for launching the RSTS to an altitude of 30 km whereupon, after staging, it returns to the launch area for soft landing. The second stage delivers the payload to orbit after which it also lands softly.

The launch complex is supposed to be located on the territory of the nuclear tests range in Nevada. The test range of Woomera (Australia) has been chosen for tests and landing sites.

Initially, the launch of the K-1 had been scheduled for 2000. The company's financial difficulties, however, make the K-1 project's chance for success almost unrealistic.

The Kelly Space and Technology company (San Bernarndino, California) is developing STS of the Eclipse family.

The most powerful model will be a partially reusable transport system, Eclipse Astrolinear, (Fig. 45), intended for delivery of 1.6 ton cargoes to a low near-earth orbit.

The basic element of this STS is a reusable winged stage with an interior payload compartment accommodating the vehicle due for launch with its upper stages. The stage is 37.5 m long, the fuselage diameter is 6 m, and the spread of the delta wing is 24 m.

The launch of the first stage is supposed to be executed at an altitude of 6 to 12 km, to which point it will be towed on a cable by Boeing-747 aircraft. Using its own sustainer engine the stage must reach a suborbital path of flight. At an altitude of about 120 km a package of payload with two solid propellant stages will be discharged from it. Those stages will deliver the package to the work orbit. After a gliding descent the first stage must land in a normal aircraft fashion on an airdrome.

Proposed for use as a sustainer LPRE of the first stage is the NK-33 oxygen and kerosene engine. Also considered as alternatives are the Russian-built NK-32 (offered by NPO Trud), RD-180 (NPO Energomash), Aerospice, and RS-56SA (offered by Rocketdyne, USA). Proposed for use on the second and third stages are SPREs of Thiokol, respectively, Star-48B and Star-63F.

The total cost of building the Eclipse Astrolinear system whose commercial operation was scheduled for 1999 is estimated at US$ 130 million. The cost of launching the STS must be in the region of US$ 10 million.

The KST company proposes to optimize the technologies required for building this STS within the framework of two less complex projects, Eclipse Sprint and Eclipse Express.

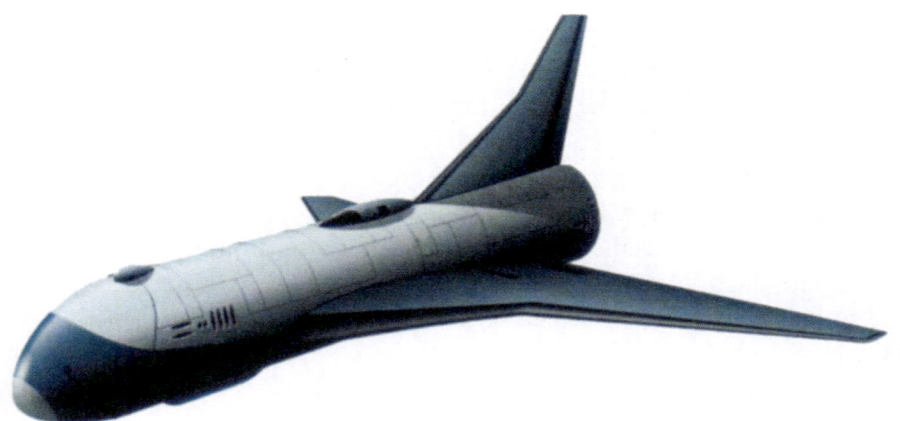

**Fig. 45.** Eclipse Astrolinear RSTS

The Eclipse Sprint project was to develop a transport system for launching purpose-oriented instruments to suborbital trajectories. Its principal element must be a modified F-106 fighter plane. Its ascent to the launch altitude will be accomplished also by towing performed by the C-141A military transport aircraft. The cost of retrofitting NASA's F-106 plane is estimated at US$ 7 million. The federal reserve has made available US$ 110 million to the KST company for optimization of the towing technique.

The Eclipse Express project proposed to equip the F-106 plane with a rocket engine in order to increase its power performance. This must provide a possibility to launch small satellites weighing 100 kg. The project is estimated at US$ 17 million. After realization of the two above said projects the KST company planned to undertake the development of the Eclipse Astrolinear system.

The Pioneer Rocketplane company (Lakewood, Colorado) is engaged in design and development of a partially reusable STS, the Pathfinder[1] intended to launch into space small and medium weight vehicles.

The principal element of the system is a piloted (two seat) transatmospheric vehicle 25 m long shaped like the Space Shuttle orbitter. The power plant of the Pathfinder is comprised of two F-100-PW-200 turbojet engines manufactured by Pratt and Whitney and one RD-120 oxygen and kerosene engine manufactured by Russia's NPO Energomash.

As far s the aerodynamic configuration is concerned, the Pathfinder vehicle is a low-wing aircraft having a delta-shaped wing with an extended fore front edge.

---

[1] This is an independent project, irrelevant to NASA's Future–X Pathfinder Project

The outboard wing panel will accommodate fuel tanks for kerosene, while the fuselage central part will receive a cryogenic fuel tank for liquid oxygen.

The load-bearing structures of the vehicle must be manufactured from composite materials with the use of heat resistant resins.

A metallic thermal protection system (TPS) has been chosen for the vehicle. The tiles manufactured on the basis of aluminum oxide will be fitted directly to load bearing structures.

The Pathfinder spaceplane takes off from an airdrome with an empty (for weight reduction) oxidizer tank by using for take-off and acceleration two F-100 turbojet engines. It's only in flight that liquid oxygen is pumped from a tanker plane into the Pathfinder. After that, the spaceplane is brought into a ballistic trajectory by means the RD-120. At the apogee the load is delivered to an orbit with the help of its own solid propellant engine.

**Fig. 46.** Roton RSTS demonstration prototype

Of all the RSTSs being considered, the one offered by Roton is the most exotic. To realize this concept, the Rotary Rocket company has been established. The development work as such had been entrusted to the Sealed Composites company (California). The developers presumed that Roton would be one of the cheapest RSTSs (the cost of flight is under US$ 7 million) since that would be a single stage vehicle similar to the VentureStar.

Roton's design allows to deliver to a low NEO up to 3.2 tons of payload. The piloted Roton is launched vertically with the help of a specially developed LPRE with ring nozzle and using a lifting rotor of the type used by helicopters. Fitted on the blade tips of such rotor are LPREs manufactured by RocketJet.

Upon delivery of a payload to an orbit the Roton returns to the atmosphere and lands with the help of an autorotating rotor like a helicopter.

The authors of the project believe that this will resolve the problem of building launch and landing sites.

The Roton carrier will be 20 m high and 6.7 m in diameter. Its rotor blades will be each 8 m long.

The first flight of the experimental vehicle (Fig. 46) took place in 1999. The vehicle reached an altitude of 23 m and developed a speed of 86 km/h. The main goal of the experiments was to demonstrate the viability of the concept according to which a vehicle has rocket engines installed on blades' tips.

According to initial plans, the Roton piloted space vehicle was to go into service in 2001. However, the financial difficulties of the Rotary Rocket company not only precluded this from happening, but have cast doubt on realization of the project altogether.

Fig. 47. HOPE-X RSTS, Japan

Theoretical research into RSTS began in Japan as far back as the late 1970s. Ten years subsequently, a special scientific committee, upon consideration of projects submitted by the National Space Development Agency (NASDA), National Aeronautical Laboratory (NAL), and Institute of Space and Aeronautical Science (ISAS), came to a conclusion about the feasibility of creating a piloted RSTS.

Initially, the HOPE program envisaged realization of three phases of work:

1. Building an experimental winged vehicle, HOPE-X (Fig.47) intended for optimization of required technologies in suborbital flights at a maximum altitude of 110 km. The first launch of the product was scheduled in 2001 with the help of the H-2 vehicle.

2. Development by 2005 of a standard reusable maintenance spaceship which in terms of size would be a close replica of the HOPE-X vehicle (16 m long, 10 m wingspan).

3. Building in a longer term in the case of adequate funding of a heavy lift craft weighing 22 tons.

However, in 1997 the project was suspended because of financial cuts. The task of carrying the loads to the Japanese sector of the ISS was shifted onto the HATV unmanned spacecraft.

The Japanese may start the full-scale development of the reusable transportation system after 2010.

India is developing a miniature reusable single-stage orbital vehicle called AVATAR (Aerobic Vehicle for Advanced Transatmospheric Research). It is not bigger than the MiG-25 fighter plane. The vehicle is designed to make 100 flights and can deliver to a low orbit a payload weighing from 0.5 to 1 ton with a specific cost of US$ 67/kg. The AVATAR is equipped with ramjets fueled with hydrogen and takes off like a usual airplane. On gathering cruise speed, the vehicle is boosted up to M=7 whereupon LPRE activates which sends the vehicle into orbit. There is no oxidizer on board the vehicle during the launch. The oxygen required for the LPRE will be gathered during the cruise flight by on-board systems that liquefy atmospheric air while separating and accumulating liquid oxygen. However, in view of India's limited financial capability (allocations to space amount to US$ 280 million per year) the AVATAR project is highly unlikely to be brought to successful completion.

The European space research organizations considered at various times various projects of the RSTSs, such as Hotol (UK), Hermes (France) and others. In 1998, ESA completed the FESTIP program under which research had been carried out into advanced STS, including reusable types. The reusability or expendability of ESA's advanced transportation system will be determined by the results of work on creating key technologies (FLTA program).

Overall, it can be said that ever more countries seek to obtain their own facilities for sending payloads into space in order to be independent in exploration and exploitation of space. In the 21st century there may be 20 of them.

## 1.3 Spaceports of the world

Currently, apart from Russia a number of other countries are in the possession spaceports, which they intensely use to launch spacecraft for their space programs. Such countries primarily are USA, France, China, India, Japan.

It is characteristic of foreign spaceports (test sites) that they are located on ocean shores and much farther to the south as compared to Russia's sites. This eases the problem of choosing an area for jettisoned parts of carrier rockets and provides an opportunity to launch spacecraft into orbits with a wide range of inclinations and a lesser power consumption.

The main characteristics of foreign spaceports (test sites) are given in Table 13.

**Table 13**. Key characteristics of foreign spaceports (test sites)

| Spaceport (test site) name | Owner | Coordinates | | Launch azimuth |
|---|---|---|---|---|
| | | Latitude | Longitude | |
| Eastern Test Site | USA | 28° 30′NL | 80° 30′ WL | 44° – 100° |
| J.Kennedy Space Center | USA | 28°30′ NL | 80° 30′ WL | 44° – 100° |
| Western Test Site | USA | 34° 40′ NL | 120° 40′WL | 140° – 307° |
| Wallops Island Flight Center | USA | 37° 50′ NL | 75° 20′WL | 85° – 129° |
| Kourou | France(ESA) | 5° 14′NL | 52° 46′ WL | 120° – 330° |
| Tsyuysan | China | 41° 20′ NL | 100° 20′ EL | 130° – 250° |
| Xi Chang | China | 28° 12′NL | 102° 02′ EL | 90° – 220° |
| Taiyuan | China | 37° 05′ NL | 112° 06′ EL | |
| Hainan | China | 19° NL | 109° 30′ EL | |
| Uchinoura | Japan | 31° 15′ NL | 131° 04′EL | 80° – 155° |
| Tanegashima | Japan | 30° 23′ NL | 130° 58′ EL | 80° – 155° |
| Shrikharikota | India | 13° 47′ NL. | 80° 15′ EL | 120° – 160° |
| Woomera | Australia | 31° 10′ SL | 137° EL | 10° – 290° |
| Christmas isl. | Australia | 10° 25′ SL | 105°143′ EL | |
| Alcantara | Brazil | 2°12′ SL | 44° 12′ WL | |
| San Marco | Italy | 2° 56′ SL | 40° 12′ EL | 50° – 130° |
| Palmahim | Israel | 34° 46′ SL | 31° 54′ EL | 279° – 292° |
| Al-Akbar | Iraq | 33° 30′ SL | 43° EL | |

### 1.3.1 Spaceports and test sites of the USA

The USA has the most ramified and developed network of spaceports. This is determined primarily by the tasks and objectives set by the USA before its space industry, by the funds allocated to this industry and by the scientific and technological capability achieved by the country.

The development of spaceports (test sites) in the USA is primarily governed by the plans to create multi-purpose launch complexes that could send into space carrier rockets of various lifting capacity.

Four spaceports (test sites) form the foundation of the US ground-based space research infrastructure:

- two sites, the Western Test Site (WTS) and the Eastern Test Site (ETS) of the Department of Defense, both subordinate to the US Air Force;
- two test sites of the National Aeronautics and Space Administration (NASA), (J. Kennedy Space Center and Wallops Island Flight Center).

The proximity of many space rocket engineering and production centers to test sites contributed to building large test and production facilities in those regions. The general location of the US space centers is shown in Fig. 48.

**Fig. 48.** USA space centers

In spite of being subordinate to different agencies, **the Eastern Test Site and the J. Kennedy Space Center** are located in the same area, on the eastern coast of the North American continent at Cape Canaveral and on Merrit Island (Florida).

The J. Kennedy Space Center (spaceport) (Fig. 49) is NASA's mainspaceport. Here, America's first space vehicle, Explorer-1, had been launched on February 1, 1958. Later on, the port was used to launch the Thore, Atlas and Titan carrier rockets, to send into space Mercury and Gemini piloted spaceships and to operate the Space Shuttle Transportation System. Currently, it is used to launch spacecraft with the help of Atlas, Agena, Saturn-1B, Atlas-Centaur and Delta carrier rockets under its own programs and those being implemented jointly with the Eastern Test Site (ETS). Also, it carries on research, conducts scientific experiments, and provides training for research workers.

Fig. 49.
J. Kennedy Space Center

The J. Kennedy Space Center comprises more than 50 buildings, structures, and technology research sites. Around 3,000 people service the Center. The launch complexes are grouped by types of carrier rockets and are placed in order of increasing power of the carriers along 18 km from North to South. There is a rather large area of barren land on the Island of Merrit that could be used for building new launch complexes.

The Eastern Test Site had been set up in 1963 on the basis the US Air Force's largest Atlantic Test Site that had been used since 1950. The total area of the ETS is around 400 sq km. Forty eight launch complexes are being maintained at the spaceport in various states of readiness. The maintenance staff of the spaceport numbers more than 20,000 people.

The test facilities of the ETS allow to launch Atlas, Atlas-Agena, Titan-1, Titan-2, Saturn, Titan-3C and other carrier rockets.

The trajectories of the carrier rockets from the Eastern Test Site and J. Kennedy Space Center stretch to the south east over 20,000 km above the Atlantic and Indian Oceans. On the trajectories there are three main areas for jettisoned hardware, a network of 15 instrumentation sites and hundreds of monitoring posts based on ships, aircraft and land.

The climatic conditions at the spaceports are satisfactory, though hurricanes and typhoons occur here rather often in which case the wind speed can rise to be as high 55 m/sec. The air temperature varies annually from 0 to +50°C. The place is swampy, located on a plain. The rock lies at a depth of 50 m. The spaceports can be accessed by all types of transports, namely, by motor- and railroads, by aerial and maritime routes.

**The Western Test Site** is the USA's spaceport second in size and significance. It is situated 250 km away from Los Angeles (California) on the western coat of the Pacific Ocean and covers an area of roughly 400 sq km. The spaceport stretches along the shoreline for about 40 km.

The first launch of the Discoverer-1 spacecraft from the Western Test Site took place in February of 1959.

The Western Test Site comprises the USA Vandenberg Air Force Military Base, the test sites of Point Mugu, Point Arguello, and its interior test site. Concentrated here are 11 launch complexes with 20 launch pads. Figure 50 shows one of the launch complexes.

**Fig. 50.** Western Test Site

The Vandenberg Air Force Military Base and the test site of Point Arguello are used to launch military spacecraft. The base has 3 launch complexes for Atlas and 2 for Titan carrier rockets.

The flight trajectories of the carrier rockets pass over water area of the Pacific Ocean which considerably eases the problem of safety and environment protection in places where the first stages of the rocket are jettisoned. This also cuts the costs of land appropriation and conservation of territories exposed to drops of discarded hardware. The range of inclination of orbits into which spacecraft are launched from this spaceport varies between 34° and 90° (in western direction) and between 81° and 125° (in eastern direction).

Launches at the Western Test Site (WTS) are executed mostly to the west, i.e. in direction contrary to rotation of the Earth since launches to the east send vehicles flying over densely populated areas of the USA. The WTS is

USA's only spaceport that can launch SC into orbits passing over the Earth's poles in which case the craft do not fly over populated areas of the country in the initial phase of the flight.

The total staff of the spaceport is over 17,000 people, of which 10,000 are military servicemen. The staff are selected for jobs so that to make the spaceport suited for use by all armed forces.

In 1984, a launch and landing complex for reusable transportation system, Space Shuttle, had been commissioned at the Vandenberg Air Force Military Base. Its landing strip is 4,500 m long, 90 m wide. However, not one launch of the Space Shuttle has been executed at the Vandenberg Military Base.

**The Wallops Island Flight Center** (Fig. 51) had been established in 1945 by the Langley Scientific Research Center and National Consultative Committee on Aviation. This is one of NASA's major scientific test bases for development and launching of experimental rockets and Earth's small artificial satellites. The spaceport is partially located on the eastern coast of Virginia and on the Wallops Island, 260 km off the US capital. In 1974 the spaceport had been renamed the NASA Wallop Flight Center. The USA uses it for joint space research programs carried out with Italy, Japan, Canada, Australia, the UK.

Fig. 51.
Wallops Island Flight Center

The spaceport consists of three main areas: the former Air Force base, areas on Wallops Island and areas on the continent 3 km west of the island. The bulk of the spaceports territory is the former Air Force base containing research and test facilities, design offices, launch complex control posts, telemetric information reception and transmission stations, airdrome.

The narrow strip of the Wallops Island, only 8 km long and 0.8 km wide, accommodates 6 launch complexes equipped with all the required facilities for assembly, testing and launching of carrier rockets with space equipment.

Located on the continental part of the Center are instrumentation sites, radar station, and test flights base. The island is connected with the mainland by means of a dike with a motor road.

The flight path of carrier rockets passes in the direction of the Bermuda Islands and is fitted with the required measuring instruments and telemetric stations. The admissible launch sector of rockets is restricted by angles from 670 through 1450. The craft are injected by the Scout carrier rocket into orbits with inclination from 370 to 540. The maintenance staff of the spaceport numbers around 600 people.

A number of states in the USA (Florida, Texas, Alaska, Virginia, Hawaii and others) make strenuous efforts to create commercial spaceports on their territories. By doing so they plan to improve their economies thanks to commercial utilization of space. Thus, in 1991 construction began of a launch complex on Kodiak Island near the Pacific coast of the southern part of Alaska. That island had been selected for its favorable location enabling launches to be performed in the direction of South. This is an ideal place for launching SC into low near polar orbits. The trajectory of the flight passes far off from densely populated areas. The spaceport can be used to launch both commercial and state-ordered cargoes. The first launch from Kodiak Island took place in 2001.

A commercial spaceport is being established on Wallops Island in the state of Virginia where NASA's test site is located. The new launch complex will be able to lift-off lightweight privately-owned carriers like Athena and Taurus.

Administration of the state of Florida has established the commercial Spaceport Florida on the Eastern Test Site. Plans are made to use the old Titan's launching pad and to build two new firing tables for small lift carriers and a work station for small spacecraft.

## 1.3.2 Spaceports of other countries

Over the last decades, Western European and Asian-Pacific nations have generously contributed towards the space exploration effort. As regards Western Europe, its countries strove from the very first to pool their financial resources as well as scientific, technological and industrial potential. Initially, it took the form of the European Space Research Organization and then the European Space Agency (ESA) established in 1975.

The prime target of ESA is creation and operation of space equipment on a commercial basis. Each member state of ESA has its own national program under which it creates its own space exploration equipment or participates in programs of its partners in the agency.

**France.** France was the third country to have launched on its own in 1965 an artificial satellite into near-earth orbit.

The guidance and coordination of work aimed to explore and exploit the outer space are provided in France by the National Space Research Center.

Until 1967 France had been using for implementation of its space programs the spaceport at Hamaguir, Algeria, built on a rocky plateau in the Sahara Desert.

In 1967, as the first phase of the spaceport at Kourou was put in operation, the spaceport at Hamaguir was closed down and all the equipment of launch complexes was dismounted.

The spaceport of Kourou (Fig. 52) is located in French Guiana on the north-western part of South America and occupies a shoreline strip 60 km long and 20 km wide. Currently, it is co-owned by France and ESA. The spaceport is equipped with three launch and technical support complexes for assembly, testing and launching of spacecraft with the help of carrier rockets like Diamond, Europa, Ariane.

**Fig. 52.** Spaceport at Kourou

In 1965 France declared the establishment of the spaceport and on 9 April 1968 performed the first launch of a supersonic rocket, Veeronique. In December 1979 the Ariane launch vehicle had been lifted off from the ELA-1 launch complex that had been in operation for ten years. In 1985 a second launch site, ELA-2, was built. In succeeding years (since 1987) ESA decided to build a new launch site, ELA-3, for launching Ariane-5 vehicles. Plans are made to build in the foreseeable future an ELA-4 launch complex designed to launch the Ariane-5 rockets for yet another 25 years.

The French government opened the spaceport at Kourou to any state wishing to launch its spacecraft from that area.

The Kourou spaceport provides two adequately equipped flight trajectories – one 4,000 long in the direction of the Azores, another 3,000 km long in the direction of the Bermudas.

The permanent maintenance staff of the spaceport numbers 600 to 700 people, of which around 55% are drawn from locals.

To optimize some stages of the carrier rockets and their structural elements France also uses a test site at Biscarosse on the Atlantic cost of the country.

**China.** China's space program is being implemented under the guidance of the Academy of Space Technology. China put its first artificial earth satellite, China-1, with the help of the Long March-1 launcher in April 1970.

A test site known in the West as Shuang Cheng-Tze is considered China's first spaceport. The Chinese themselves, however, refer to it as Jiuquan.

Since using the spaceport of Shuang Cheng-Tze for launching spacecraft into GTO is economically prohibitive, a spaceport at Xi Chang had been built for such launches.

A third spaceport, at Tai Yuan, (known as Wuzhai in western classification) is used to launch spacecraft into near-polar and solar-synchronous orbits.

The spaceport at Shuang Cheng-Tze (Jiuquan) built in 1965-1970 is located 1,470 km west of Beijing in the Gobi Desert near Xiangchanghe. It was from there that China's first artificial earth satellite had been launched and its first ballistics missiles fired. The spaceport is equipped with three launch complexes. Carrier rockets are launched toward the north-east.

To develop its piloted flights program China decided to build at this spaceport a totally different launch complex and a facility for assembly and integration of heavy duty carrier rockets whose construction was originally slated for the early 21st century.

The spaceport at Xi Chang (Fig. 53) built in 1984, lies in the south-west of China 1,300 km off the spaceport of Shuang Cheng-Tze. Being much closer to the equator than the spaceport of Shuang Cheng-Tze, it specializes in launching SC into geostationary orbit.

Located in the tropical rainfalls zone, Xi Chang sustained in summer an onslaught of mud streams. The launch sites, however, were not damaged. In 1999 the spaceport's facilities were repaired and significantly enlarged.

The Tai Yuan (Wuzhai) spaceport had been built in 1988 as a test site for testing ballistic missiles. It has a single launching facility used for launching the basic CZ-2C (Long March-2C) carrier rocket. Six carrier rockets had been launched from here under the Iridium program (two spacecraft in each launch). The activity of the spaceport is now at a very low ebb.

On the island of Hainan, used previously for launching only sounding rockets, construction is now underway aiming to erect a new launching complex, maintenance facilities, and a recreation park for tourists. The launch sites will be built near the town of Wencheng 3 km off the coast. The complex will have two channels. Two smaller islands in the east, 35 and 70 km off the coast, can be used as monitoring stations.

**Fig. 53**. Xi Chang spaceport

The prospect of using China's spaceports looks as follows:

Hainan and Xi Chang will be used to launch SC into geostationary transfer orbits, Tainan for near-polar orbits while Jiuquan can focus exclusively on piloted vehicles programs.

**Japan.** Japan was the fourth country to launch in February 1970 its own artificial earth satellite, Osumi, with the help of its own carrier rocket and from its own spaceport.

In its space work Japan sticks by its own programs which are being implemented under the guidance of the National Aerospace Development Agency (NASDA) and the Institute of Space and Aeronautical Science (ISAS) of the Tokyo University.

To implement its national space program Japan uses two spaceports fitted with modern manufacturing and testing equipment, Uchinoura and Tanegashima plus several scientific research centers.

The Uchinoura spaceport had been built in 1963. It is situated on the island of Kyushu near the shore of the Pacific Ocean. Its area is 51 sq km. Japan's first artificial earth satellite had been launched from there.

Located to the south of Kyushu is a small island of Tanegashima on which Japan's second spaceport of the same name had been built (Fig. 54).

Japan's two scientific research centers, Kakuda and Tsukuba, are in fact component parts of the spaceport. Both are directly involved in tests.

The Kakuda center located 260 km south-east of Tokyo is intended for testing rocket engines.

The Tsukuba center is located 64 km north-east of Tokyo. The center works on the reusable piloted spaceship. It comprises a large number of scientific research institutes, two universities and a test base.

**Fig. 54.** Tanegashima spaceport

**India.** India's space program is being implemented under the guidance of the state's special department of space exploration.

To test its space rocketry and to launch spacecraft by its own carrier rockets, India had built in 1971 the spaceport of Shriharikota situated on an island of the same name near the eastern coast of the state of Andhra Pradesh about 80 km north of the town of Madras. The first launch of the Indian artificial earth satellite, Rohini, from that spaceport with the help of a domestic-made carrier took place in July 1980.

The objects put in orbit of the artificial earth satellite have an orbital inclination relative to the equator plane in the range from 44° to 47°.

The spaceport has launching and manufacturing facilities for preparation and launches of the Indian carrier rockets (SLV, ASLV, PSLV). It is duly outfitted with test rigs and production equipment, such as a solid propellants plant, a test complex with a high-level flight simulator to evaluate the performance of rocket and spacecraft engines and subsystems. Also, it has installations for assembling, testing and launching of carrier rockets, a plant for manufacture of rocket propellants and systems used in telemetry, tracking, remote control, information acquisition, etc.

Some storage facilities with equipment for transfer and thermostatic control of fuel components (nonsymmetrical dimethylhydrazine, nitrogen tetraoxide, liquid hydrogen, and liquid oxygen) are located around the launch site. The fueling of various stages and the pre-launch inspection of the fully assembled carrier are performed from the Launch Control Center located far from the launch site.

For safety reasons the flight control center is 6 km distant from the launch site. The communication between the two is accomplished by means of optic fiber cables. The computer center operating in the real time mode provides information about the carrier's whereabouts, its speed and flight altitude and processes any other trajectory related data. The real time service registers information about events in flight and gives a quick access to it. During the launch procedure the computer center provides its computing facilities to all systems and services of the spaceport and land-based stations located in and outside the country. This allows to confirm the launch results just an hour after the carrier has been launched.

**Australia and Oceania.** In November 1967, the Vresat satellite had been launched from the Woomera space rocket launch site near a town of the same name in the state of South Australia. The launch had been executed under the Sparta military program with the help of the modified Redstone carrier rocket of American manufacture. In 1971 the Prospero, a satellite of the UK, had been launched after which the site had not been used for launching any more spacecraft into orbit.

The administration of South Australia had repeatedly considered the projects of converting the Woomera test site into a spaceport, which could also use, among other carrier rockets, those produced by Russia, like Start, Soyuz, and Kosmos-3M. But the sharp cuts in Australia's space program allocations suspended the implementation of those projects.

There are several places in Australia suited for creating spaceports on them. At various times, several projects had been considered of building spaceports that would use Russian-made carrier rockets. For example, a possibility was studied of building an international near-equator spaceport close to the cape of York on the coast of Queensland. The port was supposed to launch craft in the direction of the Pacific Ocean. Also, a possibility was considered of using at that port the Proton-K and Zenith carrier rockets. The financial difficulties, however, thwarted the plans.

Shown in Figure 55 are the supposed locations of spaceports in Australia and Oceania.

Also, a possibility was studied of building spaceports on the territory of Papua New Guinea in Oceania and in the north of Australia, about 30 km off Darwin. Such spaceports, it was supposed, would use the Proton family of carrier rockets.

For lack funding required to commence the construction, things in Papua New Guinea did not go beyond a verbal statement of the project feasibility.

Darwin held much promise for the project since it's a conveniently situated sea port having an airdrome and a developed industrial infrastructure. The local population, however, opposed the plan of building a spaceport there as in that case the orbit insertion trajectory would pass over Arnhemland, a place of Australia's aboriginals' hunting preserve and worship. The Australian administration is very sensitive about any infringement of rights of the indigenous people which is why the project disappeared in the deeps of administrative offices.

The financial constraints precluded the construction of the spaceport on the Hummock Hill Island near the eastern coast of Australia. It was supposed to use the Yedinstvo (Unity) space rocket complex developed by GRTs (the Makeyev State Rocket Engineering Center).

In 1995, the Australian company  Asia Pacific Space Center (APSC) had been formed. The company aimed to create and operate a space launch complex that would render services on a commercial basis.

In 1998–1999 the APSC company negotiated with the Khrunichev State Space Research and Production Center for building a spaceport on the Christmas Island in the eastern part of the Indian Ocean and for launching from it the Angara family of space vehicles. The negotiations, however, bore no fruit. More fruitful were negotiations for construction and operation of the Avrora medium lift space rocket system, which was to be launched from the Christmas Island. It was the sole project that had been realized practically. Even though the negotiations are still in progress, the construction of the launch complex by specialists of the general engineering office has already started.

The main structures of the spaceport will be erected in the southern part of the island near the Southern Cape occupied now by phosphate pits. One and a half kilometers to the north from them a carrier rocket workstation with an assembly and test complex will be set up. Those will include a storage facility for fuel components and liquid oxygen. The cargo preparation building with two rooms, the launch control post and the observation platform will be farther off, about 2.5 km distant from the launch site. Also, there will be built a complex of test benches, laboratories, a fire brigade station, a mess hall, and a housing complex with a recreation zone.

The trajectory of launches performed from the Christmas Island will pass mostly over the water area of the Pacific Ocean, with the rocket flying over the island only the initial 25 seconds (of which 10 are contributed by the vertical lift-off).

The burnout stages will only fall on the ocean. However, during launches into near-equatorial orbits the fall areas will be places with a fairly intense navigation which will necessitate extra precautions.

The first launch of the Russian Avrora carrier rocket from the international spaceport on the Christmas Island may take place late in 2004 – early in 2005.

Fig. 55. Location of supposed spaceports in Australia and Oceania
1 – Cape York;
2 – Papua New Guinea;
3 – Darwin;
4 – Woomera;
5 – Hummock Hill;
6 – Christmas Island

**Great Britain.** Great Britain carries on its space work under the national program and international ones with the USA and ESA.

For launching its sounding rockets, experimental ballistic missiles and carrier rockets the UK has been renting until 1976 the Woomera spaceport in the southern part of Australia. It was used to test the Black Arrow and Europa-1 carrier rockets. Also from it, the Prospero, Britain's first artificial earth satellite had been launched in October 1971.

**Germany.** Germany carries on its space work under bi-lateral arrangements with the USA, France and jointly with ESA under the guidance of the Aerospace Scientific Experiment Center. Germany has no rockets of its own, though it was actively involved in creation of Western Europe's Ariane launcher using a German-made third stage fueled with "oxygen-hydrogen" components.

In 1947-1977 the privately-owned OTRAG company built in Zaire with government sponsorship a test site having a launch complex for carrier rockets. It was used to launch several rockets along a ballistic trajectory. However, following Zaire's requirement in the late 1970s the site was scrapped. Later, the similar fate awaited the site in Libya which had been built by the same company and from which several experimental launches had already been performed along a ballistic trajectory.

**Israel.** In September 1988, Israel launched with the help of the Shavit carrier rocket from the Palmahim launch site in the Negev Desert its first artificial earth satellite, Ofeq-1, weighing 155 kg.

It was intended for launching ballistic missiles and carrier rockets. Objects inserted into orbit of the artificial earth satellites have an orbital inclination relative to the equator plane in the range between 142° and 144°.

**Brazil.** The spaceport of Alcantara is situated in the northern part of Brazil near the Atlantic coast. It has been in operation since 1997. It is intended for launching artificial earth satellites into orbits with inclination relative to the equator plane in the range between 2° and 100°. An international commercial spaceport is expected to be built in the future.

The first and up to date the only launch of the carrier rocket from the spaceport took place on November 2, 1997. On that day the Brazilian carrier rocket with the SCD-2A satellite took off but failed.

**Italy.** The Italian space program uses carrier rockets made by the USA (Scout), the European Organization for Development of Carrier Rockets (Europa-1) and European Space Agency (Ariane). The Commission on Space Research and the Aerospace Research Center have been charged with the management of Italy's space programs. In 1988 the Italian Space Agency was established.

The first Italian earth satellite, San Marco-1 was launched by the American Scout carrier rocket in December 1964 from a unique floating platform, San Marco.

The San Marco floating platform with the required equipment had been created on the basis of a maritime drilling platform and towed away to the Formosa Bay in the Indian Ocean 5 km off the coast of Kenya.

The San Marco platform is 90 m long. Placed on it are a 36 m long hangar to assemble and test rockets and a launching device. During launches the platform is held stationary by twenty steel supports.

The launch complex includes one more sea platform, Santa Rita, located 500 m off the launch platform and serving as a command instrumentation post. The platforms are communicated by two tens of underwater cables.

In December 1989, Abid carrier rocket had been launched from the Al-Akbar Space Research Center 50 km west of Baghdad. Iraq had created it jointly with Argentina. The third stage of the rocket entered near-earth orbit and made 6 circuits around the earth.

The spaceport was intended for launching ballistic missiles and carrier rockets. The objects inserted into an orbit of the artificial earth satellite have an orbital inclination of 34° to 50° relative to the equator plane. During the Desert Storm operation in 1991, the spaceport sustained severe damages and has been out of operation ever since.

## 1.4 Spacecraft control systems abroad. The present state of the art and the prospect for the future

### 1.4.1 General description of foreign control systems for spacecraft

The spacecraft tracking networks (ground-based control systems, GBCS, known in the Russian terminology as instrumentation and command complexes, ICC) began to be created abroad in the late 1950s when the first American SC were launched into space. Until the mid- 1960s the GBCS had been only in the USA and the USSR. As time went by, the GBCS were built by other countries, international consortiums and some private organizations.

Currently, more than 30 countries, international consortiums and private commercial companies are in possession of SC and GBCS for them.

In foreign countries, the USA and the ESA are undisputed leaders in terms of expertise in controlling SC for a variety of applications. Considerable capabilities are demonstrated by the International Telecommunications Satellite Consortium Intelsat (ITSC), a number of US companies, space agencies and organizations of France, FRG, Great Britain, Italy, Canada, Japan, China, and India. Brazil, Israel, Australia and other countries expand their space programs.

The above said European countries are members of ESA though they have their own space projects under which various SC and their control systems have been built and still more are being developed.

The GBCSs of the USA and ESA outperform those of other countries and organizations in quantity and intensity of operations, in the number and variety of spacecraft being monitored, in the stability of their structure and in other characteristics. At the same time, the GBCSs of some countries boast up-to-date ways and means of controlling the SC.

In some countries (USA, France, UK) the military and civil space programs are being tackled separately. This influences the general management of the SC control, i.e. the means of controlling military SC are singled out as separate control systems. However, work for military organizations can proceed without obvious separation from the general space exploration programs (China, Israel, Japan).

Controlling the SC, regardless of their application and characteristics, is essentially the performance of operations like measuring the trajectory, calculating the orbit, monitoring the state of the on-board instruments (telemetry), development of on-board equipment control software, transmission of commands, software and other information to the SC, reception and processing of special purpose information fed by the SC. That is why the functions and structure of the GBCS of SC of various countries have much in common. Whatever differences exist result from different quantity and assigned missions of the controlled SC, from the geographical peculiarities of the countries (opportunity to cooperate), the amount of allocations and other factors.

Given below is the composition and structure of GBCS of a number of countries (USA, France, Japan, China, European Space Agency and others) which gives a general idea about the state of foreign SC control systems.

## SC control systems in USA

In the US the control systems of military and dual use SC are operated by the DoD, NASA, and the National Oceanic and Atmospheric Administration (NOAA) of the Department of Commerce. The distribution of the SC control systems among the US agencies is shown in Fig. 56.

In the US, the formal categorization of space programs into civil and military began in the early 1960s. By 1964, a Transit GBCS had been produced. The launch of the first reconnaissance SC like Samos and Ferret urged the production of the Air Force's satellite control facility (AFSCF) as part of GBCS tracking network. It went into standard service in 1972 (the abbreviation changed for AFSCN).

Table 14. shows characteristics of the US SC GBCSs.

**Table 14.** Characteristics of US SC GBCS

| SC GBCS, agency | GBCS function | GBCS component package | GBCS general characteristics |
|---|---|---|---|
| AFSCN, DoD/AF | Military SC control | 2 SC control centers; 9 stationary command and instrumentation points; 1-2 mobile CIP; also used are SDS and TDRSS | Number of SC serviced in orbits: 80...90. SC visibility zone (% time in 24 hours): 15–20%– for low orbital SC 50–70%– for GPS SC and in high elliptical orbits; for SC 100% for SC in GEO; 50–100% – for low orbital SC (in transmission of data via 1–3 SDS and TDRSS relay satellites) |
| NSCN, DoD/Navy | UFO, Fleetsatcom and NOSS SC technological control | Fleetsatcom and UFO SC control center – 1 common sector in AF CSOC CC; NOSS system: 1 sector in AF CSTC CC (USRC); CIP: AFSCN GBCS are resorted to. | Number of SC being serviced in orbits: 10...15 (included above in AFSCN GBCS). SC visibility zone (in % of time over 24 hours): |

| SC GBCS, agency | GBCS function | GBCS component package | GBCS general characteristics |
|---|---|---|---|
| | | | 15–20% – for low orbital NOSS SC; 100% – for Fleetsatcom and UFO SC in GEO |
| Army/DoD SC control systems | DSCS control Network. of laser range finders (SC of various agencies) | DSCS SC CC: 1 sector in AF CSOC CC; CIP: AFSCN AF GBCS are resorted to; approx. 10 mobile laser stations. | Number of DSCS SC being serviced: 7–10 (included above in AFSCN GBCS). SC visibility zone (in % of time over 24 hours): 100% – for DSCS Fleetsatcom SC in GEO. |
| STDN+ ground-based GN+TDRSS | Control of scientific, applied, piloted SC of NASA and other states. Carriers' launch control | 2 SC CC; 6 CIS in ground-GN subnet; TDRSS: 3 terminals, 3 auxiliary CIS, 4 range finders, 6 TDRSS relay satellites. | Number of SC being serviced in orbits: 25-30: SC visibility zone (in % of time over 24 hours): 5–10% – for low SC via GN net (without TDRSS); 5–100% – for SC in orbits up to H=12000 km via TDRSS (with 1-3 relay satellites) |
| STDN+ ground-based GN+TDRSS | Control of scientific, applied, piloted SC of NASA and other states. Carriers' launch control | 2 SC CC; 6 CIS in ground-GN subnet; TDRSS: 3 terminals, 3 auxiliary CIS, 4 range finders, 6 TDRSS relay satellites. | Number of SC being serviced in orbits: 25-30: SC visibility zone (in % of time over 24 hours): 5–10% – for low SC via GN net (without TDRSS); 5–100% – for SC in orbits up to H=12000 km via TDRSS (with 1-3 relay satellites) |
| DSN, NASA | Medium reach (above 12,000km) and far-out SC control. Geodynamic measurements | 2 SC CC; 3 stationary CIP | Number of SC being serviced: 10-15 (subject to change). SC visibility zone (in % of time over 24 hours): 100% – on SC trajectories distant from Earth; 50 – 100% – for high SC on TDRSS users) |
| NOAA, control | Meteorological survey and Earth remote sensing SC control | 1 SC CC; sector in MSOCC CC of NASA, 1 Greenbelt; 2 CIS (+1 borrowed from GBCS of CNES Center, France) | Number of SC serviced in orbit: approx.10. SC visibility zone (in % of time over 24 hours): 100% - for SC in GEO; 85-100% - for low SC (in work via TDRSS); 5-10% - for low SC (without TDRSS) |

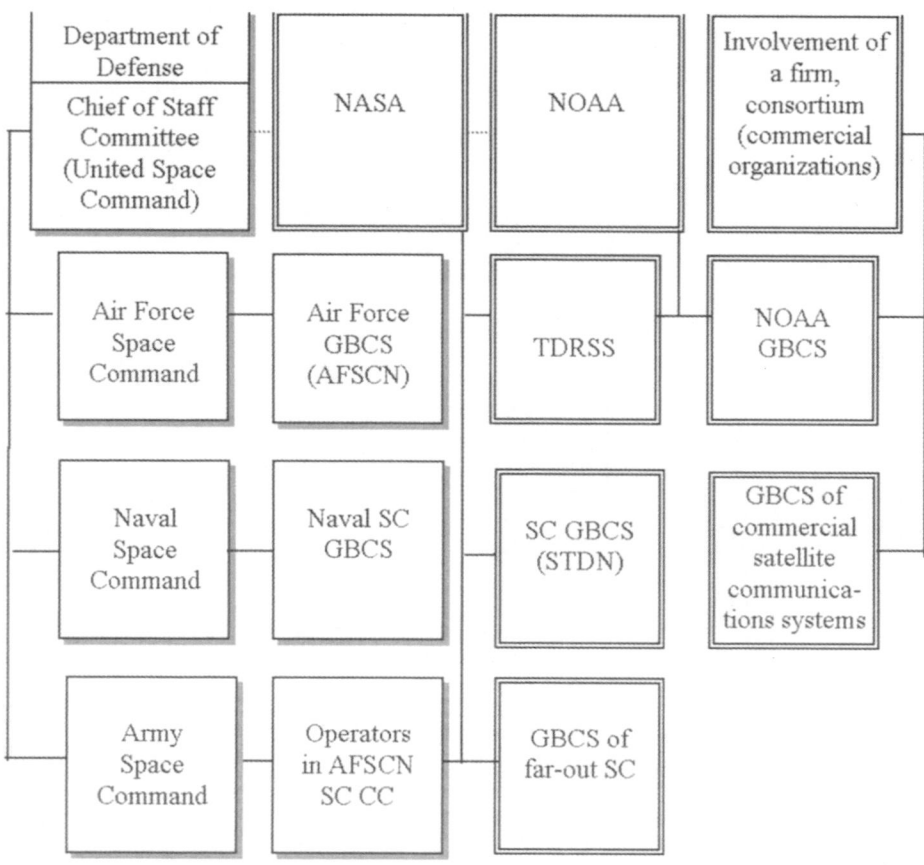

**Fig. 56.** Distribution of SC control systems among US agencies

**SC ground-based control systems of DoD.** Currently, the control systems of SC that are in charge of DoD form several instrumentation and command complexes or ground-based control systems (GBCS) of SC. Now these GBCS belong to space commands established in all armed services, namely: Air Force Space Command, Navy Space Command, Army Space Command (ground forces). Structurally, the Navy Space Command and Army Space Command are subdivisions of other "commands" comprised by the Navy and Army. Operationally, the space commands of armed services are subordinate to US Space Command, often referred to as the United Space Command (USC).

*AFSCN GBCS of AF Space Command.* The Air Force Space Command is in control of DoD's main multi-point AFSCN GBCS (AFSCN is an AF network for controlling satellites and 9 command and instrumentation points (CIP) (Fig. 57 and 58).

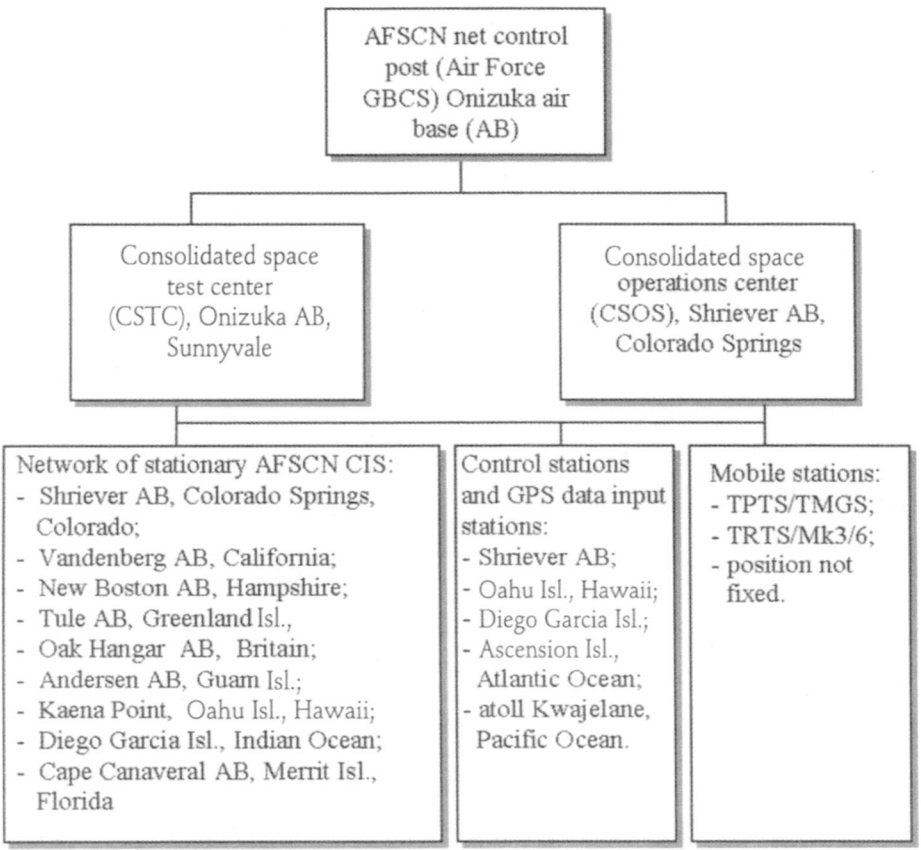

**Fig. 57.** Component package of AFSCN GBCS

Geostationary orbit fully embraces the total visibility zone of AFSCN GBCS. For SC of the GPS system (height 20,000 km, period 12 h) and for high elliptical SC (500x36,000 km, 12 h) the AFSCN GBCS provides visibility over the greater part of both circuits completed 24 hours.

Overall, 16 to 17 command and instrumentation stations (CIS) of the ARTS type are used on AFSCN GBCS as the main facilities. General purpose CIS operate in S range (1.7...2.2 GHz) and Ku range (13...15 GHz).

To meet the demands of the GBCS, a satellite data system (SDS) is used with one or two relay satellites (RS) on high elliptical orbits. The system makes it possible to relay to earth the information regarding low orbit reconnaissance SC in real time mode from any observation point. In addition to it, the AF Space Command rents TDRSS space relay transmission channels from NASA.

The operation and servicing of tracking stations and control centers of AFSCN ground-based control systems for spacecraft are provided by the personnel of 50th space wing of the US Air Force Space Command of 14th air army (by the end of 1996 about 3,300 people, of which 2,600 were military and 700 civil; 500 men staff reduction was planned towards 2,000).

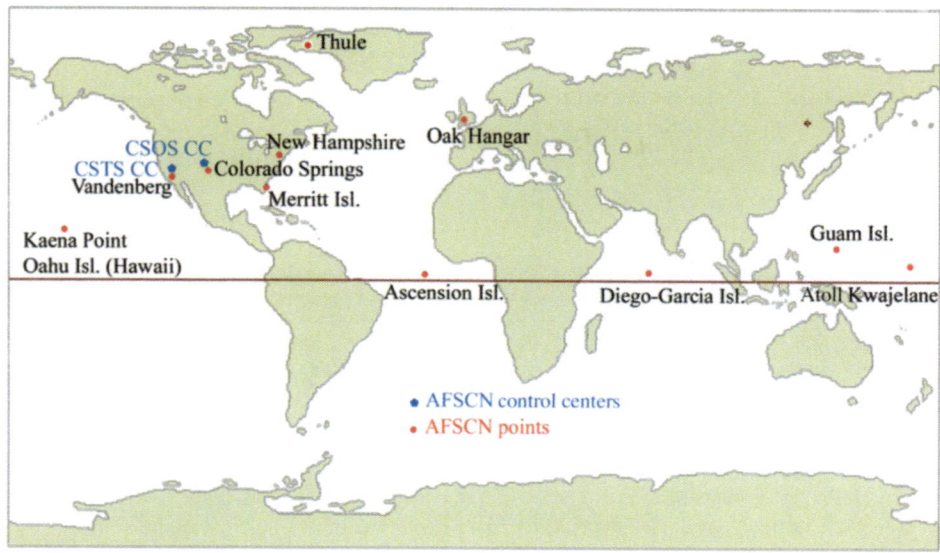

**Fig.58.** Location of US Air Force GBCS centers and points (AFSCN)

The consolidated space operations center (CSOC) controls the military systems of navigation, meteorological survey, communications, and missile assault warning. Officially responsible for some of them, e.g. FLT and UFO, are the space commands of the Air Force and the Army, which is why the Air Force CSOC is involved in servicing their spacecraft as a common control center using the personnel of both Navy and Army space commands.

The CSOC has 7 to 8 SC control sectors and a planning sector to organize the operation of AFSCN GBCS facilities (a similar sector-center, the basic component of GBCS, is comprised by CSOC). The CSOC staff numbers about 2,000 people.

The consolidated space tests center (CSTC) controls the SC of specific reconnaissance (optoelectronic and radar), SC of radar reconnaissance and radar battlefield surveillance (including SC of Navy's NOSS), some experimental military SC, IUS booster unit. Also, CSTC plans and controls the operation of facilities at AFSCN GBCS.

CSTC has several SC control sectors and a sector-center for controlling the facilities of AFSCN GBCS. The personnel of the Air Force Space Command at CSTC is around 700 people.

The following jobs are expected to be done at the AFSCN GBCS in the near future and in the longer term till 2020:

1. enhancement of AFSCN GBCS throughput capacity;
2. improvement of SC controllability;
3. wider use of space relay system, satellite date system, TDRSS and separate inter-satellite command and instrumentation lines;
4. continuation of standardization in AFSCN GBCS;
5. systems automation at SC control points and centers;
6. improvement of command and instrumentation systems;
7. wider use of mobile systems;
8. improvement of controllability of space-based GPS;
9. use of GPS signals on board spacecraft, carrier rockets and on GBCS facilities;
10. enhancement of inter-operation between AFSCN GBCS and test sites;
11. commercialization and privatization of AFSCN GBCS.

The above trends in development of AFSCN GBCS are explained below.

By 2010 – 2015 it will be required to provide control of up to 140 – 145 military SC or even a greater number (in 1998 there were 80 – 90 SC) depending on decisions to deploy SC systems for antiballistic missile (ABM) defense and military small SC.

Provisions are made to reduce the delay in delivery of command and program information and data reception to SC, in reception from SC of purpose-oriented and operational telemetric data and of on-board malfunction signal. Such a reduction must make those operations proceed in a close to real time mode. The main means to achieve this will be inter-satellite communications channels, time reduction in data exchange between SC, tracking stations and SC control centers combined with time reduction in their processing and decision making. For this purpose, automation of exchange processes will be continued, information support of communications lines between GBCS elements will be built up (now it's up to 1.5...5 Mb/s, with 45 Mb/s in the longer term), and the computing power of SC control centers will be increased.

It is believed that in the interests of AFSCS GBCS it is essential to use the SDS and TDRSS's relay satellites not only for relaying data from SC but also for controlling them. The introduction of lightweight user-oriented equipment for operating SC via relay satellite (such a system is currently in production) will contribute to this. Also envisaged are other means of inter-satellite communication, including separate radio- or laser lines between military stationary SC (of communication, warning) and inside constellations of low orbit small spacecraft. Block-2R satellites of GPS are fitted with VHF radio communications line between communications SC, including those for mutual inter-satellite trajectory measurements. Those radio lines are not yet used because of small number of such SC in orbit. The inter-satellite communication is also envisaged for the Block-2F type of GPS SC (roughly in 2003).

The standardization of technical facilities, procedures (techniques) of control and software support is required for standardization of control of various types of SC in AF space complexes. It is also necessary for the mutual support and interaction of various agencies (up to merger in the future of their SC control systems, especially in the state-owned sector).

The automation of reception, gathering, processing, evaluation and distribution of information will enhance the efficiency of SC control, will retard the growth of personnel at GBCS in spite of increase in the growth of operations. Requirements made of skills of the personnel and of their job duties will also be eased (officers will be replaced by lower personnel, engineers by mechanics, etc.) Plans are made to reduce the officers / lower personnel ratio on CSOC duty shifts to 1:2 (formerly 2:1, now approx. 1:1).

Transition will be continued at SC control centers from the architecture of information-computing complexes based on "large" computers to workstations networks based on personal computers. For example, a DPS automated system has been introduced into CSOC and more advanced MAGIC system is being currently introduced on a step-by-step basis. More sophisticated software will be used, including the types that contain "artificial intellect", for servicing expert systems, data bases, etc.

It is also anticipated to use a greater number of radio channels, to strengthen the information support, increase resistance to interference and the stealth capability of command and instrumentation systems, including the channels of space relay systems. Specifically, transition will continue to higher frequency bands (13...15 GHz, 20...40 GHz, 60 GHz, optical). It is recommended to abandon before 2010 the use of the S-band in CIS and to retain until then the two band CIS.

It should be noted that the use of the 60 GHz band in SC-SC radio lines or, with a still greater effect, of an optic band will rid the SC of both organized and unorganized interference originating from ground-based facilities of various applications since the Earth atmosphere significantly weakens the signals of those bands. Especially promising is the optical band for which the weight and power consumption of transmission devices can be reduced by several times as compared to the millimeter band (about 60 GHz).

The component package of GBCS can be changed so as to increase the number of mobile (especially transportable) facilities, including transportable stations for reception and processing of data received from SC (meteorological survey, missile assault warning system, reconnaissance and others) intended to support troops in the theater of war. Standardization of such stations is envisaged in order to reduce the number of satellite system tactical terminals and to change their types.

Within the framework of AFSCN GBCS integration will continue of GPS control systems and improvements will be introduced into the navigation field integrity control. New techniques are now being developed aimed to improve noise resistance of all links of GPS system. The continued launches of the GPS Block-2R series of SC with increased duration of autonomous inter-satellite communications, which SC are supposed to replace Block2A SC, along with introduction of inter-satellite instrumentation equipment into SC of the subsequent Block model will cut down the amount of operations related to controlling the GPS satellites which will reduce the workload of AFSCN GBCS.

Introduction of the GPS user's navigation equipment into the component package of on-board SC control systems, including those serviced by AFSCN GBCS, will allow to improve the accuracy and efficiency of SC navigation (coordinates, speed, time, orientation), to reduce the amount of operations or the amount of means of changing the trajectory, and to lessen the workload of GBCS. Studied now are new methods of controlling SC by means of users' navigation equipment integrated with the telemetric, communications and relay equipment via a relay satellite or within a SC constellation.

Optimization will continue of GPS users' navigation equipment (UNE) along with the techniques of using it on carrier rockets. In doing so, account will be taken of the previous experience.

With reference to GBCS the introduction of UNE will make it possible to improve the coordinates reference control of tracking stations, especially mobile ones. The establishing of orienting reference of GBCS and test sites' facilities in a single time scale will be simplified.

The interaction of AFSCN GBCS and space test sites in the way of instrumentation and data gathering will improve since TDRSS will be used both to control SC and to control launches of carrier rockets and their booster units. Plans are made to fit advanced carrier rockets and booster units with small-sized transceivers of a new generation, including those with phased antenna arrays functioning in S-, Ku- and Ka- bands for operation via TDRSS. It is also regarded as a promising idea to control rockets in flight by combining a relay control and control of the rocket trajectory via GPS signals. A Magic Box component package has been developed that integrates users' equipment of TDRSS and GPS systems.

Reception and rendering of commercial services is regarded as a promising means of cutting down the operational costs of SC. Massive acquisitions are planned of commercially available off-the-shelf hardware (or subassemblies) and software. Also, maintenance of facilities and other jobs are expected. Conversely, technologies and idle equipment of GBCS can be made available to various users (e.g. private companies) on a commercial basis

It is possible to privatize some GBCS facilities with their subsequent servicing (the customer duly pays for services rendered). TDRSS, for instance was expected to be privatized.

*GBCS of Navy Space Command.* The Navy Space Command is in charge of satellite control facilities generically designated as NSCN, i.e. Navy's satellite control network. The foundation of the network until the end of 1997 used to be GBCS of NNSS/Transit complex comprised of one SC control center and four CIS (command and instrumentation station). Because of cessation of operation of Transit on January 1, 1998, NSCN GBCS is being restructured. The Navy planned to use that SC to monitor the ionosphere in prediction of conditions for ultra-short wave communications, which is why some elements of NNSS can remain here.

Since the end of 1996 the Navy Space Command has been engaged on its own in technological (in the area of service systems) and purpose-oriented (communications relay systems) control of Fleetsatcom SC, and since October 1997 in control of UFO SC that are replacing them. The technological control of Fleetsatcom and UFO SC had always been the domain of the Air Force. In recent years it was accomplished from USOC via stations located in AFSCN GBCS points. After the systems changed hands, the technological control began to be exercised by the Navy Space Command personnel located (probably, temporarily) in the CSOC sectors and in AFSCN GBCS points engaged for a particular SC. Work still proceeds on organizing information exchange channels between AFSCN GBCS points and Fleetsatcom and UFO systems' communications network control centers. Therefore, later on those centers could be used to accomplish a full control of those systems.

The technological control of SC engaged in navy ocean surveillance satellite (NOSS) is the domain of the Navy and accomplished by Air Force Space Command from CSOC via GBCS of the AFSCN.

The GBCS of NSCN Navy Space Command is now being restructured and its development depends on decisions concerning amalgamation of control facilities used in above said space systems within the framework of agencies (Navy and Air Force) and on a national scale.

*Space command facilities of the Army.* The Space Command of the Army does not have its own SC control facilities forming a single GBCS.

Since 1987 the space command of the Army is fully responsible for operation of DSCS communications system. However, up to the present the technological control of the DSCS SC has been accomplished from the Air Force control center via AFSCN GBCS points (now with participation of the Army space command personnel in control centers and points. The Army space command continues to effect the purpose-oriented control of the system via 5 control centers of DSCS SC communications relay stations and through communication in the system, including the communications subsystem. As noted above, there is a stand-by mobile mixed station to control defense satellite communications system (DSCS) and to organize communication via those SC in the station's operating zone (belongs to the Army Space Command).

The Army Space Command has a number of laser and optic stations (including mobile ones) used when necessary by other agencies for precision monitoring of orbits of some SC (geodesic, navigation, reconnaissance, natural resources prospecting, etc.)

Thus, at the moment the Army Space Command does not have a GBCS organized into one entity. Therefore, the control of SC that are in charge of the Army Space Command will be maintained in close cooperation with the Air Force Space Command.

**Civil use complexes.** In the US, the principal control facilities of civil use SC (applied and scientific, including piloted SC and deep space vehicle) are in the possession of NASA, NOAA, and private companies.

NASA controls SC belonging to state-owned organizations. On the basis of agreements assistance can be provided to private companies, international organizations and foreign countries in controlling and reception of data from SC of other US agencies (DoD, NOAA and others).

Specifically, in the mid 1960s "A Memorandum of Understanding" was signed between NASA and DoD. According to it, NASA and DoD could resort on a mutually beneficial basis to each other's assets. Technically this was achieved due to the fact that both agencies' tracking stations use the same standard frequency bands. Signals from SC could be received both by DoD's and NASA's facilities with their subsequent transmission to SC control center (normally in a recorded form, without decoding and processing). Such cooperation was envisaged under Apollo spaceship program of moon trips. Currently, this is practiced under Space Shuttle program. Joint use is possible thanks to the tracking and data relay satellite system (TDRSS).

NOAA operates jointly with NASA meteorological satellites (civil) and natural resources exploration satellites. NOAA has several tracking points as well as points to receive information from such SC.

Private companies control commercial satellite communications systems belonging to them.

For more than 25 years the Intelsat international consortium (or ITSC) has been using a commercial satellite communications system of the same name. In this consortium the USA is represented by the COMSAT company which is the core of the organization and plays a leading role in controlling the system and its SC.

*NASA's SC control facilities.* Currently, within NASA the facilities for controlling SC and gathering data are concentrated mainly in two GBCS (tracking networks) (Fig. 59 and 60).

One of them, designated by abbreviations STDN, GSTDN or GN, provides for flights of near-earth SC (among them are high orbital ones flying up to the Moon), unmanned and manned SC, including the Space Shuttle transportation system. The second GBCS is a deep space network designed to control a SC in deep space. In addition, NASA owns the tactical data relay satellite system (TDRSS) considered as "the national asset". It is available to other US agencies, e.g. DoD and NOAA. Functionally, TDRSS is fully integrated with STDN GBCS and forms with it a single complex.

The division of NASA's facilities into two above said GBCS is in a way a matter of convenience since both are closely interconnected and subordinate to one and the same department of NASA, the department of space operations. Within the framework of interdepartmental and international cooperation NASA can use the facilities of DoD, radio telescopes of scientific research organizations, tracking and data reception stations of some companies, countries and organizations.

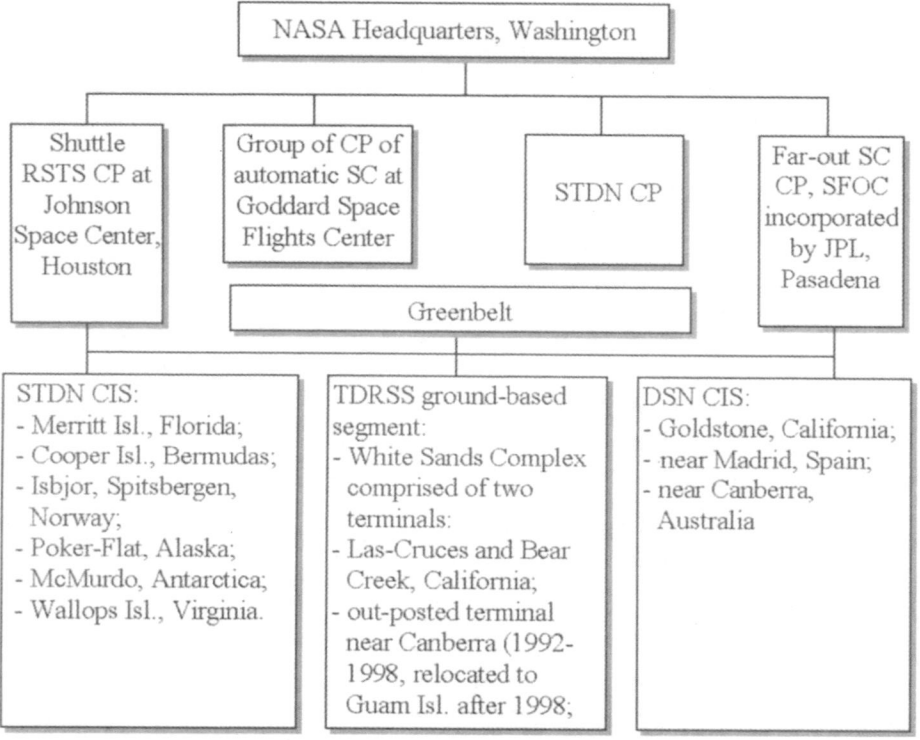

**Fig. 59.** NASA's ground-based control system

Presently, the STDN complex numbers 6 tracking points (stations), which are command and instrumentation points. Using those points the control centers perform all the operations related to function and use of SC.

The points are located both on the territory of the USA and outside it, in Spain, Australia, on the Bermudas and Ascension Island. The main and best equipped are the points in the Goldstone Desert (California, USA), near Madrid (Spain) and not far from Canberra (Australia). It is these tracking points (stations) that after introduction of TDRSS in 1989 are mainly used to control NASA's SC, including those with orbits above 12,000 km which cannot be served by TDRSS. The remaining points – Merritt Isl., Cyprus, the Bermudas, Greenbelt (Maryland) and Ascension Isl. (Atlantic) – are regarded as auxiliary with the former two planned for use mainly during SC launches.

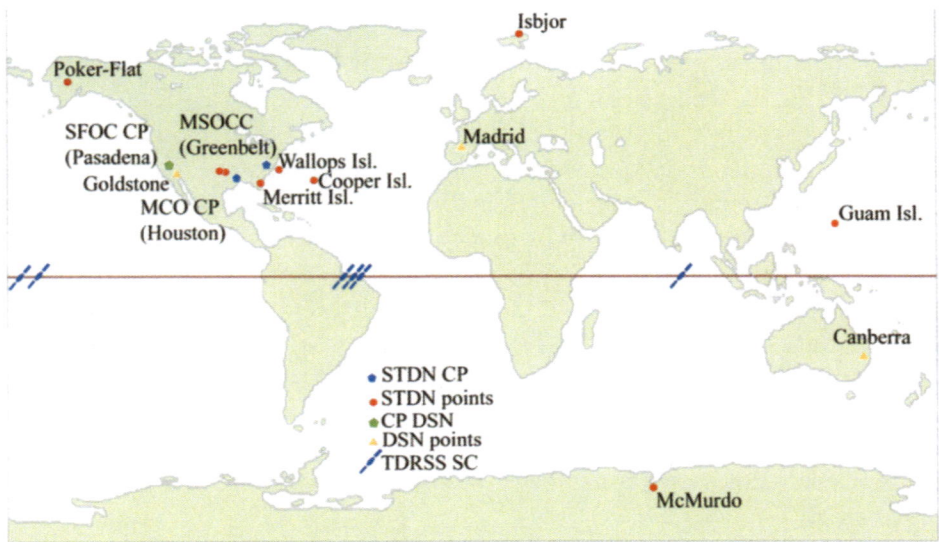

**Fig. 60.** Location of NASA's GBCS control centers and points

In STDN GBCS, two main SC CC can be singled out. One of them, MSOCC CC is situated at the Goddard Space Flights Center (Greenbelt, Maryland, near Washington), another MCC (Mission Control Center) at the Johnson Manned Space Flights Center (vicinity of Houston, Texas). Actually, each of those CC is a combination of interconnected control centers which cooperate in serving separate SC or series of them.

It so happened historically that the Goddard Space Flights Center is a base of NASA's department of scientific and applied near Earth SC. For the period of implementation of a new spacecraft program a center used to be formed on the territory of the Center aimed to control the craft. Such center was called "Project operations control center". Similar centers are established both within the framework of the common use CC and as independent CC if operations with SC proceed practically without intermission and require processing of massive information flows (for example, Hubble space telescope – NHST SC and other programs). Upon completion of the project, the specialized control center, which could be accommodated in a separate building, is dismantled and re-equipped for a new SC. Currently, operating at the Goddard Space Flights Center are two common use control centers (sectors, halls) and up to 5 special purpose control centers (one for NHST SC). Should necessity arise to analyze situations on board SC or to change their operating programs, extra personnel of the flight control center could be put on the job.

The Johnson Manned Space Flights Center comprised by NASA works on manned flights programs including Gemini, Apollo, Shuttle, and ISS. The availability of specialists, simulators and simulation complexes are the factors that determine the choice of this particular space center as a flight control center for above mentioned programs.

In 1967-1968, in view of a massive growth of amount of operations and their complexity, with regard for a certain wear and obsolescence of primarily CC, as well as for reasons of prestige the most up-to-date flights control center was built for the Apollo space program at Houston. At the moment, in spite of two essential upgrades in the course of the Shuttle program, this center is considered inadequate. Under the ISS program, a new control center has been built that is intended to control the station for 30 years.

There is a separate center (3 sectors) at the Johnson Control Center with an assigned duty of controlling cargoes delivered to space in the Shuttle's cargo compartment but not separated from it during the flight. The same sub-center controls operations with those cargoes. Upon separation of the autonomous cargo, control over it is passed on to the control center at the Goddard Space Flights Center unless no return of the cargo to the Shuttle's compartment is anticipated.

During the Space Shuttle flights with orbital laboratories of the Skylab series, the control over operation of the latter, the planning of and supervision over experiments in them is accomplished via the control center established in 1991 at the Marshall Space Flight Center (Huntsville, Alabama). Also during such flights, NASA cooperates with ESA's SC Control Center (Darmstadt, Germany) that had created those laboratories and uses them to place its cargoes for experiments.

The STDN complex is capable of controlling no less than 100 low orbit SC (the limit of SC users in TDRSS) and a certain number of high orbit SC. The actual number of SC supported by STDN GBCS did not exceed 50...60, of which up to 10 SC with the aid of TDRSS.

By the late 1970s, the STDN GBCS personnel reached 3,100 people. The introduction of TDRSS and the closure of many stations brought about staff reduction whose estimated number can now vary between 1,500 and 2,000 people (about 200 people on the ground-based station of TDRSS, 100-500 people on each of the main CIS, 200-300 people in each of the SC group control centers.

The STDN GBCS CIS use unified command and instrumentation system (USBS). This is a multi-functional combined (except for some operations and modes) S-band radio station (sub-band 2.1/2.2...2.3 GHz). In frequency band, specifically, in the reverse channel (SC-Earth) the USBS station is compatible with the DSS CIS in DCN GBCS, with SGLS CIS in AFSCN USAF GBCS, with TDRSS and with many CIS in GBCS of other countries and organizations.

Also, there are combined S band transceivers capable of operating both with USBS CIS and ASGLS CIS in USAF GBCS. Most sides and the antenna system in those transceivers are common and the transition to operating one or another CIS is accomplished by a command through switching the devices of signals formation and modulation (demodulation). The subsystems of the Space Shuttle system are one such example.

*TDRSS and reduction of elements of ground-based complexes.* TDRSS (Tracking and Data Relay Satellite System) was developed in 1976-1983 to the order of NASA. In the late 1970s, at the final stage of development NASA decided to develop all its new low orbit SC in the form of TDRSS users (the decision is complied with). The capability is retained to effect direct communication between such SC (including Shuttle) and NASA's ground network stations operating in S-band (part of stations also have now $X$-band channels (7-8 GHz) in which TDRSS does not operate). By contrast to GN, TDRSS is called a space network (SN). TDRSS is united with the GN sub-network to form a common tracking network of NASA, STDN.

By the late 1970s, the STDN GBCS numbered 22 tracking stations scattered across the globe. After introduction in 1983 of the first relay satellite of TDRSS, the number of GBCS gradually dwindled to 9. After the standard full-strength deployment of the system with three relay satellites (RS) in 1989 the number of stations decreased to 5-6. As this occurred, their target functions were revised and they began to be used as sub-trajectory aids in launching Space Shuttles and NASA's SC.

Currently, TDRSS permanently uses two to four RS form six being in GEO and three spaced special relay stations. For SC users the TDRSS provides a multi-user (in S-band only) and/or individual (in S- and $Ku$-bands) service modes. One RS can simultaneously serve:

a. up to 20 SC (or other objects of rocket engineering) in a multi-user mode (via phased antenna array on relay satellite in S-band) with the rate of information transfer up to 10 Kb/sec – when from user to SC; up to 100 Kb/sec when from SC to user;

b. up to 2 SC with individual servicing (via two independent two-band parabolic antennas on relay satellite) with the rate of: in S-band: up to 300 Kb/sec when to SC; up to 6 Mb/sec when from SC; in $Ku$-band: up to 25 Mb/sec when to SC; up to 300 Mb/sec when from SC.

New RS that will gradually replace in 2000-2001 the current RS will have $Ka$-bands channels which increase the rate of transfer from SC in multi-user mode up to 1.5 Mb/sec (but there will be 5 channels on RS instead of 20); in case of individual servicing – up to 300 Mb/sec in $Ku$-band and up to 600-800 Mb/sec in $Ka$-band.

The DSN GBCS are designed to control SC flights in deep space, including craft in orbits around planets and on their surfaces, as well as beyond the Solar system. The control over SC is fully passed on to DSN GBCS after its relocation from near-earth orbit to interplanetary trajectory. Until then, involved in SC control are the facilities of STDN GBCS.

The DSN GBCS is also used in international space programs (among them those with Russia's participation) and for scientific purposes (radio astronomy, geophysics etc.). If necessary, the DSN GBCS can make use of technical assets of other countries and organizations (radio telescopes, optics).

The DSN GBCS include three CIS referred to as outer space communications systems. They are spaced about 120° apart in order to maintain the continuous communication with SC during the Earth's rotation. They are situated in USA (the Goldstone Desert, California), in Spain (near Madrid) and in Australia (near Canberra). Also, used for launches are the sites on Merritt Island (USA, Florida) and near Johannesburg (South African Republic lets its station for launches).

The SC control center is located at Pasadena, California, and is incorporated by the Jet Propulsion Laboratory subordinate to NASA. The center controls all far-out SC and gathers information about them. It's noteworthy that more than 10 SC can simultaneously operate in far-out space. The data is processed, stored and distributed by the Jet Propulsion Laboratory (JPL) whose specialists and assets can also be used to control SC. The new center has replaced two SC control centers used previously, the center for flight support of various types of SC and the center for controlling the Pioneer family of SC.

The DSN GBCS tracking points use 3 types (modifications) of CIS, reciprocally compatible but different in time of manufacture (introduction), in diameter and in component package. The main CIS having the greatest capability to control SC in deep space is the Mars modification DSS station with an antenna 70 m in diameter and equipment operating in S- (2.1/2.2...2.3 GHz) and X- (7...8 GHz) bands. The DSN GBCS CIS enables communication to be maintained with SC during their flight to the border of the Solar system ($5.95 \cdot 10^9$ km) and Pioneer/Voyager SC that have gone beyond it.

Better suited for near-earth areas are the stations with antennas 30 m in diameter (some of them have no X-band).

Stations on the points are spaced up to 15...20 km apart, but they can be united at the signal level, up to simultaneous operation of several or all stations for one SC in which case their antennas form a common system ("an array"). As noted above, plans were made to group together all antennas and the stations' equipment of each CIS of DSN as well as of the neighboring CIS of STDN in a place where the main antennas of Mars stations (70 m in diameter) are located. In integration of all stations' equipment and antennas into a single complex the required configurations of subsystems (antennas, receivers, transmitters, recorders, etc.) for each communication session is chosen by connecting them with the help of computers. One of the DSN antennas is shown in Fig. 61.

**Fig. 61.**
Deep space tracking
network antenna
(Goldstone, USA).

The number of personnel at DSN GBCS, in SC control center especially, depends on the intensity of work on current space programs. A flight of SC to a target may take many months or years. On intermediate stages only separate experiments are conducted and control is exercised of the flight trajectory and of the state of on-board equipment. The bulk of operations happens to be performed during the short periods when the flight's primary tasks are being tackled. This is what causes abrupt fluctuation in the workload of the center's personnel. The number of personnel at DSN GBCS was at its highest in the 1980s (900 people).

NASA wishes to return to the predominant research and development under projects and to gradually curtail its participation in work when the built systems are being used. This refers among other things to techniques and operations utilized to control the SC, for which purpose the so-called "space operations unified contracts" had been drawn up with validity till 2010. The contract provides for a transfer, since 2000, to a private company (Lockheed Martin) of the right to effect control over NASA's various SC and gathering of data from them, including cooperation with spacecraft or instruments' owners or experiments on such equipment.

Thus, the main trends in development of NASA's facilities for controlling SC are:
- improvement of performance of TDRSS, including its integration
  (as far as possible) into the international "space network";
- modernization of communications system for NASA (NISN);
- commercialization and privatization of NASA's GBCS.

NOAA's GBCS. NOAA ((National Oceanic and Atmospheric Administration) operates dual use SC – GOES Next SC (meteorological survey, ERS), low orbit

NOAA SC (meteorology, ERS), and Landsat (ERS) (a total of 8...10 SC, including partially operable ones).

*NOAA's GBCS* includes SC control center in Suitland, Maryland, and two CIS situated on Wallops Island (Virginia) and near Fairbanks (Alaska). A station at Lannion, France, belonging to France's CNES, can be involved in the job.

The control over SC can be exercised in close cooperation with NASA. In addition to its SC control station NOAA uses the assets of the control center provided by the Goddard Center (Greenbelt) and TDRSS (for low orbit SC).

Also, NOAA has a wide network of stations to receive data from its SC all over the world. Many of those stations can receive data from meteorological SC and observation SC (natural resources monitoring) of other countries.

There is promise in modernization of tracking hardware and software used in GBCS belonging to NOAA on account of plans to create after 2000 a united meteorological survey system based on low polar SC and in view of the general tendency for unification of control systems of SC owned by federal agencies.

## SC control systems of European Space Agency (ESA)

The European Space Agency (ESA) was established in 1975 to promote cooperation of European countries in space exploration, in development of space technology and in search for its practical application. Initially, ESA comprised 11 countries (France, FRG, UK, Italy, Spain, Sweden, Belgium, Denmark, the Netherlands, Ireland, Switzerland). Later on, ESA was joined by Norway, Australia, and Canada. Finland has been an associated member of ESA (possibly, now a full member).

Officially, ESA is an organization engaged in research and development of space rocketry only till the product is accepted for normal operation. For quite some time, however, ESA had been doing not only research and development, design and testing, but also operated a number of space equipment until it became possible to put it to commercial use. Nevertheless, ESA is still operating purely scientific SC, including deep-space vehicle. The agency does not manufacture the SC it develops, manufacture being the domain of industrial companies under the supervision of ESA's research workers and engineers. ESA develops or acquires ground-based SC control facilities and information gathering equipment under contracts with other countries' companies (members of ESA, if possible).

ESA's main research organizations are:
- ESOC (European Space Operations Center), Darmstadt, Germany;
- ESTEC (European Space Research and Technology Center), Nordvik, the Netherlands;
  European Space Research Institute (ESRIN), Frascati, Italy.

The main function of ESOC is to control the agency's SC in course of their launch and to guide them in orbit. ESOC operates CIS for EAS. Also, the center prepares and oversees contracts for on-board and ground-based control systems, formulates suggestions concerning funds allocation, participates in international programs. ESOC personnel numbers about 800 people.

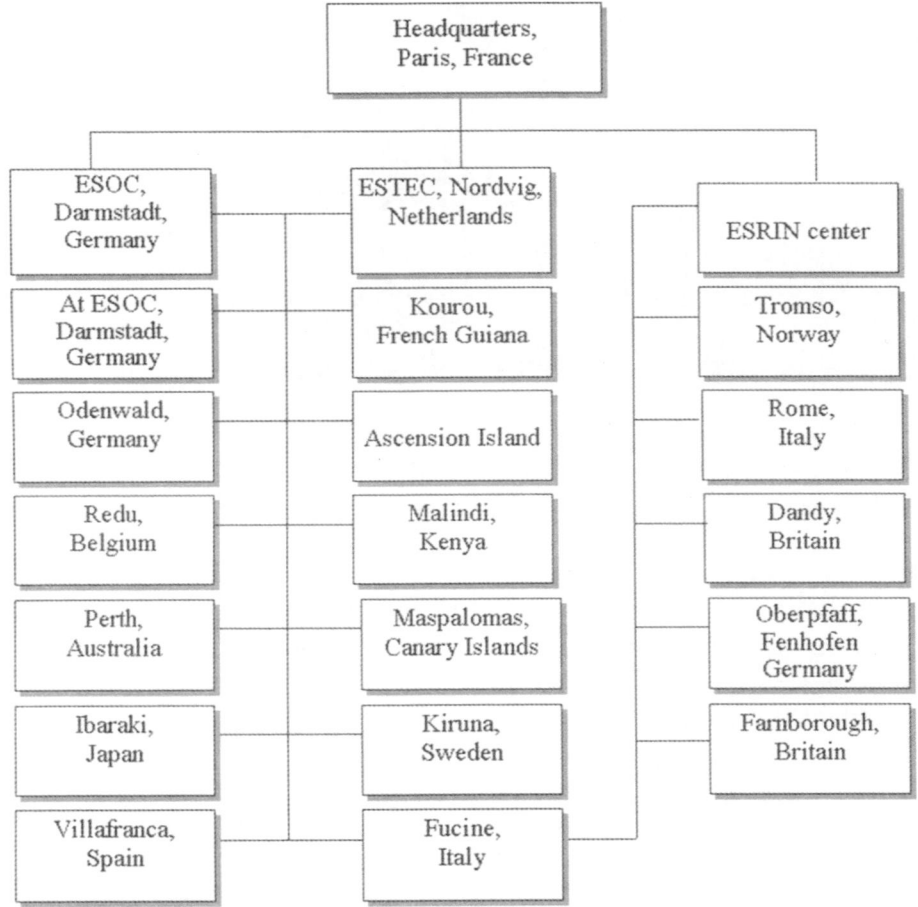

**Fig. 62.** Location of ESA's control centers and points

ESTEC works directly on scientific programs. It conducts research in advanced SC and their equipment, develops and tests some development prototypes. The center's personnel numbers 1,300 people.

The structure of ESA is shown in Fig. 62.

ESRIN supports the information service of ESA, coordinates activities of various information services of ESA and uses its Eathernet of eight ground-based stations to receive, preliminarily process, archive and distribute the data supplied by ERS.

**Fig. 63.** ESA control center (Fucino, Italy)

ESA operates the ESTRACK GBCS in its "as-is" state. The complex includes the main and several specialized and stand-by SC control centers and 10 to 11 CIS located in different countries. The main control center (ESOC) for many types of SC is in Darmstadt (near Frankfurt-am-Main, Germany), while specialized centers are in Villafranca (near Madrid, Spain) and in Fucine (near Rome, Italy). This GBCS is used to control some dual use SC of such consortiums as Eumetsat, Eutelsat and of member states of ESA. Those consortiums have put in operation their own SC control centers which eased the workload of ESOC control centers.   The dual use space relay system of DRS which is due to be introduced  circa 2000 will enlarge the operating zone of ESA's GBCS and increase the rate of data transfer from the serviced SC, which effect will also be obtained thanks to laser communications lines. The distinctive feature of the space relay system being developed is the direct access of the user (owner of SC) to relay satellite and the delivery of information from the serviced SC directly to the user. ESA plans to develop the European space navigation system (GNSS-2) which may necessitate creation of new elements of ESTRACK GBCS.

## SC control systems of France

The national space exploration center (CNES) carries out both civil and military space programs (jointly with DoD).

Military satellite communications system, Sirakus, was created in 1988. It is based on the Telecom SC. Since 1995 Helios SC derived from the Spot SC have been launched. Helios-2 SC is being developed jointly with other

European countries. The Spot dual application ERS SC continues to be operated. The Cerise experimental radiometric reconnaissance SC has been launched. Work proceeds on a number of military systems (Zenon, Osiris, missile launch warning system and others) with slated commissioning dates after 2000.

The GBCS of CNES has resulted from many years' work on civil programs. Organizationally, this is a single entity, though functionally it is divided into departments dealing with particular space systems. Now the GBCS is being restructured in order to improve the control of the above mentioned military and civil SC.

The GBCS incorporates 2 SC control centers and up to 10 CIS scattered across the globe. SC control centers are located in Toulouse, Southern France, at the USAF base in Francazal (auxiliary for Helios SC). Also, CNES has a tracking spaceship, the Henri Poincare, used mostly to launch SC from the Kourou test site (South America).

The development of CNES SC GBCS is expected to proceed in the following directions:
   a. servicing the ever growing number and variety of SC, including military and dual use types;
   b. standardization of CIS and their optimization for compliance with demands of military application;
   c. introduction of SC relay control and transfer of data from low orbit SC through a space relay system (specifically, radio and laser channels of DRS system being created under ESA program);
   d. improvement of component package and of placement of CIS and GBCS facilities;
   e. expansion of international cooperation in use of SC control systems both under ESA and with other countries.
The improvement of GBCS will make the control of and data delivery from SC global and fast to use.

## SC control systems of UK

The UK operates the Skynet military communications SC and participates in controlling NATO's communications SC. The UK is considered to be Europe's largest (second in the world) consumer of information obtained from various SC of many countries and organizations. The data processing results (including photographs from meteorological and ERS SC) accumulated over the years can be used for military purposes, for example in crisis situations.

GBCS of the Skynet system consists of a CC (Oakhangar Air Force base near Farnborough, vicinity of London) and several CIS, the main one of which shares the territory with SC CC. The tracking points of AFSCN US Air Force GBCS can be used as auxiliary ones.

Great Britain participates in such projects as ERS SC, DRS, Helios-2 SC and others.

Over time, the military space systems may grow in importance which will require expansion and centralization of GBCS. Cooperation will continue in mutual use of SC control systems in various countries under ESA's programs. After 2000, within the framework of ESA the low orbit SC of UK will also get access to DRS channels.

## SC control systems of China

China operates military and dual use SC for communications, meteorological survey, and ERS. It also launches experimental SC, military ones among them. To control those SC it has a multi-point GBCS organized as a single entity and operated by the Chinese Association for Launching, Tracking, Telemetric Measuring and Control of SC. This association is subordinate to the State Council's Committee on Defense Science, Technology and Industry.

The GBCS comprises three SC control centers and around 10 stationary CIS. There are also 10 launch and sub-trajectory instrumentation points for three space test sites. In addition, two large and two smaller instrumentation spacecraft are in operation.

Two neighboring SC control centers are in the Xi'an district of Shaanxi province. One control center is primary and provides for operation of various SC, low orbital ones among them. The third center (for meteorological satellites) is deployed at the Beijing Meteorological Survey Center. Nine GBCS points are located on the territory of China as far apart as possible. One CIS was deployed in 1997 on Tarawa atoll (a state of Kiribati, Micronesia). The points are outfitted with various CIS. There is a large number of points to receive data from SC, including foreign ones (meteorology, ERS). The above said tracking craft are used mostly to launch SC and test ICBM.

The continued work on military, navigational and smaller SC, on piloted flights and other programs makes new demands of command and instrumentation complexes (CIC):
- increase of CIC throughput capability;
- creation of new tracking points outside China in order to further expand the operating zone of CIC;
- uprating the performance of measuring and computing systems of CIC and communications channels;
- standardization of CIC facilities, including command and instrumentation stations on tracking points.

China expands cooperation with other countries, among other fields, in mutual utilization of control systems of SC and in data reception from them.

## SC control systems of Japan

The National Space Development Agency (NASDA) develops and operates SC for communications, ERS, meteorological survey and others with dual application. Scientific research SC are the domain of Institute of Space and Atmospheric Studies (ISAS).

Both organizations have their own SC control centers and CIP. It is possible, however, that on some CIP, located outside Japan, the systems of both organizations are installed and jointly used whenever necessary.

Here the GBCS of Japan can, by convention, be regarded as a single technical entity since ISAS is formally subordinate to NASDA. That GBCS has no less than three SC control centers (Tsukuba – for many SC, Kashima – for communications SC, Kamitsu – for ERS and meteorological SC) plus 8-9 CIP (part of them combined with CC): Tsukuba, Kashima, Kamitsu, Kachuura, Usuda – center and point of deep space communications, Okinawa Isl., Ogasawara Isl., Christmas Isl.

Japan's GBCS deals with the entire spectrum of jobs pertaining to SC control as well as data reception from SC and technical support of carrier rockets during their launches. Many programs are being implemented jointly with NASA.

The prospects for development of GBCS are primarily associated with creation of DRTSS which is similar to and compatible with DRS and TDRSS. The increase of GBCS capabilities is necessary above all in connection with plans to create the HOPE lightweight RSTS and JEM (Japanese experimental module) as a contributory effort in creating the ISS.

## SC control systems of international consortiums and commercial firms

A number of commercial privately owned firms operate around 10 satellite communications systems. Among such systems are Intelsat, Galaxy, Telstar, Panamsat, Inmarsat, Iridium, GlobalStar and others. The operating zone of those systems can be global or regional (territory of the US and target areas) depending on the number of SC used in the system.

As a rule, firms have their own few-points ground-based complexes that control the functioning of SC, communications channels quality, system resources distribution, etc. It is also standard practice to rent tracking stations from other firms and agencies of the US and foreign countries, e.g. from DoD, NASA, ESA, France's CNES etc.

Among the private companies that often render services in SC control to other companies, are the American Comsat corporation participating in the international communications consortium, Intelsat. The Intelsat consortium was formed in 1964 and operates a commercial communications system of the same name using 7 to 8 operating Intelsat SC of various models. The GBCS of this system controls, in addition to operating SC, the state of a number of stand-by SC and the retired SC withdrawn from the system (a total of up to 20 SC). The GBCS has 7 CIP and one SC CC.

**Fig. 64.** Astra ground-based direct TV broadcasting SC control center near Betzdorf, Luxembourg

The CC of SC and system as a whole is situated in Washington, at Comsat headquarters to whom it belongs. Apart from controlling the Intelsat SC and the system, CC provides its assets to other companies, normally for launches of SC and their commissioning (deployment, relocation to work place).

CIP of Intelsat's GBCS are scattered rather widely throughout the globe because the SC incorporated by the system are placed both in the western and eastern hemispheres. Some tracking stations of the consortium are located on CIPs of GBCSs belonging to other countries (including those comprised by the consortium). Tracking stations of those countries can also be used (on negotiated terms).

One of possible services in controlling SC is demonstrated by the Hughes Aircraft company, manufacturer of communications SC based on its standard model used by many other companies that own satellite communications systems. This company builds control stations for SC manufactured by it and then operates them in order to keep its own system working and – on a commercial basis – in order to control communications SC of other companies.

The control of SC on-board equipment after deployment of communications antennas and check-up of relays can be accomplished via radio line in the main communications channel. SC often use continuously operating transmitter beacons that serve to check the position of SC relative to ground-based stations and to transmit telemetric data about the state of SC subsystems.

The control of communications SC in a stationary orbit is facilitated by the fact that they are continuously in the tracking stations' visibility zone (except during launch and relocation to a work point). In addition, a number of SC have a system of automatic orbit maintenance (correction).

More sophisticated is the organization of control in communications systems based a large number of lightweight (small) SC in low orbits. Several projects of such systems are now being implemented, the best known of which are Iridium, GlobalStar, Teledesic systems with a number of SC in them in the region of several tens (around 280 SC in the Teledesic system). The expected advantages include, as a rule, a greater reliability, self-sufficiency of SC, lesser susceptibility of the system to the loss of part of SC, price affordability and admissible time of the system recovery, all which factors ease the requirements made of control over their functioning. The control of the system can be accomplished on the same principle as used in the ground-to-space network in which case data is delivered from the control station to SC through inter-satellite channels. Under such circumstances, several interconnected control stations can be used, including simplified mobile ones.

## 1.4.2 Evolution and trends in development of SC control complexes abroad

Until the mid-1960s, specialized tracking networks used to be created to control some or other type of SC. They were developed by various companies, had their own SC control equipment and ground-based stations (GBSs). This referred to most civil and military SP of the time. The shortfall of such organization was the rapid growth of the number of ground-based tracking stations with their rather low workload which resulted in low efficiency-to-cost ratio.

The growing number of SC in orbits urged NASA and US DoD to create a general application tracking network with unified command and instrumentation systems (GBSs with the required on-board equipment for SC). The successive integration of various control systems generated in the US the above said control complexes of NASA and DoD. For various technical and organizational reasons the complete unification and standardization of control facilities and techniques were not achieved, though attempts had been made to adapt to each other the complexes of NASA and DoD. Their further integration and unification aimed to reduce budgetary allocations for their operation and development are still regarded in the US as top priority issues.

The characteristic feature of US GBCS controlling a large number of various SC is the multi-point nature of its structure. Thus, by the early 1980s the STDN numbered up to 22 CIPs located both in the US and other countries which enabled communications sessions to be conducted on each orbital circuit. However, in spite of such a large quantity of CIPs, and massive expenditure on operation of STDN, the total zone of radio visibility of SC orbits from all CIPs did not exceed 20% of the entire trajectory of their movement which on a number of occasions impeded the performance of operations on SC in real time mode, especially in case of emergency on-board.

The drastic way to cut down the number of GBSs required to control SC, to reduce the servicing personnel and to increase the efficiency of control and heighten the effect of using SC, was the creation TDRS to receive scientific (purpose oriented) information from SC and to control them in flight as well as to gather information from carrier rockets, booster units and spacecraft during launch and return of spaceships. TDRSS provides an information link with low orbit SC belonging to NASA and is used to relay reconnaissance information from US DoD's SC. Furthermore, TDRSS radio channels are rented to obtain communication with SC of foreign countries.

The creation of space relay system radically changes the structure of SC GBCS. Under such circumstances, GBCS turns into a chain of elements of mission control centers (MCC) – ground-based TDRSS – relay satellite (RS) – SC – user, that is, a single point structure. Actually, in view of the global nature of operation, it is essential to have no less than three RS spaced apart on longitude whereas the achieved throughput and the state of equipment in orbits allow to keep in orbit up to six geostationary RS.

On the ground, three terminals have been built (two in the vicinity of White Sands, California, one in the Pacific Ocean zone). Each terminal can keep continuously in touch with 2 to 4 RS. Ground-based stations with traditional CIS are only used to control NASA's SC being in orbits at an altitude above 12,000 km (for existing generation of RS) or SC not included for various reasons in RS users.

In principle, this structure is a prototype of complexes designed to control and receive information from SC in the 21st century when informational contact between ground-based centers used to control and process information from SC will be possible at all times regardless of mutual position of MCCs on Earth and SC in space.

The idea of using RS to provide global and continuous communication with SC is applicable not for near-earth satellites alone. It can bear fruit during work in deep space. For example, in exploration of Solar system's planets with a wide use of smaller satellites, penetrators, various equipment for the study of planets' surface and atmosphere, the transmission of scientific data and control of above said objects can be most effectively accomplished with the least power consumption by relaying signals via base artificial satellite of the planet, which satellite will serve as a link with a research center on earth that organizes research in deep space. Such projects of a common RS are already mentioned in special literature abroad and some experience of such relay is also available. NASA plans more experiments in this area while implementing the Mars program.

Much work is now underway abroad aimed to create low orbit multi-satellite communications systems and systems for transmission of wideband information. Conceptually, those systems are to work with low power terminals like a telephone receiver or with mobile and stationary small-sized transceivers thus providing them with cheap global continuous communication with any zone on Earth. In the coming decade (until 2010) such systems will be widely used along with high orbit systems of satellite communications. To obtain the global capability, low orbit systems will use inter-satellite relay between SC of the system inside the orbital plane and between SC of different planes. The use of communications channels simultaneously for transfer of command information will enable single point complexes to be built for control of those low orbit systems.

The use of standardized information exchange interfaces via satellite communications systems makes redundant the creation of specialized communications and data transmission systems which now operate in tracking networks.

Tendencies for creation and use of the space relay system are also characteristic of other countries leading in space research and international consortiums that are in possession of various space systems (ESA, Japan, China, organizations like Intelsat, Inmarsat and others). Because of limited territories, those countries and organizations use foreign territories for placing their CIPs or locate them on-board ships (China).

ESA and Japan, as economically and technologically most advanced space powers that have a high scientific and technological potential and much experience in using SC, also create their own space relay systems (SRS) in order to maintain the global communication with scientific SC and advanced piloted modules of ISS.

NASA, ESA and NASDA work on the compatibility of all three SRS for reservation and improvement of capabilities of national SRS and for creation in the longer term of the world's single space relay system.

The international cooperation in space, the implementation of major international space projects (ISS, space systems of global monitoring, creation of global space communications systems embracing many countries, etc.) require large-scale integration of space technology systems, including means of controlling them.

The principal required condition for this is the unification and international standardization of hard- and software automation systems, means of communication and information exchange with SC.

Since the 1990s, this standardization has been widely used under projects supervised by Consultative Committee for Space Data Systems (CCSDS) in addition to efforts of the International Standardization Organization (ISO), International Telecommunications Union (ITU), Inter Range Instrumentation Group (IRIG standards).

Objects subject to standardization:

- radio frequency bands, functions and structure of Earth – onboard line;
- parameters of receivers and transmitters;
- standard units of formatted data;
- command radio lines procedures;
- data processing and compression;
- interfaces and protocols of data exchange at various levels;
- decision-making logic etc.

Over recent years, Super MOCA-700 standards of the CCSDS group have been developed. They define requirements imposed on planning and using of control facilities in a FCC, in ground-based terminals and in SC for all stations of the life cycle and for various degrees of SC self-sufficiency (autonomy)..

The introduction of all these standards ensures compatibility of control facilities and complexes being developed by various countries' companies. It can be said that even now, if necessary, a global system can be organized for interconnected control systems of SC belonging to space agencies of various countries. Under such circumstances, the wide use of microprocessor technology and the formation of signals structure along with processing techniques exclusively by software make it possible to fully emulate various command and instrumentation systems when CIS of one tracking system can be tuned in on command of FCC for work with SC controlled by another network having a signals structure of its own.

Most likely, in the coming years the problem will be resolved as to how to make uniform the standards of space data transfer, in which case the terminal time of entry into communication, Doppler frequency shift, zones of mutual radio visibility etc. would conform to standards of on-earth communication that use the concept of multi-level interaction of open systems. With due regard to inter-satellite relay the global earth-to-space information network will be established (a complete analog of the Internet) that continuously provides users of space information, managing operators, astronauts, research workers etc. with an access to information resources, SC instruments and so on. The creation of such an earth-to-space network will generate a space "telescience", i.e. a system giving an opportunity to a scientist, a technologist, a designer to carry out experiments and do research with the use of equipment onboard SC without leaving his work place at a laboratory, a university, or doing it even at home in real time and without the rigmarole of form-filling, choosing the time for experiments, participation in elaboration of command software for SC.

The development of SC automatic control systems will go on through refinement of control technology with transfer to SC of many functions currently performed at FCC.

In the way of navigational and ballistic support (NBS) there is promise in creating an autonomous (self-sufficient) navigation system (ANS) by using navigation equipment of users (NEU) of space navigation systems like GPS (USA) and its subsequent advanced modifications.

The GPS system offers great opportunities to many users of SC, launchers, and other products of rocket engineering. NEU of GPS system is now installed on more than 100 SC of many types and belonging to various countries.

NEU of GPS enables the following functions to be performed (part of them are used practically):

• determination of coordinates with an accuracy not worse than several to tens of meters and the speed of objects with an accuracy of 0.01 m/sec with results obtained on-board either via transmission of measuring results directly to earth through a telemetric channel or through a radio channel via relay satellite; in this case it is considered desirable to unite NEU with inertial measuring units in order to reduce information loss through interruptions in reception of the navigation signal.

• determination of object's spatial orientation;
control of mutual position of elements of large space structures (beam curvature, position of manipulator's arm, fluctuation of solar batteries as engines begin to operate and others);

• docking of two and more objects in space in various phases of the mission, monitoring of astronauts' movement outside the craft etc.

The creation of ANS will make it possible to fully abandon measuring the trajectories from ground-based points that will be utilized to obtain NBS only during the test flight period and in emergencies on board SC. Advanced ANS can also be created on the basis of other principles: the use of ground-based radio beacons; of mutual inter-satellite measurements systems; of astronavigation system (which will be aided by improved accuracy of star catalogues by two to three orders); the employment of principles of a large base interferometry with spaced radio beacons, including not only those on Earth, but also on the Moon and, possibly, on other planets in case of deep space flights.

Rapid development and wide use are expected for on-board telemetric systems that control the state of on-board equipment, perform its technical diagnostics and aid its restoration to working capacity. There will be performed on board such operations as: processing and compression of information; the summed-up information will be transmitted to the ground-based FCC on receipt of command or as scheduled only in emergencies for which ňo remedy is provided on board SC, in which case decisions will be made by operators on the ground.

The component package of telemetric equipment will include, apart from traditional sensors, optoelectronic sensing instruments providing for receipt of video information about the movement of launcher, booster and spacecraft units, about the opening of solar panels, antennas, manipulators, etc., about coordinates of such movement and, finally, information about signal and spectral characteristics of various physical and chemical processes, including those that occur in case of explosion on board SC.

The transfer to SC of navigational and ballistic support functions, along with functions of controlling and diagnosing the on-board equipment, of planning its functions on commands from GBCS, plus the use on board of highly stable generators will make it possible to create fully autonomous SC.

The US Air Force plans to manufacture such autonomous SC towards 2015 – 2020.

The building of autonomous SC and adequate FCC is only possible if such a problem as intellectualization of SC automatic control systems (ACS) is solved through introduction of artificial intellect (AI) facilities and techniques of controlling them into on-board and ground-based hard- and software of ACS and through the use of the latest achievements in computing and communications facilities and their software.

The AI systems, robotics and neurocomputers are in the US top priority space technologies. Work in this area began over 20 years ago. Expenditure on AI increased from US\$ 217 million in 1987 to US\$ 520 million in 1992. The AI systems are based on use of models of a particular type, the rules of conclusion and decision-making arranged as knowledge and data bases with a suitable man-to-machine interface.

The first operating system of AI developed to order of NASA and representing a specialized expert system (ES) has been created at the Goddard Center in the mid-1980s. This ES controlled 67 problems in data exchange with COBE SC via TDRS.

ES are known to be used at the Goddard Space Flights Center for work with SC-based gamma ray observatory (GRO). Also, they are used by extre-me ultraviolet Explorer satellite (EUVE). The Jet Propulsion Laboratory (NASA, USA) created SHARP (Spacecraft Health Automated Reasoning Prototype) for automated control of on-board systems of Voyager–1 and –2 SC. This ES tracks the flow of data that overfills the screen, effects integration with a high speed system of presenting full color graphic information, generates and promptly interprets images, detects changes in a data flow, etc. Such ES are also created by other NASA centers as well as US Air Force for AFSCN complex, for instance MAGIC ES.

One of today's examples is the Remote Agent software for far-out SC. NASA developed this software for the first automatic interplanetary station, Deep Space-1 (DS-1) under the New Millennium program. The aim of the software created on the basis of the artificial intellect will be to control SC (not only those in deep space) with a minimum involvement of a human operator.

The software has been developed jointly with the Ames Research Center and NASA's JPL to decrease the cost and increase the spatial scale of space exploration through "computer autonomy". This software can serve as a basis for development of software for other SC.

One of the promising ways to radically reduce the cost of SC flight is to reduce the personnel involved in SC control from hundreds of people to some ten. The estimated cost of flight of SC (e.g. automatic interplanetary station) can be reduced by 60% due to the use the Remote Agent type of software.

The Remote Agent software is built on models. Models of SC components are introduced into the software which enables it to independently compute the particulars of control procedure for achieving the set target. What is needed for a new SC is just the refining of models.

The Remote Agent software consists of three jointly operating components:

1. high level planning and scheduling unit;
2. model-based fault protection;
3. "smart executive".

The planning unit, as a craft commander, scans the operating schedule several weeks in advance, planning the operations and distributing the resources, such as power supply. A small ground-based group directs to it tasks (assignments) instead of detailed instructions (commands) as is the case now.

A fault protection unit (designated Livingstone) acts as a virtual chief engineer of the SC. Using the computer-aided pattern of correct behavior, the unit must detect the fault and suggest a means of rectifying it.

The executive unit realizes plans prepared by the planning and protection unit, bringing them up to the details level. It can receive an operating plan directly from operators on Earth or, should the plan appear inadmissible, refuse to execute it. This is essential to the management with their considerable effort in double checking of each command which, however, does not preclude occasional errors.

If for whichever reason the Remote Agent software refuses to interact with Earth, a provision is made for direct interference in SC control.

It is believed that the New Millennium software has hastened the progress of SC automation, giving it a roughly 10 year's lead. Plans are made to produce after DS-1 new spacecraft with a greater degree of autonomy that will enable them to change their configuration. Should any part of SC in flight start functioning contrary to design, the craft will be able to detect and change models in the software and algorithms, i.e. to adapt itself.

The specialists of ESA and its ESOC, who have a broad experience in controlling a variety of SC predict further development of the center for another 30 years (till 2023). The development is expected to yield new technologies in controlling SC.

In approximately 2020, SC are expected to become highly automated systems capable of performing robotic functions with only occasional interference of ground-based services. In case of regular operation of SC the ESOC will be relieved of the task to continuously track and control SC. There will be left to it only formulation of plans and schedules of flights. Safety will be achieved through enhancing the fault-tolerance of computing, expert and neuron networks.

The structure of ESOC will wholly rest on distributed systems based on automated work stations (AWS). Expert systems will be merged with neuron systems in which case "indistinct logic circuits" will be used with the prospect of building fully automated SC control systems.

The ESOC will participate in SC control in the event of malfunctions onboard for whose elimination procedure will be executed stored in the memory of AWS. All actions related to controlling the SC will involve re-programming of on-board software.

It is anticipated that in the coming 30 years the computing power of ESOC will multiply by 100 times (the same increase is expected onboard SC). It is pointed out that the limiting could be caused not by the computing resource, but by the technology of AI and attempts to put it into operation.

ESA's prediction also shows that by 2023 there may be neuron networks onboard. An on-board control complex may contain up to four computers, three operating and one stand-by.

The prospective way to complete automation of FCC is determined by the principle : "virtual computer" + expert system + neuron network.

The merger of ES with neuron nets has a complementary effect on either. ES can operate in a dialog mode, can have an interpretation system, use and amend the rules of decision making, etc. Neuron networks are teach-yourself system, have the highest speed of processing the information, recognizing and classifying situations and images even when information is incomplete or indistinct. The high operating speed will be particularly perceptible in optic neuron networks, in wμhich decision-making may approach the time during which a light beam passes through multi-layer optic neuron masks.

Attempts at military application of neuron computers started in the US under DARPA Neural network Atud program in 1987-1988. Beginning in 1992, the US worked under a five-year program funded by the government and aimed to develop and introduce neurmputers in military use. By the end of 1992 the world's market of neurocomputers was estimated at US$ 120 million (hardware) and US$ 50 million (software). According to forecasts, by 2000 the market was to grow to US$ 790 million (hardware) and US$ 260 million (software). The number of companies in neurocomputer business was to reach 150 entities.

In the US, the main centers engaged in neurocomputer research are the Carnegie-Mellon University, California Institute of Technology, NASA's Goddard Research Center, Massachusetts Institute of Technology, the companies like TRW, Nestor and HNC. Also, taking part in funding is the Defense Advanced Research Projects Agency (DARPA). DARPA aims to build in 3 to 5 years machines with a performance of $10^9 - 10^{11}$ neural switches per second, and in 6 years' time – machines with a performance of $10^{12}$ neural switches per second which comes close to performance of the human brain. A seven year program is prepared with a US$ 400 million funding. Also, other agencies are expected to contribute financially to this government supported project with a total value of US$ 1 billion.

In the longer term neural computers will be readily embraced by the space technology, particularly, in robot and manipulator guidance; in SC docking; in processing of images received from SC in which case the processing time can be reduced by two orders; in evaluation and forecast of damages on space stations; in stellar navigation; in solving the problem of virtual reality which requires super computing performance associated with three dimensional graphics, eye-to-hand system, creation of animation packages; in solving the tasks of coding and decoding; in processing of texts; in automatic translation; in processing of voice-presented information; in connections in complex communications systems, etc.

A totally different level of automation will characterize SC control centers in the future. It will save the personnel all the monotony of processing various data flows and will relieve them of the burden of lengthy preliminary study of numerous instructions for controlling SC.

SC control will be accomplished between FCC and SC in an end-to-end exchange mode without interference of operators of ground-based points. The information exchange facilities on the SC will not, as a rule, be serviced.

The implementation of the above described and other intellectual information technologies used in space systems control of the future necessitates automation and communication facilities that would be radically new in terms of performance ($10^{11}...10^{12}$ and more operations per second) and data rate (tens and hundreds Gb/sec) with reduction by several orders of size and increase of failure-free operation time to tens of years.

In 1999, processors Pentium-III were manufactured using 0.18 micrometer manufacturing techniques. By 2002-2003, the precision is expected to refine to 0.13 micrometer.

It is anticipated that by 2017 the silicon-based manufacturing processes will reach their physical limit, though multi-processor chips may come along. Among the technologies that can exponentially increase computers' performance are molecular or atomic technologies, DNA and other biological materials based on photons in place of electrons and quantum technologies that use elementary particles.

So far those technologies exist in research laboratories only.

However, Hewlett-Packard announced recently its first success in creating the components of future molecular computers which, according to the company's estimates, will be 100 billion times more economic than today's microprocessors. A molecular computer the size of a grain of sand can contain billions of molecules. The production of molecular computers may start by 2015. The use of biological materials will enable computers to be slimmed down to the size of a live cell. Bill Ditto of Georgia Institute of Technology, intends to use biological computers made of neuron-like elements for creation of brains of robots that solve problems by self-programming techniques.

Even though optic computers will appear in a few decades, optic elements today, including optic fiber, play an increasingly important role in high speed communications systems. While in electric cables the speed of data transfer is about 140 Mb/sec, that in optic cables reaches 10 Gb/sec. The use of optic solitons technology – light pulses capable of spreading in dispersion media over large areas without changing their shape – can eventually increase the rate of data transmission to 320...400 Gb/sec per channel with the length of regeneration stretch up to 1,000 km or more.

The above shows that the advancement in space vehicles control systems is unlikely to be held up by technical characteristics of computers and communications systems which, according to some estimates, will reach as soon as 2020 the performance of the human brain (20,000,000 billion operations per second) and then will outperform it. The capabilities of those control systems will be determined by economic resources made available for their development, by the needs of astronautics and by mathematical methods of solving the long term tasks.

## 1.5 Trends in restructuring Aerospace Industry

Aerospace industry is one of the world's burgeoning and promising high-tech businesses with a high rate of return. Over the last years the industry has been through events and developments that drew attention of a variety of specialists, political scientists, economists, legal experts, etc. The interest in this comes from the desire to understand those developments and, if possible, to forecast their progress and consequences.

Various authors resort to various terms to define the developments. Those are "consolidation', "merger", "integration". All of them, indisputably, reflect the characteristic feature of the developments, but focus on only one side of the phenomena. A more suitable term seems to be "restructuring", that is, the changing of the established structure of aerospace industry.

That it is not a mere merger is testified by the fact that the Loral company became incorporated by the Lockheed Martin corporation. The assets of Loral which in 1994 was one of USA's ten leaders in military sales were acquired by Lockheed Martin for US$ 10 billion. As this happened, the executive director of Loral became vice-president of Lockheed Martin and simultaneously headed Loral Space and Communication, a newly formed company with Lockheed Martin's 20% stake.

Noteworthy is also the fact that the Loral Space and Communication company is a holder of stocks of Globalstar International, which is creating a low orbital satellite communications system competing with the Iridium system, in the building of which Lockheed Martin is taking part. Facts like these are many.

Attaining the effect accorded by a big scale business that arises out of merger of like enterprises with ensuing reduction of production costs and a growth of a segment occupied in the market, is not the prime target in the present-day situation, though a target it definitely is. Efforts are being made not merely to increase the quantity, but to heighten the quality of assets acquired and put to use.

The processes that are now underway are not by a long chalk new. The first wave of mergers and fusions took place in the 1980s. The specialists, however, note the voluntary nature of the deals that characterizes the current restructuring.

All this shows that those processes are far from simple.

The complexity of the processes reflects the complexity of causes that triggered them. The obvious and, hence, universally quoted as the initial cause is the sharp drop of defense orders in the wake of changes in the global military and political situation. This, however, does not reflect the real state of affairs.

The restructuring is important not only for the interested companies, but also for the national economy in general. This is the sole reason why the US anti-monopoly committee adopted such a passive stance that contributed to the emergence of such monsters as Lockheed Martin, Boeing McDonnell Douglas, and Northrop Grumman which leads to believe that that is not a spontaneous occurrence, but a process controlled by federal agencies.

### 1.5.1 Why restructure?

The causes that generated restructuring of aerospace companies abroad and keep up the process can be divided for convenience sake into three groups: military-political, economic and resource related.

### Military-political causes

The military-political causes are determined by the new conditions that emerged once "the cold war" was over and the cuts in military spending cut down federal defense orders.

For example, during 1987-1994 period the global military spending slid from US$ 1,300 billion to 800 billion. Notably, the cuts of military allocations in absolute figures are accompanied by still sharper cuts of their share both in the total expenditures and in the gross domestic product (GDP).

While in fiscal years 1951-1961 the US national defense spending accounted for 50% of the total federal budget, in the early 1970s it slid to less than 40%, in 1981-1992 FY – to 30%, and to a mere 15% in 1998.

During the 1985-1998 period the defense budget of the US slid from US$ 343 to 250.7 billion. Over recent years, a tendency is observed for some growth in absolute figures of the defense budget. Its share in the GDP, however, remains practically unchanged.

Changes in the US defense budget and its share in the GDP in the 1990s are shown in Fig. 65.

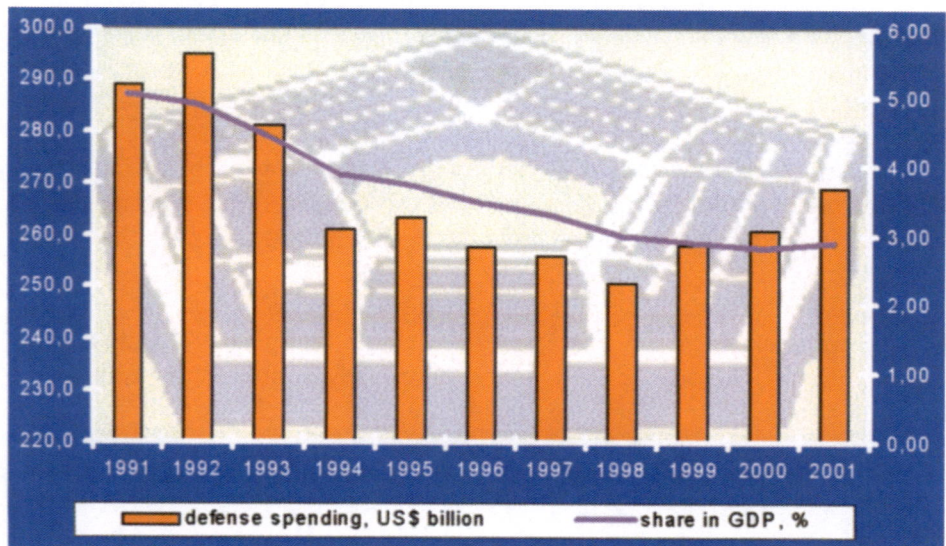

**Fig. 65.** US defense spending and its share in GDP

As can be seen from Fig. 65., as early as 1998 the share defense spending in GDP decreased in the US by nearly 2.0% compared to 1991 and amounted to 3.0%. By 2001 it remained practically unchanged.

The reduction of defense spending resulted in reduction in military procurements. During the 1987-1994 period the scope of arms procurements by the Pentagon dropped in real terms by 67%.

The military space programs' funding has changed over the past years very little, if at all. However, with allowance made for the annual 3% inflation, their real value in 1991-1996 decreased by US$ 1.5 billion.

The reduction of military space programs' funding reduced the amount of orders awarded by DoD to aerospace companies for development and manufacture of new space equipment. Thus this most advanced sector of the US defense industry (in the early 1990s it accounted for more than 2% of the country's GDP) suffered more than others since the companies it incorporated had been massively involved in defense production.

For example, in the first ten US companies with the largest military sales in the early 1990s, General Dynamics had a military sales share equal to 86%, Lockheed Martin – 85%, Martin Marietta – 76%, Boeing – 42%, Raytheon – 41%.

The new conditions are characterized not only by reductions in defense spending, but also by its restructuring in allocations for development and production of military equipment. According to high ranking officials at the DoD, "in the new conditions the US armed forces will mainly use types of arms with a long and a proven record". This implies the cuts in orders for manufacture of new types of weapons and military equipment. The changes in arms procurement policy slated back in the early 1990s seek to achieve the self-same goal. The new policy, DoD's officials believe, will ensure the development of the newest arms up to the prototype manufacture and enable, if necessary, their production in the required quantities. This will make it possible to choose the necessary technologies from those available, to manufacture development prototypes of the weapon systems and to conduct demonstration trials, that is, to obtain the initial equipment without expenditure on large scale procurements. According to an official statement of DoD's spokesman, the US can now afford to postpone the production of some or other weapon without detriment to its national security or technological leadership.

Sure enough, such policy hits really hard the profits of manufacturing companies, since it is production and operation of systems that consumes the bulk of allocations and thus provides profits for manufacturing companies. During the 1990-1995 period, the funding of military equipment procurement programs decreased by US$ 36.8 billion which accounted for 45%, while the funding for procurement of products of aerospace companies accounted for 59%.

## Economic factors

**What is happening on the Stock Exchange.** Not only politics, but also economy urges the restructuring of finance-industrial groups in defense industry. The USA has an efficient system of putting severe financial pressure on the defense industry. The processes on the stock exchange are in this respect especially important.

The first impulse for consolidation came with the fall of the Berlin wall in 1989. The stock exchange players under the influence of the talk of the day about the inevitable demilitarization and peace dividends immediately went bearish with defense plants' shares. But the major military customers, laden with still bigger order books and in possession of moneys used the players' bearish stand to embark on a massive buy-up of companies in order to consolidate their own position on the market. As a result, both those who were buying and those who were selling defense assets managed to store up huge reserves of free cash later used both to pay the dividends and to make further acquisitions. But the expectation of an increased cash flow generated a situation in which over a fairly short period of time in the early 1990s the shares of leading contractors leapt up in price the way they did during the Reagan military boom in the early 1980s. With every new transaction the companies' stocks rose still higher. As some foreign specialists later explained "the stock exchange players, initially convinced that defense companies are monsters destined to die out like dinosaurs, later changed their mind and now perceive the defense industry as a machine for making free cash".

But since the order books of all defense contractors are already depleted and there is no chance to increase either the rate of return or the spare cash supplies, the defense industry on the whole is facing a second wave of consolidation which, foreign experts believe, will be determined by serious strategic factors of long term survival.

The major diversified military industrial groups engaged in finance must finally delineate the key areas of their defense business and sell off non-core assets while stock exchange quotations are sufficiently high. The companies that will outrace others in "discarding" unwanted assets will be at an advantage since they will make profit before the collapse of quotations and conversion of the market into a shopping mall controlled by buyers.

On the wave of restructuring of defense industry most transactions will be made by way of exchanging stocks which will make it possible not only to obtain tax privileges but will also reduce the gap between sellers' and buyers' prices that comes without fail in the wake of sharp rises in supply of defense assets.

Thus, the forecasts of the stock market indicate that the restructuring of contractors' defense production is only entering the strategic phase of its development and the number of companies by the beginning of the 21th century will decrease to the same degree as in the early 1990s during the first prearranged wave of mergers and fusions.

**The formation of the market of space products and services.** The dwindling share of federal orders for products of the aerospace industry on the one hand and the increasing commercialization of space activity on the other, have generated the world's market of space products and services that is fraught with all attending circumstances.

Therefore, whatever happens in the aerospace industry is not in any way new and uniquely specific. In other industries with a longer record situations like this had probably arisen more than once and still keep arising.

The search for strategic partners in the atmosphere of stiffening competition is dictated by the companies' natural desire to survive. Alliances between various manufacturers provide for realization of this desire by giving access to new markets, new technologies and to more efficient methods of industrial management.

The wide use of the strategy of alliance formation is especially characteristic of the car building industry. The stiffening competition drives even big industrial groups to form unions. There are many ways and means of carrying on a cooperative business, from complete control of one developer and manufacturer over another to mere joint work of two partners under a commercial agreement.

Financial alliances imply close technical and commercial cooperation in research and development of new products as well as elaboration of marketing and selling strategies. This allows to access new technologies and to create mechanisms that keep competitors off the strategic segments of the market.

In the case of a full financial control of some firm over another, it becomes a matter of an overt acquisition, not of alliance. In aerospace industry this is exemplified by the formation of the Northrop Grumman corporation that came about after the Northrop corporation acquired the Grumman corporation (the estimated cost of assets was US$ 2 billion).

Alliances are also formed in order to gain control over certain segments of the market and to cut the manufacturing costs while turning out products in small batches (this is typical of production of space equipment). Under such circumstances agreements for cooperation may be concluded and joint ventures established. One such example is the acquisition of Loral's divisions specializing in production of quality electronic equipment by the Lockheed Martin cooperation and its participation (with a 20% stake) in creating Loral Space and Communications, a new independent company.

Should profitability be reached, each partner may enter into separate agreements for joint production of costly examples of equipment.

The cooperation in separate special purpose programs could be exemplified by creation of small commercial SC for remote sounding of

Earth from space. Lockheed Martin and Raytheon set up specifically for this purpose the Space Imaging Inc. firm. Also, similar SC designated Eyeglass had been jointly developed by Orbital Sciences Corp., GDE Systems Inc., and Itek Optical Systems, the companies that established the Eyeglass International firm to do specifically that job.

The rather complex nature of relationship within such formations is worth a special note. Occasionally situations arise in which some firm or other finds itself involved in a rival project. For example, the Earth-Watch company has been deploying since 1997 a remote sounding system of the same name. The company had been formed by Ball Aerospace and Technologies Corp. and WorldView Imaging Corp. Its partners are Hitachi of Japan, Nuova Telespazio of Italy, and others. Spacecraft are manufactured by Defense System Inc., established by CTA Inc. In this case Defense System Inc. is the investor of WorldView Imaging Corp. The CTS company is now incorporated by Orbital Sciences Corp. which is involved in work on creation of the Eyeglass SC.

Thus, one of the reasons of what has been happening in recent years in the aerospace industry is the growing commercialization of space industry and the formation of the world's market of space products and services.

## Resources related reasons

In 1990-1992 the US sustained a slump in the high-tech sector of industrial production (which also includes aerospace industry). The cause of the slump was the reduced federal investments in new scientific and technological developments that were being introduced by the industry under state programs. While in 1987 those investments amounted to US$ 57.9 billion, in 1996 they reached US$ 47.4 billion. In the early 1990s there was a stunt in growth of total funding for research and development (R&D). In addition, the growth of federal expenditure on R&D had been suspended and sharp cuts had been and are still being made in the share of federal expenditure on R&D in the US GDP.

This recession, according to some foreign analysts, reflected a special situation in which the state's reduction of capital investments in the futures (so is termed the expenditure on R&D in the official documents) had a rapid impact on the current production. In its turn the reduction of federal investments in science and technology results from the economic policy pursued by the US and reflects the general tendency for reduction of federal expenditures, which tendency is generated by the government's attempt to balance the federal budget whose deficit over the period between 1981 and 1995 amounted on the average to 4% of the GDP and roughly to 17% of federal annual expenses. In 1996 the budget deficit amounted to only 1.4% of the GDP.

The reduction of budget deficit proceeded in parallel with the general reduction of the share of the federal sector in the US GDP which is reflected in the processes that go on in the aerospace industry.

## 1.5.2 Analysis of restructuring of aerospace companies

The current restructuring of the US aerospace industry is the most significant phenomenon experienced by the entire US industry after the last war. By 1990 it had resulted in over 30 mergers and fusions. Those imply both major appropriations of divisions of some firms, which appropriations were made by other companies and mergers of corporations themselves with formation of firms having a new name.

However, it would be wrong to regard the processes that go on in foreign aerospace industry as mere mergers and fusions. The business strategy of the leading aerospace companies is differentiated. Some companies put more emphasis on penetrating the markets of related products while others focus on strengthening their own assets in order to preserve the key technologies of their business, whereas still others concentrate on foreign sales. The interaction of all three strategies boosted supply and demand of aerospace industry divisions: there was an increase in the number of both those wishing to sell non-core assets and those wishing to buy engineering facilities with the purpose of consolidating their position on the market or gaining a new one. The predominantly voluntary nature of transactions makes the current wave of restructuring specifically different from the previous process (in the 1980s), which abounded in hostile absorptions frequently motivated by outright profiteering that reflected overheating of the stock market in course of the economic boom of the 1980s.

One of the biggest transactions in the US aerospace industry (to the tune of US$ 10 billion) was the merger in 1994 of two major companies, Lockheed and Martin Marietta. Notably, that was a voluntary deal whereas in the late 1980s Lockheed had to repeatedly rebuff the attempts to illegally appropriate its majority stake. Before the merger both companies in the early 1990s had made big acquisitions. Martin Marietta had acquired large divisions of General Electric and General Dynamics whereas Lockheed had acquired for US$ 1.5 billion an aerospace division of General Dynamics.

Before the merger the Lockheed corporation was engaged in development and production of space rocketry and rocket weaponry (up to 45% of its total sales), aviation equipment (30%) and electronic equipment (around 25%). The employed workforce exceeded 83,000 people. In 1993 the total sales reached US$ 13.1 billion, the cost of assets – US$ 8.9 billion, in 1992 respectively US$ 10.1 and 6.7 billion. In 1993 the company's earnings amounted to US$ 0.4 billion. In 1992 in terms of contracts awarded by the government (US$ 6.1 billion) the company was placed second, and third among contractors of the DoD (US$ 4.65 billion). In 1991 Lockheed was placed eighth in the first category and ninth in the second (respectively US$ 3.7 and 2.3 billion).

Martin Marietta, too, was one of the US leaders in rocketry, space and electronic engineering as well as a prime organization in development and production of ICBM, space carrier rockets, rocket solid propellant boosters, US advanced heavy launchers.

The joining in 1994 of the Aerospace production group incorporated by General Electric – the main supplier of satellite communications systems, radars, simulation systems and other aerospace equipment – turned Martin Marietta into the world's largest company in the field of military electronics and aviation equipment.

**Fig. 66.** Top managers of Lockheed Martin and the Khrunichev Research center at the international aerospace show in Le Bourget (left to right: N. Augustin, A.Kiselev, D.Telepp)

The corporation's sales between 1990 and 1991 amounted to US$ 6.1 billion, in FY 1993 it was US$ 9.4 billion. In terms of cost of military contracts among 100 largest contractors of US DoD, the company was third in 1993 (US$ 4.7 billion), 10th in 1992 (US$ 2.4 billion), 5th in 1991, 6th in 1990. The defense orders accounted for 74% of the company's total sales. The corporation's military R&D funding in 1989 received US$ 1.9 billion, in 1990 – US$ 2.01 billion (the first place among US DoD's contractors during the said period). The workforce of the corporation in the early 1993 amounted to 65,000 people.

In the entire Lockheed Martin corporation the employed workforce amounted to 165,000 people. The estimated sales in 1997 were US$ 30 billion.

Another big event in the restructuring was the merger of Northrop and Grumman in 1993 and the formation of the Northrop-Grumman corporation.

Before the merger, the Northrop company consisted of a subsidiary firm, Northrop Woldvid Aircraft Services, and three divisions: B-2 Division, Military Aircraft Division and Electric Systems Division. The number of Northrop personnel as of the end of 1993 reached 30,000 people, the estimated cost of assets was US$ 2.9 billion, sales amounted to US$ 5.1 billion, earnings to US$ 100 million (in 1992 the latter two amounted respectively to US$ 5.5 billion and US$ 120 million). In terms of the government's contracts received in 1992 (US$ 4.9 billion) Northrop was sixth, and 2nd among contractors of DoD (total value of contracts was US$ 4.85 billion). In 1993 Northrop received from DoD contracts with a total value of US$ 3 billion which placed it 7th.

Before Northrop acquired Grumman, the latter comprised six subsidiary firms and two divisions (Grumman Aircraft System Division and Grumman Space Systems), manufacturers of aerospace equipment.

In 1992 Grumman concluded contracts worth US$ 2.3 billion with federal agencies (17th place among US companies), and DoD to the tune of US$ 2.2 billion (12th place). With these achievements (US$ 1.7 billion) the corporation was placed 10th in 1993 among DoD's contractors.

In 1993 the workforce at Grumman's facilities numbered 17,900 people. The corporation's sales in those times amounted to US$ 3.2 billion, the revenues – to US$ 580 million. The cost of assets by the end of that same year reached US$ 2 billion.

After formation of the Northrop-Grumman corporation the workforce in it reached 47,500 people. By the end of 1994 a forthcoming reorganization of the corporation was announced along with impending lays-off. By the end of 1995 9,000 personnel were due to be dismissed, half of whom worked at the facilities located in California: at Pico Rivera and Palmdale (B-2 Division) – 2,400 people, at Hawthorne (Military Aircraft Division) – 1,600 and at Hawthorne (Electric Systems Division) – 500 people. Also, 3,500 people were due to be laid-off at the facilities in the state of New York.

In the fourth quarter of 1994 DoD contracted with Northrop-Grumman for US$ 900 million worth of work.

A still more significant event in the restructuring was the merger of Boeing with the McDonnell Douglas Corp.

During the period of 1985-1991 Boeing acquired several smaller firms. This enabled it to take the 6th place among DoD's contractors in amount of sales (US$ 5.6 billion). However, already by 1994 the company's military sales plummeted (down to US$ 1.7 billion) and it slid to the 9th place. Further steps in search of the way out of predicament were the acquisition in 1996 of space division of Rockwell International and the merger in 1997 with McDonnell Douglas Corp. that is also DoD's big contractor (military sales in 1994 amounted US$ 7.5 billion). This enabled the joint company, that had retained the Boeing name, to more than double its annual turnover (from US$ 22.7 to 48 billion) and to virtually monopolize the market of aircraft production, replacement parts, and maintenance.

The consultative Committee of the European Union (the chief body overseeing the international competition) tried to thwart the merger of those two aerospace industry giants, considering that such a fusion would violate the principles of free competition in this segment of the world's market. According to the Committee's experts, Boeing and McDonnell Douglas Corp, had been producing up to 84% of all jet planes in operation across the world. The newly formed industrial group would control around 70% of the world's market of civil aircraft (specifically, up to 90% of transport aircraft).

**Fig. 67.** Discussing the prospects of cooperation. D. Albow, President of Boeing's space division, (center), and A.A.Medvedev General Director of the Khrunichev Center (left).

However, the US Federal Commerce Commission, that was also investigating the merger of those giants, came to the conclusion that the merger would not impede competition and not violate the US anti-monopoly act.

As noted above, apart from merging, the companies sell off and buy up others' divisions in order to gain access to or strengthen position on certain segments of the market. Thus General Dynamics is realizing its program aimed at a deeper penetration into the military business while striving to remain a major arms manufacturer. To this end it sells its fairly valuable and large divisions to other contractors and at the same time actively seeks and acquires smaller contractors whose order books are not empty yet. Having done so, the company channels the earnings obtained from sales into consolidation of its main businesses. General Dynamics has now defined four main areas of its activity: tactical aircraft, nuclear-powered submarines, tanks, and space equipment. By doing so the company has considerably narrowed down the field of its defense industry operations (abandoned the production of missiles and fighter bombers) and at the same time secured its foothold in the chosen businesses.

It is of interest how Lockheed Martin had acquired the Loral company. In 1996 the management of Loral and Lockheed Martin signed an agreement for sale of property by Loral to Lockheed Martin and a number of sale related contracts worth US$ 9.1 billion. Notably, the agreement was regarded not as an acquisition or a merger, but as strategic unification of two companies.

Loral was one of the leaders in production of military electronic communication systems. During the period between 1985 and 1995 it had acquired a number of US companies, including Ford Aerospace and IBM Federal Systems. The annual sales of Loral during 1973-1996 increased from US$ 27 million to US$ 6.7 billion. The company employed 38,000 people.

The divisions of Loral involved in development of space systems were merged into Loral Space and Communications, an independent public limited company in which Lockheed Martin got a 20% stake worth US$ 344 million.

The acquisition of stocks of Loral's divisions specializing in manufacture of quality electronic products was yet another step taken by Lockheed Martin towards monopolization of the space sector of the US economy.

The striving of big aerospace business for monopolization was especially pronounced in the scheduled for 1998 unification of Lockheed Martin and Northrop-Grumman. However, in March 1998 the companies were notified that the Department of Justice disagrees in principle with the merger of the two companies. According to foreign experts, the merger would save yearly around US$ 1 billion, most of which comes from federal orders. As per terms of agreement concluded back in July 1997, Lockheed Martin intended to acquire Northrop-Grumman for US$ 8.3 billion. Should the deal take place, a very powerful corporation, Lockheed-Northrop, would emerge in the US economy with an employed workforce of 230,000 people and estimated annual revenues of US$ 37 billion.

At the same time, the quest of gigantic aerospace corporations to monopolize space business does not mean that there is no room here for smaller firms. The analysis of the data available shows that the rightly chosen business strategy enables smaller companies not only to survive in the new conditions but also to build up their capabilities.

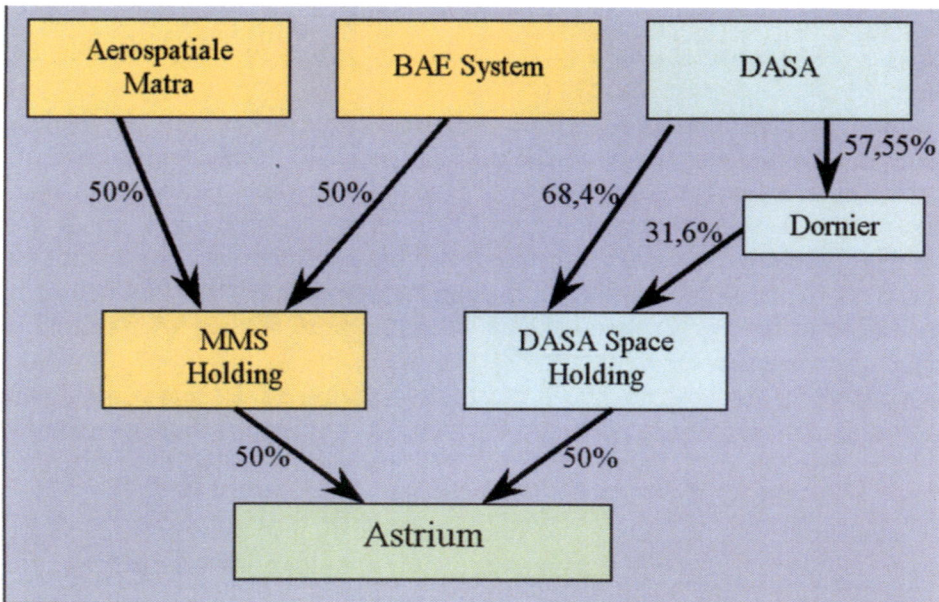

**Fig. 68.** How Astrium company was formed

An illustrative example of such an approach is demonstrated by Orbital Sciences Corp (OSC). It was established in 1982 and was concerned with development and manufacture of sounding rockets and sub-systems of exploration SC. The company concentrated its efforts on development of small SC and launch facilities for them. This enabled the company, that numbered 1,150 of employed workforce in the early 1990s (the annual revenues in the region of US$ 200 million), to increase its workforce to 4,000 people. The positions of OSC became particularly strong following the acquisition of the Canadian firm Macdonald, Dettwiler and Associates that specialized in production of ground-based stations' equipment for systems remotely sensing the Earth from space and the CTA Inc. company that was engaged in production of small SC and information systems (1,300 employed, US$ 140 million annual revenues).

Some companies quit the aerospace market to focus on other businesses. For example, Rockwell International, a former prime contractor in creating the SSTS rids itself of aerospace divisions and channels its funds into the best paying of commercial enterprises, the automation of industrial production and electronic engineering that promises rapid growth and high profitability.

Since 1995, the time when Reliance Electric Co. had been acquired for US$ 1.9 billion, that business became the chief economic activity of Rockwell International. The production of industrial automated systems and electronic equipment accounted in 1996 for 28% of the entire trade turnover of Rockwell International, whereas the aerospace segment in 1995 yielded only US$ 2.45 billion.

Similar processes go on in Europe. Thus in 1990 Marconi of Britain and Matra of France merged to form Matra Marconi Space, which in turn, after merger in 2000 with the space division of Germany's Daimler Chrysler Aerospace (DASA) formed a new space company, Astrium (Fig. 68, 69).

**Fig. 69.** Chief executives of Astrium

Thus Astrium came into being thanks to the efforts of three leading European nations – Britain, Germany and France. The company numbers 8,000 white and blue collar workers, nearly half of whom work at the German factories. Astrium is resolved to establish itself as one of the leaders in practically all areas of space work. The expected distribution of the company's business activity is as follows:

- telecommunication and navigation – 34%;
- monitoring the Earth from space and space research – 34%;
- space infrastructure (carriers and piloted space vehicles) – 32%.

Restructuring and enlargement of manufacturing facilities is also characteristic of Russia's space industry. Thus in 1994, the merger of KB Salyut (Salyut Design Bureau) with the Khrunichev Rocket Production Plant resulted in the Khrunichev State Space Research and Production Center engaged both in development and production of space rocketry. Later on, it incorporated the

Research Institute of Space Systems (the town of Yubileyny) and KB Armatura (Armatura Design Bureau, the town of Kovrov). Make-up of the enlarged Khrunichev Center is shown in Fig. 70.

Currently, the Center carries out all work required to maintain the complete life cycle of space rocket systems, including development, manufacture and operation of launch and orbital systems and ground-based infrastructure.

The Research Institute of Space Systems conducts exploration aimed to develop new launch facilities, space-based monitoring and communication systems, software support systems for testing space equipment, to improve ground-based instrumentation complexes, elaborate operational and maintenance equipment for newly created examples of space rocketry.

The KB Salyut designs, develops and tests carrier rockets, booster units, spacecraft and other space engineering products.

The Space Rockets Plant manufactures space equipment such as carrier rockets, booster units, spacecraft and their elements.

The KB Armatura develops and manufactures equipment for launch complexes designed to launch carriers produced by the Khrunichev Center and other plants of space industry.

The space rocketry operation plant supplies spaceports with equipment needed to prepare and launch Khrunichev carriers and provides its maintenance.

The Khrunichev-Telekom enterprise organizes communications networks and transmits data required to secure launches of carrier rockets.

The medical equipment and consumer goods facility turns out competitive products with the help of advanced space technologies.

Such a reorganization of production, though on a smaller scale, is being implemented at other plants of the industry. Thus in 1997 the merger of TsSKB (Central Special Product Design Bureau) with the Progress Plant resulted in the State Research and Production Center of Space Rocketry (TsSKB-Progress) engaged both in development and production of space rockets.

Overall, the merging processes in Europe are more complex than in the US. In spite of the successful formation of Astrium "the border problem" still remains a serious obstacle. There had been, for instance, protracted talks on the fusion of satellite and rocket businesses owned by Aerospatiale of France and DASA of Germany. An agreement was signed for establishment of two joint subsidiaries: for satellite systems – the European Satellite Industries (ESI) and for rocket systems – the European Missile Systems (EMS). The participation of both firms in joint ventures was envisioned on an equal basis (50% of each). Foreign experts believe that the resolution of this issue was to become a breakthrough in the restructuring of Europe's aerospace industry. However, the process got stuck and no information is available as to how those agreements fare.

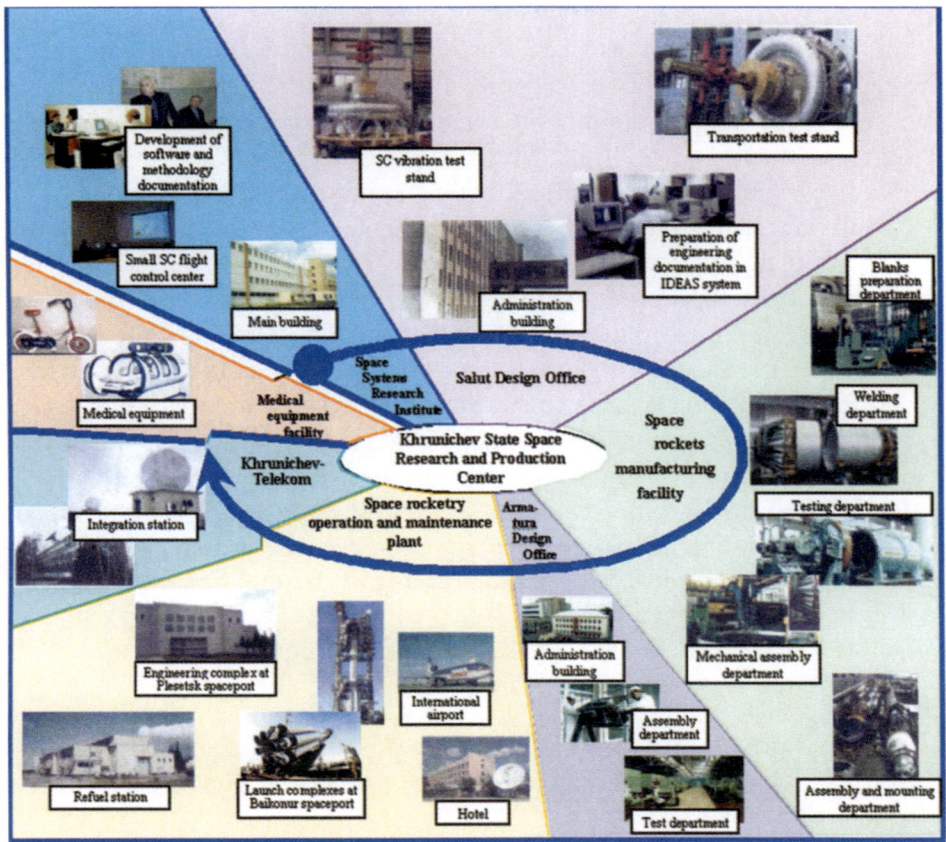

**Fig. 70.** Organization chart of the Khrunichev State Space Research and Production Center

Another hallmark of the European aerospace industry is its initially lower level of militarization. This necessitated the market orientation of space business as it was taking its first steps. Therefore, pooling the efforts for accessing the market of space services started in Europe before it did in the US. For instance, Arianespace consortium, now the leading commercial organization on the market of space payloads delivery services was formed back in 1980.

It should be noted that consortiums, i.e. temporary amalgamations on a negotiated basis for implementation of a particular project, received wide acceptance in space work of the European countries. Such a form of collective effort provides on the one hand for implementation of large scale space programs without affecting the national industrial independence in space and, on the other, it successfully resists the assault of the US aerospace companies, usually bigger ones.

On the whole, thanks to restructuring both American and European aerospace companies acquire qualities that enable them not only to retain their positions but also to enlarge their capabilities.

## 1.5.3 The aims of restructuring

The restructuring of aerospace industry abroad aims to turn the companies into radically different entities capable of not only surviving in the changing environment but also of building up their competing potential on the market of space products and services. The advantages obtained by companies by virtue of restructuring can be categorized by convention into four groups.

**Group one – current economizing** on elements of fixed costs. The in-house consolidation in this case means reduction of overhead expenses through elimination of duplicating component parts, release of territories, sell-off of infrastructure facilities, reduction of duplicating investments in R&D etc. Practically all transactions conducted in course of restructuring of aerospace industry enable the companies to cut drastically current expenses which is a decisive factor in improving the financial performance, accumulation of free cash and, hence, in further strategic acquisitions.

Thus Lockheed Martin announced a large scale in-house economy plan aimed to reduce the total expenses by 10% in five years. By 1999 the economic effect reached US$ 1.8 billion. Among the main saving measures are:
- release of the occupied territories as the result of closure of 12 plants and laboratories, 26 test sites (up to 16% of the total area in use);
- unification of five research laboratories into three;
- annual reduction of investments in fixed capital by 10%;
- reduction of expenditure on duplicating R&D;
- reduction of the 130,000 workforce in the company's aerospace sector by 12,000;
- bringing together the company's administrative, managerial and informational systems in a single center.

Reduction alone of financial support of the company's structures is expected to save US$ 500 million every year.

A separate sizable source of saving is the conclusion of new contracts with sub-contractors. Compared to two companies as separate entities, the united Lockheed Martin company will be able to order from subcontractors larger batches of sub-assemblies, component parts, and semi-finished products, thus reducing the cost of both single order and the total cost of products ordered. By creating such a scale effect for the supplier, the company saves not only funds allocated by the government for purchases of aerospace equipment (the expected annual economy of the state budget obtained through consolidation of Lockheed Martin is in the region of US$ 1 billion), but also increases employment, be it even partial, at its own plants (100,000 jobs extra) due to growing work.

Boeing also plans to obtain a considerable economic effect due to the inhouse restructuring. Before the end of 2000 the production area of the company's divisions was expected to be reduced by 1.67 million sq m (by more than 15%). The company will close down all laboratories and factories that perform duplicating functions.

A series of various types of key centers is being formed, namely, software control center, assembly centers, integration and test centers, strategic components manufacturing centers. It's worth noting that all of them will be established on the basis of already existing facilities.

Software control centers are responsible for controlling production lines. Such centers will also be established in Canoga Park and De Soto for controlling the production of rocket engines, space power supply systems and laser installations. Satellites will be supervised by the Seal Beach Center. In Downey a center is being established that will be responsible for reusable space systems. Expendable carrier rockets and defense systems will be the domain of a center located in Huntington Beach. The production of navigation systems will be supervised by a center at Anaheim. In Seattle surveillance and reconnaissance centers will be established along with those for the Sea Launch and communications systems projects.

The assembly, integration and test centers are engaged in final assembly, system integration and product testing. Technical support centers for International Space Station are being established in Huntsville, Huntington Beach and Canoga Parkn and for carrier rockets in Decatur. Their construction will provide 2000 jobs. A center at Long Beach has been eastablished for the Sea Launch project. In addition, 1,000 specialists in aircraft and rocket systems have been transferred from Seal Beach to Long Beach, and 1,100 specialists in reusable space systems and satellites from Downey to Seal Beach. The plant at Downey is closed. A center at Palmdale will take care of modifications of Space Shuttle and X-series spacecraft. The establishment of those centers cut 6,200 jobs by 2000.

The strategic components manufacturing centers produce various elements and carry out special manufacturing processes. Such centers are now also being established in a number of towns on the basis of existing facilities.

**Group two – utilization of the scale effect.** In this case a company that buys manufacturing facilities for production of similar items and consolidates such facilities on its production basis, anticipates a growth of this market segment, an increase in serial production of goods, a reduction in manufacturing cost of a unit of commodity, and hence an increase in its share on the market with resultant revenues.

The scale effect, for instance, urged the OSC to acquire CTA Inc.which will enable the former to virtually double its capability of producing small spacecraft. This is precisely what the OSC seeks to achieve by expanding its production facilities. Fifty million dollars have been allocated to implementation of a four year plan for construction of a research and production center. According to the plan the number of production buildings is to grow from two to seven while the floor area occupied by production facilities is to increase four-fold. The new facilities will enable the company to increase the rate of production of SC, to reduce the production cycle time and simultaneously to improve the quality and reliability of SC.

**Group three – attaining synergism effect** in complementary industries. A company acquiring research and production facilities aimed for use in complementary industries (e.g. electronics and information support) anticipates not a mere growth in quantity, but also an improvement in quality of assets being acquired. Unification of the research potential of two or more separate divisions enables a company to develop new technologies, products and services, to gain new positions on a particular market which adds to the company's competing capability.

The classical example of such an approach is the acquisition by Loral Space and Communication of companies and divisions specializing in provision of satellite communications services. Thus in 1996 Loral Space and Communication acquired for US$ 712.5 million the Skynet division of ATT and in 1998 the Orion Network System Inc. company.

The single-handed control of production of communications SC and provision of information support will enable the company to considerably strengthen its positions and successfully compete against such giants as Hughes and others.

The acquisition of E-Systems company made by Raytheon for US$ 2.3 billion aimed to improve the quality of assets being consolidated. The acquisition enabled Raytheon to strengthen its positions in military electronics and to take lead roles in such sectors of space products and services market as Earth remote sounding and satellite communications.

OSC, the corporation that manufactures and launches ERS spacecraft plans to expand its ERS segment of the market through acquisition of the Canadian company McDonald Datwiler & Associates which specializes in a complementary area, the production of ground stations for ERS systems.

Group four – diversification of research and industrial potential and acquisition of new markets. Through enlargement and consolidation of its research and production assets the company plans to expedite the implementation of research projects on the new markets. Pooling the efforts and creating a technologically new product with its subsequent proposal on other players' markets – such is the goal of such consolidation strategy.

One of the typical manifestations of this kind of activity is the situation on the satellite-provided image information.

Here, Lockheed Martin with its huge experience in building optic-electronic surveillance equipment for military application made headway toward the market of image information. To this end, in 1994 Lockheed Martin jointly with Raytheon E-Systems established a company, Space Imaging, that received from the government a license for work with high resolution images (up to 1 m).

Lockheed Martin Missiles and Space company created for the Space Imaging two spacecraft, Ikonos (other name is CRSSC), while Raytheon E-Systems created communication equipment and ground-based components of the system. The first SC was launched in 1998.

The Space Imaging company was united with the EOSAT firm, the seller of images from ERS SC (such as Landsat, SPOT, ERS, JERS, Radarsat). The firm had been acquired in 1996 by Lockheed Martin. The united company has the name of Space Imaging Eosat and is firmly positioned on the market of satellite imaging information.

The Orbital Sciences Corporation specializing in production of the Pegasus type of lightweight carrier rockets and small spacecraft has created a subsidiary company, Orbital Imaging. Currently, Orbital Imaging operates the OrbView-2 SC launched in 1997. A higher resolution SC, OrbView-3, was slated for launch in 1999, and an OrbView-4 in 2001.

The Eyeglass International company, actually a consortiums of firms, expected to be the first on the imaging market with the resolution of 1 m. The firms incorporated by the consortium planned to jointly develop the Eyeglass SC as per specialization of each of them: GDE-Systems – creation of image processing equipment, OSC – development of SC and carriers and launches of them, Itek Optical Systems – manufacture of optic-electronic equipment. The consortium was joined by the Eirad company of Saudi Arabia with a contribution of US$ 200 million.

The first launch of the Eyeglass type of SC was slated for 1996. Also, plans were made to heighten the capabilities of the next generation SC by fitting them with multi-spectrum equipment and even radars. However, the available sources of information give no evidence as to the practical implementation of those plans.

The WorldView firm founded in 1993 by W. Scott, the former project manager of the Brilliant Pebbles program, was the first to receive in 1993 from the US Department of Commerce a license for selling high resolution images. Together with Ball Aerospace and Technologies Corp. it established the EarthWatch Inc. company.

To the order of EarthWatch Inc., the Defense Systems Inc. which is a subsidiary of CTA Inc. (which in turn had been incorporated by OSC) created the EarlyBird spacecraft that enabled images to be received with the resolution of 3 m (Defense Systems Inc. is simultaneously investor of the project). The EarlyBird SC was launched from Svobodny spaceport with the help of Start-1 carrier rocket built on the basis of a ICBM manufactured by the Kompleks-MIT Research and Engineering Center. The partners of EarthWatch Inc. are Hitachi of Japan, Nuova Telespazio of Italy and several American companies.

Of interest is the fact that out of four commercial ERS SC projects, the OSC is involved to some extent or other in three of them. This evidence of the company's furious activity is verified on the other segment of the space products and services market – the provision of satellite communications.

In May 1998, Orbital Sciences Corp. and CCI International N.V., an off-shore company, announced their cooperation in an effort to create ECCO satellite communications system.

CCI International N.V. is a subsidiary of Constellation Communications Inc. (CCI) registered on the Dutch Antilles. The stockholders of the company are American firms Bell Atlantic Global Wireless Inc., Raytheon E-Systems Inc., and SpaceVest.

Under a US$ 450 million contract OSC manufactures and launches 12 low orbit SC making up the first constellation. In addition, having invested US$ 150 million OSC became one of the project investors.

Another project in which Orbital Sciences Corp. is actively involved is the creation of the Orbcomm satellite communications system. To implement this project, OSC, Teleglobe Inc. (Canada) and Technology Resources Industries Bhd. (Malaysia) formed a joint company, Orbcomm Global L.P.

Orbital constellations of the Orbcomm system comprised of 30 spacecraft operate successfully. However, Orbcomm Global L.P. went bankrupt and in 2001 its assets were sold to International Licensees LLC.

Still more foreign participants are involved in a project aimed to create another satellite communications system, Globalstar. This project is proposed by Space Systems/Loral jointly with the Qualcomm Inc. company.

To carry out the project the Globalstar Limited Partnership (L.P.) was established which received in 1995 a license from the US Federal Communications Commission for creation and operation of the personal satellite communications system with the use of a constellation of low orbit satellites.

The partnership of Globalstar includes key participants in development, production, deployment and operation of the system as well as the main companies that will be engaged in marketing and selling the system's services.

Among the former are Space Systems/Loral and Qualcomm Inc. (both from the US), Hyundai (South Korea), Daimler Benz Aerospace (Germany), Alenia (Italy). The latter include AirTouch Communications, Dacom, Elsag Bailey, France Telecom, Loral Space and Communications and Vodafone. Alcatel of France is in both groups. Space Systems/Loral controls 39% of the fixed capital (51% of Space Systems/Loral stocks belong in its turn to Loral Space and Communication, 20% of whose capital is controlled by Lockheed Martin). Early in 2000, Clobalstar L.P. began to operate the system.

Yet another project of satellite communications system that pools the efforts of major companies is Teledesic. In 1994, Bill Gates, the founder and proprietor of Microsoft, and Craig McCow, a baron of cellular communications, announced an establishment of a new company, Teledesic LLC (Kirkland, Washington) that set its sights on creation by 2002 of the global telecommunications network based on low orbit satellites system.

The Motorola company is assigned the role of a general contractor. According to the reached agreement Motorola is to receive 26% of Teledesic stocks in exchange for the financial contribution and the cost of engineering work that will be redirected from the Celestri program. The total value of Motorola's contribution is estimated at US$ 750 million, with 10% belonging to Boeing and the rest to Bill Gates and Craig McCow.

Also well known is another project with participation of Motorola, the Iridium satellite communications system proposed back in 1987 and "translated into reality" via the Iridium LLC consortium that was established for this particular purpose and brought together a number of companies from different countries. However, the project was doomed to failure for a variety of economic reasons.

The processes that go on in the US aerospace industry are similar to those that occur in aerospace industry of other countries. For example, a number of Canadian firms have pooled their efforts for winning positions on satellite imaging market. For this purpose the Spar Aerospace, ComDev and McDonald Dettwiler & Associates established the Radarsat International, a company concerned with operation and marketing of images received from the Radarsat SC.

The Radarsat SC launched in 1995 is equipped with a synthetic aperture radar capable of operating in all weathers and all lightening conditions.

The European aerospace companies not only participate in the American projects but also come up with initiatives, offering space services provision projects. For example, Alkatel of France, a company already involved in the Globalstar project, joined as the leading partner the SkyBridge Limited Partnership (L.P.) established in order to create and operate the SkyBridge communications system based on 80 low orbit SC.

Included in the company are the American firm Loral Space and Communication, also involved in the Globalstar project, Spar Aerospace Ltd. of Canada, the prime developer of the Radarsat SC, CNES and Aerospatiale of France, SRIW of Belgium and three major Japanese companies: Toshiba, Mitsubishi and Sharp. A workable orbital constellation with a limited component package (40% of SC) is due to be deployed by 2001.

So there are several basic causes that brought about the restructuring processes in foreign aerospace industries. One of them was the changes in political-military environment after the end of "the cold war" that entailed reductions in defense budgets and cuts in military orders. Another reason of restructuring lies in economy, even though the initial impetus came from changes in the political-military environment. The case in point are the processes that take place on the stock exchange, the most powerful financial tool to wield the defense industry.

In the US, the processes that occur in aerospace industry were largely influenced by the economy drive pursued by the government pressed by the need to balance the federal budget. The pursuit of such policy resulted in cuts in federal spending that reduced the share of federal sector in GDP.

The main cause of restructuring is the commercialization of space work which is manifest in emergence of the market of space products and services and in the stiffening competition of aerospace companies "for a place under the sun" on that market.

All these factors in the aggregate drove foreign aerospace companies to search for ways of retaining their positions, which eventually resulted in restructuring of the industry.

The restructuring of the 1990s is different from the previous similar phenomena in the large scale of the processes underway. During the period since 1990 30 odd transactions have been made associated with merger, purchase and sale of various companies and their divisions. The biggest of those deals were the mergers of such companies as Grumman and Northrop, Lockheed and Martin Marietta, Boeing and McDonnell Douglas. The distinctive feature of practically all those transactions was their voluntariness. Such transactions are often more than a mere buy-and-sell. For example, Lockheed Martin acquired Loral, thus becoming one of the co-founders of the new firm, Loral Space and Communication, which in turn established on a share basis another few firms.

The merger of the major aerospace companies did not entail monopolization of the entire industry. The experience of OSC shows that if business strategy is chosen correctly, small firms can not only survive but can also build up their capabilities.

The restructuring gave aerospace firms a number of advantages enabling them to successfully develop and compete in the new conditions.

**Firstly,** thanks to the in-house consolidation the current expenses dramatically decrease. For example, Lockheed Martin plans thus to save every year up to US$ 500 million.

**Secondly,** consolidation of similar businesses generates a scale effect, which increases the series production and decreases production costs.

**Thirdly**, consolidation of complementary manufacturing facilities provides new capabilities which ensures manufacture of qualitatively new products. Fourthly, economic diversification of aerospace companies enables them to access new segments of the market of space products and services.

Diversification is one of the most important areas of activity of aerospace companies in the new conditions. The diversification trend is especially pronounced in such fields of space work as communications and Earth remote sounding. This is quite natural since it's these areas of work in space that provide steady profits at the current stage of progress in astronautics. Once the profitability of other kinds of space work becomes evident (for example, materials manufacture), the same processes will start up there.

Under such circumstances, plain and evident is the desire of aerospace companies to be not just executors of loose orders awaiting them and struggling for contracts, but to be among those who offer final products (communications services, satellite imaging information, etc.) to consumers and be thus in control of the entire spectrum of production services – from the development of the system concept up to its marketing, as is the case with Orbital Sciences Corporation which develops launch facilities, communications spacecraft, ERS spacecraft and operates those systems via companies established jointly with other firms.

Some firms participate in several associations engaged in realizing similar or even rival projects, for example Motorola (Iridium – Teledesic), Orbital Sciences Corp. (Orbcomm – ECCO in communications, OrbView – EarlyBird in ERS).

The formation of joint companies by industrial enterprises for marketing their products is a characteristic trait of today's restructuring. Mergers for diversification are largely international with partner companies coming from various states.

The restructuring of the European aerospace companies has some specifics resulting from the specific conditions in which those companies carry on their business. One such specific feature is "the border problem" which is a serious obstacle to company mergers. Another peculiarity derives from the lesser militarization of the European aerospace industry and, hence, lesser commercialization of space work from its very start.

That is why consortiums were readily accepted by Europe. Such a form of consol dating the efforts insures on the one hand implementation of large scale space programs without detriment to the national industrial base, and, on the other, provides effective opposition to the pressure of US aerospace companies, usually larger ones.

## 1.6 Reliability as the basis of efficient functioning of space systems in the future

One of the factors governing the efficient use of space systems is their reliability. During the period between 1950 and 1980 the inadequate reliability of space equipment, primarily carrier rockets and SC, used to cause many accidents (Fig. 71) and dramatically restrained the progress of space rocket engineering and the use of its achievements in research and applied programs. Also, it thwarted the development of international space products and services market.

In the early 1990s the situation changed radically. The development of multi-satellite space systems of various applications boosted the market of space products and services.

The space technologies will continue to develop and find ever wider application in the 21st century. One of the essential preconditions for this to occur is the insurance of high quality and reliability of space hardware.

The quality and reliability of SC and their components is most closely associated nowadays with the following issues:
- insuring higher reliability of advanced launch facilities;
- insuring ever longer active service life (operational longevity) of full size and small spacecraft for communications, navigation, meteorological survey and ERS;
- insuring reliability and safety of ISS;
- deployment and replacement of low orbit multi-satellite systems;
- insuring the quality and reliability of Russia's segment in international space programs;
- insuring the reliability of carriers and spacecraft through the use of failure-free on-board equipment;
- development and improvement of methods of setting assignments
to spacecraft and their subsystems, evaluation and checking of their quality and reliability;
- improvement of international standardization of commercial space rocketry.

### 1.6.1 Insuring reliability of advanced launch systems

Intense work is now underway in the countries leading in space rocket engineering aimed to create advanced launch systems.

The protracted 40 year period (1957-1997) of operating carrier rockets made on the basis of combat missiles has drawn to a close. A considerable number of modifications of basic models developed during that period had come about through modernization of separate elements of carriers within the ramework of established lay-out patterns and without altering the essential nature of equipment. The evolutionary period of carrier improvement is coming to an end, the capabilities of old basic variants being all but exhausted.

**Fig. 71.** Accident on Titan-4A launch vehicle

The current period of creating new launch systems is characterized by the following:

• lay-out (structural) diagrams of carrier rockets can be best elaborated based on the experience of previous operation of carriers;

• to reduce the time and cost of development and operation of carrier rockets, a modular principle of carrier construction is resorted to, in which case rocket modules or basic elements are used as versatile structural components;

• in development of modules and basic elements the component parts are used (engines, control system, structural elements) that have at the current moment the optimal flight characteristics.

The reliability reached on the existing carrier rockets (Fig. 72 – 73) is inadequate for advanced carriers, due to which:

• the percentage of successful launches throughout all the period of operating carriers for all rocket families does not exceed 97% (the Sputnik family), while for basic variants of carrier rockets the guaranteed reliability does not exceed the level of 0.969;

• series of successful launches exceeding the fairly high level of 50 (which corresponds to the potential reliability level of 0.98) is demonstrated by only a limited number of carriers, e.g. Soyuz-U, Proton-K, Cosmos-3M, Cyclone-2, Delta-2, and Ariane-4;

• the time required to bring up the carrier to a period of high quality and reliable operation varies between 10 and 15 years which at the moment is inadmissible;

• production failures resulting from violations of manufacturing techniques and combined with operational malfunctions seriously affect the actual level of reliability of fairly well optimized carriers that have been in operation for a long time;

• the insurance fees that essentially influence the cost of a launch are now at the high level of 17 to 20%;

• the accident rate of the world's fleet of carrier rockets is still rather high, which is emphatically demonstrated by the results of 1998 and 1999.

The abortive launches of such carriers as N-2 (Japan), Titan-4, Delta-3 (USA), Proton-K (Russia), Zenit (Ukraine) did much damage and adversely affected the deadlines of programs and the course of their implementation:

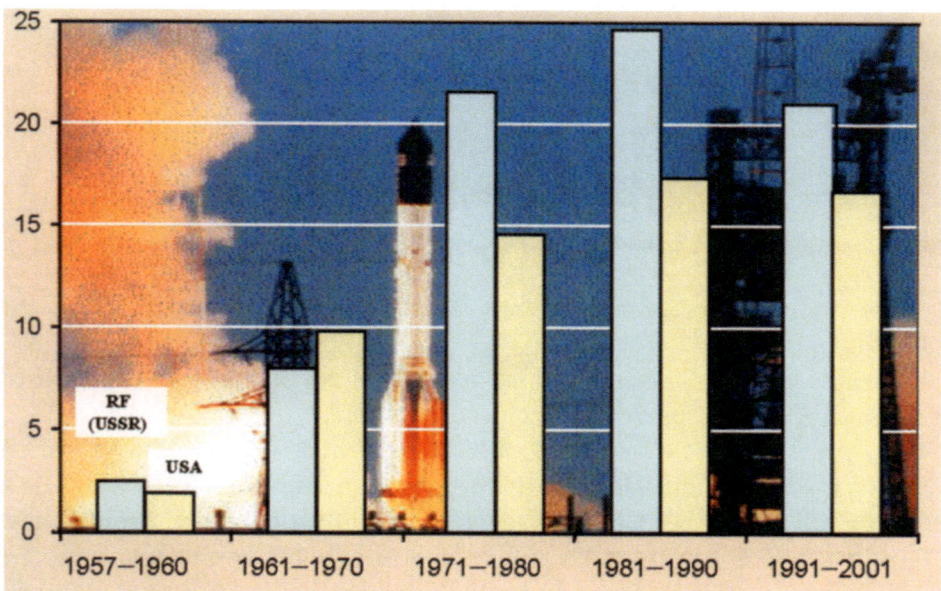

**Fig. 72.** Number of launches per carrier accident in USSR (Russia) and USA during 1957-2001

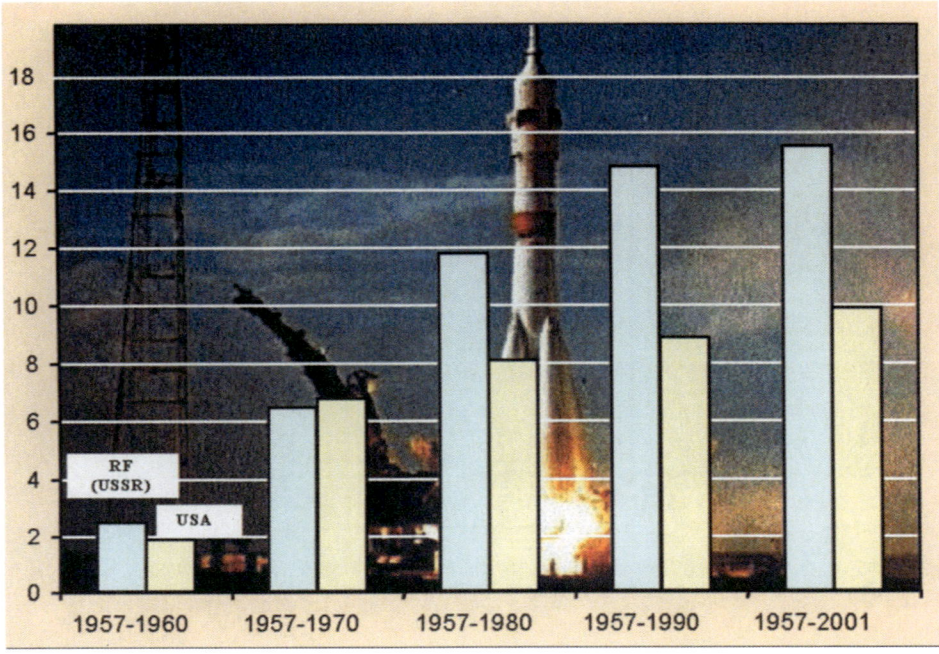

Fig. 73.
Aggregate number of carrier launches per one accident in USSR (Russia) and USA during 1957-2001

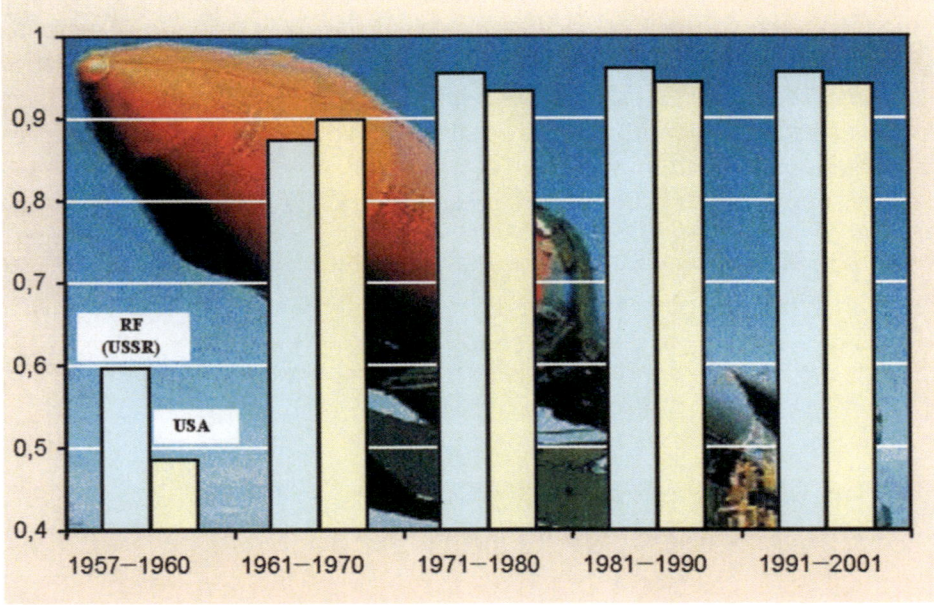

Fig. 74. Frequency of successful launches of carriers in USSR (Russia) and USA during 1957-2001

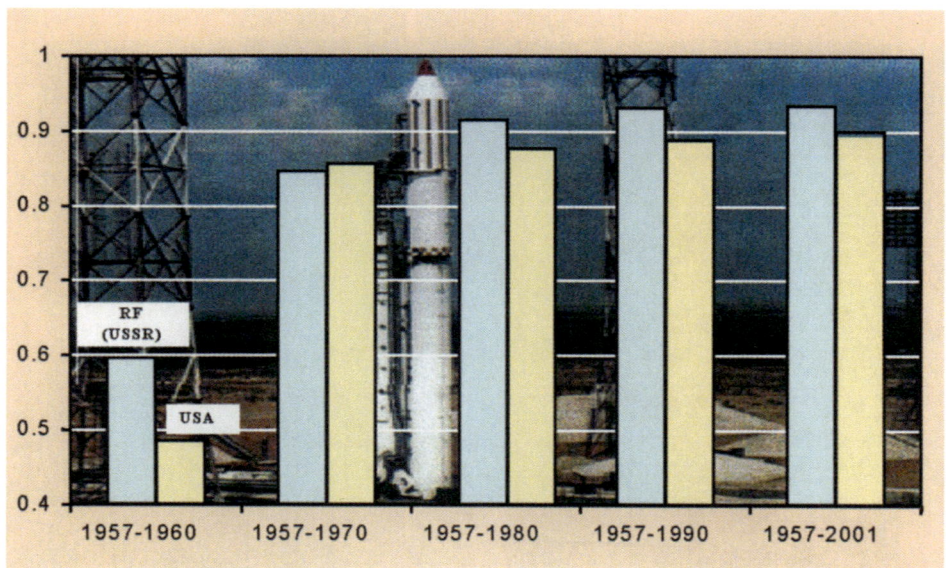

**Fig.75.** Aggregate frequency of successful launches of carriers in USSR (Russia) and USA during 1957–2001

• the damage done by the Titan-4 accidents (2.08.93, 12.08.98, 9.04.99, 30.04.99) amounted in each accident to more than a billion dollars;
• 2 accidents of the new Delta-3 launch vehicle (17.01.98, 5.05.99) seriously affected the plans to launch the carrier rocket (the satellite and its launch in May 1999 were insured for US\$ 265 million);
• the Globalstar consortium decided to replace the Zenit carrier by Soyuz in deploying its Globalstar system. Six out of 7 planned launches of Soyuz have been already successfully performed.
In order to obtain competitive and efficient use of advanced carrier rockets, the demands made of them must be of a higher order:
• the designed value of the carrier's reliability in flight $R^R_{CR}$ must be chosen from the range of values between 0.985 and 0.995. It means practically that the designed value of successful launches varies between 65 and 200;
• the reference level of reliability of the carrier in flight, $R^R_{CR}$, is established at the level of 0.975-0.99 at confidence probability $g = 0.9$. It means in practice that the values of successful launches must vary from 45 to 100.

The tendencies in changes of non-failure operating time of the carrier, *Na*, are shown in Fig. 76.

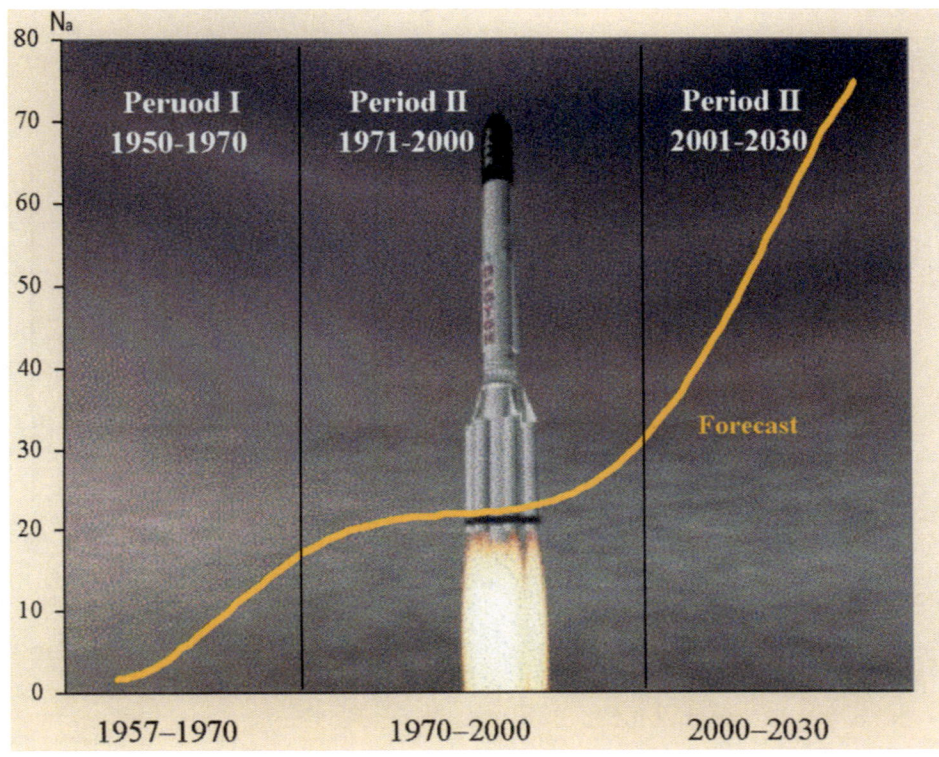

**Fig. 76.** Tendencies in change of non-failure operating time of the carrier rockets during 1957-2030.

Based on the generalized experience in creating national and foreign carrier rockets and in implementing such programs as Apollo, Soyuz-Apollo, Space Shuttle and the initial stage of Buran, some general principles can be quoted that are used to insure reliability of carriers, which principles are most appropriate in development of new carriers or modernization of existing ones:

• use of proven technical decisions, assemblies and systems;

• creation of carrier rockets on a pattern containing a minimum of elements with subsequent duplicating (in some cases trebling or use of the majority system) of critical elements;

• formulating of lists of critical elements of carrier rockets (based on the analysis of likely failures and estimation of their influence on reliability and safety of carriers) and taking extra measures to improve the reliability of those elements at the stage of on-the-ground optimization;

• providing reliability mainly by "on-the-ground" optimization in conditions as close to those of real operation as possible.

The chief factors that have been determining and in the foreseeable future will continue to determine the reliability of SC are shown in Table 14.

**Table 14.** The main factors of carrier rockets reliability

| Periods | | |
|---|---|---|
| **1950 – 1970** | **1970 – 2000** | **2000 – 2030** |
| 1. Prolonged on-the-ground tests and flight trials<br><br><br><br>2. Optimization | 1. Heightening the level of:<br> – ground processing efficiency<br> – manufacturing standard<br> – inspection authenticity<br>2. Use of redundancy at the level of units, channels, instruments | 1. Optimal lay-out-arrangements<br>2. Use of proven rocket modul<br>3. Failure-proof onboard instruments and efficient diagnostic and forecasting inspection |
| – max. diameter, m<br>– Take-off weight, t | 3.05<br>868.0 | 3.05<br>939.3 |

Reliability of engines, their type and size critically affect the reliability of the carrier rocket on the whole.

The optimal strategy of insuring the reliability of carriers' power plants implies:

• the use of minimal number of engines in power plants;

• optimization of engines in the stage of their on-the-ground adjustment in conditions as close to operational as possible and to an extent sufficient to prove the required level of reliability.

Reliability of the first stage power plant is particularly important for medium, heavy and super heavy lift classes of vehicles. Obtaining thrust on the first stage of the carrier at the level of 1,000-2,000 ton-force depends on the number and size of engines used in the first stage power plant. This is why the choice of the type and size of engine used in the first stage power plant becomes of prime importance for the safety of the entire carrier.

The families of carrier rockets being developed in the US, ESA and Japan use mainly large size engines that run on ecologically clean and high-energy types of fuel.

Analysis of the above described families of carriers makes it possible to state the following:

1. In forming the carrier family under EELV program in the US it was decided to abandon the traditional lay-out arrangement based on the principle of using the first stage of large and small solid propellant boosters (Titan and Delta family of carriers and some models of Atlas). This allows to essentially improve the reliability of the first stage power plant.

2. The use of large size engines dramatically simplifies the lay-out arrangements of carriers and cuts the number of engines used on the first stages and on the entire carrier (Table 15).

In the carriers being developed the number of engines used on the first stage and on the entire carrier is less than on the existing carriers by a factor of 2 to 5 (Table 15).

3. To obtain the same reliability of carriers, the requirements on reliability of engines of Atlas-2AS carriers must be essentially higher than those on reliability of engines RD-180 or RS-68.

**Table 15.**
Number of engines used in component package of advanced launch systems and prototype carriers

| Carrier rocket | Number of engines | |
| --- | --- | --- |
| | Stage 1 | Entire carrier rocket |
| EELV - MLV (Lockheed Martin) | 1 | 2 |
| Atlas 2AS – prototype | 7 | 9 |
| EELV-MLV (Boeing) | 1 | 2 |
| Delta III – prototype | 10 | 11 |
| EELV-HLV (Lockheed Martin) | 3 | 4 |
| Titan-4 – prototype | 4 | 7 |
| Ariane-5 | 3 | 4 |
| Ariane-4 – prototype | 8 | 10 |
| H-2 | 3 | 4 |
| H-1 – prototype | 10 | 12 |

4. The period of flight trials and initial phase of regular operation of practically all carriers now in development are characterized by a higher failure rate on engines and other subsystems of the carrier caused by insufficient on-the-ground adjustment, undue regard for peculiarities in interaction between the rocket's subsystems and by the difference between real space conditions and those on the ground. The failure-free operation rate of a multi-engine power plant is also considerably lower than that of a power plant with few engines.

## 1.6.2 Supporting active operation life of durable spacecraft (DSC)

In modern conditions characterized by reduced funding for space programs, soaring prices of launch systems, and cuts in production of carrier rockets and spacecraft, the creation of multi-satellite systems for communications, navigation, missile assault warning etc. and keeping such systems in standard

operation are associated with great difficulties and fraught with serious consequences in case of their failure.

Given below is the consideration of the following issues:

• the attained longevity of SC in foreign countries;

• the attained longevity of domestic-made SC;

• recommendations concerning reliability levels of durable SC for the period
till 2010.

## Longevity level attained by foreign spacecraft

Analysis of development of foreign rocketry between 1970 and 1990
supplies the following conclusions.

The US has created SC with designed service life of 7 to 10 years (DSCS-2,
DSCS-3, Leasat, Fleetsatcom, TDRSS, Navstar) for space stations used in
communications, relay operations, navigation, early missile assault warning
and radar battlefield surveillance. This allows to deploy space systems in their
standard configuration for the entire period of operation without subsequent
replenishment and with allocation of 2 to 5 stand-by SC (depending on the
standard number of SC units in the system) to replace losses caused by unreli-
able function of the carrier and malfunction of spacecraft. With the number of
units in standard configuration varying from 4 to 6, it is possible to provide the
following output of SC: DSCS-2 – 15, DSCS-3 – 14, Fleetsatcom – 8.

In 1994, deployment of the Navstar multi-satellite navigation system was
completed with 24 SC in standard configuration. The protracted period
(1978 – 1994) of creating the unique system had ended.

Between 1978 and 1984 10 experimental prototypes (Block-1) with
designed active service life of 5 years were put in orbit. Their actual service life,
however, lasted 7 to 11 years.

Between 1989 and 1994 a standard orbital constellation of 24 SC (Block-2)
was deployed. The designed active service life was 7.5 years. To replenish the
system, 4 examples had been manufactured.

Between 1997 and 2001 the system is planned to be replenished by SC
(Block-2R) being modernized now. Its guaranteed service life is 7.5 years.
Twenty one SC have been ordered.

Between 2001 and 2012 the system is expected to be replenished by new
generation SC (Block-2F) with guaranteed service life 12.5 years and operational
life expectancy of 15 years. Thirty three spacecraft are planned to be supplied.

During the period under consideration several generations of communica-
tions SC were created that belonged to Intelsat and Inmarsat consortiums.
The Intelsat type of SC has the designed service life of 15 years, whereas
Intelsat-8 type – 18 years, Inmarsat-3 – 13 years.

The above allows to infer that during the period till 2010 it is practical to
create a SC with service life varying between 15 and 20 years.

## The longevity reached on domestic-made spacecraft

The study of progress made by the national space rocket engineering over the period under consideration yields the following conclusions:

Multi-satellite systems of communication, relay operations, navigation, electronic surveillance and missile assault warning use SC with guaranteed service life of 1 to 3 years. In reality, however, the average operating life is 1 to 2 years longer than that covered by the guarantee. Some examples have a service life as long as 5 to 8 and even 10 years (when used in a lower technical condition or as a stand-by). The annual number of launches to replenish multi-satellite space systems is rather high. This necessitated considerable spending on deployment and replenishment of space systems during their 15 to 20 years of operating time.

In all serially produced SC it is mostly the power supply system on board the craft that determines its service life. On most SC the expiry of service life is associated with dramatic deterioration of power supply, which is why SC begins to be used with restrictions in its application.

The second in importance factor that affects the longevity of SC is the prevailing level of technology and production standards. The service life of SC and the failure-free operating time of their onboard equipment vary widely, which testifies to the manufacturer's inability to insure the stable production of high quality on-board equipment. This results in pre-timely failure of SC before the expiry of its active service life, in use of redundancy on equipment rather for elimination of faults than increase of its longevity.

The support equipment of SC developed in the early and mid 1970s (command and instrumentation system, complex power plant, heat regulation system, orientation and stabilization system, telemetric control system) has a service life of around 7 to 10 years. This testifies to the erratic reliability of serially produced SC in terms of longevity and demonstrates their capability to be generally much improved. The special purpose equipment of the SC (relays, on board instrumentation and navigation equipment) has a service life of around 7 years.

In 1995, deployment of the unique GLONASS navigation system, comprised of 24 SC, ended in Russia. Sixty eight spacecraft had been used to conduct flight trials and deploy an orbit constellation (as a reminder for comparison, 34 SC had been used to do the same job for the Navstar). The guaranteed service life of serially produced SC is 3 years, the average being around 4 years. The actual service life of SC varies widely from 2 to 5 years.

Flight trials of Express and Gals SC, whose guaranteed service life is 5 years, are drawing to an end. In development are navigation SC with guaranteed service life of 5 to 7 years. The Applied Mechanics Research and Production Association develops a support system module with service life around 10 years.

Methods and algorithms have been elaborated for setting and regulating requirements imposed on longevity. Also, algorithms have been worked out for assessing and forecasting the longevity of durable SC as well as algorithms for determination of production output in instances when durable SC characteristics are known for various operation strategies.

The above stated suggests the following conclusions:

• the longevity of SC used in multi-satellite systems needs to be essentially increased; otherwise deployment and replenishment of multi-satellite space Systems is now virtually impossible; evidence to this effect is provided by experience of deploying and replenishing the GLONASS space navigation system;

• technical and guaranteed service life of SC is determined by the power supply system;

• special purpose equipment and support on board equipment develope with the use of component parts available in the early and mid 1970s have a service life of around 7-10 years.

Considering the state of things as they are with insuring the longevity of existing SC and those now in development, the following recommendations can be given concerning the requirements on durable SC longevity for the period between 2000 and 2010.

During the initial period (2000-2005) it is necessary to resolve the problem how to develop SC with guaranteed service life from 5 to 7 years.

With guaranteed service life of 5 years this will imply the following production of SC for operation of space stations over a 10 year period;

six SC in the system. The production output – not more than 15 SC;

eight SC in the system. The production output – not more than 20 SC;

twenty four SC in the system. The production output – not more than 56 SC.

During the second period (2005-2010) it is necessary to resolve the problem how to develop SC with guaranteed service life from 7 to 10 years.

With guaranteed service life of 7.5 years this implies the following production output of SC for their operation over a 15 year period:

six SC in the system. Production output – not more than 15 SC;

eight SC in the system. Production output – not more than 20 SC.

The most important task today is to develop a navigation SC with guaranteed service life of 7 years.

Until 2010 a navigation SC must be created with a service life of 10 years.

## 1.6.3 Insuring the reliability and safety of long life space stations

Manned space exploration is hardly possible without insuring the high reliability and safety of the hardware used: carrier rockets, spaceships and orbital stations. Accidents and failures cost far too much. The history book of space rocketry contains, among others, black pages describing fatal accidents that killed Russian cosmonauts and American astronauts (Fig. 77)

**Fig. 77.** Apollo 1 crew who died in 1967 in spacecraft fire during ground tests
(left to right: V.Grissom, E.White, R.Chaffee)

Russia has a unique experience in building and operating long life orbital stations:

• during the period between 1971 and 1983 seven Salyut long life orbital stations were put in orbit;

• on 20 February 1986 a basic block (BB) of the Mir station was put in orbit. In the subsequent 10 years, 5 research modules (Kvant (quantum), Kvant-2, Kristal (crystal), Spektr (spectrum), Priroda (nature) and a docking compartment were docked to the BB. When fully deployed, the Mir station weighed over 140 tons (with two docked ships).

The unique orbital complex Mir ended its life on 23 March 2000 in a southern part of the Pacific Ocean, having marked its 15 anniversary a month before it happened.

In all, the Khrunichev State Space Research and Production Center manufactured and put in orbit for the Salyut and Mir stations 13 modules. In addition, prototypes were manufactured and their autonomous tests carried out.

The results of building and operating the Mir station testify to the fact that for the first time in the Earth's history there had been 15 successive years of successful operation of an international space research laboratory built and controlled by Russia. The orbital scientific complex functioned as if it were a ground-based research center. This became possible thanks to securing a high level of reliability and safety of the station.

The main achievements reached by the Mir station are a convincing proof of it:

• in 15 years' time the station hosted 28 main expeditions plus short Sitations; the station remained in the piloted mode 12 years and 7 months;

• overall, 104 people had worked at the station (42 from Russia, 62 from abroad); rich experience had been accumulated in collaboration of international crews; the total time of all astronauts' stay onboard the station had reached 12,000 man-days;

• the station was the site of 55 target-oriented research programs, of which 27 were international;

• a considerable number of experiments (engineering – 6,700 experiments, materials study – over 2,450 experiments, bioengineering – over 130 experiments, astrophysics – over 6,200 experiments, photography – 125 million km$^2$);

• a medical support system was created aimed to service flights lasting up to one and a half years; a record duration of a continuous space flight was reached – 438 days (V.Polyakov) and a record total duration of man's stay in space was attained – 748 days (S.Avdeyev);

• during the station's flight 142 dockings had been made;

• operability of the orbital station had been sustained for 15 years based on the use of designed longevity margin of the main elements of the long life orbital station (housing, docking assembly, solar battery, propulsion system, heat regulation system, etc.) and timely maintenance and repair;

• experience has been accumulated of how to operate the onboard operability support instruments, using for this purpose the system of evaluation and forecast of operability of the station's subsystems and units;

• experience has been accumulated of how to eliminate emergencies (fire, depressurization of the Spektr's module, failures on the thermal control subsystems, malfunction of onboard computing system, life support system failure).

## The main principles of insuring the reliability and safety of long life orbital stations

The safeguarding of high reliability and safety of the Salyut and Mir long life orbital stations became possible due to the use of:

• systematic approach to and consideration of the factors governing the reliability of long-life orbital stations (LLOS);

• high reliability carrier rockets for launching the station modules, transport and cargo spaceships;

• extra requirements imposed on the workmanship and reliability standards of component parts of manned space complexes;

• optimal design and engineering decisions;

- adaptive system of maintenance and repair;
- proven and effective system of crew training;
- high reliability automatic and manual systems of docking the modules, transport and cargo ships;
- design decisions, principles, methods and means of testing, checking and repair adjusted to the optimum during development and operation of the Salut stations.

Over the past 30 years the Khrunichev Center has been incessantly engaged in design, development and experiments aimed to attain high reliability and safety in operation of long life orbital stations and their modules.

That work brought forth principles, methods and design decisions that provided greater reliability and longer service life. While, for example, transport supply ships of the late 60s had been designed for a year's operation in space, the full function cargo unit of an ISS will be operated in orbit for 15 years.

To insure reliability and safety of LLOS, the following has been developed:
- design and engineering principles of safeguarding reliability and safety of LLOS;
- techniques for evaluating and forecasting the technical condition of LLOS and for extending the service life of its component subsystems;
- principles and methods of maintenance and repair of LLOS systems and subsystems.

## Design and engineering principles of insuring the reliability and safety of long life orbital stations (LLOS)

The design and engineering principles of insuring the reliability and safety of LLOS include:
- the modular principle of building the LLOS;
- the use of basic standardized block for all modules of LLOS regardless of each module's specific application;
- safeguarding the prescribed service life of the main systems not subject to replacement (housing, power units, etc.);
- acknowledgement on Earth of the service life of systems not subject to replacement;
- insuring the guaranteed protection of the systems not subject to replacement;
- insuring 100% security for a crew (leak-proofness, gas medium, micro-meteorite protection, emergency return system, etc.);
- design and engineering system of interconnections providing an unimpeded replacement of block units;
- development of methods and building of compact equipment for testing the systems during their operation, modernization and for block replacements;

• adaptability of systems to modernization during prolonged operation; equipping the modules with diagnosis systems that control and register the correspondence between the actual and designed performance;

• massive introduction of the systems diagnosis equipment, creation of local integrated information facilities;

• insurance of stable operation of the modular system and LLOS as a whole in case of emergency (resistance to the set levels of system failure) with replacement or follow-on repair of failed elements.

Based on those principles, decisions and arrangements have been elaborated and introduced in order to obtain higher reliability, greater capability and longer service life of LLOS. Those measures also cover design and development work, on-the-ground optimization and test flights.

The following could serve by way of example of original design and engineering decisions:

• the use of electro-mechanical gyrodyne flywheels in the orientation system. Such wheels essentially reduce the fuel consumption delivered from Earth; the daily consumption of fuel by engines alone is 25 kg, that by gyrodynes varies between 1,5 and 4 kg;

• the use of modern control algorithms in the control systems and the possibility of reprogramming them from Earth.

## Principles of maintenance and repair of long life orbital stations (LLOS)

The main principles of maintenance and repair of LLOS include:

• deployment of the orbital station and maintenance system must be carried out on agreed dates with the maintenance system being deployed first;

• most of maintenance and repair equipment must be in the state of permanent readiness for tackling suddenly arising assignments;

• the maintenance and repair system must be resistant to the change of types of some of its elements, which is inevitable in the course of prolonged operation;

• maintenance and repair must normally proceed pursuant to the principle of current technical requirement; exceptions are only made in cases directly dangerous to the crew of the space station; the realization of that principle is accomplished by choosing the parameters that determine the technical condition of the station, by using the means of monitoring, diagnosis and warning of on-the-ground repair centers;

• a priorities scale must be established in the maintenance system which determines the sequence of jobs and distributes the means of maintenance;

• maintenance and repair must not affect the station's on-duty time;

• timely provisions must be made for maintenance and repair, special devices must be prepared in the form of electrical replicas of the equipment, specially developed bench installations must be made available that could mock up various emergences.

- provisions must be made for analysis in course of maintenance of failures and their causes, information must be gathered regarding reliability and technical condition of the elements, units, systems and subsystems of the orbital station (vibration resistance of the housing, solar batteries, the leak-proofness of rubber components, etc.).

The reparability of the LLOS in flight is achieved through:

- the modular construction, i.e. the configuration of the station in the from of detachable blocks that could be replaced in orbit;
- standardization of blocks, modules, their elements and inter-modular connections;
- compatibility of the construction with the maintenance personnel, provision of habitable conditions and creation of work zones;
- optimization of technological operations used in maintenance and repair.

The above shows that the required reliability, service life and safety of LLOS are obtained through:

- modular construction of the station;
- optimal design and engineering decisions;
- range of actions directed to safeguarding the reliability and safety of LLOS component parts during such periods as design and development work, on-the-ground optimization, manufacture, flight tests and operation, maintenance and repair.

The fifteen years of successful operation of the Mir station fully verified the workability and efficiency of the above said principles of insuring the reliability and safety of LLOS and motivated their use in creating international space stations.

## Safeguarding the required reliability and safety of the international space station

For the first time in the history of space rocketry, the biggest ever international project – creation of an ISS – has been carried out. Both previously implemented programs and those in implementation, are inferior to the ISS project in scale and amount of work to be done, in the number of participating nations and co-operating organizations, in responsibility for solution of reliability and safety problems in the course of creation and prolonged operation of the ISS.

The issues of reliability and safety got their share of attention in all Soviet and American manned flights programs: Vostok, Voskhod, Soyuz, Mir, Mercury, Gemini, Apollo, Skylab, Apollo-Soyuz, Space Shuttle.

The solution of theoretical and practical problems pertaining to reliability and safety of the above mentioned programs provided much experience in Russia and the US that can be effectively utilized in resolving the problems of attaining reliability and safety in the project aimed to build a ISS.

In attaining reliability and safety of an ISS the following issues should be treated with especial care:
- development of a concept of providing the ISS reliability and safety;
- insuring the reliability of launch facilities used in building the ISS;
- the ISS reliability and safety management during operation.

The concept of assuring the reliability and safety of the ISS must determine:
- requirements made of the reliability and safety of the ISS as a whole;
- the main principles of insuring the reliability and safety of the ISS;
- requirements imposed on the principal regulatory documents related to the reliability and safety of the ISS on the whole;
- requirements made of models, algorithms, software and methods of evaluating and checking the reliability and safety of the ISS.

Given below is an evaluation of risks sustained in a program aimed to build an ISS, especially in jobs associated with launch facilities.

According to the third version of schedule of work aimed to build the ISS within 1998-2003, the following was planned (accepted now is the fifth version of assembling the ISS, Revision-E, for the 1998-2004 period):
- 34 flights of the SSTS (1 demonstration flight);
- 52 launches of the Soyuz carrier rocket;
- 3 launches of the Proton carrier rocket;
- 4 launches of the N-2 carrier rocket (1 demonstration flight);
- 2 launches of the Ariane-5 carrier rocket (1 demonstration launch).

The analysis of the proposed launch schedule allows to establish the following:
- executing 95 carrier rocket launches and Space Shuttle flights in 6 years is a tremendous scientific and technological challenge with a successful issue relying on some extra work to be done for improvement of the carrier's reliability;
- the use of proven and adequately reliable SSTS, Proton-K and Soyuz carriers arouses no doubt;
- the use of the Ariane-5 carrier (out 5 launches one was abortive, one partially successful and only three passed off according to plan) and H-2 carrier (two failures out of 8 launches) is not, in our opinion, adequately justified.
- The eventual success of the mission is primarily determined by the reliability of the Soyuz carrier and the SSTS.

To be able to build an ISS with a $P=0.7$, the reliability requirements of the carrier and the Shuttle must be set at the following levels:
- Soyuz carrier – 0.997 (0.987 in case of one failure);
- Shuttle RSTS – 0.996;
- Proton-K carrier – 0.99;
- N-2 carrier – 0.99;
- Ariane-5 carrier – 0.99.

Given below is evaluation of the possibility to obtain the required reliability on the carrier rocket and the Space Shuttle reusable system.

Reliability of the Soyuz carrier can be characterized as 0.992, which is the resultant evaluation of the carrier's reliability demonstrated over the last 15 years of operation with regard for statistics of launching the Molniya (lightning) carrier rocket.

The potentially high reliability of the Soyuz carrier is sufficient for the successful building of the ISS. However, to realize this potential it is necessary to work out a package of work to be done to secure steady quality and high reliability of the carrier rocket.

The estimated reliability of the SSTS based on operation results is 0.989. The last years' estimation is 0.995. The SSTS is of a recoverable system, the reason why obtaining the reliability of 0.996 appears quite practical.

The estimated reliability of the Proton-K carrier during 1993-1998 is 0.98. The Proton-K repeatedly boasts successful series of more than 25 launches (52 being the maximum). The required reliability can be obtained provided extra measures are taken to control the quality of products used for building the ISS.

Based on above, the following conclusions can be drawn:

1. The systematic approach to evaluation of the technical risks in implementing the program aimed to build launch equipment of an ISS supplied comparative information as to how the reliability of the Soyuz, Proton-K, H-2, Ariane-5 carriers and the Shuttle SSTS meets the system requirements.

To successfully build the ISS, heightened requirements must be made of reliability of the launch facilities used.

2. Plans (programs) must be drawn up for all launch facilities as to how to insure their reliability while building the ISS.

Actions taken as per plans above, must in the first place be aimed at insuring the quality of component parts used, the stability of workmanship and enhancement of authenticity of inspecting the technical condition of the carrier's systems and subsystems.

The accidents on the Proton-K in 1999 and serious criticism of the technical condition of the Shuttle's cable network testify to the importance of that work (in 1999 Space Shuttles Orbiters were withdrawn from service for repair).

The centerpiece of reliability and safety management on the ISS during its operation must be the automatic system that evaluates the technical condition of the ISS and predicts its reliability and safety as well as those of its main modules (Prognoz (prognosis) automatic station).

Prognoz automatic station must evaluate in aggregate exponents of the technical condition, reliability and safety of space stations, on the basis of which data decisions are taken as to how to control them.

## 1.6.4 Insuring the quality and reliability of the Russian segment in international space programs

The international cooperation in commercial space programs essentially expanded in 1980-1990. Following the establishment of the first consortiums, Intelsat and Inmarsat, quite a number of systems and programs emerged of both global and regional significance, such as Comsat, Landsat, Meteosat, Eutelsat, PanAmSat, Asiasat, Iridium, Globalstar, and others. In 1998, construction of an international space station began.

The specific features of the period are as follows:

- a considerable growth in amount of work and Russia's work share in international space programs (ISP);
- a wide range of cooperation under ISP with various countries (USA, Germany, France, Australia, Asia) and companies (Lockheed Martin, Hughes, Loral, Motorola, Alcatel and others);
- a diversity of work areas for participation in ISP (launches of communications SC to stationary orbits, building of low orbit communications systems, Earth remote sensing, piloted flights programs);
- long term nature of work on creation and replenishment of space systems with transition to the 21st century (ISS programs, Globalstar, Teledesic, Skybridge and others).

In spite of the stiff competition on the world's market of space products and services Russia is one of its key participants in international space programs. The quality and reliability control system used to insure high operational standards of space hardware, complexes and examples of rocketry had been created in Russia between the 1960s and 1980s and met on the whole the requirements of the world's technology. The methodological approach used in Russia to evaluate and control the reliability of examples of rocketry is similar to that used in the US. A considerable experience has been accumulated in development and operation of space missile complexes, space stations, carrier rockets, spacecraft, and their component parts.

The requirements made of quality and reliability of the Russian segment of international space programs (ISP) are characterized by the following features.

High (absolute) level of requirements imposed on the quality of rocketry used in ISP.

Russia, China and the Ukraine, the newly emerged figures on the world's market of space products and services, have no right to a mistake. An accident on a carrier rocket automatically abrogates contracts:

- China – accident of the CZ-3B carrier rocket (loss of Intelsat-708 SC) – abrogation of contract with consortiums Intelsat, Asiasat and the Echostar company;
- the Ukraine – accident on the Zenit carrier rocket (loss of 12 Globalstar SC) – abrogation of contracts under PanAmSat and Globalstar programs.

The Russian system of rockets quality and reliability control emerges so abundantly on the world's market for the first time in space exploration history, which is why it undergoes a thorough analysis and careful check-out.

Taking account of the specific requirements imposed on the quality and reliability criteria as set by customers in the US and ESA.

The engineering specifications elaborated by the Head Customer for the Zarya functional cargo block OR FCB imply the following quality, reliability and safety requirements that are different from those accepted for the Russian-made components:

• insuring the reliability – protection against failures ending in disasters;
• verification of readiness for launch and flight operations;
• establishment of detailed requirements for controlling the condition of onboard equipment during trials;

certification of products and quality control system according to requirements set by regulatory documentation of the US and ESA.

To insure the quality of the Russian segment of the ISP as per the latest requirements and to make it fully compatible with specifications accepted by NASA it necessary to:

• improve the methods of insuring the quality of commercial rocketry;
• optimize the Russian system of rockets quality and reliability control in design, test and production phases.

Based on analysis of quality and reliability control of the Russian-made rocketry and with due regard for modern requirements imposed on quality of the Russian segment of ISP, the following work areas were delineated for improving the methods of insuring and controlling the quality of the Russian segment of ISP:

1. Improvement of methodology, structure and subject matter of "The Quality Insurance Program", the principal regulatory document, in accordance with which work is organized and carried out aiming to insure the quality and reliability of the products being created.

2. Setting extra requirements relating to the subject matter and quality of design work aimed to insure the competitiveness of the design decisions, the required levels and margins of quality, reliability and safety of commercial space rockets.

3. Improvement of the quality control system based on the use of modern methods and newest means of control.

4. Development of algorithms and software aimed to insure the required authenticity and accuracy of quality and reliability estimates in various phases of development and trials of commercial rocketry.

5. Improvement of the verification methods and certification procedures for checking the readiness of commercial rocketry for launches and flight operations.

6. Elaboration of proposals for improvement of work on international standardization of rocketry and creation of a single normative base for insuring the quality and reliability of commercial launch facilities, spacecraft and ground infrastructure.

7. Improvement of the information support used to insure the quality and reliability of commercial rocketry being developed.

## 1.6.5 Optimization of strategies used to deploy and replenish multi-satellite space systems based on reliability and cost criteria

By the end of the 1970s both Russia and the USA embarked on development and deployment of global navigation systems, GLONASS and Navstar, which were to comprise 24 full size SC (21 basic + 3 standbys). The considerable increase of the number of SC in the system seriously complicated the timely deployment of the system.

In the 1990s, the issue of the day was development and creation of multi-satellite systems based on small spacecraft (the Gonets (messenger) program, Iridium, Globalstar, Teledesic, and others) that were supposed to contain from 45 (Gonets) to 288 (initially 840) SC (Teledesic).

The deployment of such systems in one to two years is a rather complex task.

A methodological approach is now elaborated that substantiates the optimal strategies of creating and replenishing multi-satellite space systems on the basis of full size and small spacecraft with regard to the cost and reliability criteria.

The analysis and evaluation of  basic achievements during deployment of space navigation systems Navstar and GLONASS allow to note the following:

• essential advantages of launching in groups became evident in three criteria – space navigation system (SNS) deployment time, the cost and the reliability of deploying the SNS (Fig. 78, 79). The deployment time of the Navstar SNS was double that of GLONASS (5 and 2.5 years, respectively), the cost of deploying the Navstar SNS (only the cost of launch is considered) amounted to US$ 960 million, the cost of deploying the GLONASS amounted to US$ 480 million;

• the high reliability boasted by the Proton and Delta-2 carriers insured a perfectly failure-free deployment SNS. However, the estimated reliability of deploying the Navstar SNS is 0.71, that of GLONASS – 0.78.

The proof of differing reliability of deployment is manifested by an accident of the Delta-2 carrier during replenishment of the Navstar SNS in 1997 (the first launch of the new model of SC, Block-2R, took place January 17, 1997).

The considerable reduction in cost of deploying the GLONASS SNS as compared to the Navstar SNS was attained due to the use of a carrier rocket having the best specific cost ratio:

$$C_{spec} = \frac{C_{cr}}{n_{sc} \cdot R_{cr}}$$

where
$C_{cr}$ – cost of carrier rocket;
$n_{sc}$ – number of SC put in orbit;
$R_{cr}$ – reliability of carrier rocket.

For:     Proton-K carriers    $C_{spec}$ = US\$ 20.62 million per one SC
          Delta-2   carrier     $C_{spec}$ = US\$ 40.57 million per one SC

The maximum number of SC injected annually into the system was:

for:               Navstar SNS        – 6
                    GLONASS SNS       – 9

Apart from standby SC put in orbit during deployment of the Navstar SNS, an on-the-ground standby package is provided consting of 4 SC. Such standby systems are used to replenish over $T_0 = 7$ years (the operation period of $T_0 = 7$ years is determined by the estimated longevity time of SC Block – 2A). Considering the large number of planes in SNS (6 in Navstar SNS, 3 in GLONASS), only one possible way of replenishment could be recommended, namely, launching a single SC.

Based on above, the following conclusions can be drawn:

1. The injection of groups of SC in deployment of multi-satellite space systems provides a considerable advantage in terms of cost, reliability and deployment time.

2. In deployment and replenishment of multi-satellite space systems carrier rockets are needed that have the least value of $C_{spec}$ and a high reliability level.

3. In deploying multi-satellite space systems with the use of a single SC injection plan, the deployment time is rather long and comparable to the average time of SC functioning in orbit. This makes the replenishment of the space system (SS) a continuous process; a period of replenishment of a following model of SC used in the SS is replaced by a new period of deploying a yet another model of SC. This conclusion is verified by the replenishment practice of the Navstar SS, for which SS replenishment has been planned until 2012 (successive injection into orbit of models Block-2A, Block-2R, Block-2F).

Table 16. shows the required levels of reliability of carriers for various versions of fulfilling the task.

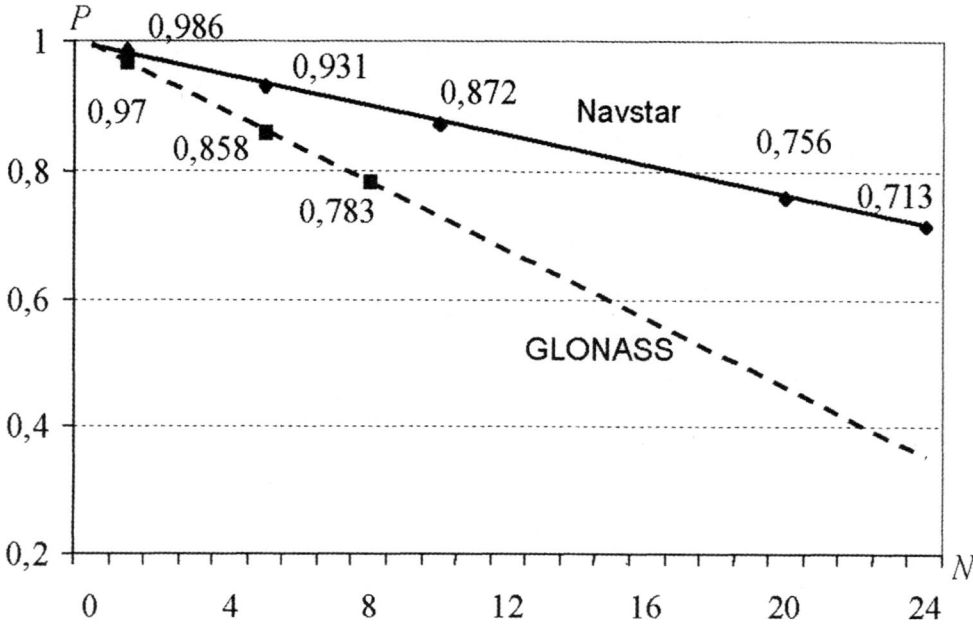

**Fig. 78.** Reliability of deploying Navstar and GLONASS SNS

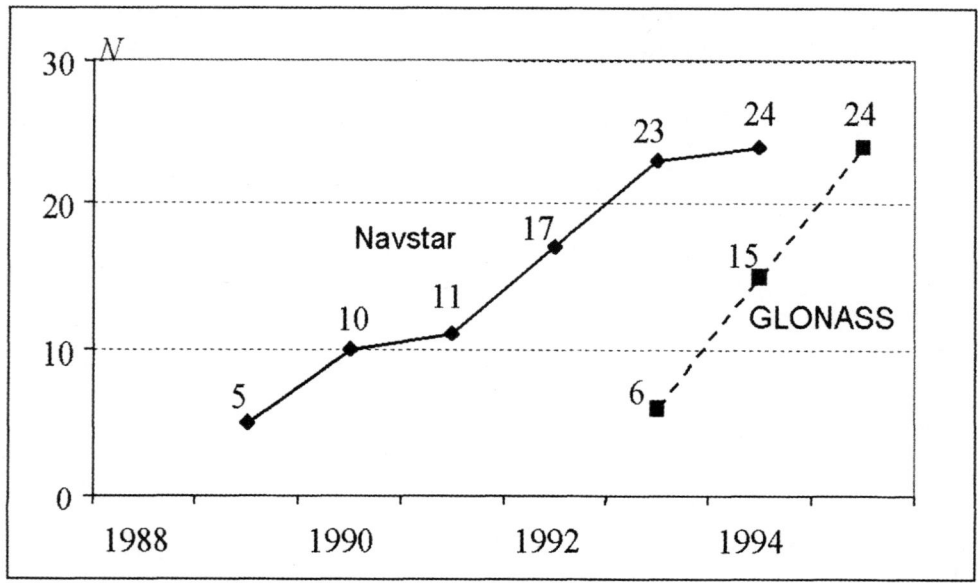

**Fig. 79.** Dynamics in deployment of Navstar and GLONASS SNS

**Table 16.** Reliability requirements imposed on carriers used to deploy multi-satellite systems

| Task fulfillment probability | Carrier reliability requirement | |
|:---:|:---:|:---:|
| | $n = 8$ | $n = 24$ |
| 0.9 | 0.987 | 0.996 |
| 0.8 | 0.972 | 0.990 |
| 0.7 | 0.956 | 0.985 |

4. Replenishment of SS in the event of failures on SC in orbit is carried out using the ground standby SC on the single injection pattern. Also used in such cases are carrier rockets having the least value of $C_{spec}$ at $n_{sc} = 1$.

The qualitative analysis of plans to deploy the Iridium and Globalstar SS (before the loss of 12 SC on January 9, 1998, during an abortive launch of the Zenit-2 carrier) allows to infer that:

• in deployment of the Iridium SS an inadequately reliable Chinese-made CZ-2C carrier had been used and the capabilities of the Proton carrier (in number of launches) were not fully utilized, which did not allow to optimally use the plan of deploying the SS;

• in deployment of the Globalstar SS an insufficiently reliable Zenit-2 carrier is used, the capabilities of the Soyuz carrier are not filly utilized, the Proton carrier is not used altogether, which also prohibits to deploy optimally the SS. After an accident on the Zenit-2 carrier with 12 SC the deployment of the Globalstar system was accomplished with the use of:

• 7 launches of the Delta-2 carrier (initially 2 launches were planned);

• 6 launches of the Soyuz carrier (initially 3 launches were planned).

### 1.6.6 Insuring failure-free operation of advanced durable spacecraft and carrier rockets

Insurance of failure-free operation of durable SC and carrier rockets involves a complex of actions aimed to obtain guaranteed operating periods of 7 to 10 years and failure-free operation of carrier rockets, including warning, protection and checking aimed to lower the failure level of components, to raise the longevity level, to detect, locate and rectify faults and malfunctions of onboard equipment.

**The principles of obtaining failure-free operation of advanced durable SC and carrier rockets.**

1. The failure-free operation of durable SC and carrier rockets is obtained and maintained due to the methodical observation of the following principles:

- concept of obtaining and maintaining the reliability of durable SC, carrier rockets and their onboard equipment;
- programs of obtaining the reliability of durable SC and carriers;
- analysis of failures on prototypes and the likely failures of SC and carriers being developed;
- choice and substantiation of using components employed in onboard equipment of durable SC and carriers;
- substantiation of types, repetition factors and standby modes of onboard equipment;
- substantiating the use of modern redundancy and standby techniques on a modular level for various types of onboard equipment;
- substantiating the use of the onboard system for diagnosis, fault checking and isolation and recuperation of SC and carrier;
- insurance of high quality production of failure-free onboard equipment;
- providing control of technical condition of onboard equipment in various phases of production and trials;
- protection of onboard equipment against adverse exposure;
- providing of failure-free operation of onboard equipment in various phases of creating SC and carriers;
- regulatory documents for development of failure-proof durable SC and carrier rockets.

2. The concept of obtaining and maintaining the reliability of durable SC and carriers must imply that the required longevity and failure-free operation are achieved without fail for each SC and carrier.

As per requirement above, in the phases of design work and experimental optimization the required margins of reliability must be demonstrated for all possible types of failures.

3. Extra measures are to taken within the framework of reliability support program (RSP) in order to insure the reliability of durable SC and carriers with due regard to their specific features.

Among the extra measures to be taken within the framework of the RSP to insure the failure free-operation of SC and carriers are:
- modeling of the service life of failure-proof durable SC and carriers;
- substantiating of redundancy in equipment, use of failure-free systems;
- exclusion of the influence of external factors on the reliability of durable SC and carriers in operation on the ground and in standard  operation;
- analysis of causes and consequences of likely failures, study of the influence of failures on the output effect of durable SC and carriers;
- drawing up of lists of critical elements of durable SC and carriers as well as elements that limit the longevity of durable SC and carriers;
- insuring the reliability of the critical elements of durable SC and carriers and elements that limit the longevity of durable SC and carriers;

- substantiating the choice, evaluation and control of the quality of component elements used in the equipment;
- evaluation of workmanship and authenticity of checking the technical condition of durable SC and carriers in acceptance trials and tests on engineering and launching complexes;
- amount and mode of running-in trials and entry control;
- simulation of typical failures on durable SC and carriers.

4. The control of failure-free operation of durable SC and their onboard equipment is carried out successively in the phases of: initial design; ground tests; flight trials and operation.

5. The initial design work must include quantitative analysis and substantiation of optimal standby (redundancy) that is required to meet the reliability requirements imposed on durable SC and carriers as established by the task assignment. All the main types of redundancy must be considered, including various types and means of structural, functional, temporal and software redundancy of the equipment and its components as well as the margins of efficiency and service life.

It is also important to develop algorithms for substantiation of optimal structural and temporal redundancy of the onboard equipment of SC and carriers used to substantiate the design decision.

6. The draft project envisions analysis of likely failures (ALF) which is a mandatory element of the Reliability Assurance Program (RAP) of the durable spacecraft (DSC) and carrier rockets' RAP. In the methodological support of ALF a tree-of-failures technique (TFT) is used that allows to conduct analysis of causes and effects of likely failures and to elaborate measures aimed to prevent them.

7. The initial design work implies the analysis of causes of likely failures which is a mandatory element of the reliability issurance program in durable SC and carriers.

The methodological package used for analysis of likely failures employs a failures-tree method that provides for analysis of causes and consequences of likely failures and enables procedures to be established for prevention of such failures.

8. The initial design must contain materials for substantiating the use of components comprised by the onboard equipment of SC and carriers and for checking the reliability of components used.

The entry control system of component parts of durable SC and carriers must provide for 100% forecasting parametric control.

9. The manufacturing quality of durable SC and carriers is determined based on results of acceptance trials carried out at the manufacturing facility of the producer of onboard equipment and at the prime manufacturer's facility. Also, account is taken of trials results obtained at the engineering and launching complexes as well as during initial period of operation.

The proof of the high quality manufacture of durable SC and carriers is the operational readiness probability of the manufactured durable SC and carriers after trials.

To evaluate the quality, information is used about the defectiveness of component elements of durable SC and carriers and about authenticity of checking the technical facilities at various times of their service.

To evaluate the defectiveness of the elements of durable SC and carriers, it is necessary to use the information about the technical condition of the onboard equipment obtained during acceptance trials at the manufacturing works as well as in course of routine maintenance and tests on the engineering and launching complexes.

## 1.6.7 Development of space systems and components quality control methods and their application for evaluation and maintenance of product quality

The massive growth in complexity of rocket engineering, the development of durable SC and high requirements for failure-free operation changed radically the methodology of obtaining and maintaining their reliability. The attention was focused on analyzing the potential and actual causes of failures that occurred during trials and in development of effective means of preventing them.

The main principles of modern methodology of obtaining and maintaining the reliability.

1. The systematic approach to obtaining the reliability based on reliability insurance program.

2. Use of probability factors included in contracts with the Customer.

3. Comprehensive tests optimized in close-to-real operating conditions.

4. Use of automatic design systems during design and engineering work which cuts development time, precludes design errors, makes comparisons of various versions of building the systems, optimizes the projects based on cost to reliability and weight to size criteria.

5. Use, during production, of automatic and automated manufacturing processes, management systems and non-destructive control techniques.

6. Creation of experimental base providing for optimization of rockets component elements in the phase of ground trials.

7. Creation of failure-proof onboard equipment of carriers and SC.

The main trends in improvement of methods of evaluating and controlling the reliability of advanced space equipment at the present time are as follows:

The first trend – the step-by-step confirmation of requirements results from the impossibility to confirm high reliability requirements (at the level of 0.98 – 0.99 and higher) at all stages of development and creation of space hardware. The strategy of step-by-step confirmation of reliability requirements that takes proper account of the planned amount of ground and flight trials makes it possible to confirm the level of reliability of carriers, booster units and spacecraft as established in the task assignment by the time their flight trials are finished. The proposed idea was materialized in engineering decisions concerning the Proton-M carrier rocket complex and Breeze-M booster unit complex. Similar decisions are planned with reference to the Angara family of carriers.

**The second trend** – complete use of information regarding the reliability and technical condition is primarily brought on by the necessity to consider at a greater length while appraising the reliability of space hardware and its components the physical parameters that govern the working capacity of carriers, booster units, and spacecraft. This fact requires development of methods of evaluating and controlling the condition of space hardware and its component parts during acceptance trials and tests as well as tests on engineering and launching complexes.

**The third trend** – taking account of optimization of space hardware being in current operation. This refers primarily to the analysis of typical failures on carriers, booster units and spacecraft and, hence, to development of models in the form of advanced systems' failures trees. The study on the basis of space systems' failures trees holds much promise.

The presented concept, principles and guidelines for provision of quality and reliability of advanced SC and carriers in case of their timely realization will make a great impact on the scientific and technological level of products being developed, on their cost, quality, reliability and, eventually, on their competitiveness in the world's markets of space technologies and services.

## 1.6.8 Space insurance: summing up the past and looking to the future

Space work is risky to people's health and life and is fraught with heavy property losses in case of damage or destruction of space hardware.

The risk reduction issue and cushioning of adverse consequences of accidents on space hardware is a complex multi-faceted business that can be managed successfully if only it is approached comprehensively within the framework of the system known as "Risk Management", one of whose elements is insurance against space risks.

Insurance of space projects is one of the most complicated types of insurance (Fig. 80). Each space project is a unique enterprise that requires an individual analysis of inherent risks to be carried out and an adequate approach to be elaborated for insuring them with due regard for other possible means of reducing the risks of project realization.

The space insurance is a new and relatively independent type of insurance with the distinctive features as follows:

• the insurance of radically new high-tech hardware, whose application, manufacturing techniques and operation are significantly different from those of traditional machines dealt with by insurers (for example, in insurance of automobiles, maritime and aerial ships, etc.).

• high risk levels associated with operation of rocketry, large insurance sums and fairly high tariff rates;

• an individual or unique approach to estimating each kind of risk and difficulties in obtaining authentic evaluations due to lack of retrospective data.

Nowadays, the Russian space risks insurance system is in a new phase of its development: the insurance of risks associated with separate stages of preparing a SC for a launch and controlling it in flight comes to be replaced by a more comprehensive and program-based protection of property interests of participants of space work. However, obtaining the protection through insurance still remains a rather difficult matter. The insufficient funds available to the project's Customer, the more so, to the Supplier of space products and services, for purchase of an adequate insurance protection on the one hand, and, on the other, the insufficient capacity of the domestic market of insurance against risks in space work, add to the complexity of the situation.

The Megaruss insurance group (an insurance company of the same name

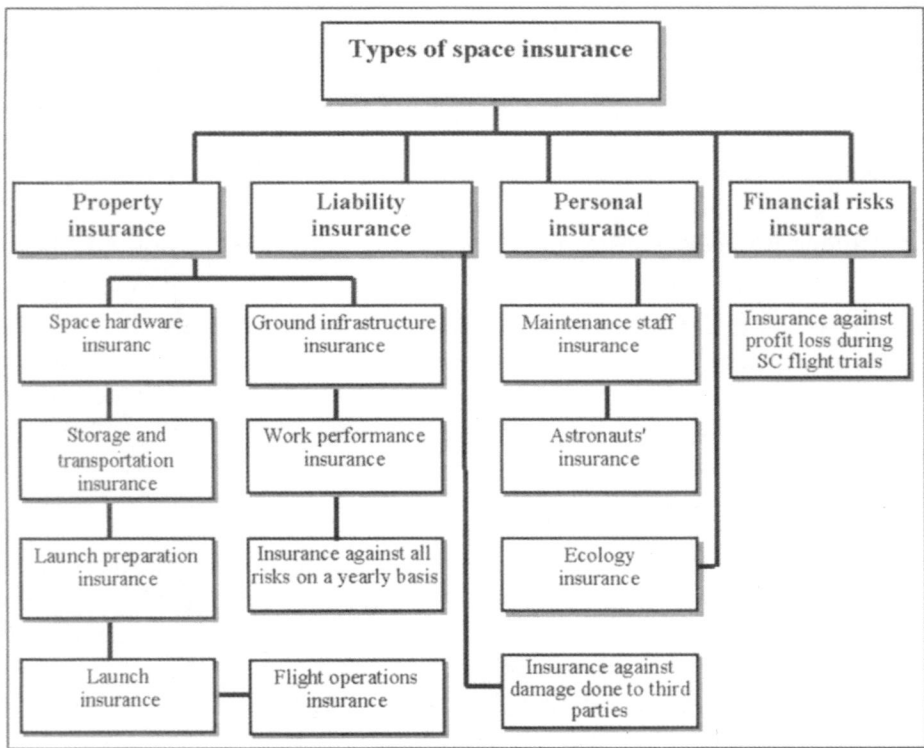

**Fig. 80.** Types of space insurance

was formed in 1992) has played an important role in establishing an efficient system of insurance against risks in space work under defense orders, federal space projects and programs of international cooperation. The efficiency of insurance against space risks was reached in the group due to:

• participation in the development of programs of insuring all participants of space projects;

• consolidated insurance of space work risks;

• package-type placement of space work risks on the Russian and the world's markets which action allows to maintain the economically most favorable and competitive rates at the same level over a protracted period ' of time;

• optimization of products subject to insurance that meet the current requirements of the international insurance and reinsurance market;

• drawing the leading insurance brokers and insurance companies, both on the Russian and foreign insurance and reinsurance markets, into reinsurance business;

• development of General Policies that simplify the procedure of drawing up an insurance contract;

• safeguarding the reliability of an insurance deal, the timely and full repayment of an insurance indemnity.

This comprehensive plan of the insurance protection has been realized by the Megaruss group through elaboration of the consolidated risk management program (CRMP) in the initial phases of implementing the space project with the execution of this program at all subsequent stages of work on the project. The elaboration of CRMP becomes one of the necessary elements (stages) of insurance service for space projects that precedes drawing up an insurance contract.

The essence of risk management is detection and evaluation of risks inherent in all cycles of a space project, plus development and realization of a package of measures aimed at diminishing the probability of risk incidents and at reducing their damage should such accidents occur.

By now, the Megaruss group has participated, as a prime insurer, in insurance of seventy odd space projects for building various carrier rockets.

Table 17. shows a list of space projects carried out using the launch facilities manufactured by the Khrunichev State Space Research and Production Center for the Federal Space Program and international cooperation programs. Those projects, executed over the last five years, have been insured by the Megaruss group.

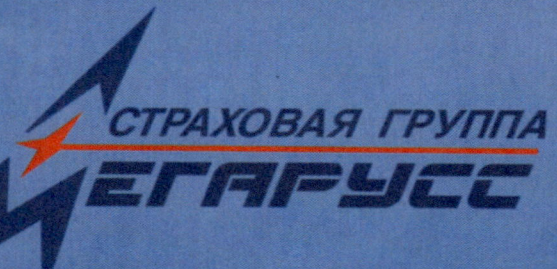

СТРАХОВАЯ ГРУППА
МЕГАРУСС

# SPACE RISKS
# INSURANCE

Table 17

| Item No. | Launch date | SC | Insured risk |
|---|---|---|---|
| 1. | 03.96 | Astra-1F | Complete destruction or damage of Proton carrier during preparation and launch of carrier with SC |
| 2. | 09.96 | Inmarsat-3 | Complete destruction or damage of Proton carrier during preparation and launch of carrier with SC |
| 3. | 05.97 | Telstar-5 | Complete destruction or damage of Proton carrier during preparation and launch of carrier with SC |
| 4. | 06.97 | Iridium-1 | Complete destruction or damage of Proton carrier during preparation and launch of carrier with 7 SC |
| 5. | 08.97 | Panamsat-5 | Complete destruction or damage of Proton carrier during preparation and launch of carrier with SC |
| 6. | 09.97 | Iridium -2 | Complete destruction or damage of Proton carrier during preparation and launch of carrier with 7 SC |
| 7. | 11.97 | Coupon | Complete destruction or damage of Proton carrier during preparation (SC starting from the forebody). |
| | | | Complete destruction or damage of Proton carrier during launch of carrier with SC |
| | | | Civil responsibility to third parties for damage incurred in preparation and launch of Proton carrier with Coupon SC |
| 8. | 12.97 | Astra-1G | Complete destruction or damage of Proton carrier during preparation and launch of carrier with SC |
| 9. | 12.97 | Asiasat-3 | Complete destruction or damage of Proton carrier during preparation and launch of carrier with SC. |
| 10. | 04.98 | Iridium -3 | Complete destruction or damage of Proton carrier during preparationand launch of carrier with SC. |
| 11. | 05.98 | Echostar-4 | Complete destruction or damage of Proton carrier with SC during preparation and launch |
| 12. | 08.98 | Astra-2A | Complete destruction or damage of Proton carrier with SC during preparation and launch |
| | | | Complete destruction or damage of Proton launching complex |
| 13. | 11.98 | Cargo block | Complete destruction or damage of Proton carrier during preparation and launch of carrier and SC in flight operations. |
| 14. | 11.98 | Panamsat-8 | Complete destruction or damage of Proton carrier with SC during preparation and launch |
| 15. | 02.99 | Telstar-6 | Complete destruction or damage of Proton carrier with SC during preparation and launch |
| | | | Complete destruction or damage of  Proton launching complex |

| Item No. | Launch date | SC | Insured risk |
|---|---|---|---|
| 16. | 03.99 | Asiasat-3C | Complete destruction or damage of Proton carrier with SC duringpreparation and launch |
| | | | Complete destruction or damage of Proton launching complex |
| 17. | 05.99 | Nimic | Complete destruction or damage of Proton carrier with SC during preparation and launch |
| | | | Complete destruction or damge of Proton launching complex |
| | | | Civil responsibility to third parties for damage incurred during launch |
| 18. | 06.99 | Astra-1H | Complete destruction or damage of Proton carrier with SC during preparation and launch |
| | | | Complete destruction or damage of Proton launching complex |
| 19. | 07.99 | Raduga-Breeze | Complete destruction or damage of Proton carrier with SC during preparation and launch |
| | | | Complete destruction or damage of Proton launching complex |
| 20. | 09.99 | Yamal | Complete destruction or damage of Proton carrier with SC during preparation and launch |
| | | | Complete destruction or damage of Proton launching complex |
| 21. | 10.99 | Express-A | Complete destruction or damage of Proton launching complex |
| 22. | 03.00 | Express-A | Complete destruction or damage of Proton launching complex |
| 23. | 04.00 | Sesat | Complete destruction or damage of carrier components during transportation and preparation |
| | | | Complete destruction or damage of carrier during launch and introduction into flight operations |
| | | | Complete destruction or damage of Proton launching complex |
| | | | Civil responsibility to third parties for damage incurred during launch |
| 24. | 07.00 | Service module | Civil responsibility to third parties for damage incurred during launch |
| 25. | 10.00 | GE-1A | Complete destruction or damage of Proton carrier |
| | | | Complete destruction or damage of Proton carrier |

| Item No. | Launch date | SC | Insured risk |
|---|---|---|---|
| 25. | 10.00 | GE-1A | Civil responsibility to third parties for damage incurred during launch in areas where carrier's jettisoned components fall |
| 26. | 10.00 | GE-6 | Complete destruction or damage of Proton carrier |
| | | | Complete destruction or damage of launching complex |
| | | | Civil responsibility to third parties for damage incurred during launch in areas where carrier's jettisoned components fall |
| 27. | 11.00 | SD Radio-3 | Complete destruction or damage of carrier |
| | | | Complete destruction or damage of launching complex |
| | | | Civil responsibility to third parties for damage incurred during launch in areas where carrier's jettisoned components fall |
| 28. | 04.01 | Ekran-M | Complete destruction or damage of carrier |
| 29. | 05.01 | Panamsat-10 | Complete destruction or damage of launching complex |
| | | | Civil responsibility to third parties for damage incurred during launch in areas where carrier's jettisoned components fall |
| 30. | 06.01 | Astra-2C | Complete destruction or damage of carrier |
| | | | Complete destruction or damage of launching complex |
| | | | Civil responsibility to third parties for damage incurred during launch of carrier |
| 30. | 05.97 | Telstar-5 | Complete destruction or damage of Proton carrier during preparation and launch of carrier with SC |

One of the important jobs in insurance work is the payment of insurance indemnity upon incident of an event insured against. The payments made by the Megaruss insurance company allow conclusions about far reaching financial consequences faced by insurance companies that insure space risks

Over the entire period of insuring space projects' risks the Megaruss group has made at agreed dates and in full amount a number of big payments to insurers of the Strategic Missile Forces of Russia's Defense Ministry (MoD SMF) and to the Khrunichev State Space Research and Production Center. So the Khrunichev Center received an indemnity for the abortive launch of the Proton carrier in lifting-off the Asiasat - 3C spacecraft – US$ 4.35 million and US$ 10.5 million for the accident on the Proton-Breeze-Raduga carrier (US$ 4.5 million to MoD SMF). Also, insurance indemnities were paid to MoD SMF for accidents on the Zenit carrier with Globalstar SC –

US$ 5.01 million and for the accident on the Cyclone carrier with Strela-3 SC – US$ 2.5 million.

It should be pointed out that the insurance of civil responsibility to third parties is accomplished on the basis of the international Convention of 1972, which regulates responsibility of the launching state for damage inflicted through space work to a third party based on the insurance sum of US$ 100 million to US$ 500 million, depending on a type of launch facility used. This type of insurance protects the Insurant against risks of arising responsibility to third parties for bodily and proprietary damage inflicted during space work.

In addition to the coverage of losses on the territory of states that are not the launching parties, also subject to insurance are the losses sustained on the territory of Russia and Kazakhstan as well as in the air space of those countries.

In 1999, the Megaruss insurance company developed for the first time and realized a new type of insurance product, which is insurance of civil responsibility to third parties in areas where jettisoned components of space rocketry fall in the case of an abortive launch.

The specific feature of launching space rockets from the Russian spaceports lies in the fact that the areas where jettisoned components fall are located on the territories of Russia and Kazakhstan, which makes their insurance an urgent necessity whereas the fall areas of jettisoned components of craft launched from foreign spaceports are located on the aquatory of the world's ocean.

The indemnity in this type of insurance is currently established at the level of US$ 50 million. In 1999, for the first time the Megaruss insurance group made payments under this type of insurance scheme for damage inflicted by the fall of parts of the Zenit carrier as it failed over the fall areas (to the Republics of Khakassia and Altay). The payments amounted to 4,064 million Russian rubles.

Participants of the market of space risk insurance are rather cautious about assuming space related liabilities. Such an attitude can be attributed both to subjective factors (consideration of the negative experience supplied by the West, an individual attitude to risks, etc.) and to other circumstances, uch as unstable development of space industry, deteriorating economic performance of provider of space products and services, the traditionally closed nature of work and difficulties in accessing authentic information about risks in space business. This narrows down the market of space insurance. Overall, however, the Russian market of space insurance is rapidly developing.

The foreign market of space insurance is a unique part of insurance business by virtue of its specific nature and due to the technocratic approach to the subject matter of insurance.

Its state is primarily characterized by the level of insurance rates, the size of premium collected, the amount of indemnities paid, losses sustained and the market capacity for one type of risk.

To better understand the state of the foreign market of space risk insurance see below the analysis of its changes between 1984 and 2001.

**Insurance rates.** The changing dynamics of insurance rates is shown in Fig. 79.

The analysis of the diagram (Fig. 81) suggests that initially the diversity in rates had been great, but starting in 1997 the rates began to slide down gradually. This is largely due to the growing market capacity which reached US$ 1.32 billion in 1999.

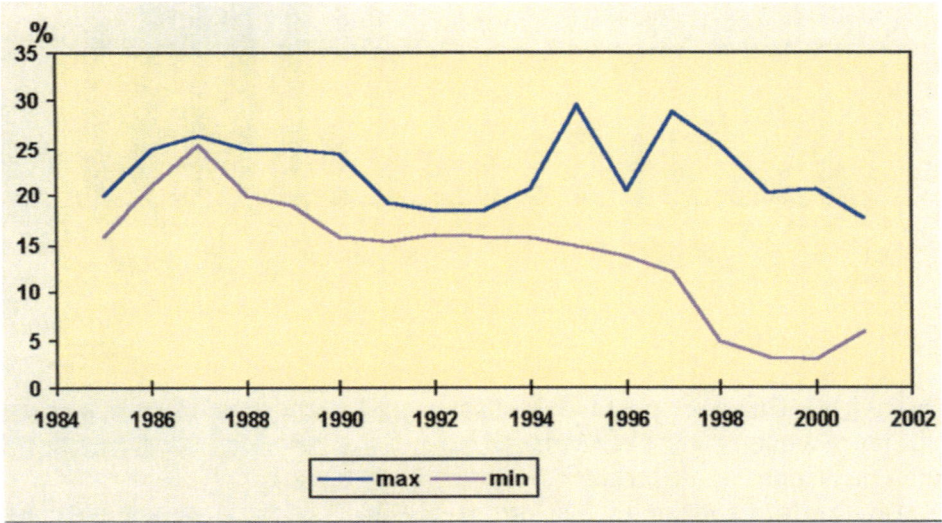

Fig. 81. The changing dynamics of insurance rates

**Market capacity.** During the period between 1984 and 1986, the market capacity for one risk was below US$ 100 million. Over the 10 years between 1986 and 1996 the capacity increased to US$ 690 million and soared up during 1997–1999 to reach US$ 1.32 billion in 1999. Such a growth of the market capacity is explained by the fact that despite the hard times in the mid 1980s when the insurance companies had to pay indemnities exceeding the total of insurance contributions, there remained on the markets some companies enabled to considerably increase the amount of insurance and improve their operability in conditions of plunted competition

However, compared to 2000 the market capacity for one risk decreased in 2001 by US$ 52.9 million for risks in carrier launches, by US$ 45.93 million for risks in flight operations of SC and by US$ 150 million for risks of responsibility to third parties because some companies, like ASG (Australia)

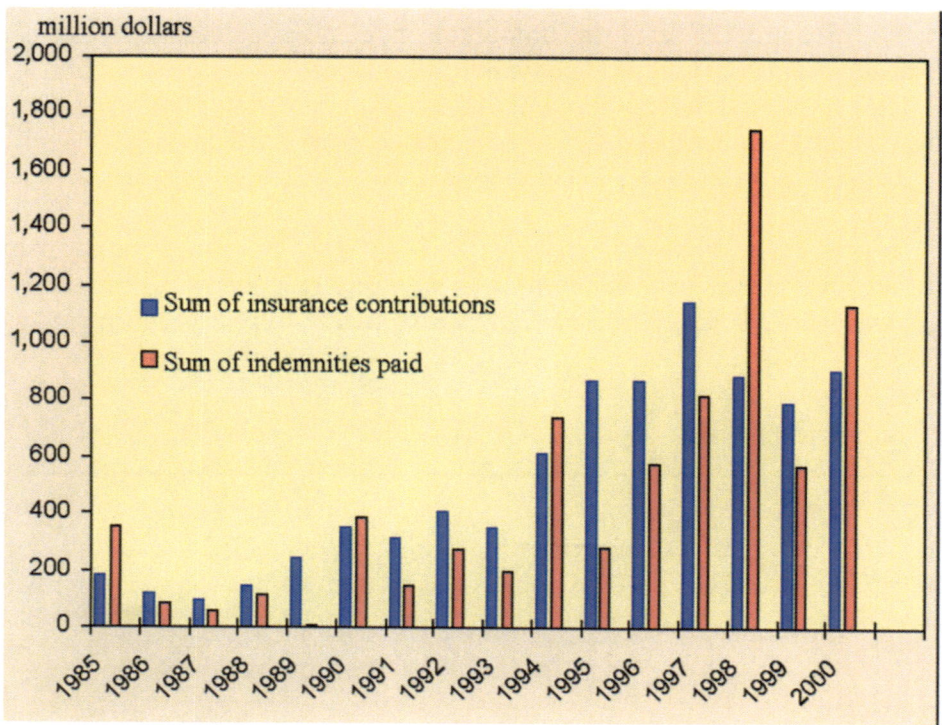

**Fig. 82.** The dynamics of insurance contributions and payments

and ACE Insurance Co Ltd (Bermudas), quitted the market of space risk insurance while others, like Marham Consortium, Hiscox and SATEC reduced their deductions in one risk.

Few are the companies that are deeply involved on a permanent basis in insurance of space risks and that have experts in this area who are capable to distinctly formulate the terms of insurance. On London's insurance market those are Lloyd's syndicates, in the US – INTEC and USAIG group, in France – LA REUNION SPATIALE, SCOR and AGF, in Germany – Deutsche Luftpul and Munich Re, in Italy – AGF, in Sweden – Skandia, in Norway – UNI-STOREBRAND, in Australia – SPACE INSURANCE group. The shares of other companies, alliances, pools and associations that follow the leaders on the space insurance market are very small indeed. It's worth noting that after the cases of large sum insurance the market capacity decreases while the rates rise.

Brokerage offices play an important role on the market. The biggest of them have experience and experts capable of putting through a good deal of organizational work aimed at managing the risks, distributing them among insurers and giving technical support to insurance programs. Standing out among them are such brokerage offices as Krouly Worren, WILLICE, Marsh S.A., Heath Lambert Group.

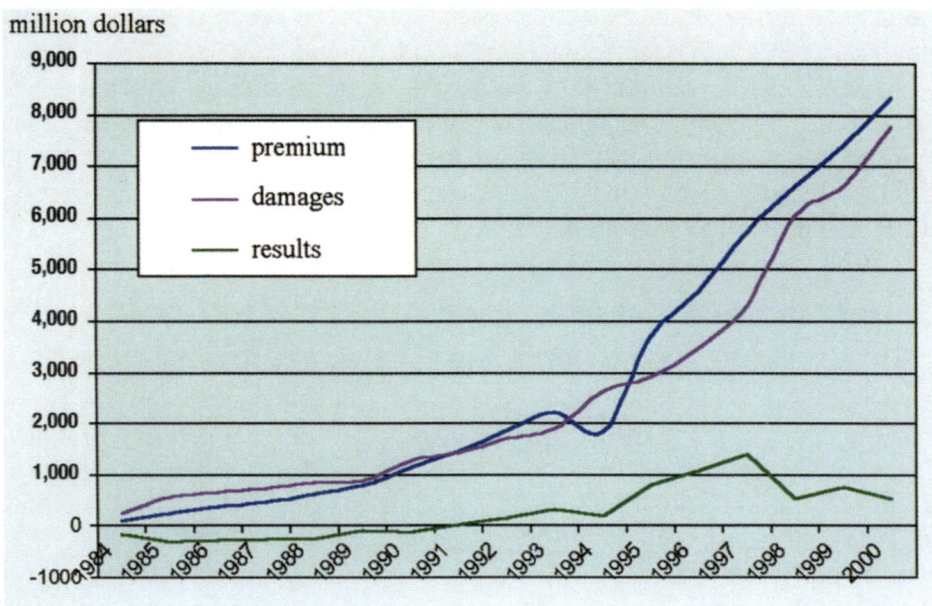

Fig. 83. The financial results of activity on the insurance market

**Premium collection and repayment of damages.** The dynamics of insurance contributions and payments made by foreign insurance companies over the period under consideration are shown in Fig. 82.

During the period between 1984 and January 1, 2001, the space insurance companies received US$ 8.33 billion worth of insurance contributions, while their expenses to repay the sustained losses amounted to US$ 7.77 billion.

The diagram (Fig. 83) shows that the insurance contributions tend to grow whereas the payment of insurance coverage is fortuitous. Nonetheless, the financial balance between the insurance contributions and indemnity payments with progressive total over the last six years has a negative balance. Fig. 83 shows as a chart the financial results of activity on the insurance market with progressive total.

The current situation on the world's insurance market has deteriorated abruptly. Because of heavy losses in all types of insurance the insurers have to revise the terms of policies in operation and to pay special attention to the quality of risks. Striving to reduce the unprofitableness over the entire spectrum of risks, they raise the rates even under such insurance agreements that have been previously loss-free. This, in the opinion of the insured, is illogical and unfair.

That is why the tendency for rates to rise that became evident in 1999 is not an exceptional feature of the foreign market of space risk insurance, but a mere reflection of the general situation that has emerged by now on the insurance market as a whole, especially in the wake of the tragic events that occurred in the US on September 11, 2001.

To keep up the operability of the world's market of space risk insurance, the sum of contributions collected must exceed the sum of indemnities paid, which is why the insurers will have to raise the insurance rates by 30 to 50%. Because of the cyclic changes that are characteristic of all markets this tendency may p=ersevere in the coming 3 to 4 years.

## 1.7 Space and national security

All the development of the Russian and foreign space exploration is inseparable from the use of space hardware for tackling defense tasks. There were objective reasons to do so. The launch facilities for delivery of space objects used to be rebuilt from combat rockets at defense facilities to orders of the military. Naturally enough, the military thought of the satellites primarily in terms of military application. As far back as January 30, 1956, even before the launch of the Earth's first artificial satellite, a decree of the Soviet Government entrusted the MoD with the task of investigating the prospects of military use of space.

During the years of confrontation between the capitalist and communist blocs known as "the cold war" the USSR had been striving to create a reliable nuclear shield based on combat intercontinental ballistic missiles. The effective use of rocketry, i.e. the precise delivery of a warhead to the desired target at the desired time implies prior detection and constant surveillance of targets over large areas of potential enemies, exact knowledge of their whereabouts and accurate "fixation" of own missiles, uninterrupted communication between leaders of the country and its armed forces, timely conveyance of orders for use of the nuclear weapon.

That is why the first defense related tasks that began to be performed using space hardware were photographic and electronic surveillance, support functions in communications, navigation, geodesic survey. This, however, was preceded by launches of experimental SC for optimization of space rocket systems and onboard equipment.

The main jobs done in various phases of military use of space are shown in Table 18.

In 1961, a new phase in exploration of space for military purposes began. The first spacecraft with an assigned defense mission was launched. That was a photographic surveillance SC, the Zenit-2. Over two years, more than ten such SC had been launched after which the first space complex went into military service.

That is why the first defense related tasks that began to be performed using space hardware were photographic and electronic surveillance, support functions in communications, navigation, geodesic survey. This, however, was

**Table 18.** Phases in military use of space

| Phases in use of space | Experimental research | Military application. Separate tasks fulfillment | Information support of services' operations. Communications support and troops handling | Equipping space as a new area of armed struggle |
|---|---|---|---|---|
| Changing the scope of missions and the fighting strength of orbital constellations of spacecraft | Experimental launches of research SC  Constellation comprised of 1...2 SC | Launches of single SC  Constellation comprised of 5...10 SC | Permanently operating space systems  Constellation comprised of 100...120 SC | New generation space systems integrated with weapon systems used in armed services  Constellation comprised of 150...200 SC |
| | **Tasks to fulfill:**  – optimization rocket systems and onboard SC equipment | **Tasks to fulfill:**  – individual tasks of photographic and electronic surveillance;  – individual tasks of navigational, communication and geodesic, support and adjustment | **Tasks to fulfill:**  – missile assault warning;  – photographic, optic-electronic, electronic and radar surveillance;  – communications and tactical control; naviga-tion, geodesic, meteor-ological survey and cartography support;  – adjustment;  – space information supply for strategic and tacticalcontrol units | **Tasks to fulfill:**  – missile assault warning;  – tactital, global, all-weather surveillance; target designation;  – communications sup-port, tactical control;  – navgation, topographic and geodesic, meteorological and cartographic support;  – adjustment;  – monitoring arms reduction agreements;  – use of space hardware in control circuits of weapon systems used in arms and services;  – space information supply for strategic, tactical and operative (on battlefield) control units in and out of pace;  – creating a network of small SC;  – creating weapons and strike means or waging battles in and dealing strikes out of space |
| | 1959 | 1961 | 1976 | 1991–2006 |

preceded by launches of experimental SC for optimization of space rocket systems and onboard equipment.

The period between 1966 and 1976 was the time when more advanced types of space hardware were test flown and introduced into service. Those include the Zenit photographic survey spacecraft; space systems for electronic

(Tselina, US-P) and radar-aided surveillance (US-A); adjustment and regulation SC (DS type), geodesic support craft (Sfera); space communications systems (Molniya, Strela); systems for meteorological survey ((Meteor) and navigation (Tsikada, Parus, and others). At the same time work began on building the space-based missile assault warning system. In 1972-1976 four experimental satellites were launched (US-K type).

In answer to the building of ABM systems and space surveillance facilities, the USSR introduced into service an enemy SC interceptor system based on IS spacecraft.

The creation and operation of space hardware abroad (in the US) began during the same period and proceeded along similar lines.

So the first experimental reconnaissance satellite, the Discoverer-1, was launched on February 28, 1959. Spacecraft of this series were used to refine the ways and means of reconnaissance conducted from space. In the 1960s, the Samos series of SC, the Ferret electronic reconnaissance SC, the Score and Syncom communications SC, and the Tiros meteorological SC began to be used for conducting imaging reconnaissance. In their development, the Samos series of SC had come through three generations (Samos-2D1; Samos-P; and Samos-M).

Special significance was attached to the space-based missile assault warning systems (initially Midas and later IMEWS) and the Vela SC based system for detecting ground nuclear explosions at high (110,000 km) circular orbits.          \

During the same period, a communications operating system was deployed at a geostationary orbit and other countries came up with indigenous spacecraft (the UK with Skynet-1A, Canada with ISIS-1).

In spite of the many space systems put in operation, the total number of the orbital constellation remained small because of the short service life of low-altitude SC in orbits.

Farther on, during transition to the new generation space systems having an essentially longer service life, more advanced onboard equipment and improved data delivery capability, there was a leap in use of space hardware for military purposes and for safeguarding national security.

Permanently operating orbital constellations of space systems of various applications were deployed for information support of Russia's armed forces. The amount of missions assigned to space systems grew considerably. Their main objects are shown in Table 19.

The use of those space systems became universally recognized and standard practice in planning both strategic operations of armed forces and tactical operations of army and naval groups.

The emergence of military space systems whose orbital elements can actively engage (defeat, suppress) potential targets, the necessity to prepare portions of space (as well as areas on the ground where space infrastructure assets are located) as a theater of hostilities – all these factors isolated space work as an independent activity which increasingly grows in importance as a means of attaining goals in armed struggle. This is shown in Table 20. The contribution of space forces and facilities to fulfillment of Russia's armed forces' tasks in different military and political periods is shown in Table 21.

**Table 19.** Tasks fulfilled using space hardware

| Space hardware application | Main periods of political – military situation | | |
|---|---|---|---|
| | Peace time | Time of threat | Periods prior and subsequent to war |
| | The aims of using space hardware | | |
| | Army and navy routine service support | Detection of enemy's preparation for attack and provision of information for military planning | Information support of Army and Navy enabling them to draw up plans of operations and use of weapons |
| **Missile assault warning** | Detection of ballistic missiles' launches and missile assault warning | | |
| | Space operations reconnaissance and information release for target designation. Detection (finer location) and determination of armed forces, their composition, characteristics and whereabouts. Detection of changes in composition, whereabouts and combat readiness of forces, monitoring the areas of local wars and major exercises | | |
| | Monitoring the observation of reaties and agreements for force reductions | | |
| | | Monitoring areas with concentrations of electronic equipment and launch installations, their finer location, issue of data to reconnaissance and weapon control posts and electronic warfare control centers. Monitoring the battle, defense and support groups of marine forces and issue of data for target designation by Navy's weapon systems. Detection of activity in engineering preparation of operations theater, in maneuvers of battle groups and reserves. | |
| | | | Monitoring the results of using the weapons |
| | Strategic nuclear forces combat control Insuring communications and data transmission in control systems Russia's of armed forces Relay of intelligence data from space-based reconnaissance equipment. | | |
| | Providing data for navigating army's mobile assets and navy's forces | | |

| Space hardware application | Main periods of political – military situation | | |
|---|---|---|---|
| | **Peace time** | **Time of threat** | **Periods prior and subsequent to war** |
| | The aims of using space hardware | | |
| | Army and navy routine service support | Detection of enemy's preparation for attack and provision of information for military planning | Information support of Army and Navy enabling them to draw up plans of operations and use of weapons |
| **Meteorology** | Gathering data for providing arms and troops control agencies with meteorologicalinformation, issue of weather forecasts and climate related info | | |
| **Cartography** | Provision of data for making and updating topographic and digital maps, urban plans and photographic documents | | Obtaining more accurate information about topographic and geodesic features of operations areas |
| **Geodesy** | Provision of data for refinement of the geodesic constants, parameters of Earth's ellipsoid and navigation field | | |
| **Special missions support** | Adjustment and calibration of anti-air and anti-space weapon systems | | |
| **Deterrents** | | | Disrupting the operation of major space support systems (ships) |

In the 1970s the US built and introduced into service more advanced LASP series of reconnaissance SC, later replaced by KH, with viewing and detailed observation capability. For the first time a satellite, the Rhyolite, featuring a large-size antenna, was sent into a GEO orbit. Its mission was to intercept radio messages issued from radio communications facilities in Europe. Rapidly developing now are space communications technology, navigation, meteorological survey services. Missile assault warning system, too, is steadily improving.

**Table 20.** Space in achieving the goals of armed struggle

| PERIOD (years) | TENDENCY | TASKS TO FULFIL |
|---|---|---|
| **1990 – 2000** | Leading role of ground and aerial spheres with a certain significance of space | – Detection of potential targets for nuclear strikes<br>– Strategic warning<br>– Combat control over nuclear forces |
| **1970 – 1990** | Increasing role of space with strategic significance of superiority in the air and on land | – Information support of armed forces operations from space<br>– Accurate fixation of mobile targets<br>– Establishment of systems and complexes for information supply and strike delivery |

| PERIOD (years) | TENDENCY | TASKS TO FULFIL |
|---|---|---|
| 2000 – 2010 | Predominant role of space in achieving the goals of armed struggle | – Information support of armed forces' operations from space<br>– Active involvement in armed forces' operations and support given to them from space<br>– Pinpoint engagement of SC, ballistic missiles, combat modules, military economic assets and army and naval battle groups performed in and from space.<br>Global non-destructive strikes against individual regions and countries. |

**Table 21.** Contributions of space hardware and forces to fulfillment of Russia's armed forces' tasks in various political-military situations

| Space systems' tasks | Results of using space hardware and forces (actual performance) | | | |
|---|---|---|---|---|
| | **Peace time** | **Local conflicts** | **Conventional large scale war** | **Nuclear war** |
| **Missile assault warning** | Enhancing the reliability of monitoring the territories nuclear powers' containing missile threats (in prospect, all countries with a missile capability) andareas in the ocean. | | | Enhancing the dependability of detecting BM Launches |
| **Strategic reconnaissance** | Enhancing the reliability of monitoring the compliance with treaties, of detecting the signs of preparation for war, signs of preparing the hostilities theater and monitoring the composition, location and preparedness of troops. | | | |
| | | Enhancing the reliability of monitoring the hostile battle groups, determination of enemy's intentions and directions of the main blow. | | |
| **Tactical reconnaissance** | Enhancing the dependability of monitoring state of enemy's strategic facilities (weapons) | Enhancing the dependability of assessing the results of using the weapons | | |
| **Communications and tactical control** | Enhancing the reliability of delivering reconnaissance data from correspondents and reconnaissance SC to troops, improving the resistance to jamming, and throughput capacity of the troop control system. | | | |
| **Navigation** | Enhancing flight and navigation safety | Enhancing the effectiveness of destroying enemy assets, battle groups and weapon systems<br>Enhancing the operabilityof aircraft in the air, ships and subs at sea, groups of troops and mobile forces in all areas, including unprepared ones. | | |

| Space systems' tasks | Results of using space hardware and forces (actual performance) | | | |
|---|---|---|---|---|
| | Peace time | Local conflicts | Conventional large scale war | Nuclear war |
| Geographic and geodesic support | Enhancing the troops provision with geodesic information about enemy territory for use of strategic nuclear forces | | Refining the data concerning topographic and geodesic features of the hostilities theater. | |
| Meteorological support | Improving the provision of troops and strategic weapon systems with support meteorological data. | | | |

According to the Russian and foreign experts, the use of space information services will improve the fighting capability of armed forces by a factor of 1.5 to 2.

It was the USA that paved the way to placing military hardware in space, primarily for attacking ground-based targets, even before the first artificial earth satellite had been sent into orbit.

The assigned task was to "neutralize" the Soviet nuclear capability by striking a preventive nuclear blow from space. Illogically, that approach ignored ground-based anti-space facilities for destroying similar Soviet assault systems should they be created. However, the experience and expertise that were available between the 1950s and 1970s, combined with obviously insufficient economic capabilities, precluded the USA and its allies from militarizing space.

Also acting as a deterrent were a number of international treaties, e.g. the Treaty of 1967 that banned placing weapons of mass destruction in space, and the Anti-Ballistic Missile (ABM) Treaty of 1972. Nonetheless, the issues of further militarization of space had always been the cornerstone in the politics of leading powers, the US in the first place. In view of he increasing importance of space for the national interests and security, the leadership in the West treated those issues with very special care.

So US President L. Johnson as far back as 1964 said: "Britons ruled over seas and controlled the world. We have been ruling over the air and leading the free world ever since we established that rule. Now this position will go to the one who will rule in space." This popular phrase, repeatedly reworded and hence ascribed to many other American politicians, became the guideline to all intents and purposes in the space politics of the world's leading powers.

**Table 22.** Advanced weapon systems

| WEAPON SYSTEMS BASED ON DIRECTED ENERGY SOURCES | | | MEANS OF ACTIVE IMPACT ON GEO-PHYSICAL PROCESSES IN SPACE INITIATED GEOPHYSICAL PROCESSES | | |
|---|---|---|---|---|---|
| LASER BEAM NUCLEAR | | | – Absorption and dispersion of electromagnetic waves in artificial medium<br>– Initiation of EM radiation of arificially modified medium | | Flows of plasma, of inert chemically active and radio active particles |
| **TASKS TO SOLVE** | | | | | |
| • Suppression of optoelectronic SC<br>• Illumination of SC targets for homing missile weapon<br>• Communication with submerged submarines<br>• Selection of combat blocks in composite ballistic target<br>• Engagement of ballistic and orbital targets | • Selection of combat block in composite ballistic target<br>• Inspection of SC for nuclear fissionable materials<br>• Engagement of ballistic and orbital targets | • Suppression of electronic facilities<br>• Engagement of ballistic and orbital targets<br>• Engagement of ballistic and orbital targets<br>• Destruction of asteroids | Protection and camouflaging of national SC against enemy anti-space weapons | Electronic suppression of hostile electronic activity | Engagement of target SC and other space rocketry |
| **ENGAGEMENT (SUPPRESSION) TECHNIQUE** | | | | | |
| • Clutter of optic-electronic equipment photo receiver<br>• Engagement and malfunction of electronic equipment<br>• Changes in targest´s radiation and reflection<br>• Distruction of structural materials througt mechanical impact and thermal exposure | • Malfunction in major integral circuit<br>• Engagement of electronic and optic electronic equipment by radiation<br>• Changes in radiation and chemical properties of mat-erials and in electronic equipment performance<br>• Destruction of rocket structural materials by thermal chemical exposure | • Creation of protracted back-ground noise<br>• Clutter of photo receiver from plasma formations on target<br>• Engagement of electronic and optic electronic equipment by radiation and ionization<br>• Destruction of electronic and optoelectronic equipment by powerful EM pulses<br>• Target destruction by mechanical impact | • Creation of background clutter for detection and guiding devices<br>• Reduction of optical and radar signature of space objects<br>• Changes of their optical and radar performance | • Disruption in data transfer channel<br>• Creation of clutter in optical and radar operating range<br>• Destruction of radio signal structure | • SC onboard equipment surface erosion<br>• Changes in performance of onboard equipment under the influence of artificial ionizing radiation<br>• Formation of methallic craters, cracks, chips on SC equipment exterior surface |

A new lap in the military space race began in the 1980s after President Reagan made his notorious speech about "the star wars" followed by development in the US of the strategic defense initiative which served as a basis of the US national politics in space for nearly a decade. The alleged object of the program was the protection of the US territory, and that of its allies, against nuclear strikes .

The principal difference of the space segment of this program lay in creation of space hardware based on radically new technologies, in development of the far-ahead engineering decisions in various fields of microelectronics, engine manufacture, structural materials, guiding and control systems, and in other developments that could be used subsequently for building not only space weaponry but also other types of weapons.

In considering the variants of implementing the Strategic Defense Initiative (SDI) many projects were proposed with various types of weapons installed on strike weapon systems (Table 22).

**Fig. 84.**
Orbital constellation of Brillant Pebbles Spacecraft in US ABM defense system

Those are such projects as "The High Frontiers" with the use of 432 spacecraft that evenly cover the entire globe and the project for building a space-based ABM system using thousands of miniature interceptors known as Brillant Pebbles (Fig. 84). Those facilities were supposed to be fitted with missile weapons. In addition, some "exotic" projects had been mapped out that proposed to engage ballistic missiles, forebodies and combat blocks by means of space and ground-based facilities using laser and beam weapons as well as weapons that neutralize targets by electromagnetic radiation, the so-called electromagnetic pulse (EMP) weapon systems (Fig. 85).

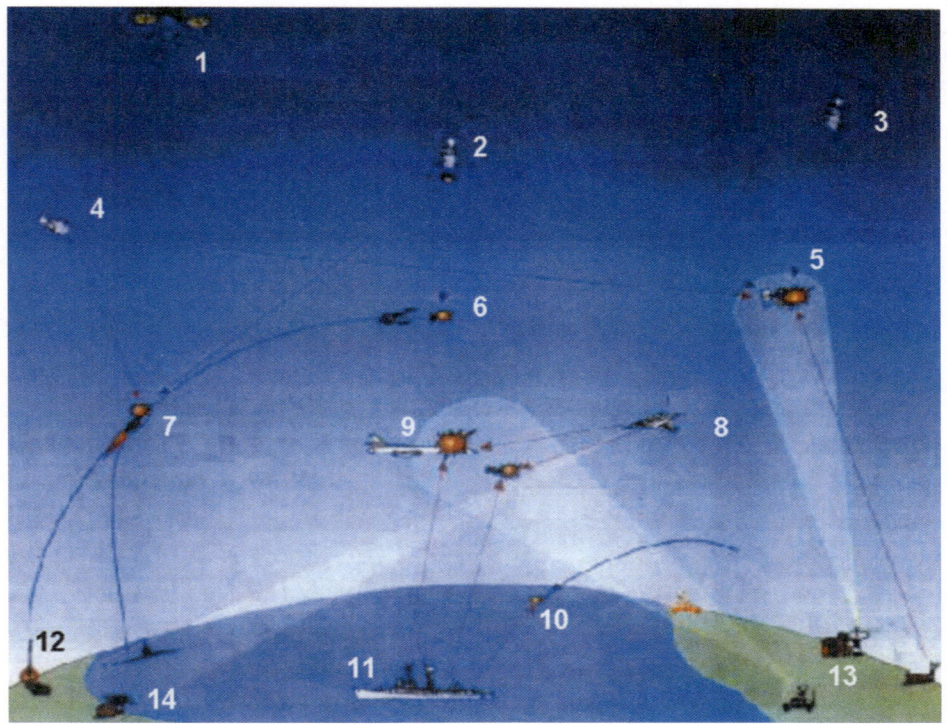

**Fig. 85.** Space, air and ground based weapons operating on new physical principles

    1 – high orbital SC target
    2 – anti-missile and anti-space SC with kinetic energy nuclear weapon
    3 – SC with anti-space laser weapon systems
    4 – SC with anti-missile laser weapon systems
    5 – low orbital SC target
    6 – composite ballistic target
    7 – ICBM
    8 – aircraft with laser weapon system
    9 – aircrafts
    10 – cruise missiles
    11 – ship with laser weapon system
    12 – launchers
    13 – ground complex with laser weapon system and microwave weapon system
    14 – air defense system radar

In addition to various projects, the USA made practical attempts to create space weapons. In the mid 1980s, for instance, an anti-satellite system, ASAT, underwent flight trials. Derived from the MXB interceptor and featuring an IR guidance, it was launched from the F-15 fighter plane.

Analysis of various projects for implementing the SDI program, which analysis was carried out by the Russian and foreign specialists, showed that their implementation, contrary to the declared objects, would disturb the established military and strategic balance of the world and that the weapons created under the program were essentially offensive in their nature.

A question often arises as to how the use of space hardware can affect the strategic stability and tip the military balance. The likely uses of space hardware for attaining the strategic stability in the world are shown in Fig. 86.

First of all, they make easily predictable the states' actions due to the monitoring of their military activities and thanks to the growing integration of Russia into the world's economic and scientific community.

No one doubts today that the space industry has a profound effect on the progress of technology and engineering in all spheres of economics and science. The hardware in orbit makes it possible not only to explore the outer space, air space, marine areas and the state of the Earth's surface, but also to use in economics the results of such exploration.

Communications and television, exploration of natural resources, navigation, topographic geodesy, meteorology, and environmental control are now activities carried on by virtually all industrialized nations. The experience of the world history proves that close economic and scientific collaboration decreases the likelihood of war between states and promotes peace.

Finally, the national security and defense capability directly depend on the operability and condition of the strategic warning facilities that warn about preparation for an aggression, about the beginning of nuclear missile strike and about the adequacy of provision with space forces.

The military strategic balance in the world and the deterrence from large scale conventional or nuclear wars are insured today primarily by a state's capability to strike an efficient counter blow at an aggressor. So far there is an approximate parity between Russia and the US in terms of strategic nuclear forces. Russia's nuclear capability, however, is above that of any other individual state.

Sure enough, to be able to deter a large scale conventional war it would be desirable to have the same ratio in conventional arms, which is not always possible for economic reasons.

It becomes then, instead of parity, a matter of maintaining Russia's Armed Forces at a level thwarting the foe's attempts to obtain superiority in several combat areas at a time: in the air, in space. at sea and on land. To make a conventional war impossible it is necessary, at least to prevent the enemy from winning superiority in any of those areas. Otherwise, defeat in war becomes inevitable. A glaring example is supplied by the victory over Iraq in the war of 1991. The multi-national forces won so quickly thanks to the indisputable superiority of their aviation in the air. A large contribution to the victory was also made by the US space reconnaissance, navigation and communications assets.

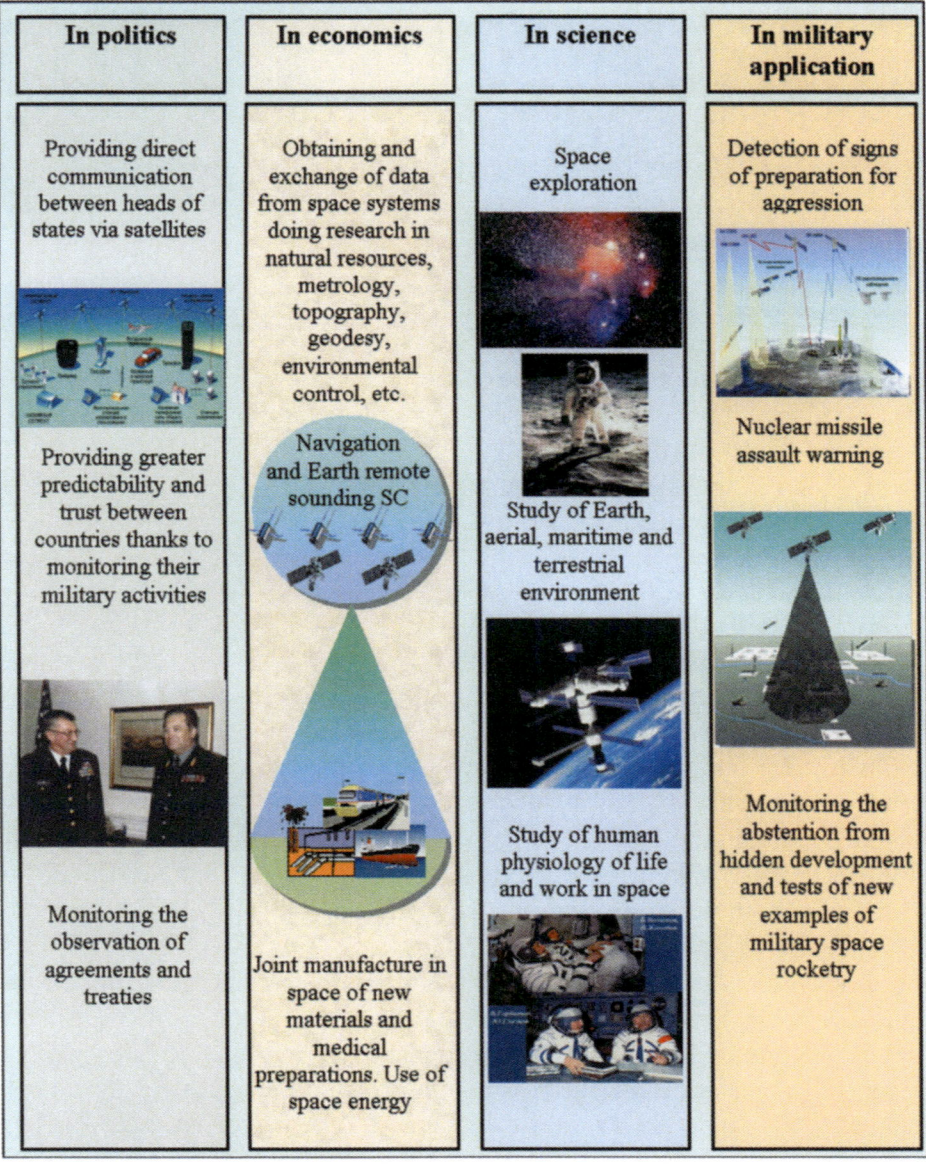

| In politics | In economics | In science | In military application |
|---|---|---|---|
| Providing direct communication between heads of states via satellites | Obtaining and exchange of data from space systems doing research in natural resources, metrology, topography, geodesy, environmental control, etc. | Space exploration | Detection of signs of preparation for aggression |
| Providing greater predictability and trust between countries thanks to monitoring their military activities | Navigation and Earth remote sounding SC | Study of Earth, aerial, maritime and terrestrial environment | Nuclear missile assault warning |
| Monitoring the observation of agreements and treaties | Joint manufacture in space of new materials and medical preparations. Use of space energy | Study of human physiology of life and work in space | Monitoring the abstention from hidden development and tests of new examples of military space rocketry |

**Fig. 86.** Uses of space hardware for safeguarding strategic stability in the world

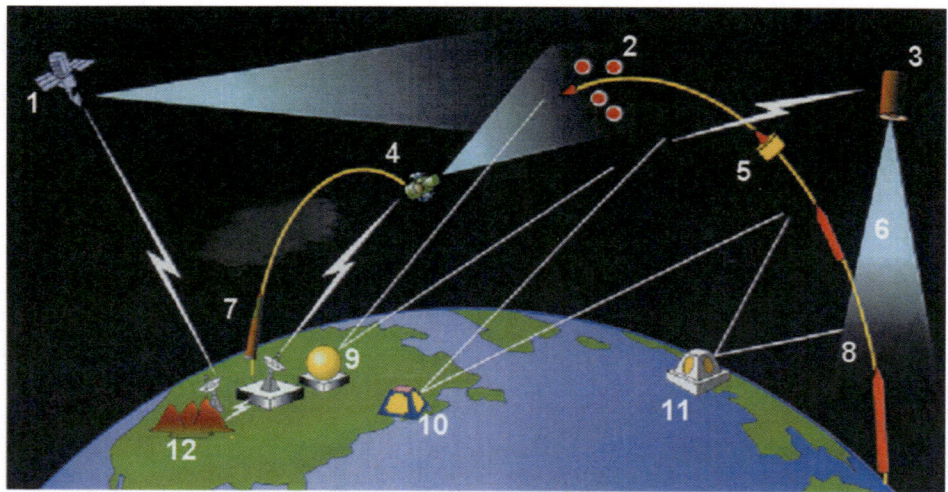

**Fig. 87**. Composition and structure of the US anti-missile defense system

1 – Detection, tracking and selection SC
2 – False targets and warhead
3 – Early missile assault warning SC;
4 – Kinetic interceptor;
5 – Warhead separation;
6 – Missile assault threat signal transmission;
7 – Interceptor missile;
8 – Active phase of missile flight;
9 – Interceptor missile control radar;
9 – Forward position radar;
11 – Early warning radar;
12 – Planning, tactical control, monitoring and communications center

Once space hardware becomes one of the armed forces' strategic components whose parity is critical for the preservation of the strategic military balance, maintaining it would be impossible should one of the parties build an ABM defense system with elements of space basing capable of warding off a counter (more so in case of a counter-head-on) nuclear missile attack or deploying an effective anti-satellite system (Fig. 87). Total superiority in space would create the required conditions for scoring a victory in any conflict or war.

For example, disabling the space reconnaissance systems leads to inability to control in real time the developments on the operations theater and to issue target designation to reconnaissance and strike complexes. This, in turn, disrupts the timely fulfillment of tasks aimed to destroy mobile nuclear weapons platforms (Table. 23).

Naturally enough, under such conditions the Soviet Union started work on opposing the SDI program in case of its implementation. Involved in such work was a massive cooperative effort of the MoD Scientific Research Establishment and manufacturing companies. The main purpose of the endeavor was not to duplicate the known projects or to carry them out sooner than others, but to find new militarily more efficient and economically cheaper means of an adequate opposition.

Table 23. Parity in space as an element of strategic military balance

| STRATEGIC MILITARY BALANCE | | |
|---|---|---|
| Means of maintaining the parity | Military parity in space – | When parity is disturbed |
| 1. Concluding international treaties that ban tests of all types of weapons in and out of space. 2. Continuation of search for creation of low-cost effective space hardware for anti-missile and anti-satellite defense 3. Preferential development of reconnaissance types, including in space, in the air and at sea. | is such a state of space forces and assets at sea in which none of the opposing parties can obtain superiority in any of combat areas (in space, in the air, at sea, on land) due to the use of space hardware. | 1. Creation of an effective ABM defense system with elements of space basing. 2. Creation of an ABM defense system assuring superiority in space. |

Thus appeared anti-ABM defense projects based on the use of passive and active means of protection installed directly on ICBM, based on creation of "launch windows" in course of striking a nuclear missile blow due to destruction of space-based ABM constellation, on creation of information processing hardware for assessing the situation in space and issue of target designation data to weapon systems, on creation of weapons striking from space at assets on the territory of the US and at sea targets. The implementation of the anti-ABM defence program completely "devalued" the claimed reliability of "the space umbrella" over the US territory.

All this contributed to the strategic stability in the world and checked the arms race in space.

That period was characterized by active work on various space weapon systems, by models of operations in and out of space that the military and scientists built in their minds, on computers and on maps. As a result, by the early 1990s the long established thesis that space is a new field of armed struggle came to be replaced by practical work on equipping the near earth space as a possible theater of military operations.

Normally, preparing the theater of war implies the construction of fortifications, building of rail and motor roads, of airdromes networks, equipping positions, depots and storage facilities with whatever is required for war, preparing the communications systems and control stations, carrying out metrological, topographic, geodesic and other work.

What does it mean to the space theater of war? First of all, it means the deployment in space of permanently operating systems of space reconnaissance, navigation, communications and tactical control, relay and other functions providing the use of space weapon systems. Further on, the building on the ground of the required elements of space infrastructure for launching spacecraft (for military use and information gathering), controlling them, receiving data from them, etc., that is the creation of infrastructure of full scale utilization and operation of space hardware.

Treating the issues of space as an area of armed struggle the Russian and foreign dedicated literature use different notions and definitions. The foreign literature prefers "the space theater of war" while the Russian literature abides by "the space theater of military operations" and, more recently, "the strategic space zone". One of the possible presentations of space as a strategic space zone is shown in Fig. 88.

The division of strategic space zone into operational zones is quite arbitrary and is characterized primarily by types of SC orbits used to fulfill tasks.

Thus by the mid 1990s, after space had been equipped for information services, radical changes took place in the use of space hardware for military purposes and for safeguarding national security.

Previously, the use of space hardware in local wars and armed conflicts was only sporadic (Vietnam, the Middle East, Afghanistan, the Falkland Islands and others) and relied on the availability of a satellite in orbit, its capability to pass quickly over the area under observation and on coordination between space communications equipment and navigation instruments. Now the situation has changed drastically.

The first experience of large scale utilization of space hardware for military purposes were the Persian Gulf crisis in 1991 in which the multinational forces resorted to space-based equipment in all phases of the operation.

The main missions assigned to control agencies in the conflict area included provision of reconnaissance and communication, assessment of losses sustained by the enemy, furnishing troops with navigational, topogeodesic and meteorological data.

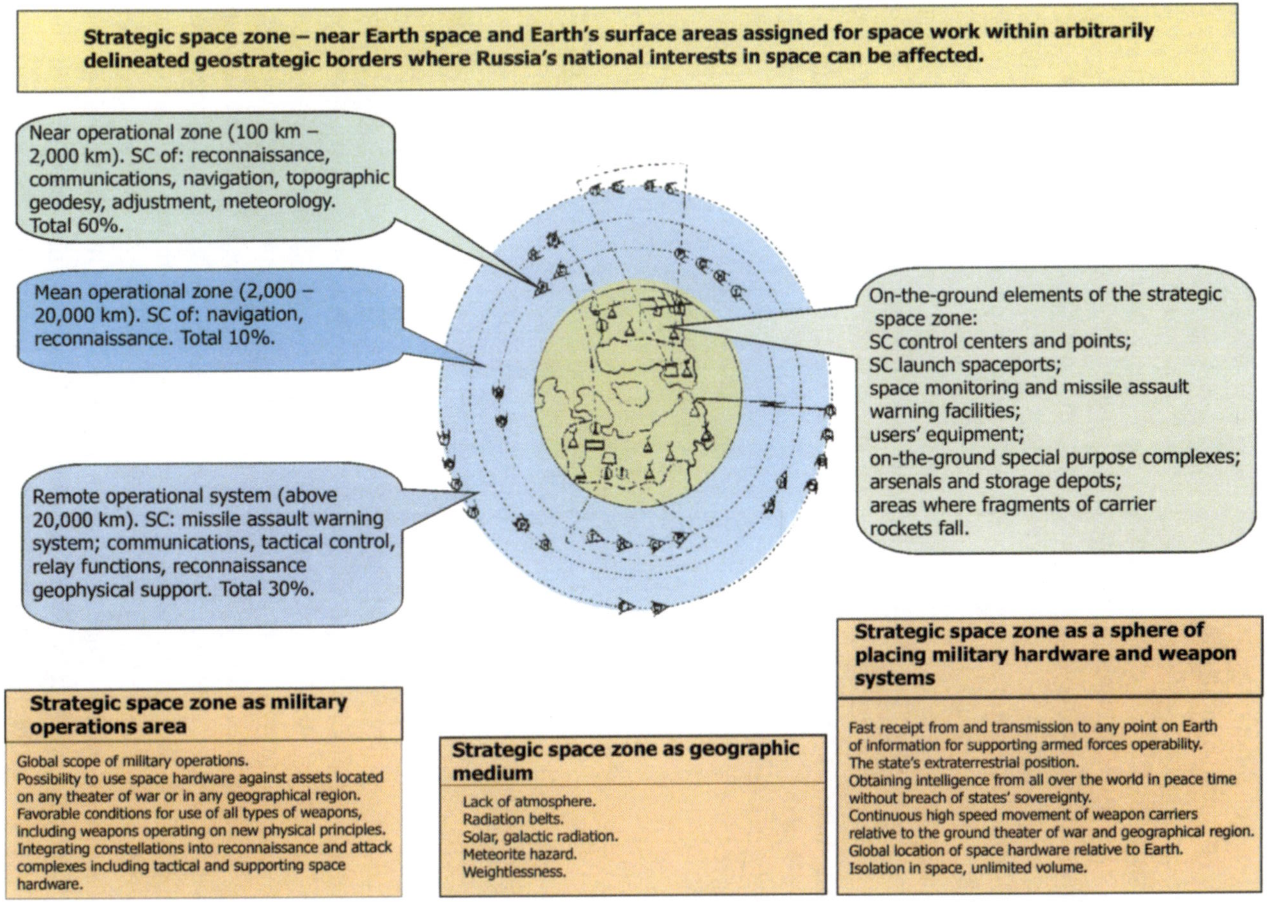

Strategic space zone – near Earth space and Earth's surface areas assigned for space work within arbitrarily delineated geostrategic borders where Russia's national interests in space can be affected.

On-the-ground elements of the strategic space zone:
SC control centers and points;
SC launch spaceports;
space monitoring and missile assault warning facilities;
users' equipment;
on-the-ground special purpose complexes; arsenals and storage depots;
areas where fragments of carrier rockets fall.

Near operational zone (100 km – 2,000 km). SC of: reconnaissance, communications, navigation, topographic geodesy, adjustment, meteorology. Total 60%.

Mean operational zone (2,000 – 20,000 km). SC of: navigation, reconnaissance. Total 10%.

Remote operational system (above 20,000 km). SC: missile assault warning system; communications, tactical control, relay functions, reconnaissance geophysical support. Total 30%.

**Strategic space zone as a sphere of placing military hardware and weapon systems**

Fast receipt from and transmission to any point on Earth of information for supporting armed forces operability.
The state's extraterrestrial position.
Obtaining intelligence from all over the world in peace time without breach of states' sovereignty.
Continuous high speed movement of weapon carriers relative to the ground theater of war and geographical region.
Global location of space hardware relative to Earth.
Isolation in space, unlimited volume.

**Strategic space zone as geographic medium**

Lack of atmosphere.
Radiation belts.
Solar, galactic radiation.
Meteorite hazard.
Weightlessness.

**Strategic space zone as military operations area**

Global scope of military operations.
Possibility to use space hardware against assets located on any theater of war or in any geographical region.
Favorable conditions for use of all types of weapons, including weapons operating on new physical principles.
Integrating constellations into reconnaissance and attack complexes including tactical and supporting space hardware.

Fig. 88. Space as strategic zone

The US space reconnaissance equipment has played the major role here. By the beginning of hostilities the US space reconnaissance orbital constellation comprised 29 spacecraft, of which 4 were imaging reconnaissance craft (optical and radar), the rest performing the functions of signals intelligence and electronic surveillance.

The performance of reconnaissance hardware provided assured detection of practically all assets of the ground forces, the bases of air forces, missile forces and their subunits as well as military industry facilities.

In course of operations the space command refined new combat techniques of using space reconnaissance equipment. A capability was checked of using the data supplied by the IMEWS (integrated missile early warning satellite) space-based ICBMs launch detection system in order to improve the operational effectiveness of the Patriot air defense missile systems. Those tasks were fulfilled by a constellation of spacecraft deployed in advance.

It was noted that the multi-national forces command used the space communications intensely even in tactical operations. The multi-national forces widely used the navigation field created by the Navstar space system. Its signals improved the accuracy of aircraft's approach to targets at night, corrected the flight trajectories of aerial and cruise missiles.

The meteorological reports prepared on the basis of information obtained from space, were used to compile and to correct, if necessary, aircraft flight planning tables.

On the whole, the military space hardware had such a strong impact on activities of the multi-national forces in the Gulf conflict that it urged the development of new techniques of using such forces.

According to experts, the war in the Gulf was "the first war in the space era" or "the first space war of our era".

Still more impressive was the  scale of using space information in Yugoslavia. The planning of missile and bomb strikes, the assessment of results obtained, topogeodesic and meteorological support of operation in all phases of its development were carried out using the data supplied by space equipment. Special attention was paid to space navigation system whose information provided for high precision weapon operability in all weathers and round-the-clock. The analysis of using space equipment in Yugoslavia and in previous conflicts allowed to finally conclude that the use of the so called space support groups created in various links of the control chain is necessary and highly effective. For instance, during the Yugoslav conflict a space systems utilization unit was formed by NATO high command in Europe in order to coordinate the activities of heterogeneous types of reconnaissance and optimize the information obtained. Around twenty tactical mobile groups were sent to the operational zone to supply space data to commanders of tactical aerial and maritime forces.

Russia's armed forces, too, have a positive experience in using space support groups in operational and tactical modes. The experience was acquired during operational training. The main tasks of those groups are the SC operability evaluation and preparation of proposals as to how to utilize the craft for obtaining the data as well as presentation of obtained information (reconnaissance, meteorology, navigation, communications) to commanders of various control links with recommendations for its use.

The results of trials allow to speak in clear terms about the component package of space support groups, their tasks, the principles of utilization, etc. Their mainstay, at least in the first phase, must be specialists in military space hardware. Depending on their composition and equipment, the space support groups can be assigned to control agencies at various levels, specifically, to military districts in peace time and to front-line headquarters and army groups (mobile forces, armies, corps) during war time with regard to the scale and nature of hostilities.

Being among the troops and having a clear-cut picture of orbital equipment capabilities, the support group officers can help the army command formulate applications for supplies of space hardware, organize personnel training with demonstration of capabilities of orbital hardware in a variety of scenarios and in various conditions.

However, the groups' principal jobs will be to process and preliminarily analyze the information coming from space information supply hardware as well as to deliver it to the users.

In war time the space support group (SSG) can be additionally charged with coordinating the operations of the tactical groups formed during a threat and pursuant to a special order within the army (division) segment and organizing their interoperability. The SSG will process and analyze the information obtained both directly from orbital hardware and the front-line SSG, evaluate and forecast the state of the navigation field in the troops operation areas, prepare their applications to the front-line SSG for issue of additional information and for provision of the personnel with required equipment and training in how to use it.

The front-line SSG will have not only to summarize the requirements for space information and formulate applications for SC launch, but also to control them. The peculiarity of tasks imposed on SSG necessitates the inclusion into them (Fig. 89) of units (groups) for coordinating the use of space forces, reconnaissance, communication, navigation and meteorological survey as well as integration of space information and logistical support services.

Under such circumstances, the coordination division will be responsible for interoperability with control agencies that oversee the use of space hardware, the timely preparation of proposals for using satellites to fulfill the troops' needs, e.g. for the required number and composition of SC, means of delivering the information etc.

The integration division is responsible for the comprehensive processing of incoming heterogeneous information and formation of its output flow for delivery to the troops (forces) control agencies. The tasks of other groups depend on the mission of orbital hardware which they will have to use. The general management, operational support and responsibility for completeness, adequacy and timeliness of information delivered rest, naturally, with the SSG command.

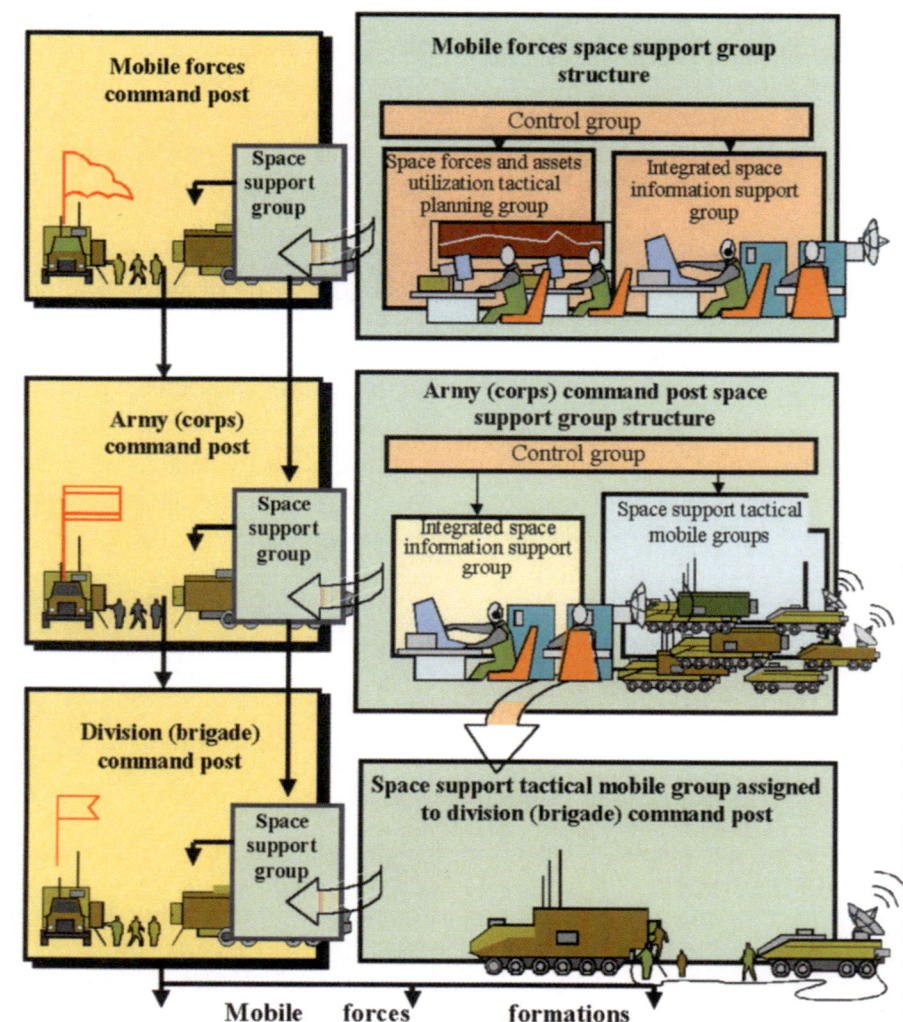

**Fig. 89.** Structural variant of the space support group in mobile forces command posts

For effective operation the space support groups will need dedicated mobile stations that receive information from satellites and issue it in the user-friendly format. A version of building one such station is shown in Fig. 90.

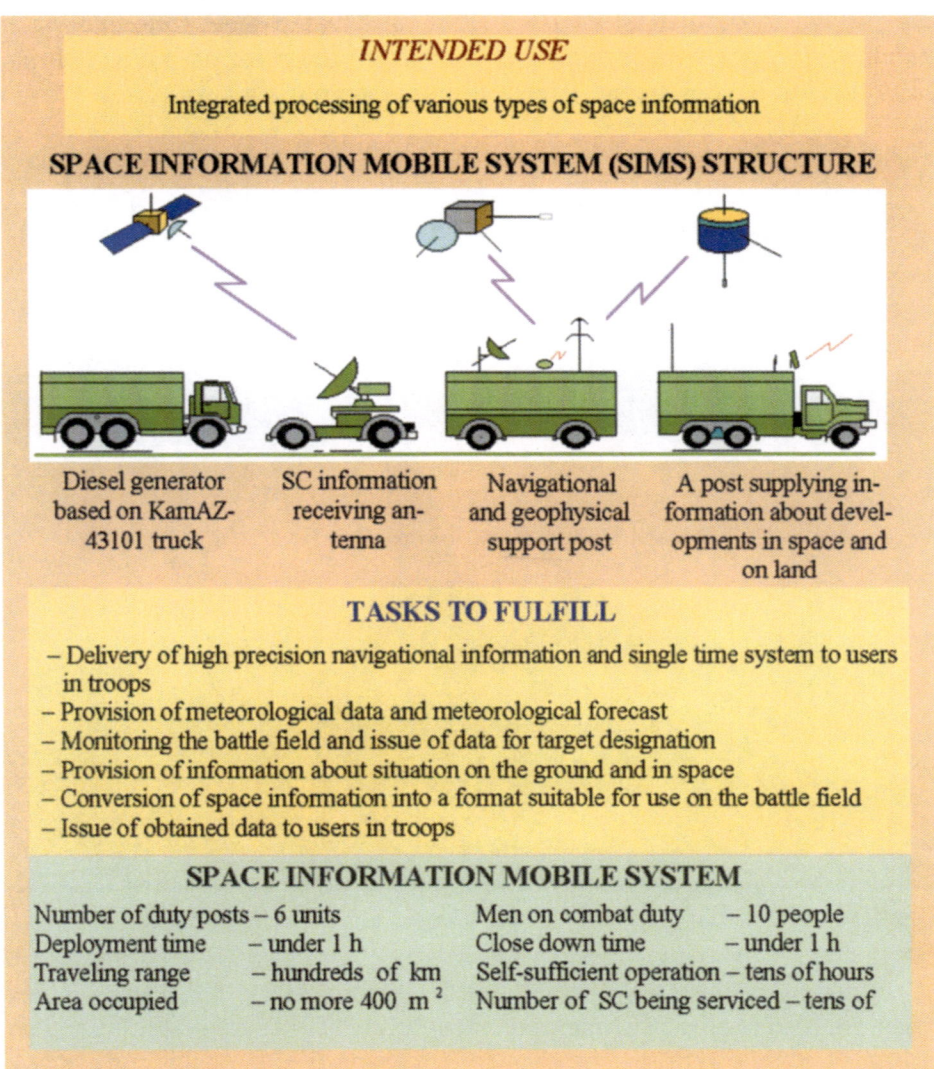

**INTENDED USE**

Integrated processing of various types of space information

## SPACE INFORMATION MOBILE SYSTEM (SIMS) STRUCTURE

| Diesel generator based on KamAZ-43101 truck | SC information receiving antenna | Navigational and geophysical support post | A post supplying information about developments in space and on land |
|---|---|---|---|

### TASKS TO FULFILL

– Delivery of high precision navigational information and single time system to users in troops
– Provision of meteorological data and meteorological forecast
– Monitoring the battle field and issue of data for target designation
– Provision of information about situation on the ground and in space
– Conversion of space information into a format suitable for use on the battle field
– Issue of obtained data to users in troops

### SPACE INFORMATION MOBILE SYSTEM

| | | | |
|---|---|---|---|
| Number of duty posts – 6 units | | Men on combat duty – 10 people | |
| Deployment time – under 1 h | | Close down time – under 1 h | |
| Traveling range – hundreds of km | | Self-sufficient operation – tens of hours | |
| Area occupied – no more 400 m$^2$ | | Number of SC being serviced – tens of | |

**Fig. 90.** Space information mobile system

This is a whole set of equipment with automated operator workstations. It can be placed on small all-terrain vehicles or helicopters. In addition to antennas and a network of personal computers its component package must include a check-and-correct station of space navigation systems' differential subsystem, a powerful server, automation and communication facilities and an independent power supply unit. This makes it possible to introduce in field conditions and in real time a space objects catalogue, to analyze errors in navigation and formulate differential corrections to range finding data as well as to structurally recover the information from SC, receive and transmit it

via external and internal local data exchange networks, decode, process
and present integrated information from automated target oriented workstations with imposition of results of semantic processing onto a digital map
of locality or for subsequent issue to users and so on.

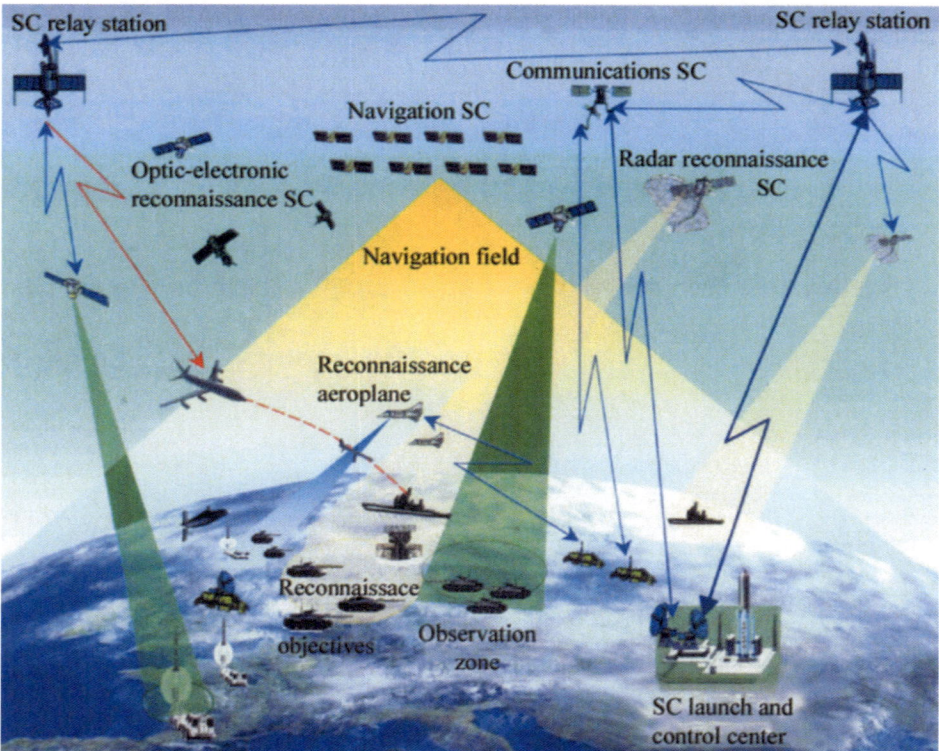

**Fig. 91.** Integrated system of space, aerial, ground reconnaissance and target designation.

In creating operationally deployable SC, thus outfitted SSG can plan their
target oriented application, operational deployment of space systems and control the SC they comprise. The considered variant of SSG is only one of possible versions. Another point is indisputable here – there is "a gap" between the
potential capabilities of space systems and their practical utilization in troops.

As for creation of SSG, it is one of the most promising ways of filling
that gap.

Over recent years, especially in times of conflicts, the US and Russia built
integrated types of reconnaissance and weapon systems. The concept of
combined application, duly coordinated in time and space, of aerial reconnaissance and weapon systems, space reconnaissance systems brought together
into a single complex represents a substantially new stage in the development
of high precision reconnaissance and weapon systems. The diagram showing
the use of such systems is presented in Fig. 91.

The algorithm of functioning of heterogeneous systems is quite simple but effective. Space reconnaissance systems (electronic and optoelectronic) having a high frequency of scanning target areas and a prompt delivery of intelligence virtually in real time mode are responsible for target location. The obtained information about target is transferred to troops and arms control centers and (or) directly to airborne weapon systems that refine intelligence and simultaneously deliver strikes. The illustrative example of realization of this scheme is an episode of Chechen hostilities in which a group of fighters headed by Dzh. Dudayev was destroyed while using a satellite communication system. Supposedly, the fact of coming on the air was detected by space electronic reconnaissance system and the target designation was transferred to aircraft on duty in the air and armed with air-to-surface missiles.

In the US the space component of the integrated reconnaissance system includes the Keyhole optoelectronic reconnaissance SC, Lacrosse radar reconnaissance SC, Magnum and Vortex electronic reconnaissance SC, DMSP meteorological satellites, and a French SC Spot. The aviation component includes piloted and pilotless reconnaissance hardware. Pilotless hardware of the Hunter, SD-289 and Predator type was widely used in Yugoslavia.

Such integrated systems have a number of unique features: first, the operational flexibility in tactical use of airborne and space components, in which case each of them can operate independently with due regard to the prevailing tactical scenario. Second, the improvement of the failure-free operation of the system thanks to its multi-component structure and the capability to conduct all-weather full-time reconnaissance due to availability of space-based systems as well as radar-assisted observation in either component. The functioning of both components and their interoperability are coordinated by the united space support groups. The scheme of functioning of integrated aviation and space reconnaissance system as used in Yugoslavia is shown in Fig. 93. Thus by the end of the 20th century the space proper became the arena of controversy between the national, including military, interests of various states. As it happens, on the frontline of the 21st century the world entered a new phase of geopolitical struggle for superiority in space.

The building and deployment in near-earth space of large-scale orbital constellations, the strategic level of tasks tackled by the systems, the emergence of space hardware capable of effective countermeasures (destruction, suppression) against various targets, the appearance in space of elements of tactical hardware – all these factors shape a new arena of armed struggle. The distinctive feature of the military application of space in the end of the 20th century was the use of space equipment for comprehensive support of troops, especially their tactical strike units, for operations on land and at sea (in the ocean).

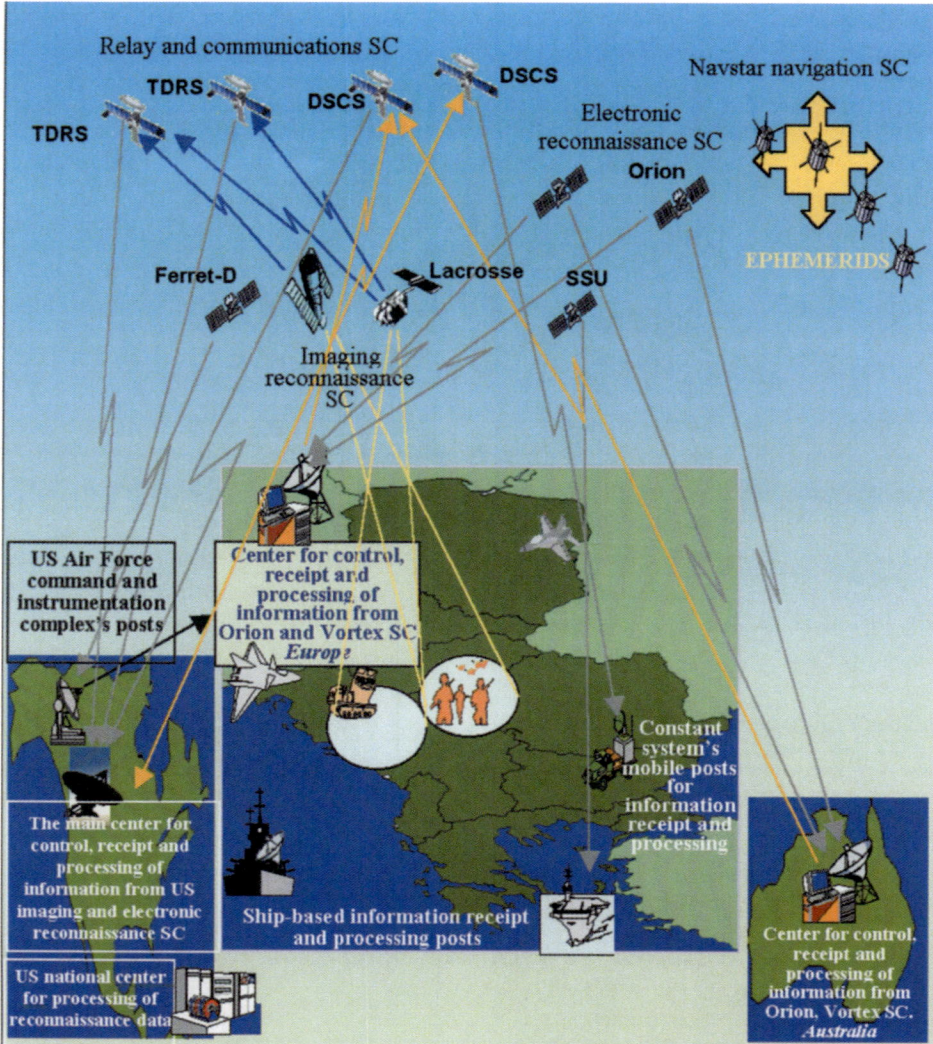

**Fig. 92.** Scheme of functioning of integrated aviation and space reconnaissance system

The integration of space information systems with weapon systems, the use of civilian SC for fulfilling military tasks and vice versa (the so-called dual use of SC), orientation towards building the war time space hardware on the basis of small SC and highly maneuverable (mobile) launch facilities, become standard practice in organizing and waging an armed struggle. As a separate fact it should be noted that by the beginning of the 21st century expertise had been accumulated that characterized the new millennium as a technologically ripe time for development and creation of weapon systems for operations in and out of space.

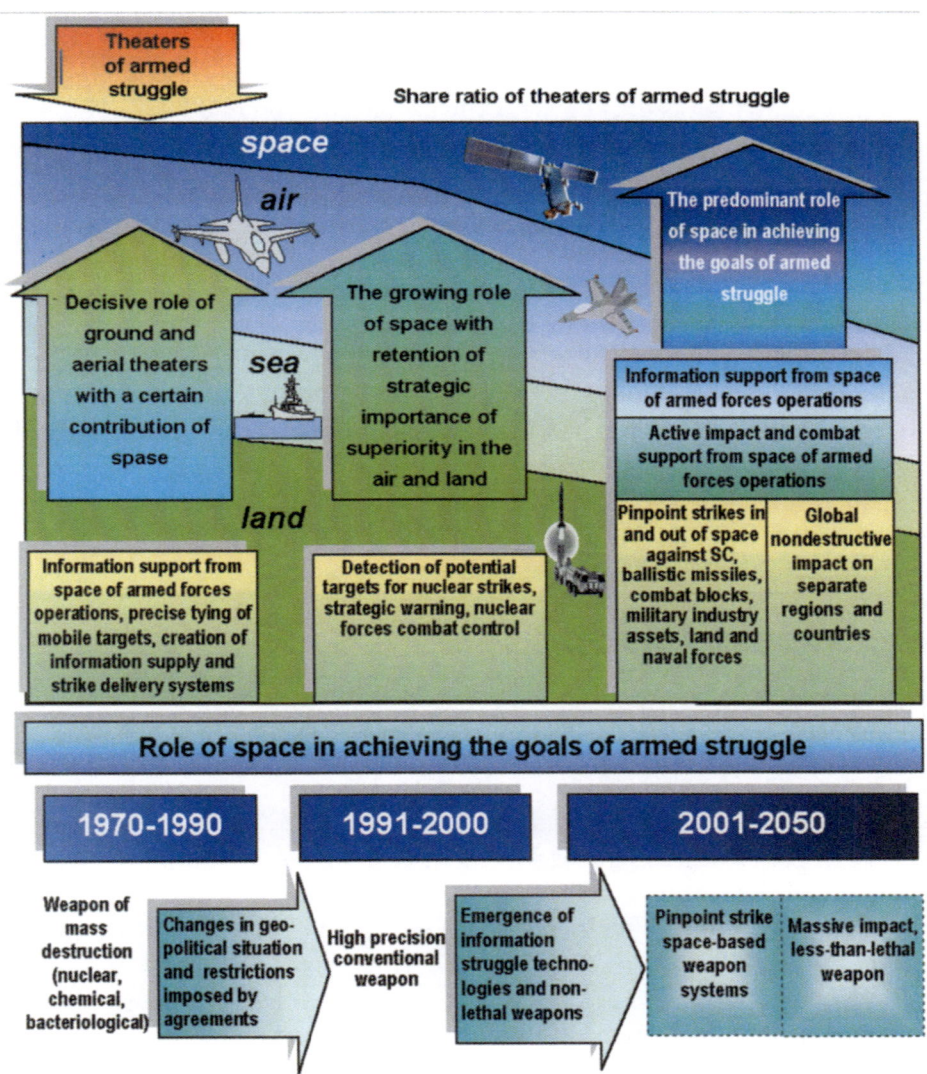

**Fig. 93.** Changes in significance of theaters of armed struggle

All this determines the main trends in development of the military space hardware in the 21st century as a key element of a state's national security. Such a development is urged by the conclusion about the changed roles of battlefields in favor of space in the armed struggle of the 21st century (Fig. 93).

As evident, one of the main tasks whose fulfillment must be assured by the military space hardware of the 21st century is the information support given to armed forces from space. The development of space hardware for fulfilling this task must proceed along two interconnected lines.

The first trend is the development of space hardware that meets war time requirements in terms of performance: components availability, high output rate, periodicity, short launch time, survivability, etc.

The second trend is the delivery of space information to the lowest links of control chain and, in the future, to each individual soldier.

The engineering base of the first trend is the work aimed to create small (lightweight) SC and carriers for injecting them. This proceeds from the transition to a new stage in advance of science and technology characterized by the increasing miniaturization of electronics.

Currently, the specific cost of manufacture of a large military SC varies between US$ 88,000/kg and US$ 220,000/kg. For a small SC these costs can be cut down to US$ 17,000/kg.

Among the advantages provided by small SC foreign experts list the following:
• smaller cost of development and manufacture;
• less time to develop and manufacture;
• smaller cost (absolute terms) of delivery to space;
• less restrictions in number of SC to be delivered;
• higher survivability due to a greater number of SC.

The period between the beginning of development and the launch of small SC may vary between 1 and 2 years which makes it possible to utilize the newest design decisions in the course of creating them.

The notion of the "small" SC stems not only from the small size (lighsat) and the low cost (cheapsat) but also from a radically new technological pattern of designing, manufacturing, testing and launching the SC combined with an untraditional application strategy.

The use of systems based on small SC can provide such basic advantages as:
• increased flexibility due to distribution of several functions currently performed by one traditional SC among several small satellites. Also, an opportunity is provided to launch small uncostly SC in crisis times;
• increased stability due to "the distributary architecture" employed in building an orbital small SC when performance of duty will not be severely affected by the loss of one or several satellites from a constellation;
• fast introduction of the newest technologies due to shortened manufacturing periods, growing number of small SC being produced and the frequency of launching them.
• wider use of commercial systems since the development of small systems utilizes more fully the opportunities of the emerging space industry which makes it possible to acquire separate elements or even whole systems manufactured on a commercial basis. This, among other things, will reduce the costs.

The creation and deployment of systems based on small SC (SSC) implies their use mainly for fulfilling tactical tasks. Such systems were supposed to provide communication on the theater of operations, to monitor enemy troops movements, to receive data for damage evaluation, to carry out separate experiments.

In countries other than Russia, SSC are already used for military purposes. As far back as the  time of NOSS (National Orbital Space Station) the USA created the Ssu SSC and used MB SSC during the Balkan crisis. Active work proceeds on creation and application of carriers that improve launch capability of SC, including SSC. Currently, the preparation of a carrier used to launch existing military hardware requires from sixty days (Atlas-2AS carrier) to 200 days (Titan-4 carrier). The best operational capabilities are demonstrated by new light class carriers of the Pegasus and Taurus families. The total time required to prepare the Taurus mobile carrier varies from 7 to 8 days. The time between  receipt of command to launch and the launch proper is 70 hours. The aircraft-launched Pegasus-XL carrier which can be stored over prolonged periods with a payload attached features a still higher launch efficiency and is determined by the time the plane needs to approach the launching site. These carriers, however, are intended for launching lightweight SC (Pegasus-XL – up to 450 kg, Taurus – up to 1,180 kg) and are practically unusable for launching the majority of military SC being now in operation. It should be noted here that the specific cost of launch grows, which invariably happens in case of using light carriers and equals three to five times as compared to the similar performance of heavier carriers. This, however, must be offset to some extent by the reduction of specific costs  obtained in development of light SC, in which case it can be as high as 5 to 10 times.

The delivery of space information to the lower links of troops control chain, down to a soldier, gained acceptance only late in the 20th century, the time when "smart" small-sized well-informed specimens of hardware came along and changed the very notion of  modern battle. Since 1993 in the US  an SMP (Soldier Modernization Plan) has been implemented. Its object is to enhance the soldier's capabilities on the battlefield.

To date, the SMP has turned out special operations individual weapons as well as soldier's combat gear and outfit. Based on the concept of "the information war", these developments use the latest achievements of science and technology and are essentially the products of revolution in defense industry at the turn of the century. In terms of its consequences it can be compared to creation of the nuclear weapon in the mid 40s of the 20th century. Introduction of space information technologies at all levels of troops management and control gives every ground to state that "digitalization of military operations" is absolutely possible.

So, according to that concept, each soldier's outfit necessarily contains control (communication), navigation and data presentation equipment. All these are not loose units, but a light and compact set (complex). Its efficiency will heavily, if not critically, rely on how space information and computer technologies are integrated. Such modern equipment as notebook computers, virtual eality headsets, pocketsize communication equipment – pagers and cellular phones – and, finally, individual space navigation instruments, are archetypes of some components of the new product. As a result, the capabilities of an individual soldier to fulfill tasks in all conditions, and a soldier's combat self-sufficiency will increase many times over. In terms of a blow dealt to the enemy he could be compared even in a conservative estimate to a squad. True, his training also needs to be adequate. A professional like this can't be trained "in a month at freshman's course". Training will be a permanent process.

The scientific research carried out over recent years and the results of optimization of practical use of space forces and hardware confirmed that the creation of a small sized transceiver for work with space information must remain one of the priority objectives. Success was achieved in determining the scope of space information to be supplied to various users on the battlefield without creating "the excessive informational redundancy". Also determined were the structure and job duties of agencies responsible for its receipt, processing and transfer as well as the preferred variants of organizing orbital and land-based facilities.

What will the soldier of the future look like?

Even supposing that in the 21st century a war or an armed conflict can be regarded as a struggle of "smart" fire and information systems, the human operator, no matter what position he holds in the army hierarchy, will continue to play the key role. No one will make a decision for him, nor will relieve him of his responsibility. But in order to "be up to the mark" in any situation each soldier must have, apart from flawless weapon, a reliable and highly efficient means of reception, presentation, processing and transmission of information about the operational situation (Fig. 95). The integral element of his outfit will be the computerized system of space communication that features a high capacity, noise immunity, protection, space navigation capability, conjugated individual data presentation devices, etc.

Separate units, or a separate soldier, if necessary, will have permanent communication with command of all levels regardless of the distance. Also, they will have a reliable control and interoperability (including fire missions) with other units, plus fast exchange of intelligence. This will enable them to receive orders and, if necessary and with due regard to the situation, to coordinate and carry out combat missions in real time.

Exact orientation on locality in all weathers at day or night will be just as normal as establishing the exact time. Paper maps, too, will become things of the past.

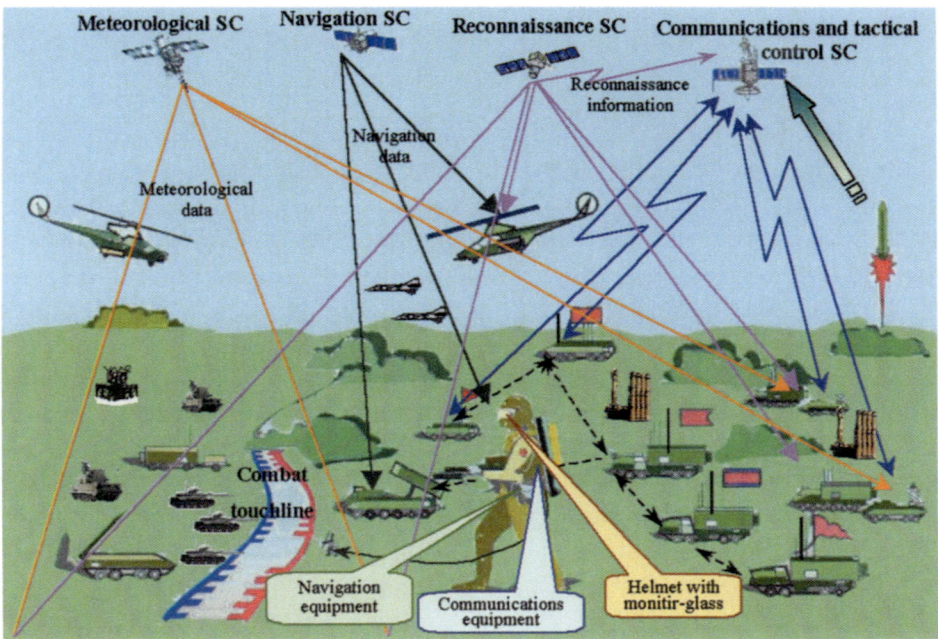

**Fig. 94.** Soldier of the future

They will be replaced by high precision digital maps transmitted directly to individual-use portable terminals that show the combat situation and the user's own whereabouts. Even "friend-or-foe" recognition will then be possible via coordinates read from the terminal. This will improve efficiency and simplify receipt of reconnaissance data concerning the hostilities. The data supplied will also contain intelligence of radar and chemical reconnaissance.

The solution of those problems will radically improve the troop management and will increase their fighting capability by several times. In previous years it would take decades and a replacement of several generations of military hardware. It's hard to believe that nowadays this can be done by a miniaturized individual-use instrument weighing a mere few hundred grams and consisting of a computer with devices for information reception and presentation and for performing communicational and navigational functions. Sure enough, changes must take place in targets handling systems of space facilities and command posts, in techniques and methods of situation evaluation and decision-making and, finally, the military must get used to the new level of control and execution of orders.

However, it would not be too much to say that this tiny, by human standards, individual use instrument will be the centerpiece of the entire system. For the military it will be like the triumphant computer for the rest of the world.

The protective monitor helmet can be one of the possible prototypes of such an instrument. Its display will receive integrated information about presence of the enemy, composition of his forces, the state of the battlefield, the location and composition of friendly forces required to oppose the foe. Naturally, the bulk of tactical orders and directions will be transmitted through it. Possibly, a problem will arise how to adapt the soldier to perception of the heterogeneous information coming from display and the real environment which is bound "to split" his attention in the case of dynamic developments on the battlefield, of selection and distribution of data for various users, of maintaining secrecy and protection etc. In addition, it would be premature to announce the creation of an advanced all-purpose outfit with elements of space–assisted functions that would suit all soldiers in all conditions. Such an outfit must be modular which would provide for fast elaboration of operating variants that would best suit the soldier's mission and conditions under which he operates. There are, however, no "insolvable" problems in sight. Similar problems have long been and are still being tackled successfully in aviation.

The predominant role of space in achieving the goals of armed struggle in the 21st century will be determined by such factors as a powerful strike from space and the capability to maintain from it the operability of friendly forces. The solution of such a task calls for creation and deployment of space-based facilities for conducting operations in and out of space. This task implies the protection of friendly satellites, provision of access to space and preventing the enemy from using the space facilities for his purposes, the destruction of ground stations, equipment and satellite communication lines, the disablement of orbital hardware. Also, the task may include the tilization of military space hardware with possible strikes from space against land-based targets. In the future, the priority in development of space weapon systems will probably shift to solution of this particular problem which will become particularly important.

Foreign experts believe that the transition to space management and strikes from space is inevitable because of their ever increasing role. It is very likely that in the future not only hostile hardware in space will be destroyed, but strikes from there will be delivered at ships, aircraft, ground targets and warheads in flight. This is precisely why some space powers carry out research in directed and kinetic energy weapon systems for destruction of such targets. They are supposed to be used both from land-based platforms and aircraft. The reference of space forces along with nuclear missiles to the category of "combat air force" is also indicative of their growing role.

The increasing importance of those systems in the future is testified by the fact that the US still continues to work on the ABM defense system which aims to achieve three following goals:

1) assure greater protection against tactical missile strikes by the end of the current decade;

2) implementation of the program aimed to deploy national ABM defense against potential threats of long range ballistic missiles;

3) development of new technological programs in the area of planned and existing defense systems.

The practical implementation of the program for creating high precision weapon to destroy pinpoint targets in and out of space is the experimental optimization of key elements and functional units of the Brilliant Pebbles small-sized space interceptor and Brilliant Eyes space observation facilities, first under the Delta-180 and then Clementina-2 research program.

In addition, the US began to optimize the elements of weapons that can also be used in space-to-surface systems.

The creation of such systems signifies the emergence of a new type of strategic weapon. All these systems are weapons for dealing pinpoint strikes at assets and troops.

However, the turn of the 21st century saw the emergence of information wars and non-lethal impact on humans. The hardware created on the basis of such technologies can be installed on spacecraft from where it is able to exert continuously or periodically enough pressure on the chosen regions in order to neutralize temporarily enemy manpower, demoralize population or commit other similar acts.

The possibility to fulfill such tasks from space will bring about qualitative and quantitative changes in the ways and means of conducting armed struggle in general. The example of building such a system is shown in Fig. 96.

It should be noted that modern times are characterized as an epoch of information wars (information struggle) in which the role and significance of space information systems are especially important.

The war in the Gulf (1991) was called the first "information war" in which the US scored a victory. In modern conditions, information is a management tool used to control all stages of the management process. Lack of information deprives decision makers of their bearings in the situation and is one of the reasons why biased decisions are taken. In other words, information is the same military resource as manpower and materiel.

Based on the analysis of results of operation Desert Storm, the US DoD elaborated new techniques of adverse affection of the enemy, among which measures based on the concept of the so-called "information war" stand out as particularly promising.

The implementation of its provisions in the context of armed struggle means a shift of priority from the traditional fighting techniques (fire, strikes, maneuvers) to the area of information supply and intellectual decision making.

The prime objective of the "information war" is disintegration and dismemberment of the enemy troops management system, thus turning them into disoriented and unmanageable elements with their subsequent neutralization through fire-assisted (physical) destruction.

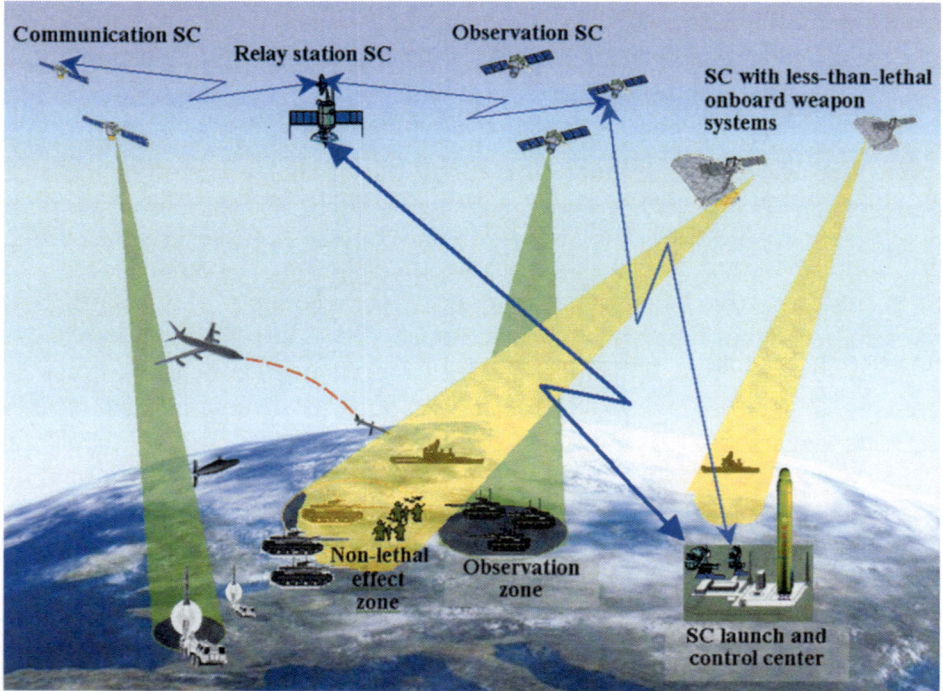

**Fig. 96.** SC with onboard less-than-lethal weapon systems

The search for new forms, ways and means of conducting armed struggle in various spheres of human activity will continue in the 21st century.

It's worth noting that space occupies a special place in modern military doctrines of national security  developed by the world's leaders in space.

For example, "The Strategy of the US National Security in the New Century" published in October 1998 outlines Washington's fundamentally new approaches to the world's foreign policy and the use of space for national security.

Space is regarded here as one of the main areas in which the developed and the prospective technologies are urgently needed to defend the US. It is also believed that space can supply the US with a capability to use more fully its power for creation of a milder international climate and to be responsible for the entire spectrum of possible threats and crises.

That is why according to "The Strategy of the US National Security in the New Century" the USA intends to retain its leadership in space, to strive for unhampered access to space and its use for national security that contributes to the country's wealth and prosperity.

It is emphasized that space has become globally important for information supply issues, which is of paramount importance to the political, diplomatic, military and economic life of the US.

Washington's politics consists in the development of all types of activities in space and in protection of vitally important interests while doing so. It is expected that threats to the US interests in space will be adequately opposed. Should this yield no results desired, measures will be taken to frustrate the efforts of an enemy that prevent the US from accessing space and using it.

A task is set to maintain the capability to fight space systems and hardware that can be used against its land and marine forces, against management and control systems or other structures that are important to national security. The US closely follows the commercial remote observation satellites so that gathering imaging intelligence from space could not impair the US national security.

The issues of Russia's national security are thoroughly considered in "The Concept of National Security" confirmed by the decree of Russian President No. 24 dated January 10, 2000. The basic point of the concept, unlike that of the Western similar documents, is to strive to create an ideology of the multi-polar world based on political consolidation of a considerable number of states and their international associations,  on improvement of mechanisms for managing international processes, on expansion of collaboration in economy, science, technology, and information systems, including those in space.

Thus with advancement of man in space and creation of new space technologies the role of space for safeguarding national security will grow. It will retain its importance and increase it with improvement of international relations, development of human civilization and strengthening of the Earth's safety as a planet.

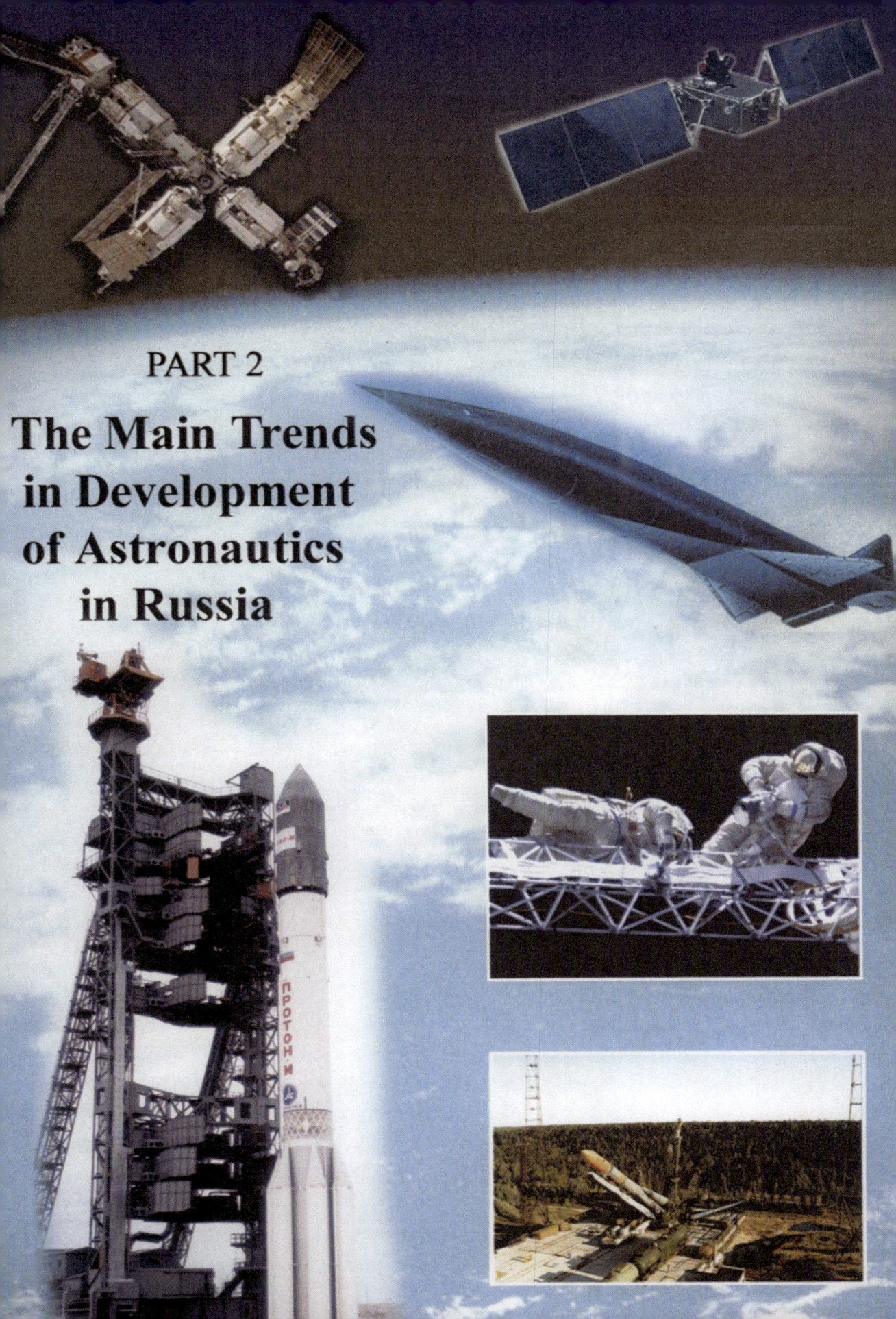

# PART 2

# The Main Trends in Development of Astronautics in Russia

Space exploration is treated in Russia as one of the national priorities regardless of social and economic reforms being carried out. That is why it should receive every kind of the state's support, political, economic and legal. Space programs should be based on a strategy that singles out priority goals in space work and elaborates tactics of achieving them. Such tactics defines the prime targets and tasks of space exploration in Russia, the ways and means of work to be done, its dates and amount of funding. Also, the tactics specifies the types of space hardware to be manufactured for social and economic development, for advance of science and defense technology, for international cooperation. All these activities should proceed with due regard to the current situation and prevailing practices in space exploration (in medium term this activity is regulated by the Federal Space Program).

It should be noted that the foundations for development of the world's astronautics in the 21st century remained practically the same as they were in the 20th century. Its purpose has been and will remain the implementation of a package of social, economic and scientific programs that insure the stable economy, national security and the implementation of global international space projects.

The main trends in the development of space hardware will remain the improvement of capabilities of the existing systems and complexes through introduction of a spectrum of advanced technologies (in design, research, manufacture, management, etc.) many of which emerge on the borderlines of various scientific, technological and industrial activities. The close connection of space industry with all fields of human activities urgently calls not only for improvement of capabilities of the existing hardware and development of its new types for tackling traditional space tasks, but also for fulfillment of other purposes for better satisfaction of demands of the ever growing number of consumers of space products and services.

The prospects of the Russian astronautics in the 21st century are inseparable from the major trends in development of the world's astronautics and Russia's international commitments in space exploration. Also, they are closely associated with maintenance of the country's potential in space research and its priority development.

In modern conditions a number of new factors emerged that govern the development of Russia's astronautics.

1. The growing international cooperation in research and utilization of space, a sizable increase in the number of states engaged in space exploration on an independent basis or using the results of space research. An important product of this process is the emergence of the world's market of space hardware (technologies) and services, which gives rise to competition that imposes new and more stringent requirements on the quality of space

equipment. This factor had a great impact on the Russian astronautics which from its inception had been "stealthy" and with a pronounced military bias. It was oriented exclusively towards domestic industry and technologies and towards the home market. Work on a project today may lead to massive introduction of advanced foreign technologies and engineering decisions into the Russian space systems which may reduce their production period and raise manufacturing to a higher international standard. However, this is often impeded by uncontrolled (or pseudo-controlled) drain of unique technologies away from Russia.

2. The economic plight of the state which traditionally has been funding and regulating the space work necessitated the shift in priorities in implementation of the state space program.

The search for more effective means of carrying out and regulating space work is a new prominent strand in the state's policy in space technology. In order to reduce spending on defense work, it is important to establish the balance in development of space systems and complexes of the new generation, mostly of dual application; to use for them a single engineering and manufacturing base and ground infrastructure that insure their development, production and utilization for adequate solution of tasks in the national science, economy, defense and social programs.

3. The high rate of return of investments in space science and technology has considerably commercialized the space work over the last decade. It was caused, on the one hand, by the state's desire to turn some space projects into self-supporting enterprises and, on the other, by emergence of both foreign market and domestic consumers with various forms of ownership.

The objective process of concentration of the private capital (including investments) in Russia may put the entire commercialization of space work on a larger scale. In addition, the necessity and opportunity to look for extra sources of funding in modern conditions along with attraction of the commercial capital to space industry can make up for the obviously insufficient funds.

This situation cannot be changed in the near 3 to 5 years for such an objective reason as the high inertia of economic development. It means that Russia's space industry enters the 21st century with an array of financial and economic problems. That is why the commercialization of space work in Russia becomes one of the important ways of preservation and development of the national astronautics. The second factor is worth a more minute consideration since this is its negative influence that causes numerous difficulties in implementation of Russia's space program because the main source of funding the space work in Russia as well as in other leading space powers (USA, European countries, China, Japan and others) has always been, still is and will be the federal budget.

In spite of the huge importance of space systems in safeguarding the national security and defense, in development of economics, science and international cooperation of Russia, the national space industry does not receive adequate support from the state. The funding provided by the federal act "On Space Exploration" is equal to 1% of the Gross Domestic Product (GDP) which is obviously insufficient (Fig. 1), jeopardizes the country's independence in space work and casts doubt on its ability to honor its international commitments in space.

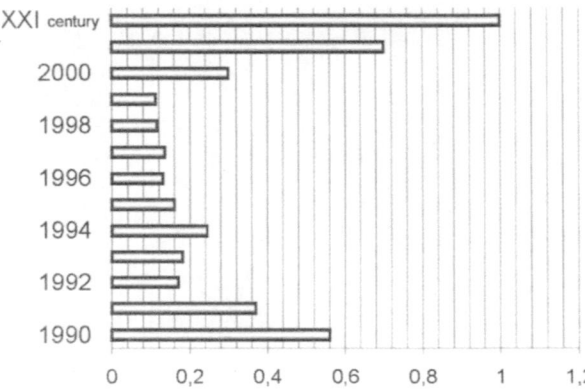

**Fig. 1.**
Russia's required funding for space work in 21st century (in % of GDP)

There is no remedy but to move quickly on to flexible commercial relations in search for stable non-budgetary sources of funding.

The extra non-budgetary sources of funding that could be used by space rocketry industry are as follows:
- commercial launches of spacecraft;
- use of space communications channels;
- space navigation;
- fulfilling tasks in Earth remote sounding;
- utilization of not used basic assets of plants and organizations of the space industry;
- sale of technologies and elements of rocket hardware on the international market;
- realization on a competitive basis of orbital equipment and frequency ranges, and so on.

For example, the Khrunichev State Space Research and Production Center that created unique specimens of rocketry, such as the Proton carrier which launched into space in a quarter century 40 types of hardware with a total weight of 4,000 tons, the orbital stations Salyut and Mir, the space modules Kvant, Kristal, Priroda and others and is actively involved in advanced projects of the new century (Fig. 2, 3).

The light class Rokot carrier rocket and the new family of carriers, the Angara, orders for 20 commercial launches under various space programs from

**Fig. 2.** A.I.Kiselev and Thomas Corcoran meeting to discuss international cooperation

The prospects of further development of the Khrunichev Center in the 21st century are linked with the expansion and improvement of services in launching various payloads into various orbits. Also, they include provision of various types of global communication, navigation, Earth remote sounding and introduction of new information technologies based on cost effective small spacecraft.

The work carried out now by the Khrunichev Center under the GMES and GES (Global Environmental Service) program exemplifies one of such latest trends.

The work under GMES and GES program is carried out on the basis of an agreement signed jointly by the leaders of Matra Marconi Space (France and Great Britain) and DASA/DSS (Germany) in December 1997. This program implies the creation of a common European service of global monitoring of the environment based on the Monitor small spacecraft (containing instruments operating in optical, radar and IR ranges). The first launches of such craft are slated for 2003. The system is intended for:

• the continuous instrumental monitoring of natural and man-caused phenomena;

• the processing of data about the Earth and issue of results to the interested customers;

• the issue of well-founded recommendations to control agencies for adequate decision making.

**Fig. 3.**
Vans Coffman, CEO and
Chairman of the Board
of Directors of Lockheed
Martin with A. A. Medvedev,
General Director of the
Khrunichev Center

What are the main trends in development of astronautics under such uneasy conditions in the 21st century? The prime target of Russia's space exploration must be the effective use and further development of the country's space potential for addressing the national most pressing needs in science and social spheres, for active participation in the international cooperation (including that on a commercial basis), for solution of mankind's global problems.

The main principles of Russia's activity in research and use of space in the 21st century must be:

1. assured protection of Russia's national interests in pursuit of its goals in space, unfailing fulfillment of Russia's commitments in space and primarily in manned flights and scientific space research;

2. independence of the state's policy in space for the purposes of national defense and security;

3. the state's management of space work, including the coordination of activities of Russia's organizations on foreign and domestic markets of space products and services;

4. well-defined distribution of functions, tasks and responsibilities for formulation and pursuit of space-related policies among legislative and executive authorities; coordination and concentration of their efforts towards achieving the prime goals of the national space policy; conducting of expert analyses of nationally significant space systems and complexes;

5. well-balanced development of military and civil space systems, unfailing completion of design and development and subsequent commissioning

of space systems and complexes of the new generation, predominantly of dual application;

6. equitable and mutually beneficial international cooperation in space, observance of requirements pertaining to monitoring space rocket technologies, prevention of discrimination against Russia on the part of other states on the world's market of space products and services, intensive use of advanced foreign technologies and engineering decisions in the Russian space systems and complexes.

## 2.1 A leap in improvement of orbital facilities

The prospects of evolutionary improvements in science and technology and the predicted demands of the national and commercial entities in use of space suggest that the characteristic feature of space work in the 21st century will be not a mere struggle for a greater number of better space systems, but also a creation of new high technology equipment for various applications that have a direct bearing on many spheres of human activities and radically change them (Fig. 4, 5).

Standing out as most promising among the advanced trends in development of space systems over the period till 2025 are primarily the developments directly linked with the building of global adaptive space networks for various applications (communication and data transmission, navigation, etc.), small, micro- and nano-satellites, neuro-network technologies, management systems based on artificial intellect, special purpose miniaturized chips and software for processing large amounts of information, high precision micro-mechanical systems, highly efficient sources of energy, etc.

During the period between 2025 and 2050 the achievements reached in technology and the basis created in science and industry will allow to use qualitatively new aerial and space vehicles of various applications, orbital power supply systems feeding via various channels both space– and land-based facilities, space laboratories and plants. Interplanetary research will continue and eventually result in piloted expeditions.

The research in the border areas of biophysics, medicine, radiobiology, electronics, etc. will bring about spacecraft that control and correct energy and information flows. This will make it possible to address the problems of monitoring the medical and biological condition of the environment and to correct the factors that adversely affect it.

The development of space communication and data transmission systems will aim at achieving the global stable and continuous communi-cation of various types of users with due regard for integration of different types of systems, increase in the network's throughput capacity and organization of multi-level telecommunication fields.

The creation of the distributed space systems, including those that use nano-technologies, will not only improve the accessibility of information channels, but will have a profound effect on the very process of organizing the informational interaction, i.e. telecommunication will embrace practically all spheres of human activities. This process will turn the orbital communication segment into an integral part of all communications systems that will allow to fully utilize not only data transmission capabilities but also those of remote control over various objects and processes.

| | Man starts to utilize near Earth space | Advanced decisions and technologies in design, development, engineering and management, massive introduction of untraditional schematic techniques, energy sources and power units | Advanced power units, inter-orbital towing vehicles, orbital stations | Reusable aviation and space hardware, lunar and orbital stations and automated plants | Global tasks of space work: Expansion of knowledge about the Solar system and the Universe. Insuring human life and safety on Earth and in extra-terrestrial settlements. Utilization of resources of planets, comets, asteroids and the Sun (materials, energy). Development of space transport and colonization of planets |
|---|---|---|---|---|---|
| 2001. Creation of conditions in science and technology for leap in quality | SC for optimizing basic technologies and conducting experiments | | | | |
| 2010. Creation of the global information field | Advanced systems of small spacecraft monitoring | | | | |
| 2015. Man starts to utilize the Moon | Distributed telecommunication integrated systems of spacecraft | | | | |
| | SC of global high precision navigation support | | | | |
| 2030. Man reaches the nearest planets, transfers production to space | SC of fundamental investigation of interplanetary space | | | | |
| | Orbital stations for manufacture of materials, preparations, etc. | | | | |
| | Orbital power supply stations to service Earth and spacecraft | | | | |
| | Families of robot SC (maintenance in orbit) | | | | |
| 2050. Man starts to utilize the Solar system | SC for monitoring dangerous natural and man-caused phenomena | | | | |

Fig. 4. Prospects in use of orbital facilites in the 21st century

In the 21st century, side by side with the existing and future communications technologies (Fig. 6 ... 7) there will appear and gain wide acceptance high rate laser channels of intersatellite exchange, video channels based on holographic image representation and transmission systems. The development of digital TV along with assuring the high quality of signals being transmitted will yield an interactive process of obtaining the information needed by the customer, in which case the computer-aided information terminal will replace the normal TV set.

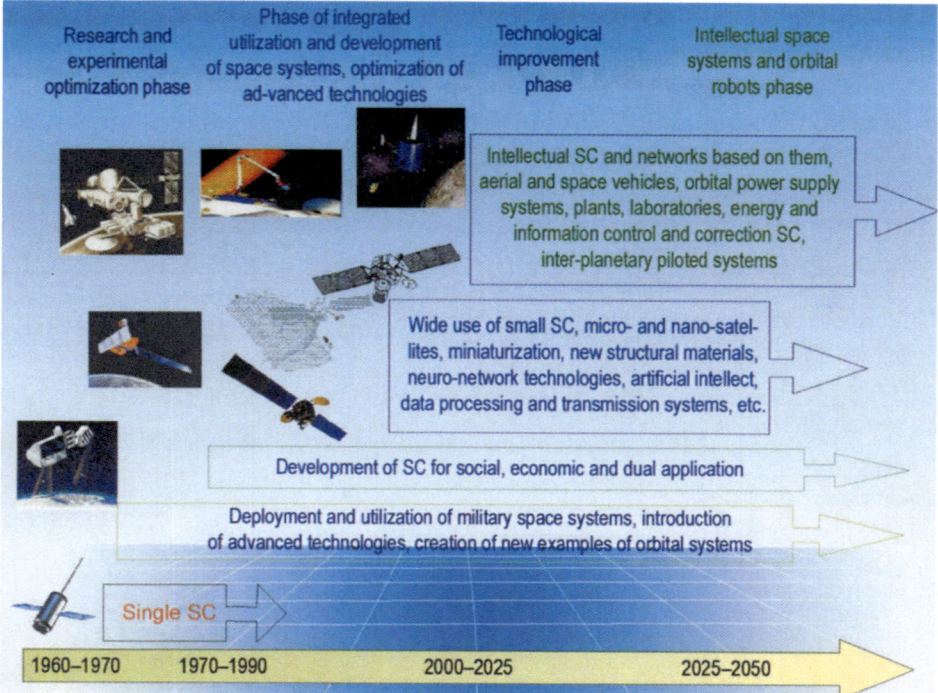

**Fig. 5.** Main phases in development of orbital systems

Also worth of note is the necessity to develop in the 21st century the networks that maintain communication with mobile objects (especially in the Northern hemisphere and remote hard-to-reach areas) compatible with the equipment of the Inmarsat international mobile communication system and similar advanced developments.

The preferred development of space communications systems is particularly important for Russia with its expansive territories and lack of landbased data transmission channels without which not only national defense and administration become inoperative, but even effective economic, industrial and commercial activities prove impossible.

However, it should be borne in mind that apart from a broad spectrum of capabilities to fulfill the social, economic and scientific needs, the energy supply systems, including those for space communications, have one serious side-effect whose study will get a great deal of attention in the 21st century. This is the so-called energy and information impact of electromagnetic fields that is manifest in changes in living organisms exposed for a considerable length of time to the effects of low intensity fields. This factor must be taken into account in designing and operating all sorts of space communications and data transmission systems. This, in turn, will give rise to and priority development of energy and information monitoring systems for use with equipment that issues energy. Also, they will play a major role in addressing the problems of urban development, industrial production, insure man's security against information and energy threats, etc.

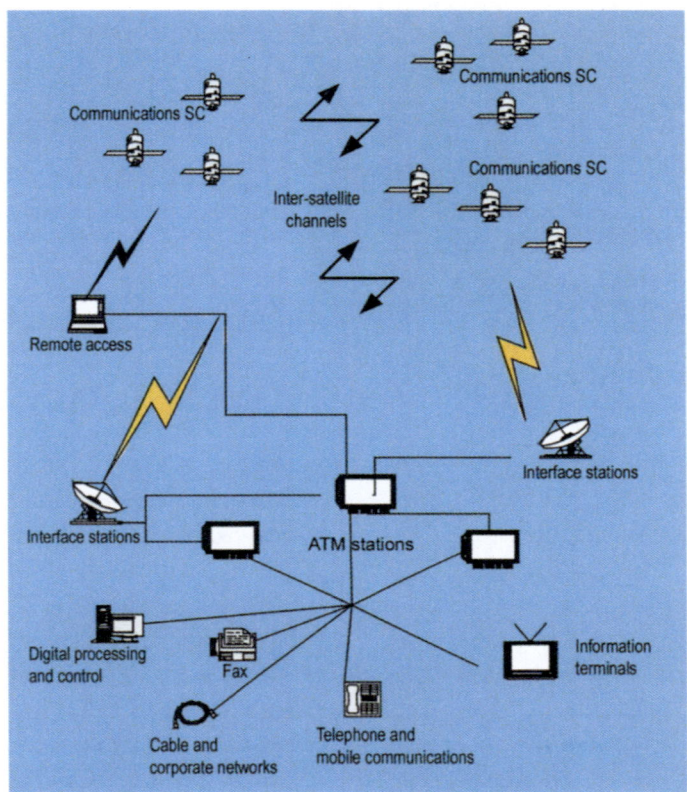

**Fig. 6.**
Prospects of development of space communications systems

All this will create conditions for the growth in the 21st century of significance of satellite communications systems in Russia's communications network to 30-50% (now it is around 6%) with the total throughput capacity reaching more than 400 channel (now it's around 60). This will be based on the increase of the throughput capacity of communications equipment installed on one SC up to 30-40 thousand equivalent telephone channels (now around 500) with securing self-sufficient operation of a communication SC for no less than 60 days (no such capability is available now).

However, the state's current funding of orbital communications systems will suffice if only extra non-budgetary allocations are made available. This virtually nullifies the state's regulation of the telecommunications field as a single whole, which can impair Russia's information security in the 21st century since the federal requirements for communication and telecasting early in the 21st century will be satisfied only by channels of commercial satellites and by leasing the channels offered by foreign communications operators.

**Fig. 7**.
Testing the Molniya
(Lightning) communications
satellite at NPO PM work-
shop (courtesy of A.G.Kozlov
and G.D.Keselman)

## 2.1.1 Space monitoring systems

In the sphere of monitoring, a number of priority development trends should be pointed out that are associated with the increase in global nature and efficiency of monitoring the Earth's surface and near-earth space by space systems as well as with improvement of prediction and management techniques, including those of advanced prediction and management of natural disasters, biological, energetic and informational condition of particular regions, towns, farming lands, pastures, plus methods of geological survey of territories (search for mineral resources, refinement of knowledge about the geological structure of the Earth) based on massive introduction of new untraditional ways and means of space exploration.

Space systems, primarily small spacecraft, will gain considerably in importance (Fig. 8) as a means of investigation and optimization of new hypotheses and theories about dynamics in development of the biosphere and noosphere. The evaluation of interaction and mutual influence of all man-caused, social, biological, chemical, physical. energetic, informational and other processes in order to build the adequate model of the Earth's dynamics and to draw up international programs of man's security will be one of the prospective challenges confronting mankind. Such an interaction is a multi-level and a multi-faceted process that affects all major activities of mankind , including those in biological, energetic and informational areas.

The adverse effect of human activity upon the environment calls for development of space systems for controlling the harmful man-caused impact on the biotope, for rapid (from 0.5 or 1 day to instantaneous) detection of disasters (fires, chemical, biological and radioactive contamination, mudflows, avalanches, floods, biosphere pollution, etc.). Much attention is being currently paid to the timely warning of such calamities and to enhancing the efficiency of

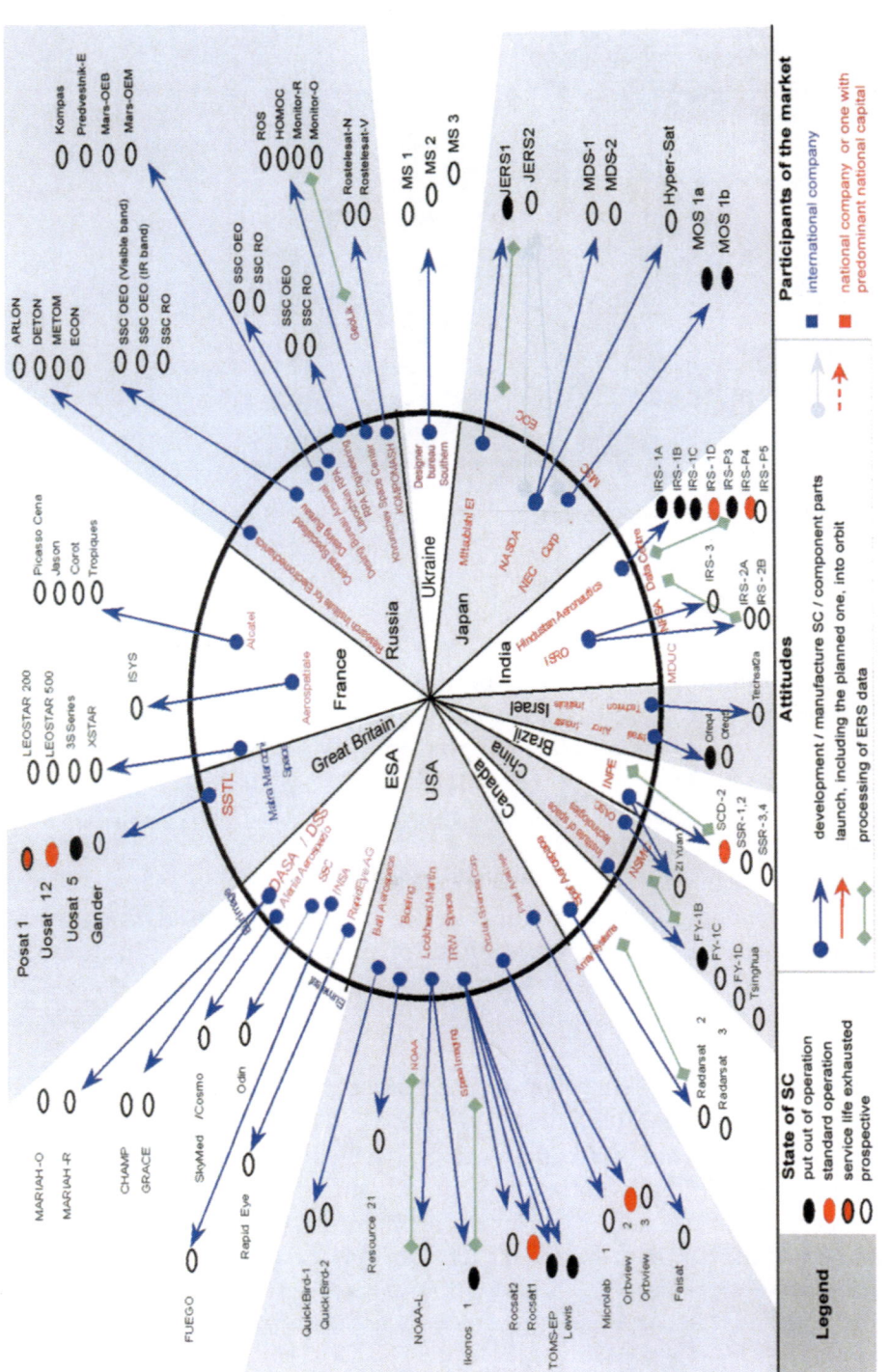

**Fig. 8.** Prospects in development of monitoring equipment

eliminating their consequences by the use of information received from space-based systems (Fig. 9).

**Fig.9.** Resurs ERS SC

The introduction of the latest achievements in microelectronics, nanotechnologies, SC power supply and heat regulation systems, lightweight components manufacture technologies, including precision optics, will serve as a base for building distributed space systems of small spacecraft aided monitoring. All this, along with the development of managing techniques, will create conditions for attaining in the 21st century:

increase by 3 to 5 times of resolution of observation of a locality from space; attainment of the ultimate periodicity of information update on ecology monitoring systems equal to 24 hours (in some dangerous areas quasi-continuous observation is performed); the use of up to 100 various spectral ranges in the course of observation.

The improvement of weather forecast, the extension of the forest credibility to 10 and more days (now it is 3 days); reduction of the possible damage caused by dangerous atmospheric phenomena (typhoons, hurricanes, storms) through finer location of disaster areas; determination of the nature of disaster and its development – these are the long-term goals set before space meteorology.

Logically, the comprehensive approach to such issues as environmental monitoring, ecological safety, the study of the Earth's development will result in building distributed space systems of small SC dedicated not only to observation and forecast, but also to control and correction at various levels of dangerous natural and man-caused phenomena (earthquakes, typhoons, epidemics, disturbance of the Earth's ozone layer, etc.)

### 2.1.2 Navigation systems

The space navigation systems are finding ever increasing use in a range of human activities (Fig. 10). They are developing into radically new systems of real time navigation of various facilities used for economic, scientific and defense purposes in order to navigate mobile objectives, to locate with a high degree of precision construction sites, to carry out geological prospecting and land survey, to monitor high value cargo transportation, to perform rescue operations, etc.

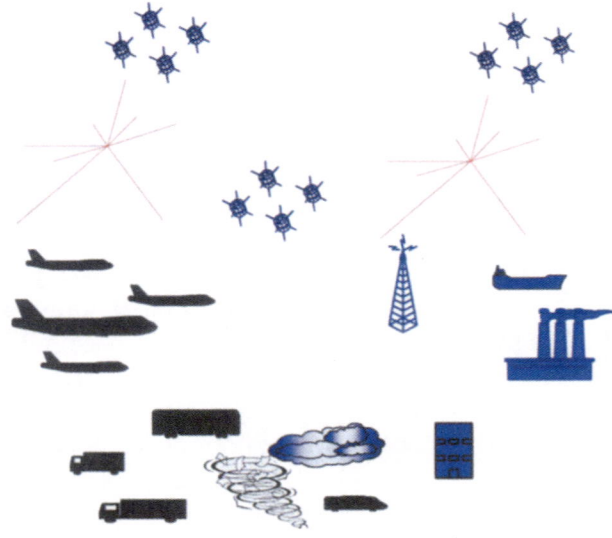

**Fig. 10.**
Prospects of development
of space navigation systems

Navigation tends to become ever more individual. Ever wider use is projected for digital maps with high precision location of current whereabouts of various objects (trucks, planes, etc.) by means of navigation systems using their own coordinate signals transmission devices.

### 2.1.3 Space energy, production and medicine

Wide use is predicted for the systems of orbital power stations that supply energy to both space- and land-based industrial facilities. According to the now developed concept of the Earth's power supply from space (which is based on the use of the Solar energy and its transfer to Earth or spacecraft in the form of radiation in microwave or optic range and anticipates the gradual transfer of a considerable part of electricity production to space), the orbital equipment will play a decisive role in power generation. The promising approach is the creation of power plants operating on new physical principles, and the search for new energy sources along with enhancing the efficiency of traditional technologies.

Starting in the second third of the 21st century, favorable conditions will emerge not only for building unique and experimental technological complexes in space, but also for removal from Earth to space of all ecologically dangerous manufacturing facilities. They could be placed both on piloted and automatic lunar production complexes. Such an approach will fully utilize all unique capabilities of the space-based production, and simultaneously will make the Earth better suited for human habitation.

The orbital medical centers can be used for research and the implementation of studies in genetics and microbiology, and for creation of new, unique medicines. This will not only provide unique conditions for developments in science, technology and production, but will also increase the reliability and safety of investigations of dangerous drugs and microorganisms.

## 2.1.4 Fundamental research

The data received from space are extremely important for understanding the profound space processes and their impact on Earth (Fig. 11). The possibility to conduct extra-atmospheric observation is absolutely essential to astronomic investigations. The Earth's atmosphere, consisting of nitrogen, oxygen and other gases, absorbs much of the stars' radiation, the reason why land-based telescopes can observe it in the narrow spectral transparency windows. The stars, however, radiate in a fairly broad electromagnetic range. The radiation of hot stars is at the maximum in the ultraviolet portion of the spectrum whereas only its negligible part is in the visible portion. The understanding of many processes of atomic excitation in stars and nebulas was oftentimes hypothetical for lack of data related to the study of stars in this range. Such data can only be obtained from observation conducted in space.

The magnetic field is one of planets' important characteristics. Its interaction with a flow of charged particles streaming from the Sun causes such natural phenomena as northern lights (Aurora Borealis) and magnetic storms. That is why the study of the magnetic field is quite important both for scientific purposes and practical applications. The data obtained allow to model magnetic hydrodynamic processes occurring in the ionosphere during magnetic storms.

The flights of spacecraft to the celestial bodies of the Solar system gave a mighty impetus to the research effort (Fig. 12). Previously, the planets were a matter of concern exclusively for astronomers. Now this also interests greatly geologists, geochemists, geophysicists, and specialists of many other sciences. A new package of sciences comes into being, including the physics of planetary resources, geology and geochemistry of the Solar system, the physics and chemistry of planetary atmospheres, the physics of magnetospheres and comets, the cosmogony, the astrobiology. It merges with the sciences of Earth and planetary astronomy as well as  with such seemingly unrelated sciences as the theory of star formation and the origin of life.

**Fig. 11.** Oreol-3 SC

Normally, three global problems are pointed out in outlining the main aims of investigating the planets and smaller bodies of the Solar system:

1. The current condition and the preceding evolution of the Solar system's bodies, the possibility to forecast their evolution in the future. Of special interest in this aspect are Venus and Mars, the two planets that resemble Earth more than others.

2. The origin and early stages of evolution of the Solar system.

The most valuable new data are expected to be obtained in the study of the Moon and the Solar system's small bodies, such as asteroids, comets, and small satellites. However, the study of big planets (including giants) can also help obtain important data.

3. The origin of life. According to scientists, Mars is the only place in the Solar system where to date there is a hope of finding traces of extra-terrestrial life or signs of its existence in the distant past.

The latter problem has always received much attention in Russia, which brought forth plenty of results acclaimed across the globe. The space rays, being a flow of atomic nuclei of different elements, literally bring into the Earth "pieces of substance" of various physical objects of the galactic and extra-galactic origin. The physics of space rays is a fundamental science that brought about a number of new trends in research. As work in space goes on, research goes ever farther and deeper. The discoveries made while studying the interaction of space rays with atomic nuclei of the matter, gave birth to the physics of high energy particles. Today, this is an independent science. With the emergence of SC the study of space radiation gave rise to the physics of near earth space closely associated with geophysics and the physics of magnetosphere. The continuous registration of space rays resulted in discovery of their generation on the Sun during its flashes. The study of this phenomenon with the help of SC together with investigation of the spread of solar rays in interplanetary space became component parts of the solar physics and the physics of solar-terrestrial bonds.

**Fig. 12.** Phobos SC

## 2.1.5 Space manufacturing technologies and materials study

The regular investigations dedicated to space manufacturing technologies and materials started roughly in 1976.

The conduct of scientific and technological experiments in space opened up radically new opportunities to study in-depth many physical phenomena, whose study on Earth is difficult or even impossible because of the gravity. The analysis of results of experiments considerably enriched the knowledge of the diffusion, crystallization and gasification. The research carried out on board orbital stations and automatic spacecraft generated a new trend in science and technology, space materials study.

In the course of applied research there came under study the crystallization of semiconductors, the impact of manufacturing techniques on the magnetic and superconducting materials, the thermodynamic stability conditions of alloy phases and of their reversibility with no-mixture areas, the influence of various modes on the properties of composite materials, the means of attaining concentrations of different types of inclusions in optic glass, etc.

The study of the processes of heat and mass exchange, of phase transitions in liquid and gaseous media necessitated the creation of special scientific equipment. For growing semiconductor crystals an array of multi-purpose dedicated devices had been designed and manufactured, capable of operating both in automatic and astronaut-controlled modes.

The experiments carried out to date showed that there is promise in continued work in weightlessness for production of new materials and obtaining of fundamental data in order to improve technologies on the ground. The most promising of these appears the production in space of high purity biological compounds, crystals, proteins, and semiconductors.

The important landmarks in research done to date are the obtaining of a large amount of experimental information about the growing of semiconductor crystals by various methods and the creation of conditions for modeling manufacturing processes in circumstances of prevailing micro-acceleration. In addition, new trends in science appeared, such as the physics of weightlessness, hydromechanics and heat and mass exchange in weightlessness. The main task of space materials study in the coming years is to carry out research for obtaining new fundamental knowledge of crystallization in substances, of heat and mass exchange in them, of conditions under which macro- and micro-heterogeneous structural irregularities occur. More thorough knowledge is sought of the composition, structure and electrophysical properties of crystals grown in micro-gravitation, which will serve as a scientific base for experimental or industrial space-based production of materials with unique properties unachievable on Earth. This, in turn, will improve manufacturing technologies used on Earth.

## Space biotechnologies and genetic engineering

New results, interesting for science and useful for industry, have been obtained in biotechnologies. Thanks to growing in space of fairly large and perfect monocrystals of proteins, determination of proteins' three-dimensional molecular structure speeds up. This makes it possible to effectively check the required restructuring of proteins by means of genetic engineering and to considerably reduce (by a factor of 3 to 5) the time and cost of development of new medicines.

The results of such research have brought to light a relatively small range of materials whose production in orbital flight is particularly promising. Nonetheless, the  data gathered is insufficient to enable the transition first to pilot manufacture and then to commercial production in weightlessness of some materials and biologically active compounds. However, it is hoped that in the near future the orbital centers of genetic engineering will resolve many of mankind's global problems.

## Space biology and medicine

Man's flights into space have become possible thanks to space rocketry and systematic research on space biology and medicine, new sciences dealing with the life of human and other organisms in space.

The biological research carried out during rocket flights and the flights of the first artificial satellites paved the way to space and largely determined the progress of piloted astronautics and, together with it, space biology and medicine. The development of these sciences in Russia proceeds along the following two lines: first, for exclusively practical purposes associated with practical tasks of medical support of space flights; and, second, the fundamental approach aiming at the study of weightlessness mechanisms, the space radiation, the electromagnetic rays, ultraviolet radiation and other factors with ill effects on living systems.

The results of medical and biological research done on board spacecraft testify to the fact that man can adapt himself rather well to space conditions and effectively work under them for one year. As this takes place, the human organism responds to short- and long-term exposure to space by adaptive adjustment of its gravity-controlled and regulatory systems. At this time substantial changes are observed in support and motion, cardiac vascular and sensor systems, in the neuropsychic sphere, in the blood and immune systems. Changes take place in the functioning of kidneys, in water/salt balance, in metabolism.

The investigations carried out resulted in the flight crew medical support system that includes crew selection and training, its medical care in flight (including emergency medical treatment) and the post-flight adaptation to Earth gravity and the customary pattern of life.

In course of medical and biological experiments with the use of various biological samples (including those at the cellular level) it was shown that weightlessness is the main ecological factor that disturbs normal functioning of the organism. However, it does not cause genetic or chromosomal mutations and generally does not disturb cell division. Nonetheless, the issue whether weightlessness directly affects the cells remains disputable to date. It is not unlikely that some physiological responses of the organism to the lack of gravity may be caused by the cells' changed behavior.

In the future, attention will be focused on the cell as a mechanical structure in the gravitational field and efforts will be made to evaluate the interdependence between the effects of weightlessness and the type of cellular differentiation and the degree of the organism's functional load. Also, it would be appropriate to study the role of the genetic apparatus in the organism's adaptation to prolonged weightlessness. Generally, it could be assumed that the lack of gravity does not impair the main functions during its life cycle.

In connection with the prospective flights of SC to other celestial bodies it is becoming ever more pressing to learn the conditions under which life in the Universe originates and spreads and to search for it or its signs on the Moon, Mars or elsewhere in space. The attempt to find life on Mars is one of the priority tasks among the planned research undertaken by automatic stations.

Over recent years, the space biotechnology has been developing by leaps and bounds. Its main aim is to elaborate the techniques of obtaining in space extra pure drugs and biologically active compounds (hormones, vitamins, enzymes).

In spite of its not yet long life, the space biology and medicine stand firm among other medical and biological sciences. This is explained by the uproarious development of these disciplines, by the novelty of tasks they face and by the impressive achievements that draw attention of specialists and the scientific community. The large amount of accumulated knowledge about the organism's activities in space affected conditions, including such factors as the flight's dynamic impacts and man-made habitation, plus achievements of space technology are essential prerequisites to the intense space exploration in the 21st century.

The trends under consideration will further gain both in quality and quantity. They will embrace all walks of life. The 21st century will be the time when the Solar system will become man's home.

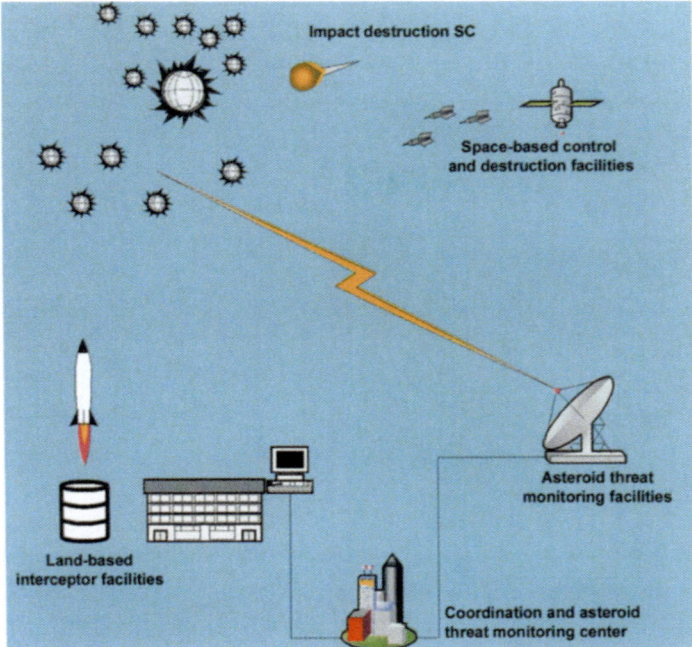

**Fig. 13.** Means of fighting the asteroid threats

## Safeguarding the asteroid safety

A collision of the Earth with celestial bodies, small as they are by the Earth's standards (measuring in diameter from hundred meters to one kilometer) may have grave ecological consequences or, in the event of their large size, may endanger mankind's life as such. The archeological investigations allow of a theory according to which occurrences like this are not new to the Earth's history. In this context, the safeguarding of asteroid safety (Fig.13) will be one of the astronautics' prime tasks in the 21st century, seeking to provide mankind's global security.

A brief review of the above stated prospects of development of orbital space hardware suggests an unequivocal indication of the increasing role of space technologies and their penetration into man's every day's life. One can safely say that mankind's future lies in the 21st century.

### 2.1.6 Creating prerequisites in science and technology

One of the most important features of space systems is the complexity of their design and development, the high standards of their engineering and operation. Specifically, the building of advanced space systems implies a fairly long period of preparatory work preceding the development of space hardware proper, during which period new advanced ideas, design decisions and technology findings are accumulated and checked out. The leap in development of orbital equipment stems directly from the package of solutions to a large number of interrelated scientific and technological problems.

The implementation of the advanced space programs of the 21st century is impossible without creating scientific and technological preconditions insuring the steady progress of space hardware. Given below are the main tasks in formulating and implementing the programs aimed to create preconditions in science and technology:

**Fig. 14.**
U-15T1 Electric arc installation at the pilot production workshop of the TsNIIMash (Central Research Institute of Machine Building). Standing in the foreground is academician N.A.Anfimov

- providing the high level of performance characteristics of advanced models of space rocketry;
- organization and carrying out of work in priority research areas related to the predicted tasks and targets of the world's and Russia's astronautics;
- sustaining at the required level of the intellectual potential, research and laboratory equipment, trial and pilot production facilities of the space industry (Fig. 14).

The targets of the prospective program for development of orbital equipment must be as follows:

- creation of scientific and technological preconditions in base technologies, primarily, of dual application for development of space hardware with required characteristics for use in defense, research and business;

preventing Russia's global lag in science and technology from the world powers in terms of space hardware for various uses and space exploration in general;

- insuring competitiveness on the world's market of domestic base technologies and space hardware for various applications built around such technologies;
- further improvement of control systems used in research and development aimed to create advanced space hardware for military, scientific and business applications.

The prime targets of the prospective program for development of Russia's orbital hardware must be as follows:

- concentration of efforts and finance for creation of top priority and pioneering technologies;
- involving the state's interested organizations and non-budgetary entities in the creation of preconditions in base technologies and their large scale use for defense, business, and international cooperation;
- the exclusion of redundancy and duplication in research and experiments carried out to various orders, the reduction of cost of developing the base technologies;
- the reduction of time needed to develop advanced models of space equipment by making full use all-inclusive solutions to scientific and technological problems and by resorting to the full spectrum of opportunities provided by the created preconditions. The reduction of the total cost of utilizing the complete life cycle of space equipment and of systems for various applications;
- organization of a competition in research and experiments carried out for the creation of base technologies and specific key elements of prospective space systems;
- preservation of the cooperative principle in the work of organizations conducting research and experiments for the creation of base technologies. Involvement in cooperation of new organizations, including commercial

ones having a high scientific and technological potential and capable of resolving technical and engineering problems pertaining to development of space systems and complexes for various applications;

• insuring standardization and unification in the development of component parts, key elements and generally advanced spacecraft and systems for business, research, and defense.

All prospective problems have to be resolved on the basis of advanced development of space base technologies used in designing, engineering and management, many of which will gain wide use across the globe.

The prospective technologies of designing orbital equipment must in the first phase provide for automatic optimization of basic elements, modules and subassemblies of spacecraft (a platform, heat regulation systems, controls, etc.) and then of serially produced SC and systems based on them with the extensive use of standardized modules, subassemblies, schematic decisions etc. Automation of the designing process and, subsequently, of the assembly procedure, will dramatically reduce the manufacturing cost of space equipment.

Of special significance will be the designing of space systems on the basis of small spacecraft with the distributed architecture. The very notion of the "small spacecraft" bespeaks the breakthrough in science and technology that is characterized by organizing the design, manufacture, trial, launch and operation at a new technological level. Now, in the course of development of small SC, technological expertise is being accumulated, which can, by right, be regarded as a base for space research in the 21$^{st}$ century. Small SC (micro- and nanosatellites in the future) will be the main cog wheel in the gearbox of high tech space systems for various uses. The chief advantages of small SC are their simple design, versatility of modules used, easy manufacturability and relatively low cost of production.

Among Russia's many space rocketry plants, the Khrunichev Center has the highest scientific and economic potential for development, production and operation of small SC.

Based on the analysis of application and operation of small SC carried out by the specialists of the Khrunichev Center, basic engineering principles have been formulated which allow to create rather quickly the required modifications built around typical elements. The main element of small SC is the standardized space platform weighing around 350 kg and fitted with service systems that maintain the conditions required for functioning of special purpose instruments and keep up the flight operating modes (Fig. 15, 16).

Having a modular construction, small SC (SSC) can be easily adapted to tackling a variety of tasks both by using special modules and by using their own component package. The modular nature of the platform design permits of an independent assembly and check-out of separate systems and units.

Also, it allows to improve various modules virtually independently of one another, thus creating new types of SSC. This, in turn, in case of non-tight fastening, will ease repair and replacement of units and assemblies in orbit.

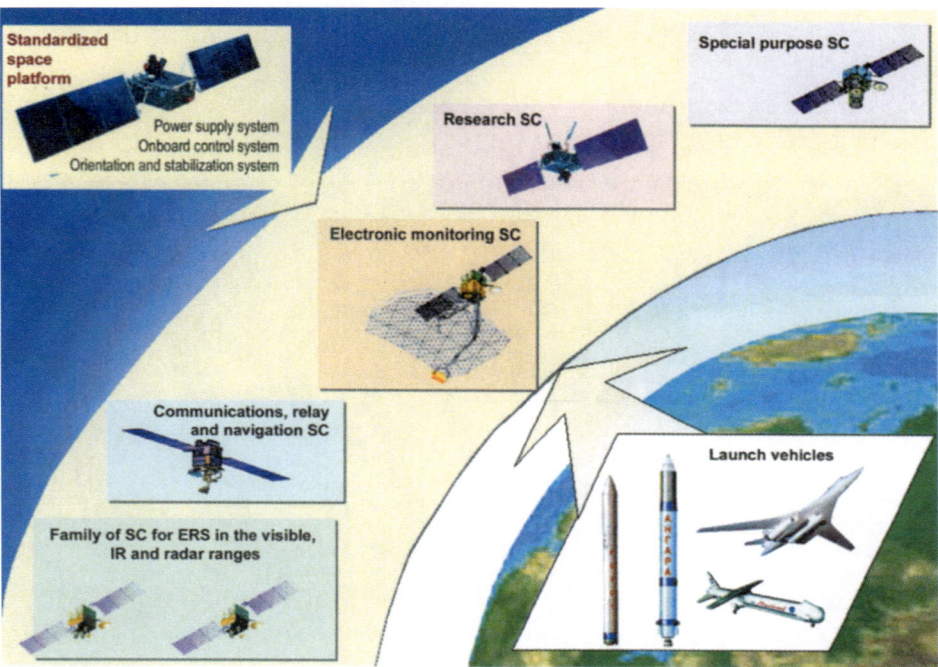

**Fig. 15.** Creation of SC for various applications on the basis of standardized space platforms (concept offered by the Khrunichev Center).

The basic variant of the platform includes task-oriented instruments and onboard systems for power supply, orientation, stabilization, and heat regulation plus corrective propulsion unit, onboard digital computer, etc.

The analysis of state-of-the-art in space rocketry shows that the development of SSC is linked with further reduction in weight, size and cost, with increase in the efficiency of energy systems, improvement of reliability and extension of the SSC service life on the whole.

The use of SSC offers a number of advantages:

1. more economical and better suited solution of a number of problems related to earth monitoring, communications, navigation and control, optimization of new technologies that require rapid response;

2. reduction of cost and time of development, manufacture and deployment of spacecraft and systems enabling advanced technologies to be introduced;

3. the possibility to materialize quickly the basic design and engineering decisions with subsequent modernization thanks to the modular configuration;

4. small own gravitational, electromagnetic and gas backgrounds;

5. low cost combined with the capability of launching by means of cheap carrier rockets obtained from conversion and by prospective aerial and space equipment;

6. higher survivability and reliability of manned space systems thanks to the higher technological reliability of SSC proper and of structures they are based on;

7. lesser financial and engineering problems in deployment of systems based on SSC;

8. possibility to use more widely on a commercial basis the existing systems and to introduce new ones due to their lower specific cost.

**Fig. 16.** Flight prototype of the Yakhta standardized platform at the assembly shop of the Khrunichev Center

According to foreign experts, the cost of manufacture of large SC varies between US$ 88,000 and US$ 220,000 per kilo with development time being 5 to 7 years, whereas for small spacecraft the respective figure is US$ 17,000 per kilo with development time of 2 years. Such SC can be launched by cheap light class carrier rockets like Rokot, Kosmos and others.

For example, the Khrunichev Center plans to launch into geostationary orbit in 2003-2004 from the Plesetsk spaceport the first small spacecraft Dialog. The launch will be accomplished with the help of the Rokot carrier rocket whereupon the craft will be placed within 140 days in the work point by means of low thrust engines.

In the development of advanced space engineering technologies and special purpose spacecraft, one can single out a number of trends associated with the currently available component packages that make up the backbone of prospective SC and ask for prompt all-inclusive optimization.

## 2.1.7 Multi-functional constructions

The combining of functions of electronics, sensors, electric power distribution systems and heat regulation devices with the use of very light modular constructions, the prospective SC won't have cables and distribution boxes linked with the bus-bar. This will reduce the weight of SC by 10 times, and halve the space it occupies. The electronic modules on many chips will be mounted directly on the bus or built into it. Flexible connecting elements or connectors will connect electric circuits without the intricate interlacing of heavy cables. Such constructions will considerably decrease the loads and vibrations occurring during SC launch and while deploying systems in space. This will lower the strength requirements of SC and cut down their weight.

**High and super-high density connecting elements, microcircuit modules, radiation-proof processors and built-in electronic devices.** Microcircuit modules are expected to be used as semiconductors. Modules will not be mounted on the surface of elements, but will be embedded in them. They will be thin enough to form a layered structure that provides the theoretical density of ~10 tera flops/m$^3$. The dimension and weight of SC can be reduced by 4 to 8 times as power consumption requirement slackens by 7 to 15%.

Based on the combination of several microcircuit modules and connecting elements, prospective receiver modules of the transmitting antenna will be developed for orbital platforms of information processing SC.

**Architecture of building the prospective space systems.** The aim of work being done in this area is to investigate the feasibility of building a distributed system of SC that performs the functions of one major SC. Such groups of SC will fulfill various information related tasks (radio-location, gathering of information concerning the near earth and outer space, the building of distributed antenna systems, etc.)

**Lightweight solar battery (SB) panels** are expected to be built on the basis of constructions manufactured from composite materials without using cables and wirings with the use of electric buses and control systems built into the panels and auxiliary elements' framework. SB will use reusable, cheaper, lightweight plastic hinges with better characteristics. In the use of thin layer elements and concentrators the specific power of SB can be as high as 116 W/kg (for the existing SB it's 40 to 50 W/kg). The special construction of SB will damp vibration and heat impact on the high performance photo-galvanic thin layer solar cells with several junctions.

**Inflated constructions** will enhance miniaturization (by placing satellites in containers), cut down the cost of manufacturing and launching of SC while providing for deployment in space of large-sized structures, e.g. antenna systems measuring up to 450 m in diameter which will dramatically improve the performance of information processing SC.

**Versatile integrated energy-signal systems.** It is anticipated that vapor-phase deposition method will be used which allows to integrate thin layer batteries and all-purpose electronics with flexible photogalvanic cells. This technique yields a layered structure of the satellite bus. Such flexible batteries could be wrapped around other subsystems which makes it possible to assign the bus functions to big modular antennas of radar stations, thus creating a sort of "a magic flying carpet".

**Microelectromechanical systems.** Such mechanisms measure around 1 mm. Millions of elements of microelectromechanical systems, each of which has a micromotor measuring from 0.1 to 1 mm in diameter and is accessible for control, can be activated individually in order to provide a certain arrangement in the group of microsatellites. Teflon can be used for fuel. After ignition with the help of a mini-charge it will generate a small pulse. A laboratory example of such a device has been activated by now around 500,000 times and demonstrated a negligent consumption of the Teflon fuel. The creation of this type of micro-motors will give birth to the concept of an expendable construction in micro-electromechanical systems. Teflon or any other fuel can also perform the function of strengthening structural elements and be used simultaneously for operating micro-motors that correct the SC orbit. .

**Large-size lightweight optic devices.** Component packages will be developed for space telescopes and laser systems based on high precision lightweight flexible and adaptable optic devices, reflective mirrors and nano-metric accuracy control systems having a highly reliable and endurable coating. The use of optic or laser communication, primarily between SC, will enable the transmission of large amounts of information.

Such technologies will be used to obtain high-resolution images from space for observation of the Earth's surface and space. That work will serve as a base for wide use of hyperspectral images for location and recognition of various objects.

**Robotics.** The most pressing task of robotics now is in all likelihood the automatic servicing and repair of various types of SC in orbit. This will dramatically cut down the capital outlays expended on replenishment of orbital constellations thanks to improved recoverability of the hardware.

**Prospective communications systems.** The building up of the energy potential of the communications SC onboard systems is unthinkable without large-size antennas. The manufacturing techniques of large size antennas with an aperture as large as hundreds of meters can already be developed in 2005-2010. During the period before 2010 single cell mirror antennas can be created with the amplification ratio up to 50 dB and multi-cell antennas measuring up to 200 m in diameter with amplification ratio up 100 dB.

**Computing assets.** In addition to the development of new technologies that insure the creation of the prospective component package of space hardware, the integration of all SC systems with computing equipment must also get special attention. Actually, the onboard computers have already become the central link that combines into a single whole a variety of onboard systems. Thus they determine the performance and efficiency of the SC used. The promise of this work lies in the massive introduction of variable structure processors that feature the properties of parallelism and form micro-miniature distributed computing systems.

**SC control systems.** It is also worth of attention how to improve the reliability and self-sufficiency of the SC control systems. The prospective SC and the systems based on them will have many properties of self-sufficient, self-regulating systems which will start to map out their own behavior for fulfilling the set tasks, perform self-diagnostics and recovery both on a single SC and within an orbital constellation of them.

The large capacity holographic memory devices will enable a great progress in the onboard systems for storage, processing and transmission of high rate flows of information, in the construction of autonomous adaptable self-learning systems for controlling SC.

**The prospective trends in development of electronic equipment.** It should be noted that along with the development of the traditional information support equipment, special attention is being currently paid to the prospective trend in electronic engineering based on the use of techniques of super-wide band short pulse electrodynamics (SPE).

The SPE techniques allow to obtain more effective, occasionally radically new, solutions of the following problems:

1. super-high resolution radiolocation (including beneath surface vision);
2. super-wide band communication with the spectrum close to "the white noise";
3. creation of distributed powerful electronic systems that are actually pulse analogs of multi-cell phased UHF arrays.

**Environment active protection and recovery facilities.** The preservation of the biotope, including near earth space, must also be one of the prime targets of space work. Quite important are the preparation and implementation of a number of programs aimed to ease and eliminate the negative consequences of man-caused effects on environment by means space rocketry. The tasks of such programs include:

- preservation and recovery of the ozone layer;
- removal of especially dangerous industrial and energy waste to space;
- clearing the near earth space of fragments of man-made facilities.

Those tasks can be fulfilled with the help of both mechanical equipment (sweep nets, reusable space waste collection stations, etc.) and energy radiating orbital facilities (laser and beam installations, devices operating in UHF and EHF ranges).

**Power motive plants of SC, interorbital towing vehicles.** One of the problems closely associated with enhancing the reliability and orbital service life of prospective SC is the development of durable and effective means of onboard energy-saving, of orbit correction, of SC maintenance and repair. Evidently, the solution of this problem will influence in many ways the solution of all prospective tasks of space work since prospective orbital hardware requires increasing power supply with its active service life varying between 10 and 15 years. Motive and power supply installations of orbital hardware are particularly complex to manufacture and optimize. Their production is a highly labor-intensive process. The adequacy of power and weight characteristics, the active service life duration, the reliability and capability of power supply systems determine largely the functional capacity and cost effectiveness of space rocketry in general. The development and optimization times of motive and power supply systems are rather long, varying between 5 and 7 years provided that sufficient preconditions are created in science and technology. This is precisely why advanced development of motive and power supply systems must be assured for the fulfillment of prospective tasks of astronautics. Only under such circumstances would it become possible to develop and build in good time complexes and systems not inferior in the main parameters to foreign counterparts and even capable of competing against them on the world's market. Ample expertise is available to further increase the specific power of power supply systems by 2 to 3 times due to improvements in the systems of power generation, accumulation and conversion.

The important means of enhancing the efficiency of using space complexes and systems and extending their service life is the employment of motive and power systems based on solar or nuclear powered plants and electrically driven rocket propulsion units. For example, the use of a towing vehicle with an electrical rocket engine will increase by 2.5 to 2 times the weight of SC (purpose-built module) in geostationary orbit.

Such prospective means of interorbital transportation are characterized by a considerable level of power consumption (40-100 kW) and by onboard energy systems with better energy and weight characteristics. One of the promising trends in the development of onboard power units is the creation of solar gas-turbine installations with electric power capacity of 10 kW and above. Those come into being thanks to the following advantages:

• higher efficiency in conversion of solar energy into electricity (25% with the prospective 50%);

• maller size as compared to that of power plants based on solar batteries;

• the option to generate high voltage alternating current;

• ong service life (thanks to the use of gas-lubricated bearings in turbo-compressor supports);

• the lesser cost of the power supply plant;

• the promising nature of gas-turbine energy converters that could be used together with nuclear and chemical energy sources;

• the option to use thermal energy accumulator.

The work on nuclear and motive energy installations aims to create advanced scientific and technological preconditions for building basic standardized elements, assemblies and units for nuclear energy installations. The priority trends in research that may demonstrate the advantages of nuclear power sources over other versions could be as follows:

a. the development of technologies that assure the creation of nuclear power plants featuring the power in the region of hundreds of kilowatts (with its prospective increase);

b. bringing the guaranteed service life of the nuclear power plants up to the level not less than that expected of the solar energy technology (including up to 10 years and more on geostationary orbit);

c. the development of technologies that assure the creation of bimodal nuclear electromotive installations (operating both in the nuclear rocket engine mode fueled by hydrogen and in the electric power generation mode for power supply of purpose-built and service instrumentation of SC or electrical rocket engine);

d. the confirmation of nuclear and radiation safety of development and operation of nuclear power plants (nuclear powered propulsion units).

The creation of advanced scientific and technological preconditions for the said trends in improvement of onboard (solar, chemical, and nuclear) power systems and equipment and the utilization of such preconditions in the development of power supply systems will enhance the efficiency of dedicated use of SC and provide the required base for implementing space programs in the 21st century.

## 2.1.8 Advanced space materials

The fulfillment of the entire variety of complex constructional, schematic and technological tasks in the creation and operation of space hardware is unthinkable without extensive development and massive introduction of the results of the space materials study.

The development of space hardware calls for new materials capable of withstanding the loads of space flights, vibratory stress during launch, low temperatures of outer space, deep vacuum, radiation exposure, microparticles impact, etc.). In addition, they must have a fairly low specific weight. All the spectrum of severe exposure, oft-times with abrupt transition of stress to the metallic and non-metallic structures and their elements, dramatically affects their in-depth structural properties, and hence the reliability and endurance of space hardware for various applications.

Metals are the main structural material for building rocketry hardware in whose "dry weight" they account for more than 90%. Therefore, the improvement of products' performance depends largely on the properties of alloys used.

Over the last few years a new generation of lithium and scandium doped aluminum alloys has been elaborated which will shortly receive wide acceptance. The replacement of traditional alloys by new ones will cut down the weight of rocketry components by 10 to 30% depending on the construction type. The techniques of manufacturing component parts from new granulated alloys combined with the possibility to raise their operating temperature to 850°C will cut down the weight of components by 10 to 30%.

The new class of structural materials called the intermetallides (chemical compounds of titanium-aluminum, nickel-aluminum and others) may revolutionize the construction of advanced rocketry of the 21st century. Those materials have a low density (3.7-6.0 g/cm$^3$) and feature high resistance against heat (up to 1,200°C), corrosion, and wear.

The titanium alloy, now in development, will equal in terms of engineering performance the traditional stainless steel (no equipment is required for welding and heat treatment in controlled atmosphere). Due to doping mainly with hafnium and niobium, the alloy will retain its non-oxidability when heated up to 850-900°C. No heat treatment will be required to remove residual stresses, which makes unnecessary the use of heat treatment ovens and welding chambers with controlled atmosphere. Should it become necessary to eliminate warpage caused by residual stresses (for example, in large size constructions like frames, girders, bottom protection screens, etc.) the heat treatment of welded units can be accomplished in air without subsequent sandblasting or etching. Machine components can be welded using only argon flow for protection without fear of seam oxidation. The seam will be workable in the wide temperature range between – 253 and + 450 °C. It opens up wide prospects for the use of titanium in rocketry instead of stainless steels and improves the product performance by nearly three times.

The enhancement of metallic materials strength by traditional means (increase of doping, improvement of thermal and mechanical properties) has to date practically depleted its capacity. Modern alloys contain a large amount of costly and rare metals, such as cobalt, tungsten, niobium, molybdenum, nickel and others which boost their cost. Furthermore, the considerable increase of doping elements in alloys brings about zonal and volumetric liquidation in ingots and consequently anisotropy of properties in semi-finished materials and products from them.

Much reserve for improving the structural properties of space rocketry is in the use of intermetallide compounds. Of special interest for the development of heat resistant structural materials based on intermetallide components are the complexes like titanium – aluminum, nickel – aluminum, iron – chrome – aluminum.

By their structure, the intermetallides (chemical compounds of metals) occupy an intermediate position between metals and ceramics. They have a complex crystalline structure with up to 30% of the covalent component in interatomic bonds, which feature determines their unique physical and mechanical properties, such as high heat resistance and thermal stability, high resistance to corrosion as compared to stainless steels (especially in oxygen), high resistance to wear. In addition, the intermetallides have a low specific density. Intermetallide alloys based on titanium can operate at temperatures up to +850°C without a protective coating. Those based on nickel can operate at temperature up to +1,500°C.

The intermetallides's properties can revolutionize the developments in many areas of technology, primarily in creation of advanced models of airborne and space equipment, including vehicles flying at hypersonic speeds (up to M = 25). The use of intermetallides in power units (rotor, stator, impellers, valve assemblies, uncooled nozzles etc.) will increase the motor specific thrust by 25 to 30% and decrease the structures weight by 40%.

**Prospective non-metallic materials.** Thermal control coatings. One of the main factors that govern the reliability and endurance of SC is the stability of its thermal operating mode since modern optoelectronic instruments of SC operate within specific temperature  range. The spacecraft thermal control instruments include various thermal control coatings that establish the balance between the heat issued inside the craft, the energy absorbed from space and the energy reemitted into outer space.

TCC are characterized by radiant heat performance which changes under the influence of various space factors (especially, ionizing radiation). This change builds up temperature  inside the spacecraft and shortens its service life. As the previous years' experience shows,  some spacecraft failed to implement the planned missions in consequence of  overheating because of the heightened ratio of solar radiation absorption by TCC in the system of passive thermoregulation. The analysis of the existing TCC shows that they are unable to assure the extension of serviceability to 15 years, especially for SC operating in high elliptical and GEO. Therefore one of the prime tasks of astronautics in the 21st century is the creation of thermoregulation coatings of "solar reflector" and "true absorbent" class which are stable heat radiators that remain antistatic during prolonged operation in space and feature only slight gas emission. The development of such coatings will reduce to the minimum the deviations from the established heat mode, cut down malfunctions and failures in the high sensitivity optic and electronic equipment which will extend the service life of spacecraft to 15 years. The promising ways of fulfilling this task are:

• the development of combined or modified heat resistant and radiation-proof components with a low gas emission (acryl, silicone, urethane resins);

• the choice or development of effective agents to counter materiel deterioration in space;

• the development of white or black pigments, including that with heightened electric conductivity and resistant to prolonged exposure;

• the development of detachable coatings to provide protection for product manufacture and storage up to 5 years.

**Advanced polymer structural composite materials. Antenna structure mirrors** made coal-plastic will gain wide acceptance for communications via satellites. At the weight of under 15 kg they will secure the destruction stress of 900 kgf with the service life of no less than 20 years.

Cellular materials (three layer) from coal-plastic in the carrying elements of constructions as compared to single layer types (monolithic) in the set operating conditions and at increased loads at the set weight of the cell will assure:

• the reduction of the structure element's weight by 40-50% and increase of its rigidity by 60-80%;

• the increase in reliability by 20-25% and extension of guaranteed operating life by 60-70%.

• In addition this type of materials will insure some special electro-physical properties (e.g. for radio locator antennas) and provide some heat stability and heat conductivity properties.

**Pressure bottles.** Pressure bottles made from polymer composite materials and operating under pressure are used successfully in space rocketry. Also, there are in operation fuel tanks, ball-bottles, rocket engine cases, pressure accumulators, breathing bottles for pilots and astronauts. The use of organic and glass fibers will allow to create durable lightweight pressure bottles.

The creation of elements of precision instruments implies the manufacture of components that retain their geometrical size (size stability) within the widely varying operating temperature range ($\pm 150$ °C). Technologies will be developed that enable the creation of polymer composite materials from coal-plastic that assures high size stability of instruments elements for a given temperature range.

**"Smart" materials.** The progress of science and technology is inseparable from the development and introduction of new materials. During the last decade, along with the constant improvement of the existing materials that insure the essential engineering and economic effect thanks to the unique combination of properties, new tendencies emerged in the materials study aimed to create new materials capable of active interaction with exterior factors. Such materials got the names "smart", "clever", "wise" and the like. They can "feel" their physical condition and exterior exposure and respond in a special way to such "sensations", i.e. they can conduct a self-diagnosis should a flaw occur and develop and take corrective actions in order to stabilize their condition in the afflicted areas.

Thanks to their diverse properties, the "smart" materials can be used in various structural elements of space rocketry (casings, fairings, compartments, friction assemblies, etc.). The use of such materials will make it possible to monitor and predict the condition of various structures and constructions at the required time even in hard-to-reach places and to considerably increase the systems' service life and reliability.

The analysis of experts' evaluations shows that in the coming 20 years 90% of modern materials used in industry will be replaced by new, specifically, "smart" ones which enable creation of elements of constructions that will shape the progress of technology in the 21st century.

**Tightening and sealing materials.** In spite of the diversity of tightening and sealing materials available, there is a need for the development of new prospective materials aimed to meet the demands of astronautics in the 21st century. It appeared because of the stiffening requirements to reduce the number of manufacturing processes in the production of items, the expansion of the temperature interval, the increase in capacity for work and service life of spacecraft and launch vehicles. The tasks are being set to create new types of rubber, sealants and compounds (including current conductive rubbers and sealants; heat-, freeze-, exposure-resistant rubbers; heat-, exposure-resistant anaerobic sealants; heat-conducting, UHF energy absorptive compounds).

Current conductive rubbers and sealants with performance improved by a factor of 1.5 to 2 thanks to the progress of manufacturing processes will remove static electricity from SC and increase their service life from 5 to 10-15 years.

**Radiation resistant lubricants** are required to assure the reliable peration of friction assemblies in various gas and liquid media in the broad temperature range on the ground and in space for 10 to 15 years. Plastic lubricants are a versatile means of protection of machines and their elements against climatic exposure during operation and storage. The lubricants being developed must be effective in all climatic zones and suitable for prolonged storage even on open sites.

**High elasticity and low gas emission constructional glue**. Vibration and impact-resistant, thermo cycling proof epoxy silicone glues are widely used now to fasten solar battery cells, brackets, and other hardware elements. Also, they are indispensable in repairing the loaded surfaces of space equipment. Their big drawback, however, is the considerable gas emission (up to 8%) in vacuum or under high temperature. The gaseous emissions pollute work surfaces of optoelectronic instruments mounted on SC and often affect their normal functioning. To insure cleanliness of instrumentation (extension of their durable operation) it is necessary to develop and use on rockets exterior surfaces such materials (including the glue) that lose not more than 1.0% of their total mass and emit not more than 0.1% of easily condensable compounds.

The gluing of heterogeneous materials operating in conditions of thermo cycling, high vibration and strong impacts requires highly elastic and strong (up to 20 MPa) types of glue.

Current conducting glues are intended for providing electrical contacts in cases when hot soldering is inadmissible or impossible, i.e. in hard-to-reach joints of screen partitions and casing.

In space rockets, the current conducting sufficiently strong structural glues are used in the control systems for:

• fastening current conducting elements, installation of electronic instruments' electrical circuits;

• screening separate units in awkwardly shaped constructions, for electrical sealing of assembly units.

Today, conditions have evolved in science and technology for creating current conducting cold solidification glues not containing precious metals, which glues are intended for making highly reliable current conducting connections in control systems of space rocketry, for screening separate places (awkwardly located for soldering) in constructions of complex shape. The creation of current conducting glues with high confection properties will remove static electricity charges from SC surfaces and hence will enhance the reliability and service life of electronic equipment and improve their fire safety.

Carbon-based materials. In the developments of new materials based on carbon, wider acceptance awaits carbon-carbon and carbon-carbide composite materials which will find wide use in space rockets (elements of power plants, thermal protection, fragments and radiation resistant screens, radio transparent constructions, etc.). Having higher performance characteristics (and a higher cost) they will cut down the product weight by 30 to 50%.

Control techniques. In work on the advanced control techniques the following should be pointed out as top priorities: controlling multi-satellite distributed space systems (including those based on micro- and nano-satellites); development of self-learning autonomous control systems based on neuro-network technologies, artificial intellect; the reduction of the land-based infrastructure of control systems; insuring safety in the use of space as it becomes ever more littered with man-made refuse and crowded by deployed SC.

The analysis of trends in development of orbital equipment in the late 20th century suggests their following peculiarities in the first half of the 21st.

The first peculiarity is attributed to the consolidated efforts in space communications systems in order to build multi-satellite low orbit communications network. Under such circumstances, till the mid 21st century the key role will belong to the orbital systems of communications and data transmission which are deployed in the area of geostationary orbit. No lesser role will be played by navigation systems located in the area of medium altitude orbits.

The second major trend in space exploration in the first half of the 21st century will be a dramatic growth in the number of orbital systems and hardware (in the first place those based on small SC as well as micro- and nano-satellites) that operate in near space (Fig. 17). Simultaneously, a considerable growth is expected in the relative number of small SC, including nano-satellites, while the share of large SC in tackling various tasks is due to decrease.

It should be noted that the priority development of the above described technologies will be the core activity in space exploration in the 21st century.

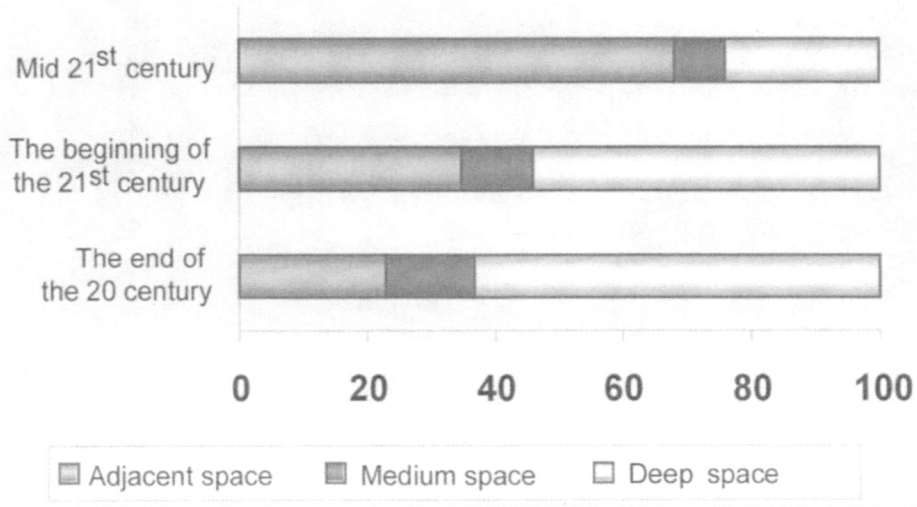

**Fig. 17.** The expected changes of relative number of orbital equipment functioning in adjacent, medium, and deep space (%)

### 2.1.9 Multi-functional space systems

The initial 10 to 15 years of the space era were the time when a tendency emerged to switch from launches of single spacecraft to the creation of permanent orbital space systems. That tendency became standard practice in an avalanche fashion (Fig. 18).

In their turn, large size space systems gradually expanded the spectrum of their tasks, embracing an ever-greater number of industries and branches of science. By contrast with the dedicated systems, multifunctional ones began to be manufactured and put in operation. This tendency was particularly pronounced in the best paying field of space work, in satellite communications.

The market situation in the mid-90s got particularly fierce in the segment of communications services with the use of space hardware. The desire to win over more and more customers had generated space systems that enable a user of a mobile communications system to access navigation related data or to transfer formalized or ERS related data, or to maintain communication between computer networks.

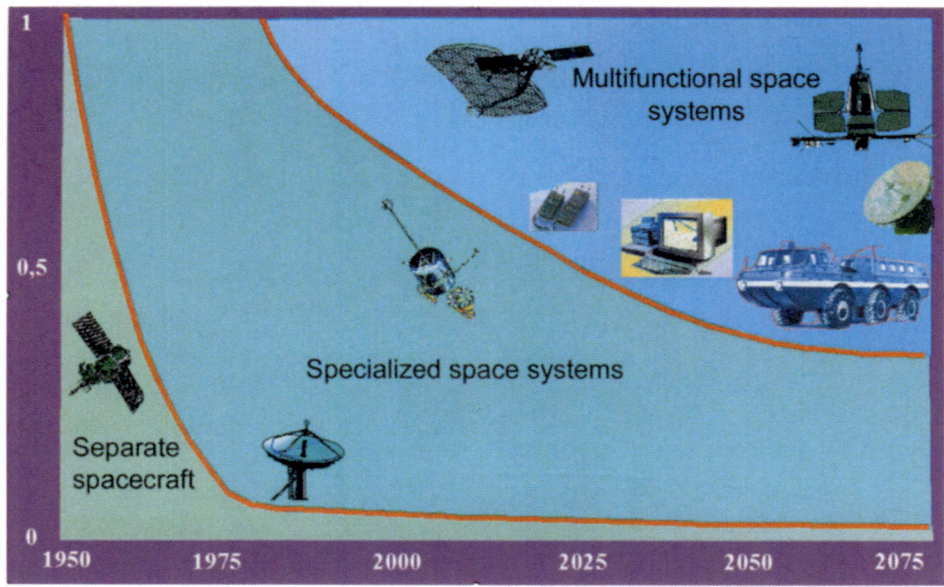

**Fig. 18.** Transition from launching separate spacecraft to multi-functional space systems

ARGOS (Fig. 19) could exemplify the system that combines the functions of earth remote sounding (ERS) and data transmission. This system's instruments placed on the NOAA weather spacecraft receive signals from users' transmitters and transfer the processed data to the monitoring center where the final calculation of their position is taking place. In addition to the data concerning the user's position, the monitoring center can receive information from ERS.

One of the weak points that impairs the efficiency of ARGOS, especially in the case of serving a big number of users, is the restricted quantity of satellites in the space segment. Generally, the distinctive feature of modern multi-functional space systems is a relatively large number (several tens) of satellites placed in low or medium altitudes or a simultaneous use of high and low orbit spacecraft within the system, i.e. the use of sophisticated orbital structure of the space segment.

Thus the space segment of the GLOBALSTAR satellite communications system comprises 48 satellites placed in eight medium altitude orbits. The orbital structure implies a global servicing zone restricted by 70° of n.l. and 70°of s.l. for most users. The multinational nature of the system manifests itself in that mobile and personal users of the system can get not only telephone networks services but also receive data (including FAX, Paging) and information concerning location.

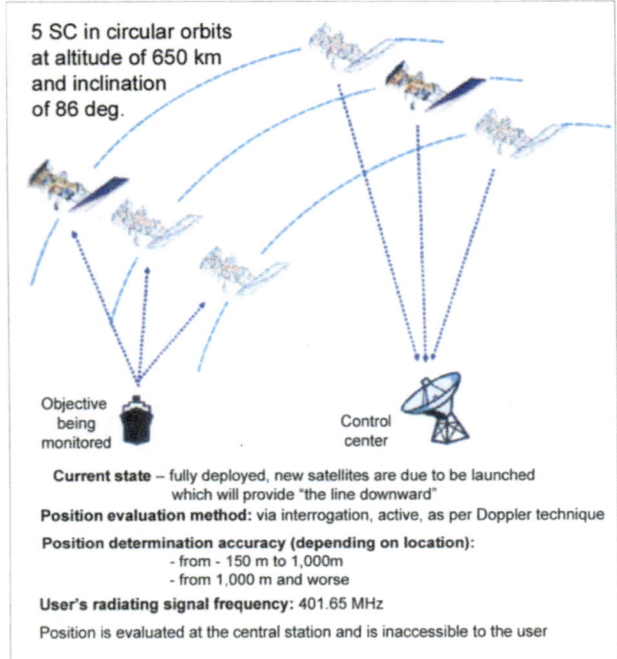

5 SC in circular orbits
at altitude of 650 km
and inclination
of 86 deg.

Objective
being
monitored

Control
center

**Current state** – fully deployed, new satellites are due to be launched
which will provide "the line downward"
**Position evaluation method:** via interrogation, active, as per Doppler technique

**Position determination accuracy (depending on location):**
- from - 150 m to 1,000m
- from 1,000 m and worse

**User's radiating signal frequency:** 401.65 MHz

Position is evaluated at the central station and is inaccessible to the user

Fig. 19.
Main characteristics of ARGOS
global navigation system

Multifunctional space systems are, as a rule, international. This is both due to the size of territories involved and the considerable amount of investments that is well above the financial capabilities of even the richest companies of the world. At the same time, a tendency still persists to develop national multi-functional systems intended primarily for safeguarding national security.

So the ORBCOMM data transmission system designed for a two-way exchange of alphanumeric information between mobile and stationary land-based consumers is intended to perform a variety of missions, including those for the US Army (Fig. 20). The use of the system abroad involves, in addition to the request for utilization, an issue of a national license.

Russia's ROSTELESAT satellite communications arrangement is a specimen of multifunctional space system. Its main aim is to organize generally accessible and specialized purpose networks for communications and data transfer between mobile and fixed users. As a multifunctional system the ROSTELESAT will provide the following services:
- formalized data transfer;
- communication between local computer networks;
- transfer of ERS data in real time mode;
- navigation support of the system's users.

The ROSTELESAT space constellation is expected to be comprised by 70 low orbit artificial earth satellites placed in seven orbital planes at the orbit altitude of 700 km and inclination of 82°.

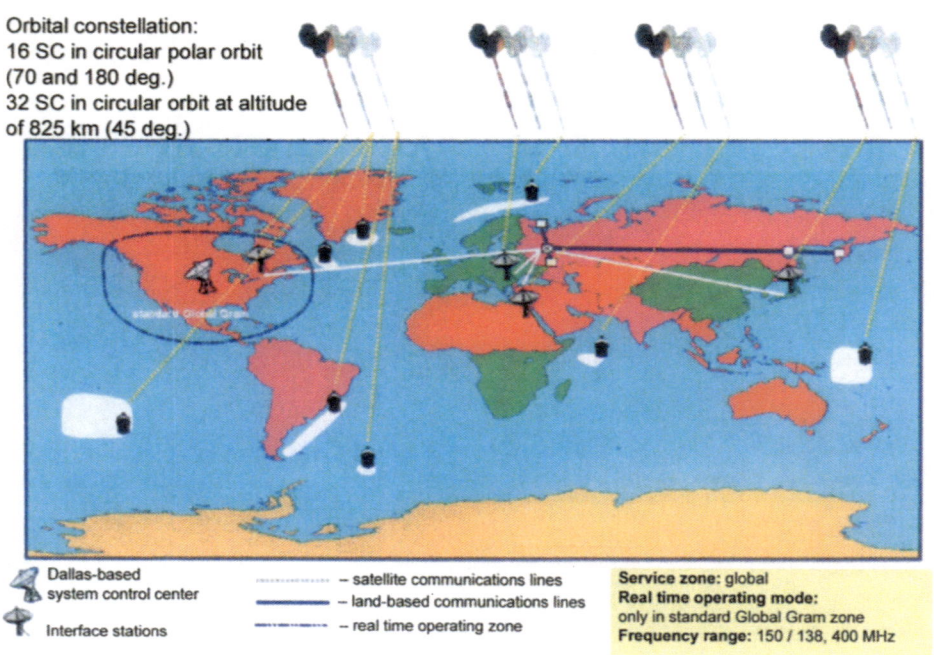

**Fig. 20.** Main characteristics of ORBCOMM satellite communications system

A totally new approach to the creation of multifunctional space systems is the principle of maximal integration of space hardware both produced or in production either under the Federal Space Program of the Russian Federation or by separate entities and organizations. This principle has been used by the Khrunichev State Space Research and Production Center in designing the Federal Monitoring System (FMS) employed to survey land-based objectives and earth resources.

Three subsystems can be singled out in the component package of FMS (Fig. 21), each of which anticipates the use of space equipment:

• 1. The sensors subsystem that locates the monitored objectives and evaluates their condition.

• 2. Communications and data transfer subsystem (telecommunications subsystem).

• 3. Monitoring information processing and analysis subsystem.

In its turn, the sensor subsystem of locating the monitored objectives' whereabouts and condition can be divided into sensors for evaluating the monitored objectives' condition and location and sensors for areas remote surveillance. The basic means of locating objectives' whereabouts' in monitoring systems are navigation sensors that can be mounted separately or jointly with function control sensors. Proceeding from the fact that the navigation support of FMS users must be global across the territory and provide information update several times per hour with an accuracy of 100 m, such demands are best of all satisfied by GLONASS and GPS global space navigation systems.

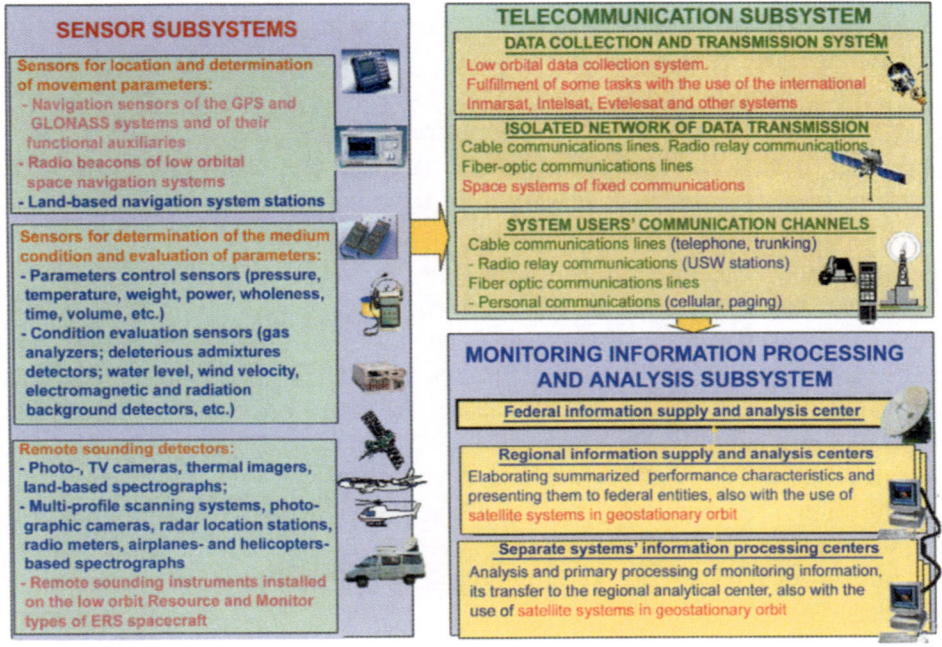

**Fig. 21.** Technological base of the federal system of monitoring objectives and resources in Russia

Other systems (land-based hyperbolic radio navigation systems, low orbit satellite navigation systems, etc) can be used in FMS as auxiliary means of locating objectives. This refers also to the European Global Navigation Satellite System (GNSS) which, as a matter of course, will expand the GPS and GLONASS, the systems with compatible basic navigation signals.

Technically, the objectives (areas) remote observation subsystems are the systems of aerospace monitoring. The component package of such equipment is rather diverse and includes various multi-channel scanning systems, photo cameras, radar location stations, radar meters, spectrographs, etc. The monitoring aviation equipment is installed both on high altitude aircraft and on small class airplanes. The space monitoring hardware implies earth remote sounding equipment, oceanic and meteorological survey systems.

The FMS anticipates to use, together with the traditional artificial observation satellites like Resource, Ocean and Meteor, the advanced small spacecraft. The Khrunichev Center is developing on the basis of the Yakhta standardized space platform a new generation space complex comprised of several small spacecraft of the Monitor type. They all are intended for operational surveillance of the Earth's surface for fulfilling the tasks pertaining to resources exploration. They are fitted, however, with different optic and electronic equipment designed for specific functions and having a rather limited application. Advanced engineering decisions are also being implemented in the creation of the distributed land-based infrastructure intended for reception, multi-level processing, storage and distribution of information about Earth and space.

The information exchange between the structural elements of the FMS will be accomplished by means of the communications and data transfer subsystem. There are several basic variants of the engineering base of the subsystem for data collection and transmission. The most promising of those is the one based on the low orbit satellite system (Fig. 22).

The main adavantages of this variant are the improved performance of the system in terms of communication periodicity in the high latitude areas, which is especially important for Russia, the existence of an orbital constellation and a network of ground stations that provide global communication as early as today. Of no lesser importance is the safeguarding of informational safety, not so stringent requirements for user radio visibility, and existence of small-sized user terminals.

The overall performance of the system engaged in the processing and analysis of monitoring information supplied by the FMS is determined primarily by the system's structure as a whole. The system is expected to be organized as a three level entity:

• at **the level of separate systems** information primary processing centers are being created that will gather information directly from sensor systems, analyze and pre-process it and transfer to the higher hierarchy levels;

• at **the regional level** regional information supply and analysis centers will be established that will be responsible for gathering information from separate groups of sensor systems, for processing it as per subject matter and for expert evaluation of situations at the regional level;

• at **the federal level** data will be gathered from the regional centers and particularly important sensor groups; formulating macro-indicators of the state of objectives and resources; expert evaluations of resources and environment; coordination of information exchange between the interested agencies.

Implementation of the principles of creation and application of the FMS will enable the use of this multifunctional space system for tackling a wide range of the federal agencies' tasks at the federal and regional levels.

The necessity to create and effectively use the Federal Monitoring System for supervision of Russia's objectives and resources results from the high priority of integrating the land-, sea-, air- and space-based monitoring facilities and reflects the tendency for expansion of the range of tasks tackled with the help of space hardware. It could be expected that this tendency will persevere in the first decades of the 21st century, which will bring about further functional standardization of space systems.

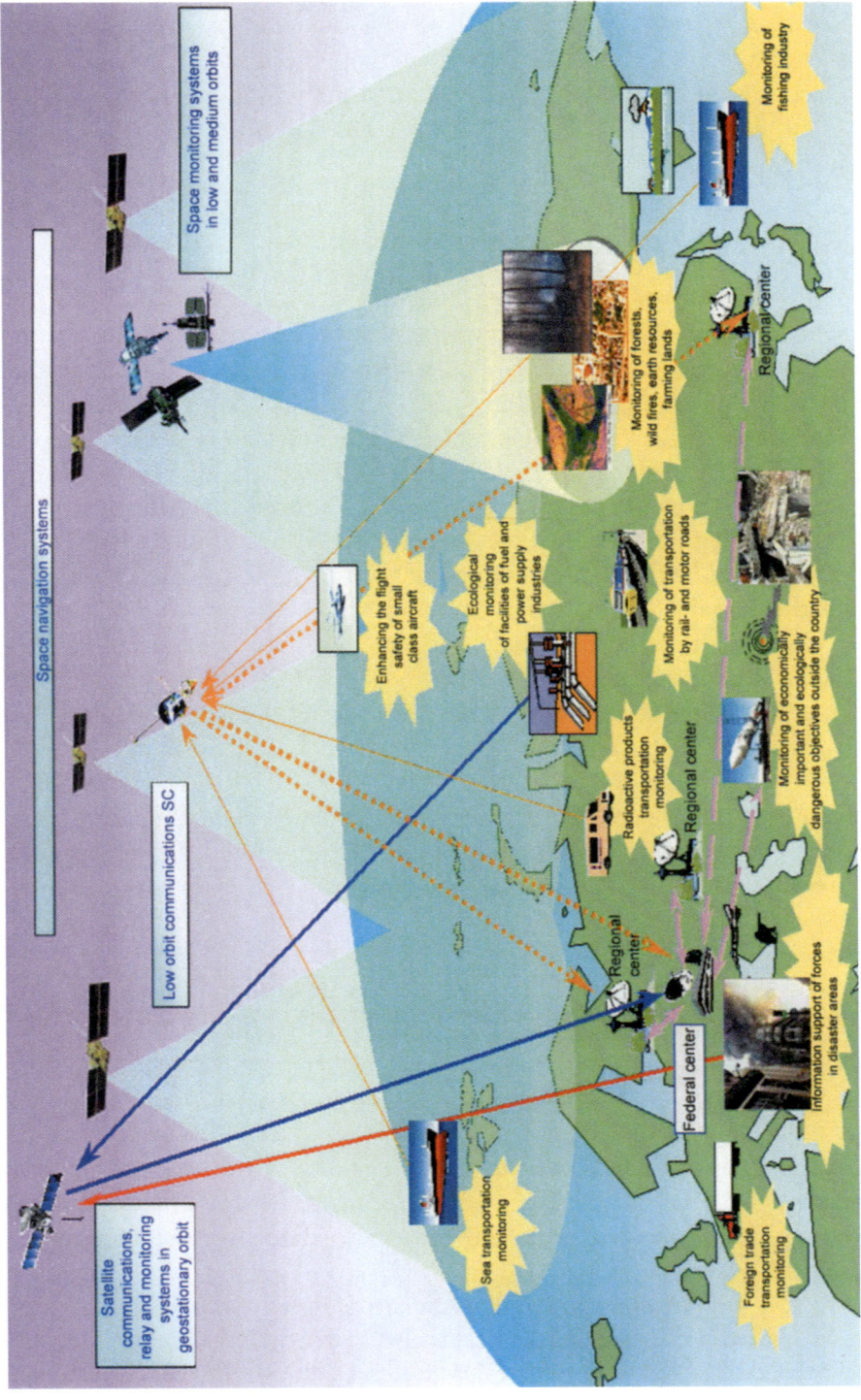

**Fig. 22.** The variant of organizing the monitoring data collection and transmission at a regional level in the interests of FMS

The multifunctional space systems are particularly sensitive to the organization of the effective joint use of monitoring systems controlled by various industries and regional agencies. Due to the fact that the military space systems traditionally possess a high potential, the priority is shifted to integration of military and civil space explorations within the framework of dual use systems. The peculiarities and main characteristics of such systems are described in chapter 2.5 of this book.

## 2.2 Russia's launch vehicles

The development of facilities for launching payloads into space (carrier rockets) had proceeded in Russia along several lines. The first line, dating from 1957 is associated with the creation of a number of carrier rockets based on the R-7 intercontinental ballistic missile, ICBM. This ICBM had been developed at the renowned OKB-1 (since 1966 referred to as the Central Design Bureau of Experimental Engineering, since 1974 as NPO Energia, and currently as the Energia Space Rocket Corporation named after Korolyev). Within that line of work more than 15 modifications of carriers had been built, e.g. Sputnik (satellite), Vostok (East), Molniya (lightening), Soyuz (union). Slated for manufacture is the Yamal (Avrora) carrier.

The new variants of carrier rockets came along normally due to the installation of new upper stages on the initial "package" and minor engineering amendments of other elements. The propitious design decisions laid down at the base rocket, R-7, its considerable technical capability and high reliability made those carriers the longest serving launchers at all Russian spaceports. Responsible for this area of development is the Progress Federal Space Center (until April 1996 the Central Specialized Design Bureau, previously a Kuibyshev-based division of the OKB-1). Currently, modernization of the Molniya and Soyuz carriers is underway which is expected to yield a new carrier rocket, Soyuz-2.

The second line, dating from 1961, was associated with the creation of light class carriers based on various strategic missiles. The crucial role here was played by the NPO Yuzhnoye (association for research and production) which built two variants of the Cosmos carrier rocket based on R-12 and R-14 ballistic missiles, as well as Cyclone-3 (11K69) and Cyclone-2 (11K68) carriers based on R-36 (8K67) and R-36M (8K69) rockets. Those carriers launch payloads weighing from 450 kg (Cosmos-2 carrier) to 3.6 tons (Cyclone-3 carrier) into low near-earth orbit and remain to date the best vehicles in their class.

In 1965, a new trend emerged in the development of Russia's carrier rockets. It involved the creation of heavy class carriers based on the UR-500 ballistic missile. Playing a key role in the trend was the Salyute Design Bureau, known for long as the First Division of the OKB-52 (now the Salyute Design Bureau is comprised by the Khrunichev State Space Research and Production Center). The Salyute has built several variants of the heavy class carriers of the Proton series (Fig. 23) whose lifting capacity for launches into low orbits is up to 20 tons. Fitted with a booster, they are capable of putting payloads into GEO from the Baikonur spaceport. Now, the Center is upgrading carriers of this series which will increase the vehicle's payload capacity by another 2 tons.

The fourth line (the late 1960s), in which the leading part belonged to the Central Design Bureau of Experimental Engineering (now Energia Space Rocket Corporation) was associated with the development and production of super-heavy class of carriers (Fig. 24). The first one in this class was the N1 carrier planned for implementation of the Soviet lunar program. Even though all launches failed and the lunar program was scrapped, it was precisely that carrier that paved the way to the building and successful testing of the super heavy carrier with lifting capacity above 100 tons. It was intended for launching into orbit the reusable orbital spaceship, Buran, and other large-sized payloads. Unfortunately, after a few successful launches the program was closed for a number of reasons.

By the mid-1970s the progress of space technology, the nature of tasks it was dealing with, the amount of spacecraft launches, and the impact of carriers on the environment reached such a degree that carriers derived from the strategic missiles could not any longer fully meet the requirements imposed by space missions. In this context a necessity arose to create carrier rockets of the new generation. This trend in the development of Russia's rocketry gave birth to the Zenit carrier whose successful launch took place in 1985. The key contributor to the creation of this carrier was the NPO Yuzhnoye (research and production association) based at Dnepropetrovsk.

The Zenit carrier boasts unique operating characteristics and high economic effectiveness. It uses ecologically pure fuel and is designed as a base model for development of a standardized series of carriers of various classes.

After disintegration of the USSR, the Zenit carrier became the Ukrainian property, though the Ukraine alone cannot manufacture it because many of its components are produced at Russia's facilities. The disintegration of the USSR made Russia face the challenge of creating purely Russian carriers launched from the Russian spaceports. In connection with this, the Khrunichev Center works on the Angara project aimed to create a family of modular carrier rockets of the light, medium and heavy classes.

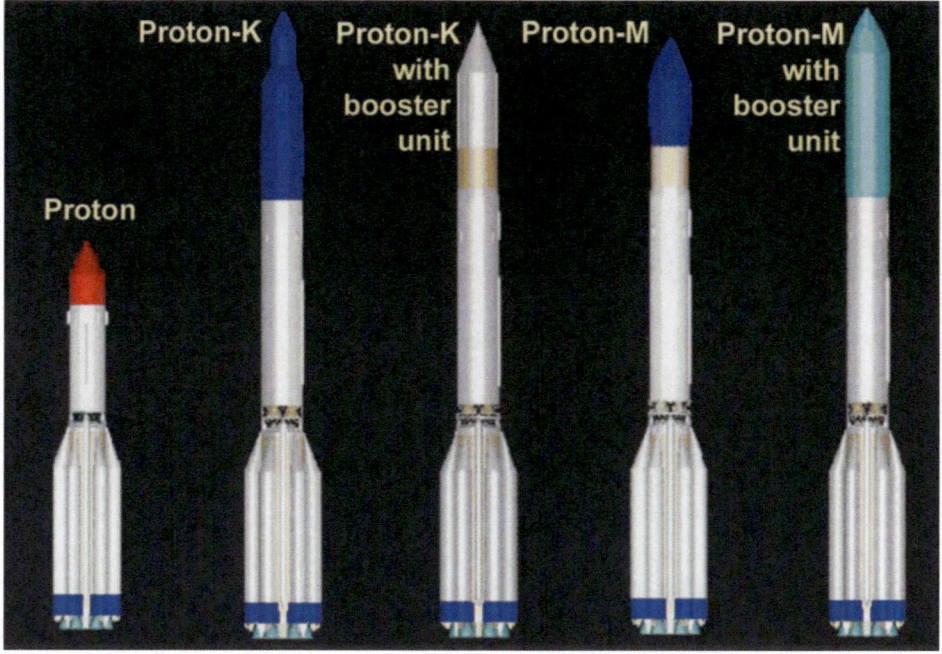

**Fig. 23.** Proton family carrier rockets

Unfortunately, over recent years Russia started losing somewhat of its status as a leading space power. This is largely attributed to Russia's changed geopolitical and economic position and to drastic cuts in funding the space work.

Russia's existing fleet of launch vehicles includes space rocket system of light, medium and heavy classes based at Russia's Plesetsk spaceport and the spaceport of Baikonur rented by Russia from Kazakhstan. A new Russian spaceport, Svobodny, is being built now. Two successful launches of Start-1 carrier have already been executed there.

Most of space rocket systems that are in current operation had been built back in the 1960-1970s on the basis of ICBMs. Those are the systems of carrier rockets Cosmos-3M, Cyclone-2, Cyclone-3, Soyuz-U, Molniya-M, and Proton-K. Such carriers, except the Soyuz and Molniya, use highly toxic types of fuel and structural components whose manufacture is nowadays rather difficult because of their obsolescence and the manufacturers' location beyond Russia's borders. The production of Cyclone-2 and Cyclone-3 in the Ukraine has practically stopped. Also, Russia is having a difficult time producing the Cosmos-3M carrier. The technically exhausted assets of the ground infrastructure, launch and engineering facilities, are not renovated. More than 19 million hectares of farming land are periodically put out of agricultural use as areas receiving jettisoned debris from overflying spacecraft.

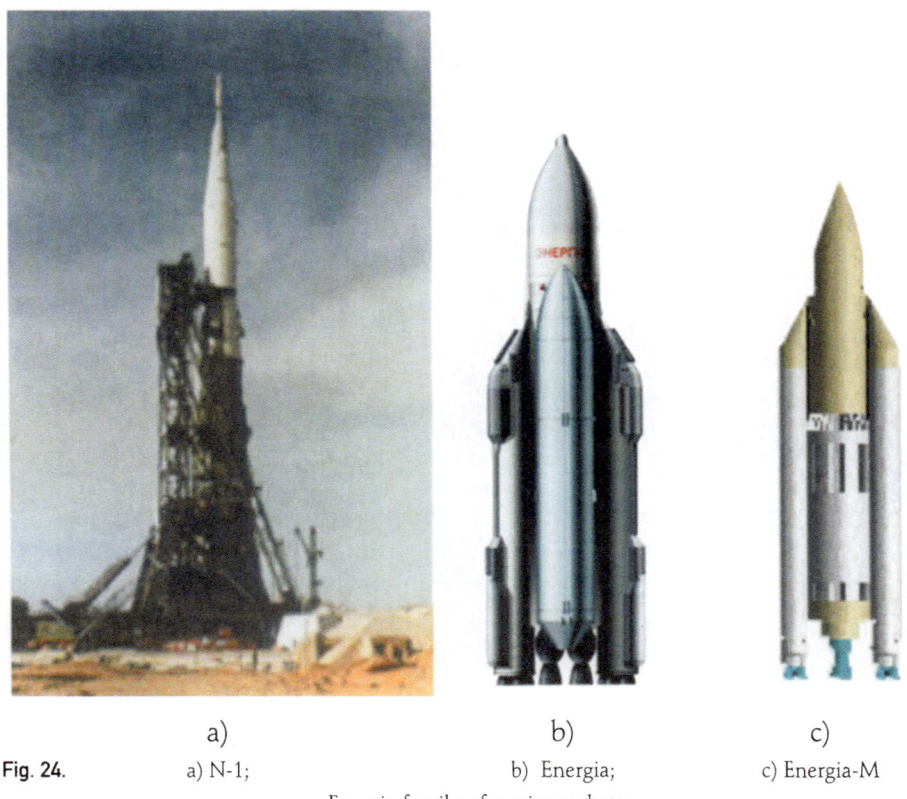

**Fig. 24.**          a) N-1;                              b) Energia;                    c) Energia-M

a)                                    b)                                c)

Energia family of carrier rockets

Under the current conditions it is of crucial importance for Russia to choose the optimum means of launching payloads into space and the adequate ground infrastructure with due regard for the country's economic capabilities and with resort to the non-budgetary sources of funding. This will back up space work for defense and national security and bolster the national economy, science and cooperation with foreign partners.

The solution of this problem involves a step-by-step transition from the existing launch system to the one of the new generation which is ecologically friendly, features high reliability and safety and can secure effective launches of SC for various uses in the 21st century. Such an approach takes account of the current state of Russia's economy and reduces the cost of all space programs due to work to be done in the following areas:

• modernization of the existing rocket systems Soyuz-U, Proton-K and the withdrawal from operation of space systems using carrier rockets Cyclone-2, Cyclone-3, Molniya-M, Cosmos-3M as the stock of those rockets is gradually depleted;

• the creation of systems using light class carrier rockets, Start-1, Strela, Dnepr, based on ICBM being phased out now (mainly at the expense of non-

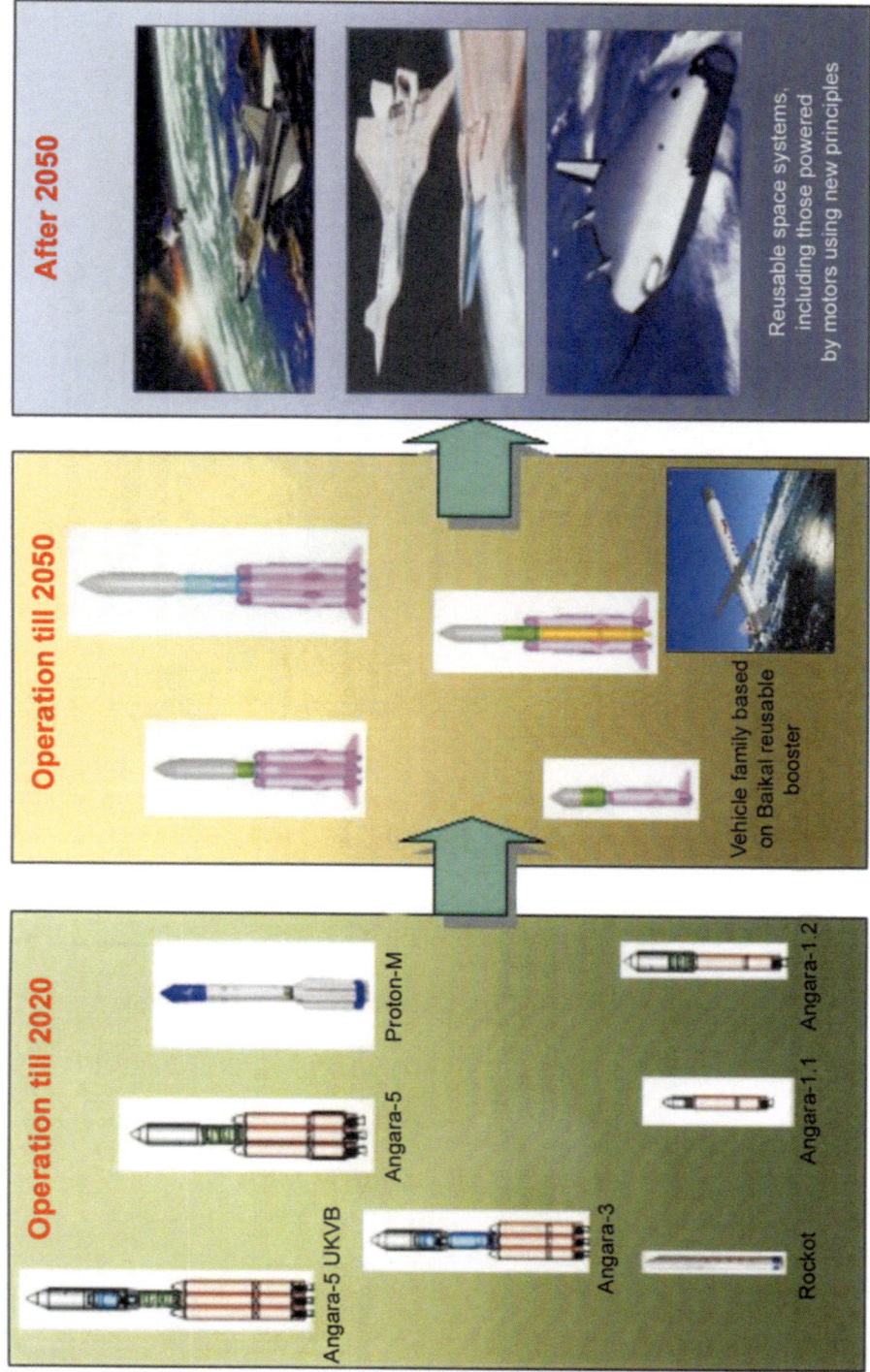

**Fig. 25.** Concept of development of launch vehicles in the 21st century on the basis af carrier rockets built by Khrunichev Center

budgetary sources) which are to become launch vehicles of small SC during the transition period, until a multi-role light class carrier rocket is created;

• the creation of the new generation systems with carrier rockets of the light, medium and heavy classes based at the Plesetsk spaceport (with Svobodny in prospect) using the Russian industry base (with utilization of budgetary and attraction of non-budgetary sources of funding).

In the longer term a transition is expected to launchers with reusable elements (return type units, power plants, etc). These are primarily all-azimuth launchers that do not require the isolation of territories onto which spacecraft jettison their spent components.

Investigations show that from the point of view of technical execution the problem of elimination of territories where spacecraft jettison their expended components, can now be most easily resolved by creating a two-stage carrier whose rocket units in the first stage are fitted with a rescue system similar to the one used on aircraft which assures their return to the launch area after separation from the carrier. In this case the nose fairing and other detachable parts, such as the adapter compartment, are launched with the help of energy supplied by the second stage into a ballistic trajectory with a subsequent fall into the world's ocean.

Still wider vistas open up for launch vehicles with the coming of a new and inevitable phase in space exploration, the transition from expendable launchers to reusable space systems. The prospective space programs are becoming ever more costly and labor-intensive due to the build-up of tasks and their growing complexity. Ever more critical is the issue of the space systems' harmful impact on the environment and the necessity to clear the adjacent space of used elements of SC, rocket stages and other debris of rocketry, the reduction of territories assigned to receive spent fragments of rockets.

One of the prospective trends in the solution of the indicated problems is the transition to reusable space systems that work on the principle of multiple use of the most sophisticated and costly elements of the system (booster units, orbital craft, sustainer motors, etc.) and their critical components. The new properties that can be used in the concept of prospective reusable space systems (launch capability for all azimuth positions, utilization of the considerable parallax at launch, the capability to change the orbit parameters by maneuvering the orbital stage with immersion into the dense layers of the atmosphere, the capability of controlled descent from orbit with the use of the developed aerodynamic surfaces of the orbital stage, etc.) make it possible to regard their use not only within the framework of a prospective launch system, but also as a means that effectively complement purpose-built complexes and systems based on permanently operating orbital constellations.

The concept of development of launch vehicles in the 21st century (exemplified by launch systems developed by the Khrunichev Center) is shown in general terms in Fig. 25.

## 2.2.1 Carrier rockets

Light class Cosmos-3M carrier rocket was created by the Yuzhnoye Design Bureau (Dnepropetrovsk) on the basis of the R-14 medium range ballistic missile introduced into service in 1961 and phased out as per Agreement of 1990 on the Reduction of Medium- and Short-Range Missiles. A new second stage has been engineered for the carrier rocket, the oxidizer tank has been somewhat changed and an adapter compartment between the stages has been built.

The new carrier took flight tests at the Baikonur spaceport under the name of Cosmos-1 (8 launches) after which it was reworked (Cosmos-3). The obtained product, however, proved unsuccessful (3 out of 6 launches at the Baikonur spaceport failed).

After modernization, the carrier got the name of Cosmos-3M and has long been used successfully to launch various SC. According to American experts who have compared 18 types of various carriers of the light class, the Cosmos-3M carrier is one of the world's best carriers in its class.

A new launch site with two take-off pads had been built at the Plesetsk spaceport to the designs proposed by the Design Bureau of Transport Engineering. The Raduga launch system, previously used to launch the Cosmos-2 carrier, had also been re-equipped. The flight tests of the Cosmos-3M carrier began in 1967, followed by standard operation in 1971. Also, it had been launched from the Kapustin Yar test site.

The series production of the Cosmos-3M was organized in Omsk at the PO Polyot (manufacturing association). The carrier rocket consists of two stages and a nose fairing. The stages are connected in a tandem fashion. The rocket measures 32.4 m in length, the diameter of its cylindrical part being 2.4 m, the launching mass 109 tons. Both stages of the carrier are powered by liquid propellant rocket motors running on self-inflammable rocket fuel mixture, AK-27I oxidizer and asymmetrical dimethyl hydrazine fuel. The flight control on the first stage is accomplished by means of gas deflectors and with the help of generator gas steering hinged nozzles on the second. The stages are separated in a semi-hot scheme.

Installed on the first stage is a four chamber RD-216 liquid propellant engine (LPRE) with a turbopump fuel supply system in which the generator gas is not afterburned. This engine has been built at the NPO Energomash (research and production association for power machines manufacture) under the supervision of V.P.Glushko. The dry weight of the engine is 1,325 kg, the operating time 131 sec. Installed on the second stage is the LPRE developed by the Design Bureau of Chemical Engineering under the supervision of A.M. Isayev. The LPRE comprises a powerful single chamber sustainer without afterburning and with pumped-up fuel supply and four-nozzle steering engine featuring a displacement supply of components to the special gas generator of the small thrust system. The characteristic features of this kind of engine is the possibility to re-activate the sustainer engine in weightlessness and the availability of three thrust modes – main, intermediate and small.

The Cosmos-3M carrier launches SC into elliptical and near-circular orbits at altitudes between 250 and 1,700 km with inclinations of 51, 66 and 83°. The weight of payload in this case varies between 1,500 kg (the orbit altitude is 250 km) and 500 kg (orbit altitude is 1,700 km). It is also possible to launch into solar synchronous orbit SC weighing up to 850 kg to the altitude of 475 km and with inclination of 97.3°. In one launch such a carrier can carry into space up to 8 spacecraft.

Introduction of SC into work orbits is accomplished with two activations of the sustainer liquid propellant rocket engine of the second stage power plant. After the first activation of the engine the stage flight proceeds along the transitory trajectory in whose estimated point the second activation of the power plant increases the speed increment (in the trajectory plane) which is required for putting the SC in the preset orbit. The stabilization of the second stage on the portion of the flight along the transitory trajectory is accomplished via a special small thrust steering engine. The nose fairing is jettisoned during the flight of the second stage at an altitude of about 75 km and speed pressure of about 14 kg f/m$^2$.

Nowadays, the Cosmos-3M is practically withdrawn from production and individual launches of SC are executed by carriers manufactured previously. The KB PO Polyot (research and production association) has attempted to modernize this carrier by replacing the obsolescent analog control system with a modern digital one, by increasing the fuel supply on the second stage, by improving its ecological characteristics, by transferring the production of all elements of the carrier to Russia's manufacturing facilities with the use of Russian component parts, etc. However, the drastic cuts in Russia's space spending which, naturally, had an impact on the PO Polyot allows of no optimistic predictions about the completion of that work.

**The development of Cyclone-2 and Cyclone-3 carrier rockets** began in August 1965 at the KB Yuzhnoye. A two-stage carrier based on the R-36 ICBM was envisioned with its subsequent modernization by installing a third stage.

The flight development tests of the two-stage carrier began in 1968 at the Baikonur spaceport. Both basic (Cyclone carrier) and modified variants (Cyclone-2) had been tested. The latter became base version. The flight tests of the three-stage carrier (Cyclone-3) took place from June 1977 through February 1979 at the Plesetsk spaceport where the engineering and launching sites had been erected with two open type take-off pads.

The main differences of those complexes from their counterparts at the Baikonur spaceport lie in the absence of service towers at the launch site, as well as in the presence of a fixed station for refueling and ampoulation of the rocket's third stage. The equipment of the Plesetsk spaceport provides for a high degree of automation of pre-launch operations with a minimum involvement of people. This makes absolutely unnecessary the presence of people at the launch site from the moment of rocket delivery till its actual launch.

The first stages of Cyclone-2 and Cyclone-3 carriers are practically fully standardized. All rocket stages are connected in tandem configuration. The length of the Cyclone-3 carrier is 39.3 m, the diameter of the cylindrical part is 3.0 m, and the take-off mass is up to 188 tons. The carrier can launch a payload weighing up to 3.6 tons (Cyclone-2 carrier – up to 2.9 tons) into an orbit at altitude of 200 km. There is a capability to send up to 6 SC in one launch. The separation of the first and second stages of the carrier is accomplished via the semi-hot scheme, that of the second and third via the cold one with the help of four spring pushers. The nose fairing is jettisoned during operation of the second stage.

The power plant of the first stage consists of two engines, the RD-218 sustainer unit and D-68M steering unit. Both run on nitrogen tetroxide and asymmetrical dimethylhydrazine. The RD-218 engine with a turbopump supply system has been developed by the NPO Energomash. It operates without afterburning of the generator gas. The D-68M steering engine also has a turbopump supply system and operates without afterburning. The launch and deactivation proceed in a single step. The chambers are turned by hydraulic drives.

The second stage sustainer engine, RD-219, has also been developed by the NPO Energomash. It has a turbopump supply system and operates without afterburning. Actually, this LPRE is a unit similar to the two-chamber unit of the RD-218 engine. The steering engine of the second stage of D-69M is identical in design and layout to the steering engine of the first stage.

The LPRE of the third stage, D-25, with a turbopump supply system operates without afterburning the generator gas. This small engine has been developed by the NPO Yuzhnoye under the supervision of V.F.Utkin. It runs on the same fuel, nitrogen tetroxide and asymmetrical dimethylhydrazine. Apart from the sustainer unit, D-25, the third stage of Cyclone-3 carrier is equipped with a special liquid jet-powered control system. It is designed to dampen the SC stages after separation, to orient and stabilize them in free flight and to ensure the sustainer engine activation in weightlessness. It runs on the same rocket fuel as the stage sustainer engine and is, essentially, a LPRE with a pressure supply of fuel components.

In creating the Cyclone space rocket system, new approaches were used to prepare the carrier for launches. This raised Russia's space rocket engineering in the mid-1960s to a new level. To date, this system outperforms all known counterparts abroad. In creating the launching and engineering complexes, the developer, (Design Bureau of Transport Engineering) succeeded in automating all the main and many auxiliary operations. The level of automation of the pre-launch and launch procedure of Cyclone-2 and Cyclone-3 carriers is 100% and no less than 80% in work on the complex in general. The sole dangerous manual operation is the reconnection of refueling communications in case the launch is cancelled. The launch site of the Cyclone employs modern management and control equipment. For the first time ever the preparation for launch and the launch proper under a single program have been fully automated starting from the carrier delivery to the take-off pad.

The production of all the main elements of the Cyclone-2 and Cyclone-3 carriers proceeded in the Ukraine: the frame and the propulsion unit were manufactured in Dnepropetrovsk with controls being manufactured in Kharkov. The Cyclone production line at the Dnepropetrovsk manufacturing facility is practically dismantled and unlikely to be reinstalled. Thus, in the current situation the prospects of Russia using the Cyclone carrier are unrealistic.

**The Soyuz-U carrier rocket** (Fig. 26) is the best known and most widely used missile out of all modifications built around the R-7A ballistic missile. This is an upgraded version of the Soyuz carrier that had been in operation between 1966 and 1973.

The first launch of the Soyuz-U carrier took place in May, 1973. In December, 1982, its modification, **Soyuz-U2**, was launched. Instead of kerosene its central block used synthine, synthetic hydrocarbon-based compound. The fuel pair of liquid oxygen + synthine gives a higher specific pulse to the propulsion unit which improved somewhat the energy performance of the carrier rocket.

Soyuz-U and Soyuz-U2 carriers launch spacecraft of various applications (including manned ones) from the Baikonur spaceport and unmanned spacecraft from the Plesetsk spaceport. The weight of payload delivered to low NEO by the Soyuz-U carrier is 6.8 tons from the spaceport of Plesetsk and 7.1 tons from Baikonur.

The Soyuz-U is a three-step carrier. All stages use liquid oxygen as an oxidizer and kerosene as a fuel. The first and the second stages of the carrier are connected in the packet configuration, in which case used as the first stage are the four lateral blocks, B, C, D, E, and acting as the second stage is the central block, A. The third stage (block I) is connected with the second stage in tandem configuration. Installed on the third stage via the adapter compartment under the nose fairing is the SC.

Depending on the type of SC the fairing may have various shape and size. The take-off mass of the carrier is around 310 tons, the maximum length – 50.7 m, the maximum cross-section size – 10.3 m. In the case of launching manned spacecraft the Soyuz carrier is fitted with a crew recovery system installed at the top of the nose fairing.

The first stage of the carrier is formed by four blocks identical in design placed around the central block on stabilization planes and independently affixed to it by two connecting units, upper and lower. To be able to separate the lateral blocks they are fitted with appropriate mechanisms. The upper connecting units are intended to transmit the axial efforts (engines' thrust) while the lower ones are for taking in the shear load. Thus the bulk of the central block's case is not loaded in flight by the lateral blocks' thrust. While on the launch site the carrier is fastened to it in the area of upper connecting units of the lateral blocks which also creates favorable conditions for carrier blocks operating under stress. Such a placement of the packet type of rocket on a launch site has no analogs.

Placed in the tail compartment of each lateral block is RD-107 LPRE developed by the NPO Energia under the supervision of V.P.Glushko. The engine has an open type of construction and comprises four main fixed chambers and two steering movable ones as well as a turbopump unit that feeds them, a hydrogen peroxide catalytic decomposition generator, control units and a power frame. The R-107 engine is mounted at an angle of 3.5° relative to the longitudinal axis of the lateral block. Under such circumstances the axes of all engines of lateral blocks end up parallel to the longitudinal axis of the carrier rocket.

**Fig. 26.** Soyuz carrier rockets in the assembly shop of the Progress Central Specialized Design Bureau (standing in the foreground is D.I.Kozlov)

The sustainer engine of the second stage, RD-108, is similar in design to the sustainer engine RD-107 of the lateral block and differs from it in the number of steering chambers (four on RD-108) and in the design of the fuel throttling device that regulates the ratio of fuel components in the engine in order to synchronize the emptying of the tanks. The engine thrust is transmitted to the housing of the block A via power frame on which all the systems of the engine's fuel supply and pneumatic reserve are mounted together with some controls.

The separation system of the first and second stages of carrier is highly original and has no analogs. It assures a reliable simultaneous separation of four lateral blocks placed around a central one. The separation system comprises the lateral blocks' oxidizer and fuel tanks' jet nozzles that provide the required force to separate the blocks by means of tanks pressurization gases passing through them; upper and lower assemblies of the mechanical connections of the lateral blocks with the central one; also it includes the separation control arrangement. The jet nozzles are fitted with lids having mechanisms for opening via a pyrotechnic device.

Installed on the third stage (block I) is the LPRE developed by the Design Bareau of Chemical Automatic Systems. This four-chamber engine operates in an open scheme with a turbopump fuel supply system. All four chambers are fed by one turboprop unit placed on the block's axis vertically between the chambers. The combustion products of the reducing gas generator fueled by the main components, perform as the actuating fluid of the gas turbine. After the turbine the generator gas moves on to four rotary steering nozzles that control the flight of the block.

After the lapse of a preset time following the separation of the central block from block I, a command is issued for opening the ball pyrotechnic locks of the separation system of the tail compartment of block I.

Currently, the Samara-based Central Specialized Design Bureau upgrades the Soyuz-U carrier under research and development program dubbed Rus. The new carrier that will be called Soyuz-2 is supposed to use only Russian-made component parts. It will replace all types of carrier rockets based on R-7A ICBM (Soyuz-U, Soyuz-U2, Molniya-M).

The main reasons of the modernization are on the one hand the obsolescent component package of the control system and telemetry (many elements are being withdrawn from production, the upcoming replacements need on-the-ground optimization and extra spending) and, on the other, the desire of the developers of payloads to increase the power capability of the carrier in order to expand the product line of SC being manufactured and enhance the range of their tasks.

The new carrier will be able to propel SC into high elliptical, geostationary transfer, solar synchronous as well as medium and high circular orbits in the wide range of altitudes and inclinations. The energy capabilities of the **Soyuz-2** carrier will significantly enhance the mass and size of payload launched into orbit thus providing for fulfillment of all tasks set now before the R-7A series of carrier rockets.

The modernization that is now underway aims at:
- placing the production of all component parts in Russia;
- replacing the existing analog control system with a new one based on an onboard computing complex which will cut down the mass of the control system by roughly 200 kg and reduce the area on which jettisoned fragments are dropped (the reduction can reach 40% since the launches will be conducted in one azimuth);
- enhancing the energy capability of the sustainer LPREs of the packet by replacing two-component jet-and-jet injectors with new single-component jet-and-centrifuge injectors;
- using the new onboard telemetry system based on modern components which will save up to 160 kg;
- equipping the Soyuz-2 carrier rocket with a Fregat booster unit developed by the NPO Lavochkin (Research and Production Association) and using high-boiling fuels (nitrogen tetroxide + asymmetrical dimethylhydrazine). The flight tests of the Fregat started in 2000. In this case the carrier will be fitted with a nose fairing having an enlarged diameter of 3.7 m;
- renovating the I block (the third stage of the carrier) by installing on it a new four-chamber LPRE with afterburning of the generator gas which will increase the payload weight by 950 kg.

It should be noted that the initial idea was to conduct modernization in two phases – the so called "minor" and "in-depth" modernization. The upgrading of the I block was slated for the phase of the "in-depth" modernization. However, the erratic and incomplete funding of the research and development results in delays and failure to do the planned work on time. Under such circumstances the phase of the "minor" modernization will also proceed in several phases – depending on the financial and manufacturing capabilities of the organizations in charge of research and development.

A series of four-stage carriers, also developed on the basis of the R-7A ICBM, are called **Molniya** (lightening). The first launch of a carrier from this series took place in October 1960. In 1965 and 1985 this carrier got a serous upgrade aimed to improve its capabilities and enhance the safety of its main-tenance. The variant in use now is the Molniya-M.

The cost effective pattern of delivery that uses an intermediate orbit with the launch of the fourth stage in weightlessness, does not allow to employ this carrier for placing payloads into fly-away trajectories and elongated elliptical orbits. The Molniya-M carrier launches SC weighing up to 2,000 kg to high elliptical orbits with the perigee altitude up to 700 km and apogee altitude up to 36,000 km with an inclination of about 63° or to orbits for flying over to other planets of the Solar system. The launches of Molniya-M spacecraft are conducted at the spaceports of Baikonur and Plesetsk from the same launch pads as for the Soyuz-U carrier. The take-off mass of the carrier is around 305 tons, the maximum length measuring 43.4 m.

The Molniya-M carrier differs from the Soyuz-U in that it has a fourth stage that forms together with the nose fairing, spacecraft and launch support unit a head block measuring 8.46 m in length and 2.7 m across the maximum diameter. There are also differences in the design of the central block's instrumentation compartment and the core girder that connects the second stage's block (block A) with the third stage (block I). Yet another difference is the absence of the control system on block I in the case of the Molniya-M carrier. The functioning of the third stage of the Molniya carrier, by contrast with the Soyuz-M carrier, is assured by the control system of the fourth stage. For the fourth stage the Molniya-M carrier uses an L booster unit consisting of an instrumentation compartment, fuel tanks package and a engine.

The fuel tanks package, in turn, consists of the torus tank for fuel (kerosene) and the torus tank for oxidizer (liquid oxygen) which are interconnected by a cylindrical ferrule. The block's sustainer, LPRE S1-5400, is a single chamber device with a turbopump fuel supply system that operates with afterburning the oxidizer gas. It has been developed in the early 1960s at the Central Bureau of Experimental Engineering. The LPRE engine is fueled by liquid oxygen and kerosene at the ratio of 1 to 2.45. The engine is attached in a gimbal suspension and provides for flight control of the L block in pitch and yaw angles. The rotation control is accomplished by nozzles operated by gas generated by a special generator. This same gas is used to pressurize the fuel tank. The oxidizer tank is pressurized by oxygen, previously gasified and heated in the heat exchanger.

The launch support unit includes a transition truss connecting blocks L and I. Installed on it are two SPRE designed to create the initial overload prior to activation of the block L engine. Mounted on the same truss are the elements of the stabilization system that functions in the phase of passive flight of the L block over the intermediate orbit and during activation of the main engine. Performing as actuators in the stabilization system are the gas nozzles integrated with the electric pneumatic valves. After activation of the block L's engine, the launch support block is separated and jettisoned.

Three stages of the carrier deliver the head block to the intermediate orbit with a perigee altitude from 200 to 250 km, apogee altitude from 400 to 700 km and with an inclination of 63°. After delivery to the intermediate orbit the head block makes a stabilized flight over orbit up to the launch point from the intermediate orbit to the estimated one. The duration of the passive stabilized flight over the intermediate orbit is 50 to 60 minutes. The propulsion unit of the carrier's fourth stage is activated at the estimated moment followed by a launch from the intermediate orbit to the estimated one. The activation of the propulsion units in weightlessness is accomplished by two powder rocket motors. As the engine achieves 75% of the nominal thrust, the truss of the launch support's block separates from the block of the carrier's fourth stage. After reaching the estimated speed, the engine deactivates and in about 8 seconds the spacecraft separates from the L block. After that, the block L is rotated and withdrawn from the direction in which SC is ejected.

**The Zenit carrier rocket** (Fig. 27) is the latest development of Yuzhnoye Design Bureau based at Dnepropetrovsk. To date, this is the world's most advanced carrier that embodies Russia's vast experience in space rocket engineering and the latest most prominent achievements. A carrier was envisioned that would possess unique operating capabilities:

**Fig. 27.** Zenit carrier rocket on the transportation and erection unit (standing in the foreground is S.N.Konyukhov).

• the possibility to prepare the vehicle in good time and keep it in operational readiness over a protracted period;
• short ready-to-launch time (not more than an hour and a half after receipt of command);
• high ecological cleanliness and non-toxicity of all fuels and gases used;
• high safety of conducting all launch related operations due to the people-free principle of the launch procedure;
• possibility to transport fully assembled stages by the railroad without stopping the oncoming traffic, and capabilities.

The decision to develop the Zenit carrier rocket was taken in 1976, almost simultaneously with the decision to create a reusable space system (RSS), Energia-Buran. Such a coincidence was caused largely by the fact that the lateral blocks of the RSS were supposed to be replaced by modified first stages of the Zenit carrier rocket. The successful and well-timed optimization of this carrier was also required for creating the RSS. In addition, the Zenit carrier was envisioned as a multi-role launch vehicle serving as a base for a series of advanced carriers of various classes.

The original idea was to build a launch complex for the Zenit carrier solely at the Plesetsk spaceport. That's where the flight tests began in the mid-1980s. The first launch took place in April, 1985. Two out of thirteen initial launches were unsuccessful, due to which fact the spaceport's council on science and technology gave a negative report concerning the results of the flight tests. The State Commission's Certificate for acceptance of the Zenit carrier for service in 1989 was passed in spite of the controversy of opinions. The next launch of the carrier showed that the experts' caution in the matter was not ungrounded. In October 1990 a major accident occurred. The failure of the sustainer LPRE of the first stage caused the rocket to fall into the gas duct of the launch system where it exploded and fully destroyed the launch pad which has not been repaired to date. The carrier needed to be reworked, primarily in the area of LPRE of the first stage which took nearly two years. The subsequent launches of the reworked carrier were conducted from the second take-off pad (which survived) of the launch complex.

Depending on the missions tackled, the Zenit carrier can be used in two-stage (Zenit-2) and three-stage (Zenit-3) configurations. Used as the third stage here is a booster unit which enables SC to be placed into high and GEO as well as into fly-away paths.

The take-off mass of the Zenit-2 carrier is 459 tons. The mass of payload launched from the Baikonur spaceport to an altitude of 200 km is up to 13.8 tons.

The Zenit carrier operates in a tandem configuration with a transverse division of stages. All the stages are fueled with non-toxic fuels such as liquid oxygen and kerosene. The stage division proceeds according to a semi-hot scheme. The control of the carrier in the flight segment of the first stage is accomplished by deflecting the sustainer engine's chambers, and in the flight segment of the second stage – by means of a special steering engine.

The interesting feature of the Zenit carrier is the non-optimal distribution of component over the stages. Such a decision was taken in order to shorten the first stage and make it transportable by railroad in a fully assembled configuration thus overcoming the restrictions applied to dimensions of goods to be hauled. This circumstance also determined such a layout of both stages, very compact for a carrier rocket.

The first stage sustainer engine, RD-170, is now the world's most powerful LPR engine. It was built at NPO Energomash under the supervision of V.N. Radovsky. The RD-170 is a four chamber LPR engine with one turbopump unit. The engine operates with afterburning the generator gas in which case the engine chambers can deflect to an angle of 6° in two planes. The engine uses chemical ignition of fuel mixture in the chambers and gas generators. The operating time of the engine is 140-150 seconds.

The propulsion unit of the second stage consists of two LPRE, RD-120 sustainer engine and a steering engine. The RD-120 has been developed by the

NPO Energomash (research and production association for power engineering) under the supervision of V.N. Radovsky. The single chamber LPRE, RD-120, is rigidly fixed and operates with afterburning the generator gas. The engine operating time in the event of a single activation is 300 seconds. The steering engine is a four-chamber system with one turbopump device. It, too, operates with afterburning the generator gas. The steering engine chambers can deflect to an angle of 31° using hydraulic drives. The engine operating time is 375 seconds. The steering engine has been developed by the NPO Yuzhnoye.

No manual maintenance of any kind is required on the launch complex of the Zenit carrier either in its preparation for launch or in withdrawing from the take-off pad if the lift did not take place. The complex component package includes some units which while executing commands of the automated control program place on their own the carrier rocket onto the take-off pad and connect to the rocket all the required communications fed by the land-based systems. Notably, they can repeat such operations many times, also on a fueled carrier for draining the fuel mixture. At the launch complex of the Zenit carrier there are no expendable items that burn up during launch. No repair is required on the launch facility after launching. This allows to launch a new carrier from the same site 5 hours after the launch of the previous one. In order to enhance the reliability of the carrier the complex is equipped with a diagnostics system that checks out the operation of the sustainer LPRE of the first stage during the rocket's launch.

The Zenit carrier is a fruit of joint effort of many organizations and factories. The main component parts of the rocket are manufactured by the factories of Russia and Ukraine. The assembly proceeds in Dnepropetrovsk, Ukraine, at the manufacturing facility of Yuzhny.

On the basis of the Zenit-2 carrier and DM booster unit (Zenit-3SL carrier) a new space rocket system has been built today for deployment on sea platforms. Dubbed Sea Launch (Fig. 28), it comprises an assembly and command vessel plus a self-propelled launch platform. To implement the Sea Launch project an international company of the same name, Sea Launch, has been formed. It includes Boeing Commercial Space (USA, 40 % of registered capital ), Energia Space Rocket Corporation (Russia, 25%), Kvaerner Maritime (Norway, 20%), GKB Yuzhnoye, PO Yuzhmash (Ukraine, 15%). The main advantages of the Sea Launch over land-based spaceports lie in the all azimuth nature of the launch, no necessity to assign territories to jettisoned fragments and the full utilization of the Earth's rotation for increasing the mass of the payload. The first capabilities demonstration launch of the Sea Launch system took place on 27 March, 1999.

**The Proton carrier rocket** has been developed by the Salyute Design Bureau on the basis of the heavy ICBM UR-500. The flight tests of the two-stage variant (Proton carrier rocket) ended in 1966. As early as 1968 a three-stage variant (Proton-K variant) has been launched. The transition from the two-stage variant to the three-stage one was fairly fast and easy. It consisted mainly in the following. Fuel tanks were enlarged on the second stage and some changes were made on the construction of the truss transitory compartment that connects the second stage with the first. The third stage was formed by shortening the initial variant of the second stage. On it, one LPRE was installed instead of four. All sustainer engines of the Proton carrier operate economically with afterburning the oxidizer generator gas.

**Fig. 28.** Sea Launch project. The Zenit-SL carrier on the launch platform

The engines of the first and second stages are installed in gimbal suspensions, which allows to control the carrier rocket with lesser losses. On the third stage the control of the thrust vector is accomplished by additional steering four-chamber LPRE without afterburning.

The Proton-K carrier uses for its fourth stage the booster unit D whose engine can activate in space many times. This made the Proton-K Russia's only carrier with a capability to deliver payloads to geostationary orbits. The first such launch was executed in March 1974. Subsequently, the block D was modernized by installing in it a special instrumentation compartment. Today, the Proton-K carrier is used mostly with the DM booster unit for boosting SC into geostationary orbit.

The energy capabilities of the Proton-K carrier provide for launching up to 20.9 tons of payload to low Earth orbit, and up to 2.3 tons to geostationary orbit (in case the DM booster unit is used).

All the stages of the carrier are connected in sequence (tandem pattern). The separation of the first and the second stages is accomplished via the hot scheme, that of the second and third – via the semi-hot one.

The first stage consists of the central block and six lateral blocks placed symmetrically around the central one. The lateral blocks are similar in design, each consisting of a front section, fuel tank and a tail section in which the engine is mounted.

The propulsion unit of the first stage consists of six autonomous sustainer LPREs RD-253 developed by NPO Energomash. Each of them is installed on two cross-beams of the tail section of the lateral block. To control the thrust vector the engine can be deflected by the hydraulic drive to an angle of 7°30'. This is possible due to the engine being fastened by special trunnions in the area of critical section of the chamber in the ball bearings of cross-beams. The RD-253 engine with a turbopump fuel supply system and afterburning of the oxidizer gas runs on nitrogen tetroxide and asymmetrical dimethylhydrazine. The engine operating time is 130 seconds.

The propulsion unit of the second stage consists of four autonomous sustainer LPREs of the same type: three RD-0210s and one RD-0211 developed by Design Bureau of Chemical Automatic Systems. Installed on the RD-0211 engine, by contrast to the RD-0210, are tank pressurization units – fuel tank pressurization gas generator and oxidizer tank pressurization mixer. These units are similar to those used on the RD-253. All LPREs motors are fastened by means of trunnions to the cross-beam so that deflection is possible of each of them to angles up to 3°15¢. The deflection is accomplished by the hydraulic drive. The second stage engines also have a turbopump system of fuel supply and operate according to a scheme with afterburning of oxidizer gas. They run on fuel mixtures of nitrogen tetroxide and asymmetrical dimethylhydrazine. The operating time is 230 seconds.

The propulsion unit of the third stage consists of the sustainer LPRE, RD-0212, and four-chamber steering engine, RD-0214. The sustainer engine is similar in design and operation to the RD-0210 engine of the second stage and is its modification. The steering engine does not afterburn the generator gas. It consists of four chambers, one turbopump unit, two gas generators and a powder starter. The  operating time of the steering engine is 250 seconds.

The separation of the second and third stages occurs due to the thrust provided by the steering LPR engine of the third stage activated before the shut-off of the sustainer LPREs of the second stage and the braking of the second stage by six powder motors placed on it. At the end of the active portion of the path the sustainer LPRE RD-0212 deactivates and leaves only the steering engine running which provides a more precise terminal speed. The payload separates after deactivation of the steering engine, RD-0214, in which case the third stage is braked by four powder engines.

In using the booster unit, DM, the lower rib of its conical adapter directly joins with the instrumentation compartment of the third stage over its upper rib by means of rods and bolts. The separation of the booster unit from the third stage proceeds along the joint of the conical and cylindrical adapters of the DM block. The conical adapter remains with the third stage. For each type of payload a suitable nose fairing is used which is jettisoned in the initial period of operation of the second stage.

Currently, it has become acutely necessary to increase the weight of payloads to be launched to geostationary orbit, to improve the ecological characteristics of carrier rockets, and to replace the control system with a more progressive one fabricated wholly by Russia's manufacturing industry. To this end the Khrunichev Research and Production Center upgrades the Proton-K carrier rocket.

The new model, **Proton-M**, (Fig. 29), will be equipped with a modern control system based on an onboard computing complex. Its first stage will be cleared of the remnants of fuel mixture after separation from the rocket and will require for receipt of jettisoned debris only one third of an area needed for the Proton-K carrier. The boosting of all sustainer motors will exceed the lifting capacity of the rocket during launch of up to 22 tons to low orbit while the use of oxygen and hydrogen booster unit will increase to 4 tons the payload launched to geostationary orbit. In addition, plans are made to equip the Proton-M carrier with a larger diameter nose fairing (5.1 m) for launching large-size spacecraft.

It should be noted that the launch complexes of the Proton carrier rocket are available only at the Baikonur spaceport.

**Fig. 29.**
Proton-M carrier with Breeze-M booster unit

## 2.2.2 Carrier rocket booster units

Booster units, also referred to as interorbital towing vehicles, are the key components of launch systems. Booster units transfer payloads from orbit to orbit or direct them to fly-away or interplanetary paths. To cope with the task, the booster must be able to perform one or several maneuvers aimed at changing the flight speed. This is expected to be achieved by activating each time the sustainer engine. The activations are alternated with lengthy (up to several hours) periods of passive flight over transfer orbits or paths. Thus, any booster unit must have a multiple activation sustainer and an additional reactive system or a power plant that provides orientation and stabilization of the booster moving with SC and creates conditions for activating the sustainer engine. The motors can be controlled via both the SC control system and the autonomous control system of the booster unit proper. In the latter case it must have a special compartment for accommodating it.

**The DM booster unit** (Fig. 30) is intended for use on the carrier rockets Proton-K, Proton-M, and Zenit-3. In 1974 the D booster unit, created in the late 1960s for the lunar expedition, underwent the first flight tests for launching SC to geostationary orbit. Later on, it was modernized and starting in 1976 its modification, the DM block, is used for launching SC to geo-stationary orbit.

**Fig. 30.** DM booster unit

In launching a SC to geostationary orbit a carrier rocket can work according to a two- or three-pulse scheme. Depending on the longitude of the spacecraft's stay in geostationary orbit, the duration of the boosters' stay in intermediate orbits changes and, hence, so does the total flight time which can be as long as 7 to 21 hours. During the flight, the booster unit can function either in a completely autonomous mode or be controlled via radio channels from Earth.

The LPRE, RD-58M, of the booster unit is a multiple activation device with a turbopump system of fuel supply. It operates with afterburning the oxidizer gas and is fueled by oxidizer (liquid oxygen) and fuel (kerosene of RG-1 grade). The engine is fastened in a gimbal suspension on an inner tier of a two-tier truss. Such an arrangement enables a control via pitching and yawing channels. To obtain the rolling control a hinged nozzle is used that is fueled by hot generator gas. The component package of the LPRE, RD-58M, also contains a multiple activation unit and pneumatically controlled automation units. To prevent the thermal impact of the outflowing gas stream on the structural elements and LPRE, a bottom protection is used which is a framework of welded pipes with shield-vacuum heat insulating material stretched over it.

The instrumentation compartment is a hermetically closed torus-like container fastened to the inner and outer tiers of the upper truss. The container is demountable. It contains the instruments of the control system and an air/liquid system of thermal regulation.

The booster unit DM is fitted with a conical and cylindrical adapters connecting it with the carrier rocket. During separation of the booster unit from the third stage of the carrier, the conical adapter is separated together with the stage. After a time, the cylindrical adapter is also jettisoned.

The weight of the dry block without discarded elements is 2,200 kg, the maximum length is 6.26 m, the maximum diameter – 4.1 m. The weight of rocket fuel and gases is 15,095 kg.

The Fregat (frigate) booster unit has been created by the S.A.Lavochkin NPO (research and production association) (Fig. 31) for use in the Soyuz-2 carrier rocket. It admits of up to 20 activations of the sustainer engine in flight and carries on board 5,350 kg of fuel. The LPRE is fueled by nitrogen tetroxide +asymmetrical dimethylhydrazine. The fuel is located in four spherical tanks. Two other such spherical tanks are used as instrumentation containers. All six spheres are placed around the sustainer engine whose chamber is installed in the gimbal suspension. The power frame of the gimbal is fastened to four brackets, each of which is welded to the appropriate fuel tank.

The Fregat booster unit also has a power plant for orientation and activation of the sustainer engine. It works on the principle of catalytic decomposition of hydrazine, a supply of which (around 85 kg) is placed in two small spherical tanks. The tank pressurization that provides for displacement-aided supply of all fuel components is accomplished with the help of helium. The first launch of the Fregat booster unit under flight test program was successfully conducted on 9 February 2000 as part of the Soyuz carrier.

**Fig. 31.**
At the assembly shop
Lavochkin Research and
Production Association
(S.D.Kulikov standing in the
foreground)

The Khrunichev Center has created a **Breeze-M booster unit** (Fig. 32) intended to replace the D/DM series blocks for use in the Proton-K and Proton-M carrier rocket. The new booster unit will increase to 3.2 tons the weight of payload launched to geostationary orbit. The Breeze-M carrier has been under flight tests since 1999.

The Breeze-M carrier consists of a central block and a jettisonable toroidal auxiliary fuel tank. The cylinder-shaped fuel compartment has an integrated bottom in case of the forward end positioning of the oxidizer tank. The upper bottom of the oxidizer tank is spherical while its lower one is of a complex shape and forms a semi-spherical niche. Passing through the fuel tank, the niche is formed by the inside conical ferrule of the tank. The conical ferrule is welded to the lower spherical bottom of the oxidizer tank at the top and to the lower spherical bottom of the fuel tank below.

The sustainer LPRE has the capability of multiple (no less than 8 times) activation. It is installed in the niche inside the fuel tank of the central block. The small thrust liquid propellant motor (LPM) running on the same fuel components as the sustainer engine provides orientation and stabilization of the

**Fig. 32**. Assembling the Breeze-M booster unit

booster unit during an autonomous flight. Also, it compresses the fuel in tanks during activation of the sustainer engine. Installed in the instrumentation compartment, the inertial control system controls the flight of the Breeze booster unit and its onboard systems. In addition, the carrier is equipped with a power supply system and instrumentation for gathering telemetric information and for extra-trajectory measurements.

In creating the Breeze-M booster unit, much emphasis was laid on improving its maintainability. Specifically, the refueling of the carrier with fuel components was envisioned at the manufacturer's site with subsequent ampoulization of the block.

The principal distinctive feature of the Breeze-M booster unit is the use of many systems and assembly units of the Breeze-KM being created for the Rockot launch vehicle. To enhance the lifting capability of the Breeze-M it was equipped with jettisonable toroidal fuel tanks in addition to the main ones on the central part of the block.

**The oxygen/hydrogen booster unit** (OHBU) is being developed by the Khrunichev Center for use with the Proton-M carrier, and, in the longer term future, with the Angara-3 and Angara-5 carrier rockets. The creation of the OHBU resulted from the need to launch into high orbits Russia's advanced spacecraft and to expand the range of services on the market of commercial launches. The block arose out of the Khrunichev Center's unrealized project for building a cryogenic booster unit, Storm, and an oxygen/hydrogen block, 12KRB, created for the Indian GSLV carrier rocket (Fig. 33).

In course of designing the OHBU, its several variants have been developed for use in the Zenit and Ariane-5 launch vehicles. However, none of them has found to date their customers.

The OHBU operates in a single-stage configuration and consists of an upper adapter, tank compartment, propulsion unit compartment and an insert between the OHBU and the carrier. The tanks of the OHBU perform as carrying structures and are laid out in sequences: liquid hydrogen on top, liquid oxygen at the bottom. The tanks have an integrated bottom. This made the OHBU more compact and slimmed it down in height as compared to the initial variant.

The control system and the onboard measuring complex of the OHBU are being created on the basis of similar systems of the Breeze-M booster unit. The electronics blocks of those systems are installed on the upper adapter. The adapter also has a docking element for installing on the OHBU either Russian or foreign spacecraft.

Being considered now are two variants of the sustainer engine for the OHBU: RD-0146 developed by the Design Bureau of Chemical Automatic Systems and KVD-1M (oxygen/hydrogen) developed by the Design Bureau of Chemical Engineering.

**Fig. 33.** Oxygen/hydrogen booster unit, 12KRB, used in the Indian GSLV carrier

The RD-0146 engine is planned to be created on the basis of the American engine RL10A-4-1 jointly with the Design Bureau of Chemical Automatic Systems and the Pratt & Whitney company. The production will proceed in Voronezh. The sustainer engine's thrust in vacuum is around 10 tf. It is fastened in a gimbal suspension for controlling the thrust vector in pitching and yawing. Two blocks of steering micromotors are installed for rotation control.

The LPRE can be reactivated a number of times for launching payloads into a target point. The insert of the propulsion unit compartment enables the block to be docked with the Proton-M, Angara and other carriers with the minimum of changes.

The in-vacuum thrust of the KVD1-M3 sustainer engine is 10 tf. The number of activations of the sustainer engine is no less than 5, the maximum duration of the autonomous flight is up to 9 hours. The  refilled supply of $(O_2 + H_3)$ fuel is up to 19 tons. In all likelihood this particular engine will be chosen for the OHBU.

The prime manufacturer of the OHBU will be Khrunichev State Space Research and Production Center. Work on the engineering draft proceeds in conjunction with manufacturing services of the facility and the Salyute Design

Bureau since part of the required technologies have already been employed by the Salyute Design Bureau during pilot production of the Indian block 12KRB.

Tanks and part of the block's structure are covered with a combined thermal insulation, the whole block being beneath the nose fairing. The space between the OHBU and the fairing is divided by the diaphragms into several zones to secure fire safety and the required temperature modes.

## 2.2.3 Carrier rockets built around the ICBM withdrawn from service

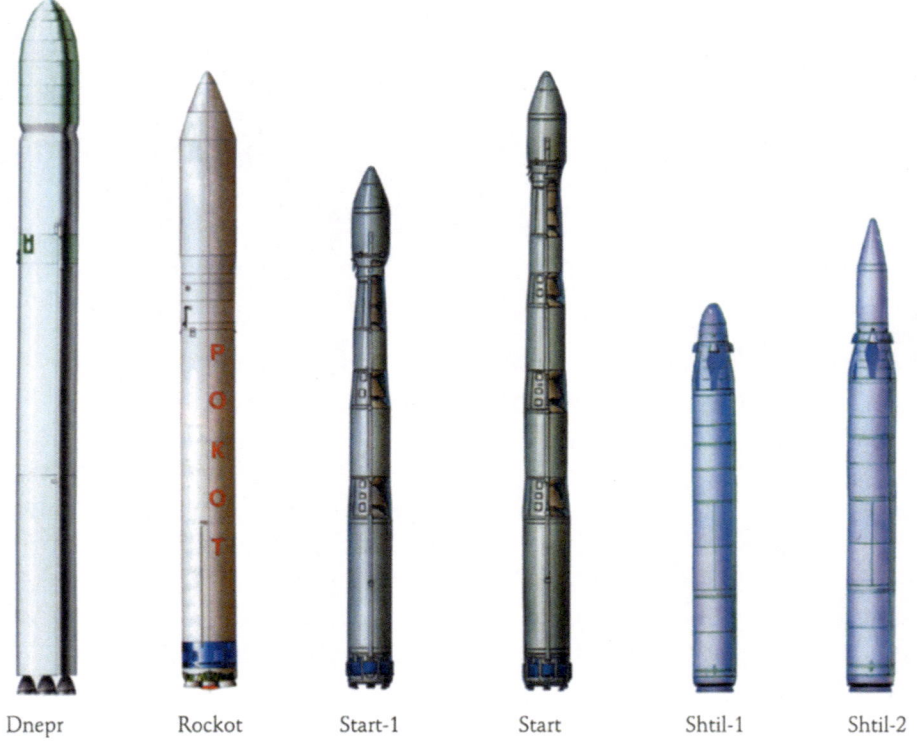

Dnepr            Rockot            Start-1            Start            Shtil-1            Shtil-2

**Fig. 34.** Strategic missiles-based carrier rockets

**The Start-1 carrier rocket** (Fig. 34) has been created by the Kompleks Science and Technology Center of the Moscow-based Institute of Heating Engineering which is well known for its ICBM, including Topol (SS-25) that gave birth to the new generation of carriers. The Start-1 carrier is intended to launch small spacecraft to low NEOs. Two successful launches of this carrier have already been performed from the Svobodny spaceport with an experimental military SC and a commercial US-manufactured spacecraft.

The Start-1 carrier rocket is a four-stage ecology-friendly solid propellant carrier whose first demonstration launch was performed in March, 1993 from the Plesetsk spaceport. The take-off weight of the rocket is 47 tons, the length is 22.7 m, the maximum diameter 1.8 m. The energy capability of the carrier enables it to launch up to 400 kg of payload to a low orbit.

**The Rockot light class carrier rocket** (Fig. 35) is being created by the Khrunichev Center under conversion program on the basis of 15A35 ICBM (RS-18 under the Strategic Arms Limitations Agreement) being currently withdrawn from service. The Rockot space rocket system has been created at the Plesetsk spaceport on the basis of 11P865P launch system. The infrastructure of the Plesetsk spaceport has been overhauled for launching the Rockot carrier. Taking part in the creation of the system is DASA company of Germany with which Eurockot joint venture has been established. Funds are already available to create an adequate infrastructure at the Plesetsk spaceport.

In creating the Rockot carrier, the existing two stages of the ICBM booster block receive a third stage, an innovative booster unit Breeze-KM with a large amount of fuel supply (5.055 tons), and a large nose fairing. This results in a light class carrier rocket with a very high commercial potential. Analysis shows that the Rockot carrier may seize up to 30% of the entire market of launches of small SC. The sustainer engine of the booster unit has a multiple activation capability which allows to launch spacecraft in various fashions, including the launch of a batch of SC into one or several various orbits. The instruments of the Breeze-KM booster unit provide for a high accuracy of putting the SC in orbit, the required orientation of payload and, if necessary, its power supply during orbital flight lasting over 7 hours.

The Rockot carrier is able to launch up to 1,950 kg of payload into orbit at an altitude of 200 km. The carrier (without payload) has the take-off weight of 107.5 tons, measures 27.7 m in length and 2.5 m in diameter. The stages are connected in tandem. The separation of the first and second stages proceeds according to a semi-hot scheme by means of the thrust provided by the second stage steering engine which is activated before the command is issued to deactivate the liquid propellant engine of the first stage. The braking of the first stage is accomplished through powder motors installed on the tail compartment. The separation of the third stage from the second occurs with the idle LPRE of the third stage by the agency of thrust of the braking SPREs of the second stage. All the stages use nitrogen tetroxide and asymmetrical dimethylhydrazine, which are fuel components with an extended storage life.

Work has been done at the Plesetsk spaceport to re-equip the launch site and update the erection and trial complex. The relevant manufacturing equipment in use has been renovated, that out of use has been dismantled.

**Fig. 35.**
Launch of Rockot carrier

The project of the Rockot carrier launch complex has been developed by the Design Bureau of Transport Engineering in 1995. The complex has been created by reconstructing the existing complex used for Cosmos-3M carrier and putting the engineering complex of preparing the Rockot carrier and SC onto the technology base of the Cyclone-3 rocket system. As this was done, the maximum use was made of the main structures and technological systems of the complex being reconstructed, without modifications or with a minimum amount of them. It has been possible to preserve the basic pattern and scheme of operations accepted for the Cosmos-3M system as well as functions of the main systems and mechanisms of the launch segment, which is quite important.

The delivery of the checked carrier without nose fairing over to the launch site is effected in a launch container. The installation of the container on the launcher is accomplished via an adapter ring that imitates the support elements of the Cosmos-3 carrier. Instead of a dismantled cable-mast the system receives a stationary support column with grasps for keeping the container with the Rockot carrier in a vertical position. This column is used to deliver technological communication lines of land-based systems to the docking points as well as for placing the carrier's controls and targeting systems. After installation on the container column, delivered to the launch complex together with the first two stages of the Rockot (the booster block of the ICBM 15A35) are the prepared nose block and the container extension. Their docking with the carrier and the launch container proceeds in a vertical position. Before the carrier with the SC is launched from the launch container, the service tower, as in case with the Cosmos-3M, is moved to safety.

The preparation of the carrier and the main block for delivery to the launch site proceeds at the engineering complex with the use of newly developed ground-based technological equipment and technical means of the base launch site. At the erection and trial complex there will be one work place organized for work with the Rockot carrier. Also a zone will be created with special conditions that meet the special requirements and the specific nature of work with the booster unit, its components and with spacecraft. The fuelling of the booster unit's engine with fuel components proceeds at the central refueling and neutralization station of the Plesetsk spaceport.

The first launch of the Rockot carrier with mockups of spacecraft was successfully performed on 16 May 2000. The launch demonstrated the unique characteristics in injection accuracy, an order above the ordinary level. In March 2002, with the help of the Rockot carrier, two US-German SC GRACE, each weighing 475 kg, were injected into orbit at an altitude of about 500 km.

Simultaneously with equipping the launch complex of the Rockot carrier at the Plesetsk spaceport, an issue is being considered of how to launch the Rockot carrier from a silo launching facility of the 15A35 ICBM at the Baikonur spaceport. The Rockot carrier has already performed from that spaceport the first test ballistic launches in 1990 and 1991. The third test launch with the Radio-ROSTO satellite took place in 1993.

The launches from the Baikonur spaceport offer their advantages in that the payload of the carrier increases due to the launch complex's closer location to the equator. In addition, there are now in the world several projects of satellite telecommunications systems oriented to low orbits with an inclination of around 50°. For systems like these the Rockot carrier could be a means of injection during launches from the Baikonur spaceport. However, as the carrier leaves the silo with running sustainers, its payload is subjected to a huge acoustic stress that modern foreign satellites cannot withstand. Therefore the Eurockot joint venture plans to rework the silo. Under study now is the feasibility of laying a gas duct  that begins at the silo bottom, meanders smoothly inside the silo and comes out at the surface. Also, a possibility is studied of creating a system of "water barrage" in the silo. Such measures are supposed to decrease the acoustic stress during launches to the level below 142 dB.

To date, practically all international restrictions imposed on launches of the Rockot carrier are lifted. The launch site at the Plesetsk spaceport has been declared as a new launch facility for injecting SC into space. The legal matters pertaining to the storage of supply of Rockot carriers at the Plesetsk spaceport have also been settled. Prior to that, similar matters had been settled concerning the launches of the Rockot SC from the spaceport of Baikonur.

The 15A35 ICBM is a prototype for building yet another carrier rocket. The NPOMash (engineering research and production association) has come up with a proposal to convert this ICBM into **a carrier rocket, Strela (arrow)**, for launching spacecraft for various applications. The Strela and Rockot carrier rockets are very similar since they use the base stages of one and the same ICBM. They differ mainly in the third stages and nose fairings. The energy capabilities of the Strela carrier provide for injection into low orbit at analtitude of 200 km payloads weighing up to 1,600 kg. The launches of the Strela carrier are expected to be conducted from silo installations at the spaceports of Baikonur and Svobodny.

In 1999 a decision was taken to build at the Svobodny spaceport a rocket system of the Strela type. The concept underlying the Strela space rocket system envisions the maximum preservation of the main features of the base system. Adopted as the base variant is the multi-role launch complex with a silo-based launching facility. The silo of the Strela carrier is a complete equivalent of the silo installation used for the RS-18 ICBM. Same as with the Rockot carrier, the first and second stages of the booster unit and the transport/launch container remain unchanged. Used as the third stage is the instrumentation block of the RS-18 ICBM comprising the standard missile control system. However, its fuel supply is smaller than that of the Breeze-KM carrier which is why the energy capability of the Strela is somewhat lower than that of the Rockot. The ground-based segment of the control and targeting systems also remains practically untouched. The one new element is the measuring instruments compartment included into space forebody and used to accommodate the instruments of telemetric and extra-trajectory measurements, the first and second stage motors emergency deactivation system, extra stabilization system for  passive flight sectors and onboard power supply sources. The space forebody can be equipped with two types of fairing – the standard fairing of the RS-18 ICBM and an enlarged fairing optimized in experimental launches of the RS-18 missiles (space forebody-2).

The NPOMash undertook to find extra-budgetary sources of funding to implement this project.

**The Dnepr space rocket system.** In 1998, a decision was taken that the international Kosmotras company (space transportation systems) would work without recourse to budgetary funds on creation and subsequent commercial utilization of the Dnepr space rocket system based on the RS-20 ICBM currently being withdrawn from service with Russia's armed forces pursuant to the strategic offensive arms reduction plan (Fig. 36).

Fig. 36.
Launch of the Dnepr carrier

The prototype of the Dnepr carrier is the heaviest and most powerful RS-20 ICBM (15A14). This liquid propellant two-stage ICBM had been developed in 1973 by KB Yuzhnoye and serially manufactured by Yuzhny machine building facility. According to the Strategic Arms Reduction Treaty (START-2), all RS-20s are to be withdrawn from service before the end of 2003 and disposed of before the end 2007. In service to date are 168 ICBMs. All the 104 missiles that had been deployed in Kazakhstan were removed to Russia after the breakup of the USSR and are now partially destroyed.

The Dnepr carrier is a three-stage rocket whose first two stages are standard booster units of the RS-20 ICBM. The original idea was to use for the third stage the standard autonomous warheads separation block. However, to increase the lifting capacity and expand the range of applications (specifically, for launching SC into medium altitude, solar-synchronous orbits) instead of an autonomous separation block other different options for the upper stage are being considered, for example, the third stage of the Cyclone-3 carrier, Lift booster unit developed by NPO Lavochkin, or the Breeze-KM booster unit adapted to this rocket.

The energy capabilities of the Dnepr carrier provide for injection into low Earth orbit a payload weighing up to 4 tons. The launches are supposed to be conducted from a silo-based launch facility of the Baikonur spaceport. One of the possible variants of using the Dnepr carrier for commercial launches of SC is participation in deployment of the low orbit system of global satellite communications, Teledesic (approximately 150 launches, two SC in each). The first commercial launch of the Dnepr carrier (Fig. 36.) took place in 1999 at the spaceport of Baikonur.

The implementation of the Dnepr project in collaboration with Ukraine creates conditions for re-establishment of the crippled technological and economic links between the two countries, for development of corporate interests, which in the final analysis will help the space industries of Russia and Ukraine come on the foreign market with a full range of space services.

For the first time in the history of astronautics, on 7 July 1998, a spacecraft was launched to orbit from a submarine, being at that in a submerged position. Two satellites of the Berlin Technical University had been put in orbit, Tubsat-N and Tubsat-N1 weighing respectively 8.5 and 3 kg. The launch was conducted with the help of the converted variant of the submarine-launched ballistic missile (SLBM) RSM-29M, **the Shtil-1 carrier rocket** (see Fig. 34).

The Shtil-1 carrier is one of the Russian-built SLBMs developed by the State Rocket Engineering Center dubbed Academician Makeyev Design Bureau (based at Miass, Chelyabinsk region). To enable the launch of SC the standard three-stage liquid propellant SLBM has been slightly modified. A special frame was added for installation of a spacecraft to be launched and the flight control software was changed. In addition, a special telemetric container with service instrumentation was mounted on the third stage to provide monitoring by land-based services.

Modifications are envisioned of the Shtil-1, Shtil-2N and Shtil-3 carriers. **The Shtil-2N carrier** (see Fig. 37) differs only in the larger size of payload compartment and is capable, depending on the prevailing conditions, of putting in low orbit up to 300 kg of payload. Installed on the Shtil-3 carrier is another boosting unit which provides for placing in low orbit payloads weighing up to 1 ton.

## 2.2.4 Advanced carrier rockets

In 1993, the Khrunichev State Space Research and Production Center, Energia Rocket Space Corporation, Academician Makeyev State Rocket Engineering Center and the Progress Central Specialized Design Bureau participated in bidding for the development of a new heavy indigenous carrier. Acknowledged as the best one was the project proposed by the Khrunichev Center. It was based on the many year research and development of carrier rockets, their creation and operation with regard to the projected requirements and the actual capability to satisfy them. Currently, the Khrunichev Center develops under the Angara program a series of carrier rockets. The key element of the effort is the creation of a heavy class carrier rocket that will be the base of Russia's space transportation program in the 21st century.

The cost effectiveness was achieved largely due to the use of oxygen and hydrogen fuel on the second stage and the oxygen/hydrogen booster unit. Compared to the competing variants using kerosene and oxygen fuel on the second stage, this approach allows to reduce by roughly 40% the take-off weight of the rocket and, hence, the weight of its structure and the cost of the entire system. The cost of hydrogen in this case accounts for less than 1% of the launch cost. All this (along with a somewhat higher cost of the hydrogen powered propulsion unit, tanks, fuel and storage system, etc.) slashes the specific cost of the launch by 30% to 35%.

The project proposed to use at the first stage of the Angara heavy class carrier an RD-174 engine that embodies unique design decisions and has been repeatedly tested in flight on the first stages of the Zenit and Energia carrier rockets. Developed by the NPO Energomash, it features the thrust of 740 tf. Used on the second stage is the hydrogen/oxygen engine, RD-0120, developed by the Design Bureau of Automatic Chemical Systems.

The manufacture of the Angara carrier envisioned the use of multi-purpose welding equipment and resorting to the experience accumulated by the Khrunichev Center in the production of the Proton carrier. The layout of the Angara carrier, as that of the Proton in its time, was arranged according to the customers' requirements: i.e. it provided for transportation by railroad with elementary assembly operations and checking at the spaceport.

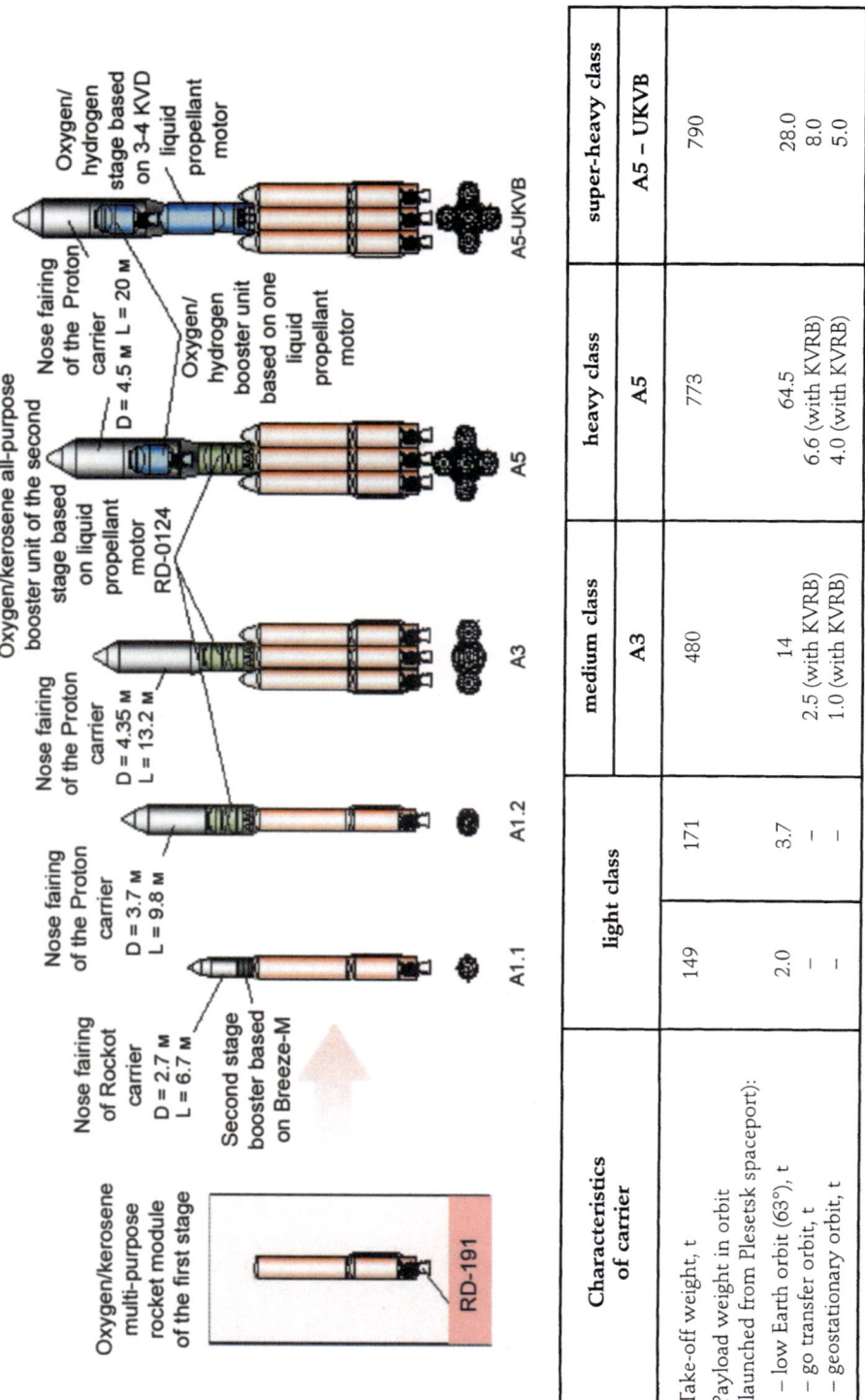

| Characteristics of carrier | light class | | medium class | heavy class | super-heavy class |
|---|---|---|---|---|---|
| | A1.1 | A1.2 | A3 | A5 | A5 – UKVB |
| Take-off weight, t | 149 | 171 | 480 | 773 | 790 |
| Payload weight in orbit (launched from Plesetsk spaceport): | | | | | |
| – low Earth orbit (63°), t | 2.0 | 3.7 | 14 | 64.5 | 28.0 |
| – go transfer orbit, t | – | – | 2.5 (with KVRB) | 6.6 (with KVRB) | 8.0 |
| – geostationary orbit, t | – | – | 1.0 (with KVRB) | 4.0 (with KVRB) | 5.0 |

Fig. 37. The famaly of Angara carrier rockets based on multi-purpose rocket module

The stages of the Angara carrier are arranged in a tandem configuration. With such an arrangement, expected for use on both stages was the package principle of laying out the fuel tanks. Hitched to the central fuel (kerosene) tank on the first stage are two lateral tanks containing oxidizer (liquid oxygen). On the second stage the oxidizer (liquid oxygen) tank is central while two fuel (liquid hydrogen) tanks are lateral. The separation scheme is "hot", i.e. the stages are connected by a truss (between central tanks). Later on (in the second phase) the layout of the Angara carrier envisioned the installation of additional devices for returning the first stage to the spaceport neighborhood without intermediate flight stop. This enabled multiple use of the system components and made redundant the areas for dropping the spent first stages (the second stages enter the suborbital trajectory and drop after the first half-circuit in a remote area of the world's ocean).

Such a variant of the Angara carrier must launch up to 27 tons of payload into low support orbits (altitude of 200 km) with an inclination of 63° (the latitude of the Plesetsk spaceport) and up to 4.5 tons into geostationary orbit in case of using the oxygen/hydrogen booster unit (OHBU). Along with OHBU it was also envisioned to use the Breeze-M booster unit.

After a detailed consideration by the Interdepartmental Commission a decision was taken to further develop the Angara carrier under the project proposed by the Khrunichev Center. In course of the design work the Angara was substantially reworked and improved. Considering the prevailing conditions in the country, the Khrunichev Center proposed the strategy of a steep-by-step construction of the heavy class carrier with the use of general application modules in its component package. The new concept preserves all the key ideas of the original variant of the Angara carrier and develops new advanced capabilities.

Nowadays, the Angara family of carrier rockets embraces the carriers from the light to super-heavy class. The main characteristics of the Angara carriers are shown in Fig. 37. At the base of this family is the multi-purpose rocket module (MPRM). It includes oxidizer and fuel tanks and the RD-191 engine. The MPRM tanks perform as carrying elements with the oxidizer tank placed in the forward end of the system. The RD-191 engine, being created by NPO Energomash is fueled by kerosene and liquid oxygen. This single chamber engine is being built on the basis of four-chamber engines RD-170 and RD-171 and a two-chamber LPRE RD-180 being created for the Atlas-2AR carrier. The close to Earth thrust of the RD-191 reaches 196 tf, in vacuum – 216 tf; the specific thrust on Earth is 309.5 s, in vacuum – 337.5 s. To obtain the control of the carrier rocket in flight the engine is fastened in a gimbal suspension. The length of the MPRM is 28.5 m, the diameter is 2.9 m. These dimensions had been chosen based on the manufacturing rigging available at the space rocket manufacturing facility.

**Fig. 38.** Top managers of Khrunichev Center at Le Bourget (1999). Left to right: A. V. Lebedev, A. S. Kondratyev, A.I. Kiselev, A. A. Medvedev

One such multi-purpose rocket module (MPRM) is the first stage of two light class carriers being created under the Angara-1 program.

Used as the second stages on those two variants of the carrier (Angara-1.1 and Angara-1.2) are respectively the central part of the Breeze-M booster unit and the I type of rocket block of the Soyuz-2 carrier.

The medium class carrier rocket Angara-3 is built by adding multi-purpose modules (as the first stage) to the light class carrier Angara-1.2. The stages of the Angara-3 carrier are arranged in a tandem fconfiguration. Three MPRM are used as the first stage. Installed on the medium MPRM via a truss adapter is the second stage (I type of block). Used as the third stage is a small size booster unit or a central block, Breeze-M, which is intended for forming the work orbit.

Its inclusion into the carrier variants with a stage of the I type is caused by the fact that the RD-0124 engine installed on this stage is designed for only one activation.

The Angara-5 heavy class carrier rocket is built by adding two lateral modules to the Angara-3 carrier. The super-heavy class of carrier is formed by replacing the second stage (block of type I) of the Angara-5 heavy class carrier with an oxygen/hydrogen stage having four KVD1 motors.

The energy capabilities of the Angara-3 and Angara-5 carriers launch into low orbit payloads weighing respectively 14 tons and 24.5 tons.

Used as booster units on the Angara-3 and Angara-5 carriers are Breeze-M booster unit and oxygen/hydrogen booster unit.

The principal launch site of the Angara family of carriers is the spaceport of Plesetsk. In building the launch complex of the Angara carrier expertise is utilized obtained while working on the Zenit carrier. The unique engineering decisions make it possible to launch from one take-off pad all carriers of the Angara family. To reduce the areas allotted to receive jettisoned fragments of space vehicles, a provision is made as early as building the Angara-1 for taking special measures.

Three sources of funding the Angara project are expected: the Rosaviakosmos, the Defense Ministry, and revenues yielded by the commercial activity of the Khrunichev Center.

On the basis of the  main variants of the Angara carriers shown in Fig. 37, it is possible to build other models. Thus, variants are being considered of installing extra solid propellant boosters on light class carriers. This will allow to select a suitable carrier for a specific SC instead of creating a SC with a regard for the available carrier.

**Fig. 39.** T. Corcoran, A.S.Kondratyev, A.I.Kiselev, A.A. Medvedev, A.K. Nedaivoda, D.V. Pivnyuk after signing an agreement for cooperation in commercial use of the Angara

So the Khrunichev Center has developed and proposed under the Angara program a strategy enabling creation of a number of advanced carriers of various classes despite financial constrains and rigid deadlines. The manufacturing period of the Angara carriers are quite stringent. For example the first launch of the Angara-1.1 carrier is slated for as early as 2003. The launches of all types of the Angara carriers are planned to be conducted from the spaceport of Plesetsk. Proposed here for launching the carriers is the modified launch installation of the Zenit starting complex. The first start of the Angara-1.2 carrier is to take place in 2004. The first launch of the Angara-5 carrier is also planned for 2004. Foreign specialists have already registered interest in the new carrier (Fig. 39). °

The Khrunichev Center intends to improve on the performance of the carrier and, most importantly, to reduce the cost of its launching not only through standardization of the Angara carriers' first stages but also through introduction of advanced and well proven technologies such as high-performance oxygen/hydrogen LPREs, automated launch preparation, the innovative booster units and nose fairings. The Angara family of carriers uses such cutting-edge technologies as the employment of reusable elements (booster stages) in the carrier structure. This is just one of the engineering decisions that dramatically improves the economic performance of launch vehicles.

The Korolev Rocket Space Corporation Energia works jointly with other organizations on the Avrora space rocket system. The project has become practically international since launches of the Avrora carrier will be conducted from the Christmas Island (Australia) in the Indian Ocean and the project customer is Asia Pacific Space Center. The Avrora project is based on the well known "seven" (simplicity of the entire system, very high reliability, a great number of launches).

The Avrora carrier will be operated in two variants: three-stage carrier (without booster unit) – for launching SC to low near Earth orbits and four-stage carrier (with the Korvet booster unit) – for placing SC in geo transfer and geostationary orbits.

The take-off weight of the Avrora carrier with the booster unit is 379 tons. From the Christmas Island the Avrora carrier with booster can deliver to geotransfer orbit with an inclination of 11.30 a payload weighing between 4.35-4.5 tons and 2.1 tons to geostationary orbit. It can deliver 12 tons of payload to low orbit at an altitude of 200 km and with an inclination of $11.3°$. All Avrora carriers' motors are fueled by liquid oxygen and naphthyl RG-1. Installed on the central block will be a combination of one liquid propellant engine NK-33 in the middle and one four-chamber steering engine RD-0124R. On four blocks of the first stage will be liquid propellant engine 117A with new centrifugal one-component nozzles. Installed on the third stage of the Avrora carrier will be one four-chamber engine RD-0124E.

## 2.2.5 New generation launch vehicles based on multi-purpose rocket booster

The issue of ecology and the striving to improve the competitiveness of the Angara carrier call for, albeit in the future, all-azimuth carrier rockets that need no allotted areas on land to receive their jettisoned components.

To this end, the rocket assemblies of the carrier's first stage must be equipped with a recovery system similar to that used on airplanes that would enable them to return to the launch area after separation from the carrier. The nose fairing and other jettisonable components, e.g. the adapter compartment, are launched using the energy of the second stage into the ballistic trajectory with a subsequent fall into the water of the world's ocean.

The optimum variant of the aerodynamic scheme of the reusable rocket booster (RRB) that minimizes the carrier's energy loss caused by installing the recovery system, is based on the concept of the pilotless single-mode flying booster optimized for performance of the subsonic cruise flight to the launch area and equipped with a two-position high-aspect wing that opens after separation of the RRB from the carrier, with an turbofanengine and with a fin assembly arranged in a normal aerodynamic configuration. Contrary to other variants of all-azimuth launch vehicles, e.g. aerospace flying vehicles or a single-stage carrier, the two stage carrier with RRB has no critical technologies and can be implemented even now with the maximum utilization of expertise available to date in rocket and aircraft engineering. In the case of comprehensive optimization of structural and ballistic performance and amendment of control software, the carrier's losses in terms of weight and energy caused by installing the recovery system of the first stage and determined by the chosen launch trajectory, do not exceed 35%–50% of the weight of payload placed in low circular orbit.

The study of the light class RRB powered by the RD-191 engine shows that the proposed variant of layout for the first stage has a multi-purpose capability and permits, if necessary, to build on its base a number of advanced carriers of light, medium and heavy classes with lifting capacity between 2 and 22 tons that use oxygen and kerosene for fuel components.

The Khrunichev Center develops now jointly with NPO Molniya (research and production association) the Baikal reusable booster unit for use in the Angara family of carriers instead of expendable multi-purpose modular boosters (Fig. 40, Table 1).

Such reusable boosters are created with the maximum utilization of expertise accumulated in work on expendable modular boosters whose design originally provides for their multiple use. The utilization of such reusable boosters will make it possible, in addition to reducing the specific cost of launch:

- to fully or partially do without jettison areas;
- to provide all-azimuth capability in launches to orbits with various inclinations;
- to release manufacturing facilities for other jobs.

The reusable booster unit is created according to the "fly-back" concept. Its centerpiece is a liquid propellant rocket engine whose design is in many ways similar to that of the modular booster of the Angara carrier.

**Table 1.** Main characteristics of carriers based on Baikal reusable booster unit

| Characteristics | Angara A1-V | Angara A3-V | Angara A5-V | Angara A4-V |
|---|---|---|---|---|
| Take-off weight, t | 168.9 | 466 | 709 | 700 |
| Weight of payload        in launches from Plesetsk spaceport, t | | | | |
| – to low orbit | 1.9 | 9.3 | 18.4 | 22.0 |
| – to geo stationary transfer orbit | – | 1.0 | 4.4 | 5.66 |
| – to geostationary orbit | – | – | 2.5 | 3.2 |

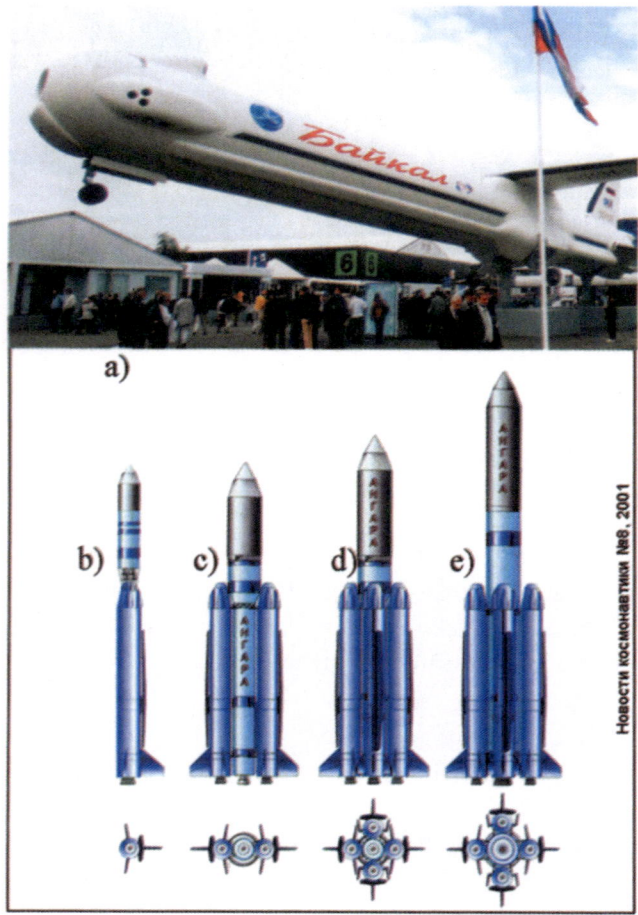

**Fig. 40.**
Development of the Angara family of carriers on the Baikal reusable booster base:

a – Baikal reusable
    booster
b – Angara A1-V carrier
c – Angara A3-V carrier
d – Angara A5-V carrier
e – Angara A4-V carrier

Parameters and characteristics of the carrier and trajectory are chosen so that to preserve as much as possible the design of the multi-purpose rocket module. The choice of the stages separation speed corresponding to M=6...7 made redundant the heat insulation layer which had a dramatic impact on the cost of turnaround servicing.

To orientate the spent booster before it enters the atmosphere's dense layer, it is equipped with a reaction control system (RCS). On entering the atmosphere and after initial braking, the booster starts gliding with the help of a rotary wing. The flight control is accomplished by aerodynamic control mechanisms. The gliding passes into a engine-sustained flight when two turbojets installed in the nose end of the reusable booster go into action. Controlled by its own onboard control system, the booster returns to the airdrome near the launch complex where it touches down on a retractable wheeled undercarriage of the type used in aircraft. The power in flight is supplied by the power supply system. The requisite information about the condition and functioning of the booster is gathered and transmitted to Earth by an onboard measuring complex.

A number of base elements and systems of the reusable booster are borrowed or developed on the basis of existing designs. This, undoubtedly, reduces the technical risks and the amount of required tests. The amount of experiments and trials to be conducted remains, however, rather large. Primarily, this refers to determination and certification of aerodynamic characteristics of the reusable booster's flight after its separation from the carrier rocket before landing at the airdrome. Such trials have already been conducted by TsAGI (Central Aero Hydrodynamics Institute), the Khrunichev Center and NPO Molniya on scale models of the reusable booster.

Taking off from the launch complex of Russia's Plesetsk spaceport, the reusable booster as part of the Angara carrier can on return land at the airdrome of Pero incorporated by this spaceport. After neutralization of the residual of fuel at the filling and neutralization station with the subsequent delivery to an erection and trial facility, the booster will be prepared for flight and will be ready for reuse in the component package of the carrier.

Test flights of the reusable booster unit are slated for 2004 in the component package of the Angara-1.2 light class carrier.

Thus thanks to the use of reusable boosters, the Angara carriers will also become partially-reusable since they will have reusable first stages comprised of modules (reusable boosters) and reusable upper stages. Such carrier rockets will be quite competitive on the world's space market. According to estimates, in terms of specific cost of launch the partially-reusable Angara carriers will surpass not only expendable analogs, but also fully-reusable carriers that are now in development.

## 2.2.6 Reusable space systems

Today's achievements in science and technology, the expertise available in development of component packages for reusable space systems (RSS) and the experimental research base available to date plus the manufacturing technology adopted in space rocket and aircraft industries will make it possible to tackle in the very near future the issues of designing and developing advanced RSS.

Currently, three distinct major trends are evident in the evolution of RSS:

1. space rocket systems using the rockets for launching the orbital winged stage;

2. aerospace systems using heavy subsonic transport aircraft as boosting stages;

3. vertical take-off aerospace systems comprising aerospace flying vehicle with a combined propulsion unit based on various types of airbreather and rocket engines.

The extreme energy capabilities of RSS, their dimension, the possibility to maneuver during launch into and descend from orbit are determined by design features and layout peculiarities.

Studied now in each of the said trends are various variants of RSS. Also, a possibility is being investigated as to the feasibility of their practical application.

The first of those trends has progressed more than others and produced RSS. By the late 1980s **a reusable space system, Energia – Buran**, was created (Fig. 41).

**Fig. 41.** Reusable space system Energia – Buran

In terms of performance and capabilities the Energia – Buran RSS (NPO Energia as prime developer) is similar to the UmjS project Space Shuttle. It included the Energia heavy class carrier rocket weighing 2,270 tons and an orbital space ship (OS), Buran, whose dry weight was about 70 tons.

The Buran OS is able to inject 30 tons of payload into a low orbit at an altitude of 200 km and bring back from the orbit 20 tons. The OS could perform up to 100 flights, with the expected use of the first stage of the Energia carrier up to 10 times. The Buran orbital spaceship made its first and, regrettably, last flight on 15 November 1988 in an unmanned mode. Having completed two orbital circuits, the ship landed automatically on Earth.

The subsequent collapse of the Soviet Union and the resulting economic plight of Russia precluded the adequate funding of the Energia – Buran program which gradually terminated it.

In spite of the fact that in creating the Buran OS it was resolved to expedite the work by using as reference parameters the characteristics of its US counterpart, the major design decisions taken with reference to the Buran can be regarded as Russia's know-how that found application in other industries. The automatic landing of the orbital ship on Earth still remains a luxury unavailable to the Americans.

Work in other lines aimed at building reusable space systems are in various phases of engineering development.

**The Burlak aerospace system** is intended for commercial launches of small size spacecraft to low Earth orbits. The prime developer of the system is MKB Raduga (machine design and development bureau).

The Burlak system is comprised of:
• liquid propellant two-stage space rocket Burlak;
• two carrying aircraft, Tu-160SK, equipped with a special launch unit and apparatuses for preparation and launch;
• two Il-76SK aircraft-based command and instrumentation posts;
• ground-based servicing complex;
• automated system for preparation of rockets and carrying aircraft's mission tasks.

Energy capabilities of the system permit to launch to low equatorial orbits payloads weighing up to 1,100 kg. The flight range of the carrying aircraft in the launch area of the space rocket is 5,000 km.

On 19 January 2000 the Progress Specialized Design Bureau and Vozdushnyy Start (aerial start) Aerospace Corporation signed an agreement for cooperation in building **an aerospace rocket system, Vozdushny Start** (Fig. 42). The core idea of the project is to use the heavy transport aircraft An-124-100, Ruslan, as a launch platform for the Polyot light class carrier rocket. The carrying aircraft equipped with systems that provide for loading the carrier with payload placed in transport and launch container into the freight cabin, for controlling and monitoring of the system in flight, jettisons the carrier in the launch area, conducts navigational and telemetric measurements of the rocket flight parameters and transmits information to the control center.

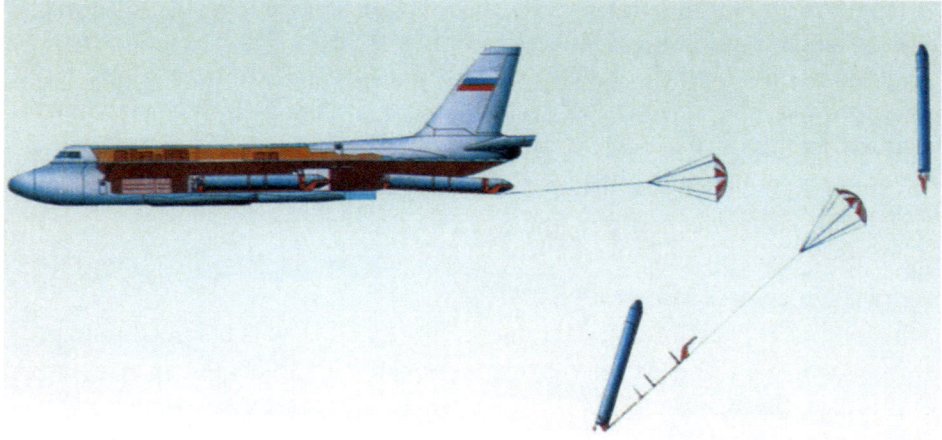

**Fig. 42.** Vozdushny Start aerospace rocket system

Vozdushny Start aerospace system includes:
- Ruslan An-124-100 carrying aircraft;
- Polyot carrier rocket;
- ground-based preparation and launch complex;
- automated systems for controlling the preparation, launch and flight.

The release of the Polyot carrier at an altitude of 8 to 11 km at the carrying aircraft's speed of 600 to 650 km/h is performed from the freight cabin so that the rocket starts off with the speed of 20 m/s (relative to the aircraft) and falls back at a safe distance prior to activation of the first stage.

The Polyot carrier has a two-stage configuration with a tandem-type arrangement of the stages. Chosen for sustainer motors were the available liquid oxyden-kerosine LPREs: at the first stage – NK-43 built by the Samara-based Kuznetsov Science and Technology Complex; at the second – a modified LPRE, 11D58M, used on the DM booster unit of the Proton carrier. The choice of those engines for the aerospace system was accounted for by their high readiness and by the plans of the OAO Motorstroitel (Samara-based joint stock company) and the Federal State-owned Unitary Enterprise Voronezh Mechanical Engineering Works to support the advanced programs. The take-off weight of the Polyot carrier is 100 tons. The Vozdushny Start system allows to launch 3 tons of payload to low Earth orbit.

The Vozdushny Start can perform launches from practically any place on Earth where at least 3 km of a runway can be made available. The possibility to install a satellite on the carrier rocket on a customer's territory resolves the problem of restrictions imposed on export of space technologies.

The Ruslan carrying aircraft is essentially a returnable and reusable first stage of the system used to launch satellite to low Earth orbit. Its use reduces the specific cost of launching a payload (roughly by 30% as compared to the cost of launching by the existing carrier rockets).

The Vozdushnyy Start aerospace system is a highly reliable (no less than 0.99), ecology-friendly (due to non-toxic fuel components), fast-to-build craft. This has been achieved thanks to the wide use of ready-made assemblies and systems. The biggest problems now are the dropping of large size cargoes (carrier rockets) and the activation of the liquid propellant sustainer engine in conditions close to weightlessness. Both tasks can be successfully fulfilled by resorting to the experience obtained in building Russia's ICBM launched from silos.

For a number of years the NPO Molniya has been working on the **multi-purpose aerospace system** (MPAS) consisting of the An-225 Mriya carrying aircraft (the first stage) and a double seat orbital plane with a jettisonable fuel tank.

The orbital plane can be used in an unmanned mode or replaced with a disposable rocket stage which gives the systems a greater application flexibility. The orbital plane uses two sustainer engines running on three-component fuel and can endure 10 flights. The plane itself is designed to with-stand 100 launches. At the initial portion of the flight, the engines run on kerosene and oxygen, and then on hydrogen and oxygen. The excellent aerodynamic performance of the orbital plane gives it during its descent from the orbit the lateral range of up to 2,000 km which makes it possible to land the craft on a variety of airdromes across Russia. The MPAS can be launched from any first class airdrome equipped with facilities for turnakound servicing and refueling of the orbital stage.

The **reusable space rocket glider** (RSRG) has been studied as one of the advanced concepts that will get special attention in the coming years. The RSRG is a vertical take-off and horizontal landing flying vehicle for launching into orbit light and medium class payloads. The RSRG performs the entire flight using standard LPREs. The studies show that the development of the rocket glider makes sense if only it decreases the specific cost of launching the payload by 5 to 7 times as compared to traditional expendable carriers, increases by 5 times the reliability of task execution, meets the ecological safety requirements and improves the serviceability. The current level of science and technology does not afford creation of such a system which is why preference is now given to partially-reusable all-azimuth carrier rockets.

Very promising for the longer term future are **reusable single-stage carriers** (RSSC) and **aerospace flying vehicles** (ASFV) (Fig. 43). Studies have shown that the advantages of the single-stage craft with a LPRE over the

double-stage one in terms of development cost and the specific cost of launch
come into effect when the "dry" weight of the craft is cut down by roughly
30% compared to the systems like Space Shuttle or Energia-Buran. This,
however, is to date unrealistic even though some Russian firms come up with
RSSC projects even these days. For example, the Salyute Design Bureau has
prepared proposals for a vertical take-off horizontal landing carrier similar to
the US project Venture Star. The Makeyev Design Bureau has developed on its
own a light single-stage rocket, Korona (crown), with vertical take-off and landing. The rocket is similar to the US flying vehicle Delta Clipper. The work
under the latter program was closed recently. No feasibility study has been
conducted for either of the above said Russian projects  Nor is it clear where
the funding will come from.

**Fig. 43.** MG-19 (project)

More realistic is the concept of the aerospace vehicle with horizontal take-
off and landing and with a propulsion unit that uses air. Since the inception of
the space era the idea of using the air gave no rest to designers. Yet the
seemingly simple idea proved quite difficult to convert into reality.

A special challenge is the propulsion unit. Each portion of the flight (take-
off, boosting, acceleration and altitude pick-up, entry into orbit) requires an
optimally running engine. But the requirements made of such a propulsion unit
often contradict the general concept of the craft. The vague prospect of the
development of the multi-mode air jet propulsion unit and the need for
substantial funding did not allow to build in the 1970s the Russian Spiral aerospace system. For similar reasons the English concept HOTOL was not
materialized in the 1980s. In the early 1990s, on the same grounds the US discarded the high-tech NASP system. Eventually, all developments boiled down
to a mere mating of a rocket with an aircraft as is the case with the Space
Shuttle and Energia-Buran systems.

The building of the aerospace vehicle is economically feasible only when
new structural materials become available and multi-mode airbreather propul-

sion units are developed. The creation of such systems on a purely commercial basis looks unrealistic.

Modern turbojet engines widely used by aircraft assure flights at M slightly above 3, which is plainly insufficient for boosting the aerospace vehicle to optimum velocities. One of the most promising ways is to build supersonic combustion ramjet (SCRAMJET). Russia is the sole country that conducts flight trials with SCRAMJET. In the 1990s, the Baranov Central Institute of Aircraft Engine Building had prepared and conducted flight tests of SCRAMJET on a hypersonic flying laboratory. The studies of foreign specialists in this field (USA, France, Germany, UK) did not go beyond testing miniaturized examples in hypersonic velocity wind tunnels and creating computerized models.

Russia's best known projects dedicated to aerospace vehicle are those concerned with the Tu-2000 proposed by the Tupolev Aviation Research and Engineering Complex and the Neva/Ajax from the St.Petersburg-based Leninets holding company.

The Tu-2000 is a delta tailles flying vehicle with leading edge extension wing. It must be equipped with liquid hydrogen ramjet engines. The vehicle wingspan is 14 m, the length is 55 to 60 m, maximum take-off weight is between 70 and 90 tons.

The goals set by this project can be realized provided that technology research is carried out in the field of advanced structural materials and wide range SCRAMJET as well as in super-cooled (slush type) hydrogen production and storage techniques.

In 1995 the estimated cost of the required research and development reached US$ 5.3 billion, the figure Russia could not cope with even at the relatively low declared cost of launch, US$ 13.6 million (at the rate of 20 launches per year). The investigation results yielded a conclusion according to which the Tu-2000 project implementation date cannot be established for certain even in a stable economy because of the excessively high characteristics claimed.

Specialists believe that flight test must be continued with increased speed and ceiling of the hypersonic flying laboratory. To this end, the specialists of the Central Research Institute of Aircraft Engine Building jointly with the Gromov Flight Research Institute have developed an experimental flying vehicle dubbed IGLA that simulates the layout, outline and flight conditions of the advanced aerospace flying vehicles (ASFV). A large number of experiments are to be conducted using the IGLA to enable the creation of a reliable, high speed, ecology-friendly SCRAMJET.

Still more exotic is the ASFV project proposed by Ajax. This is quite a daring project of an open aerothermodynamic system that converts into work the energy of the hypersonic incoming flow. The authors of the concept suggest to revise many development related aspects of the ASFV of the future. In general terms, the energy exchange proceeds as follows: during its passage through the channels at the hottest points of the flying vehicle (fore end of the fuselage, wing leading edge and the engine combustion chamber) hydrocarbon fuel (kerosene) heats up and decomposes with the help of catalysts to hydrogen and hydrocarbons having a smaller molecular weight. The hydrogen is used in the magnetohydrodynamic converter (MHD generator) to generate electricity utilized to regulate the air flow in the entry circuit of the SCRAM-JET in which hydrocarbons are burned. The thrust to weight ratio of such an engine can surpass many times over the similar parameter of conventional airbreather engines. The electric power obtained is also utilized by the plasma system controlling the lift and drag of the ASFV. The creation of the Ajax involves utilization of the following technologies:

• active heat protection of the hypersonic flying vehicle with the help of chemical heat recovery;
• magnetoplasmachemical engine (MPCE);
• a device for plasma-aided control of aerodynamic performance of the flow.

The Ajax main propulsion unit ramjet operating in the range of M between 6 and 16) is based on the concept of the magnetoplasmachemical engine. Its cycle includes the MHD circuit of power conversion enabling to control the parameters of the incoming flow and its profile within a wide range of speeds at a fixed geometrical configuration of the air intake unit. The electric power of the MHD generator is used to create plasma beams that reduce the aerodynamic drag of the ASFV, which in turn increases the thrust of the engine in the MHD-accelerator installed behind the combustion chamber.

The plasma beams fed to various zones of the incoming flow can be used to control aerodynamic parameters of the Ajax with an increase of its aerodynamic quality to 5 and above. This is required for a cost effective cruise flight and for boosting to an orbital speed. Laboratory experiments have been conducted to determine the impact of plasma on incoming airflow.

According to its authors, the Ajax offers better mobility and higher efficiency (has no onboard cryogenic liquids, oxygen and hydrogen). Also, it is capable of sustained flight in the atmosphere at hypersonic speed generating in the process large flows of energy (in the region of tens of megawatts).

The ASFV is going to be based on the existing infrastructure, i.e. airports, spaceports and ground-based control posts. Relatively low in cost, it will be able to go into orbits with various parameters and perform wide maneuvers in the air. At the same time, many experts believe that the Ajax concept is disputable since none of the engineering decisions required for its creation has gained general acceptance.

## 2.3 Manned austronautics as the trend line

The history of manned astronautics began on 12 April 1961 when the Soviet pilot-cosmonaut Yuri Gagarin made the first 108 minute space flight and went down in history of the civilization (Fig. 44). That event was the peak of the titanic effort of the USSR in space exploration and embodied the national achievements in space rocket engineering.

**Fig. 44.**
The Earth's first cosmonaut Yuri Gagarin

The manned exploration of space was a formidable challenge. Success and achievements alternated with failures and tragedies. The first in the death toll were US astronauts. In January 1967, V.Grissom, E.White and R.Chaffee burned up in the oxygen atmosphere of the Apollo spaceship during its groundtest. Three months later, cosmonaut V.M.Komarov died in testing a new transport spaceship, Soyuz (Fig. 45).

In 1971, the first crew of the Salyute orbital station, G.T.Dobrovolsky, V.N.Volkov and V.I.Patsayev, died during the return trip after a successful completion of their mission (Fig. 46). The space kept taking its toll. In 1986, the catastrophe on the US Space Shuttle reusable orbiter Challenger, ended the lives of seven astronauts.

**Fig. 45.**
Cosmonaut V.M.Komarov,
the first tester of the Soyuz

One of the landmarks, rather sad though not tragic, on this thorny path was Russia's program of manned flights to the Moon. It was started in 1964 and from the outset lagged behind its American counterpart announced in 1964 and made the nation's top priority. The success of this program became the matter of concern to each and every American. By contrast, the Russian public could only guess about the existence of any such national program. The centerpiece of both Russian and American manned lunar flights programs was the super-heavy carrier rocket. The successful flight to the Moon, landing on it and the return to Earth asked for launching into low Earth orbit more than 100 tons of payload.

The Americans started to develop their super-heavy carrier under the Saturn program in 1958. As early as 1961 a double-stage prototype of such a carrier was launched. In 1963 the final variant of a flight to the Moon was mapped out and a three-stage carrier rocket, Saturn, was picked out for the mission. The carrier enabled 139 tons of payload to be injected into low Earth orbit and 65 tons into the flight path to the Moon.

The tests of the Russian carrier N1 picked out for implementation of Russia's program of manned flight to the Moon began late in February 1969. The weight of payload to be injected by that carrier into low Earth orbit was 70 tons.

The more than four years' lunar race was won by the Americans (Fig. 47). In December 1968 the American astronauts made an orbital flight round the

Moon in the Apollo-8 spaceship. The Russian endeavor to do the same in 1969 in an unmanned mode failed (the carrier rocket fell because the engines went out of action). After the American astronauts landed on the Moon in July 1969 the Soviet leaders lost interest in the lunar program. Four subsequent – and all abortive – launches of the program's principal "mover", the N-1 super-heavy carrier rocket, buried the last hopes for Russia's manned lunar missions.

**Fig. 46**
The first crew of the Salyute, cosmonauts G.T.Dobrovolsky, V.N.Volkov, V.I.Patsayev

The manned expedition to Mars in the 20th century was technically unrealistic. However, both the USA and USSR had been studying various versions of carrying out such an expedition since the 1960s. So one of the projects envisioned the use of an electric reactive unit as a motor. The weight of the entire complex sent on a mission to Mars could reach several hundreds tons. Though unclaimed for, those projects were a step forward in conquering of space by man. The scientific and technological experience gained in such work will, undoubtedly, be used in preparation of the future flights to Mars.

After Yu.A.Gagarin's flight, Russia's manned astronautics gathered momentum and quickly went the way from solitary brief flights to permanent stay of crews in orbits.

The legendary Vostoks and Voskhods were shortly replaced by the first generation Salyute space stations that assured life support and operating conditions for the crew over a long period of time restricted only by the amount of supplies delivered to the space station. That was when it became possible to move from the deliberation of the kind "is it at all worthwhile to launch man into space?" to the problem study of the kind "will a man be able to fly as far as Mars and farther off to stars and what will he need to do so?", not a rhetoric question raised at one point by K.E.Tsiolkovsky.

Fig.47.
First people on the Moon
(lunar module's pilot E.Aldrin)

The evolution of science and technology brought forth the second generation Salyute stations (Fig. 48) that boasted an improved system of transportation services enabling organization of sustained space flights.

The following step in the development of Russia's space technology was
the next generation orbital station, the manned space complex Mir. It was built
at the Khrunichev Plant under the technical supervision and administrative
guidance of A.I.Kiselev, the head of the said fabricating facility (Fig. 49). The
Mir is a sophisticated construction of modules and assembly units that could
be adapted in flight even to sharply changing conditions. For example, in
the initial engineering and during the first years of its flight there could be
no talk about docking the complex with the orbital ship Space Shuttle (considered as the main version was the docking of the complex with the Buran).
It was only during space flight that the station was reworked and got re-equipment enabling it to carry out that task.

It should be noted that one of the results of developing the manned flights
in the 20th century was the logical conclusion that no further advance is
possible without extensive international collaboration. Therefore the next

phase in the evolution of manned flights that falls on the 21st century will be marked by the joining of efforts of various nations in work on a common project. Manned space flights programs envision large scale gradual merger of Russia's space effort with the related programs of the USA, Western Europe, Japan and Canada. The Federal Space Program provides for Russia's step-by-step introduction into international manned flights programs with the wideuse of experience in creation and operation of the Russian manned orbital station Mir. The main steps on the way to becoming such a partner were the following:

**Fig. 48.**
Salyute orbital station in flight

1. Programs of foreign astronauts' flights as crew members of the Russian complexes Salyute and Mir.

2. The Mir – Shuttle program (1994-1995) that included joint work on the Russian Mir station and the US Shuttle spaceship as well as flights of the Russian cosmonauts on the Shuttle and the visitation of the Mir station by the American astronauts (Fig. 50).

It should be noted that Russia's desire to lead in manned flights stems, undoubtedly, from the capabilities offered by the Mir. The Mir complex, whose first module (the base unit) was injected into orbit on 20 February 1986 is a major achievement in technology of manned space flights and in exploration of near Earth space. In all, 102 successful launches of various types of ships and modules  have been conducted (including launches of the US Shuttle spaceship).

**Fig. 49**. Soviet cosmonauts and A.I.Kiselev, head of the Khrunichev Engineering Works, on the plant's premises before sending the Mir station to Baikonur

The Mir complex has no counterparts and holds the world's absolute records in:

duration of service life in orbit;

• cosmonauts' total flight time on board the complex;

• versatility and amount of research work and investigations carried out on board;

• number of programs implemented as international deals and the amount of work done on a commercial basis.

The capabilities of the Mir station and the level of international cooperation on it are in keeping with the designed performance of international space stations. The nearly 15 years' operation of the Mir complex had yielded a unique research laboratory that included the natur study system comprised ofa package of spectroradiometric instruments, astrophysical laboratory equipped with six powerful telescopes and spectrometers, processing ovens, medical diagnostics equipment.

**Fig. 50.**
Mir orbital station
and Space Shuttle

Around 18,000 experiments have been conducted based on the research complex in such critical areas of investigation as production engineering, biotechnologies, geophysics, Earth mineral resources and environmental study, astrophysics, medicine, biology, materials study, technologies tests and others.

The implementation of the program was assured by integration of many Russian and CIS companies involved in science-intensive industries. In course of operating the Mir complex unique experience has been gained based on long term forecasting of technical condition, periodic extension of service life and special continually improving methods of repair and recovery including those pursued in open space.

The projects of Mir and ISS can under no circumstances be regarded as isolated programs since Russia willingly shares with its partners in ISS its experience in organization, support and performance of orbital flights. Over recent years, due to Russia's involvement in creating the ISS a question arose as to the feasibility of continued operation of the Mir because the state's insufficient funding precludes simultaneous work on two major programs. In addition, operating the Mir well above its rated service life impaired its safety. So the government took a decision to cease the existence of the station, to remove it in a controlled mode from orbit and to splash down on the ocean. In March 2001 the decision was put into effect.

The principle of international cooperation in space implies Russia's full scale involvement in the International Space Station program (Fig. 51). In the 21st century there is practically no alternative to this trend of development since expenses incurred by space exploration began to exceed the financial capabilities of any single state.

**Fig. 51.** Assumed appearance of full scale International Space Station in the 21st century

The use of ISS will address the fundamental scientific issues, will enable applied investigations and experiments in the area of fundamental science, will promote social and economic progress and foster the international cooperation. The main tasks tackled by the International Space Station will be as follows:

• conduct of fundamental research in order to expand and deepen the knowledge of the Universe and the world around us;

• carrying out applied investigations in order to obtain onboard a SC geophysical information for utilization in farming, forestry, fishing, geology, oceanography and environment protection; manufacture of pilot lots of semiconductor materials, alloys, gradient glass for research and application in electronics industry, nuclear power production, laser technology, projection television; production of biologically active compounds and medicinal preparations for pharmaceutical industry, molecular electronics, stock breeding;

• work under international cooperation programs, including that on an international basis;

• full scale optimization of elements and systems of space rocketry in a near-operational environment.

• The creation of this station is expected to:

• expand fundamental scientific knowledge of astrophysics, geophysics, ecology, materials study, medicine and biology;

• provide samples of new high quality materials, biologically active compounds and drugs for use in electronics engineering, optics, medicine and biology;

• enhance the effectiveness of research and development in creation and optimization of new types of hardware for various space systems;

• assure an increment to the national product due to the use of new space technologies in industry and thanks to utilization of information about the Earth's mineral resources and the ecological situation in agriculture, forestry, geology;

• bring hard currency earnings from sales of programs of international cooperation on a commercial basis;

• provide expertise in science and technology for advanced programs of investigation of Mars and the Moon in cooperation with foreign countries.

In September 1988, the governments of the USA, member states of ESA, Japan and Canada signed an intergovernmental agreement for cooperation in development and operation of the ISS. In the late 1993, the Russian government received an invitation from the signatory countries to cooperate in the ISS program which it accepted.

The project for building the ISS has been in development since 1980s. Its initial name was Freedom. By 1993 the work under the project had consumed US$ 11.2 billion. However, the lack in it of proven hardware and optimized operating procedures (which is readily available to Russia) that provide for the crew's extended stay and work in space flight conditions, the deficiency of emergency equipment, means of delivery of fuel and cargo to the station made the project almost unrealizable.

**Fig. 52.** D.Goldin (right) and A.I.Kiselev in the assembly and test building of the Baikonur spaceport

Russia's participation in the project for creation and operation of ISS adds stability to the program and makes it more realistic. The Russian key elements and technologies that essentially expedite the ISS assembly are as follows: the service module (SM) that ensures life support for a crew of 3 to 6; Progress-M cargo ships and their modifications that provide the station with consumable materials including fuel; manned spaceships of TM series that deliver the crew and bring it back to Earth, assure its rescue in case of emergency. No other partners under the ISS program (including the USA) have to date the equivalent equipment.

Overall, the Russian segment of the International Space Station comprises the following elements: Zarya module, Zvezda service module, docking compartments, multi-purpose docking modules, docking-for-storage modules, research and power supply platform, research modules, Soyuz TM and Progress craft. The base modules of the Russian segment of the ISS are delivered to orbit by the Proton carrier rocket.

The USA, ESA member states, Canada and Japan – Russia's partners under the ISS program – are interested in its participation in the project since they realize that otherwise the cost of the project will soar up and the creation of the station will become problematic. This conclusion is in line with the opinion of he US experts. On 7 October 1998, at NASA session Daniel Goldin publicly announced for the first time that NASA might request Congress for extra funding to preserve Russia's role in the program of creating the space tation and, simultaneously, to take measures aimed to reduce the program's dependence on the

Russian hardware. Also, Goldin said that such message had been submitted to the White House during discussion of NASA's claim for budgetary allocations in 2000.

**Fig. 53.** Press conference prior to shipping the Zarya functional cargo unit to Baikonur (shown in the pictures are V.Barnes, S.K.Shayevich, Yu.N.Koptev, A.I.Kiselev, R.Mitchell)

According to NASA's estimates, another US$ 1.2 billion will be needed to trim down Russia's role in the program. In the near future, NASA will be buying Russia's goods and services. In the longer term, the NASA plans' to produce its own goods and services, e.g. to modify the reusable transport Space Shuttle in order to do without launching several Russian cargo spaceships Progress. In the foreseeable future, Russia's participation in the project of creating the ISS is the cheapest option.

Including Russia in 1998 in the ISS partnership strengthened somewhat the country's position on the post-Soviet economic field. One of its principal partners in space research within the framework of CIS, Ukraine, expressed a wish also to participate in the project. Ukraine turned to Russia with a proposal for cooperation in building a Ukrainian research module and integrating it into the Russian component package of the ISS.

It is also anticipated to use the Russian segment of the ISS on a commercial basis. The goal of commercial space work is to refund part of expenses incurred in creation of the Russian segment of the ISS, to minimize the operating costs, to use the products of research and engineering obtained in the development and utilization of the ISS and in other branches of economy in order to build advanced competitive hardware and promote it on the market.

The following could also be of commercial interest for business in the 21st century:

• products obtained in research and development of ISS based on the latest achievements of space science and technology;

• comprehensive and timely training of crews for ISS (in addition to Russian ones) at the Gagarin Research and Astronauts Training Center;

• fulfillment of applications made by partners under ISS programs for delivery of payloads;

• preparation of ground-based equipment and training of personnel for carrying out the scheduled experiments (work) on the ISS;

• carrying out commercial orders for development and fabrication of materiel for the projects realized on the basis of Russia's segment.

Russia's integration into the international space exploration strengthens its positions in the world's community, fosters its authority and influence and helps other countries better understand Russia's interests. In analyzing the relations with leading powers in the area of space exploration it should be always borne in mind that joint scientific projects, the realization of Russia's capabilities on the market of space services and Russia's fulfillment of the obligations it assumed with reference to limiting and controlling the proliferation of rocket technologies are regarded by its foreign partners as a single whole. The breach of any constituent invariably results in reduction (or cessation) of joint work not only in space but also in areas of economic cooperation. Under such conditions, in order to preserve and further develop Russia's space potential, to expand the international cooperation and attract significant foreign investments in the country's space rocket engineering, it is necessary to honor on time the commitments under space related programs (including those in the creation of the ISS).

The projected service life of ISS is till 2013. Building it requires US$ 100 billion, in which sum Russia will account for US$ 6.5–6.8 billion. By contributing its share to the creation of the station Russia will obtain the right to a third of its resources, including: 43% of the crew's stay in space and its number, 20% – energy resources, 35% – the volume of the pressurized compartments and 44% – work places.

The ISS flight program consists of two phases: the ISS assembly phase – from 1998 till 2006 and the utilization phase starting since 2006. The starting date of station deployment is considered 20 November 1998, the day when the functional cargo unit (FCU) developed by the Khrunichev Center in co–operation with more than 240 organizations, was successfully launched. It was dubbed Zarya (dawn) which is supposed to symbolize the beginning of a new era in the international space cooperation (Fig. 53). The creation of a module which can be safely called "the transitory compartment leading into the 21st century" proceeded in difficult conditions when the ISS configuration and its requirements were being worked out. Out of 1,100 initially formulated requirements a third was amended in the course of design work, manufacture and trials. As work proceeded, the specialists of the Khrunichev Center resolved quite a few engineering and organizational problems related to adapting the FCU to international standards and to performing the functions that create the required conditions for deployment and operation of the ISS by means of:

• supporting the orbit and controlling the orientation of the ISS in the initial phases of deployment;

• supplying power to the ISS in the initial phase of its deployment;

• docking;

• performing the functions of the consumable materials storage facility; assuring life support functions.

**Fig. 54.** The likely appearance of lunar base (NASDA concept)

Currently, the ISS operates in orbit as part of two segments – the Russian segment (Zarya FCU, Zvezda (star) service module (SM), docking compartment SO1 Pirs, rescue ship of the Soyuz series) and the American segment (connecting module Node 1 Unity, laboratory compartment Lab Destiny, lock chamber Quest, Canadian-made Space Shuttle remote manipulator system SSRMS) controlled by the joint Russian-American crew of the already third main expedition of the ISS. The crew and cargoes are delivered to the station by the American Space Shuttle and Russian cargo transportation ship, the Progress. The space crews are trained under the ISS program at Russia's Gagarin State Research and Astronauts Training Center and Lyndon B. Johnson Space Center (USA). The ISS flight control from Earth is accomplished via two flight control centers, the Russian one located in the town of Korolev (near Moscow) and the American one based at Houston.

It is expected that the development of technologies and technical facilities for "small" orbital flights will get much attention in the 21st century. Such a program is exemplified by the Oryol (eagle) program which envisions creation of a small size orbital ship for small space crews (one to two men) for emergency rescue of astronauts, maintenance of orbit-based equipment and other missions.

Out of all celestial bodies the exploration of the Moon looks most feasible in the coming years. This is due to its proximity, the possibility to locate on its surface lunar bases for various applications: fabrication, repair, extraction, astrophysical survey, anti-asteroid protection etc. (Fig. 54). In this context, there is every ground to believe that manned flights to the Moon will resume in the 21st century.

**Fig. 55.**
The likely view of a base on Mars (NASDA concept)

Also, it can be assumed that the manned flights will be undertaken to the planets of the Solar system, primarily to Mars, whose temperature conditions come closest to those on Earth. The expedition to Mars is possible as soon as the first quarter of the 21st century (Fig. 55).

It should be noted that manned flights to other planets look rather problematic in view of their high cost, the difficulty of realization and the sharp aggravation of the Earth's global problems predicted for the mid 21st century. Therefore the investigation of planets of the Solar system and deep space will probably continue with the help of automatic interplanetary spacecraft and probes.

## 2.4 Power plants and propulsion units of space rocketry

The Russian designers have generously contributed to the creation of propulsion units enabling injection of payloads into space. Sentenced to death, Nikolai Ivanovich Kibalchich (1853 – 1881), a student of the medical surgery academy, submitted to the Bar not a petition for mercy, but his "Aeronautic Craft Project". In it, he expounded at length his idea of a rockett type flying vehicle. N.I.Kibalchich wrote that once dedicated researchers thoroughly consider his project and find it feasible he would feel happy for having thus served his nation and mankind.

His message, however, remained unanswered. Only 40 years later the scientific exploit of the revolutionary became known to the public at large.

K.E.Tsiolkovsky appreciated greatly this deed of outstanding heroism for the sake of science and singled out Kibalchich as the most notable of his predecessors. K.E.Tsiolkovsky had a gift of astute insight that enabled him to look into the future of space exploration although that was the time when even airplanes couldn't fly properly. "As I've been working on the jet engine I thought that rambles over Mars would start in hundreds of years. But time brings its corrections. Now I 'believe that many of you will witness the first trip beyond the Earth's atmosphere". K.E.Tsiolkovsky said those words on the 1st of May, 1935. Twenty two years later, the Earth's first satellite was launched into orbit.

Fridrikh Arturovich Tsander, Yuri Vasilyevich Kondratyuk are by right considered the pillars of Russia's jet-propelled rocketry.

The work of researchers like S. P. Korolev, V. P. Vetchinkin, B. S. Stechkin, M. K. Tikhonravov, Yu. A. Pobedonostsev, V. P. Glushko and others set the stage for the launch that took place 1957. The pioneer of design and development of space equipment is M. K. Tikhonravov with a group of enthusiasts.

The project of N. I. Kibalchich introduced S. P. Korolev to space rocketry. He headed space engineering that brought about the launch of the first satellite, man's first spacewalk, the first circling of the Moon. S. P. Korolev had lived only 60 years but they sufficed to realize the dream of mankind. He paved the way to space and his name became immortal.

All jet-propelled engines (JPE) are divided into two big groups:
small thrust engines (highly cost-effective) having the high rate of gas efflux but a small mass of ejected actuating liquid, and large thrust engines in which the gas efflux rate is relatively low while the mass of gas efflux fairly high. The former find use in generating small accelerations in open space. Only large thrust engines are used to overcome the Earth's gravity, the atmospheric resistance and to obtain the initial boost.

The prospects of development of rocket engines are shown in Fig. 56.

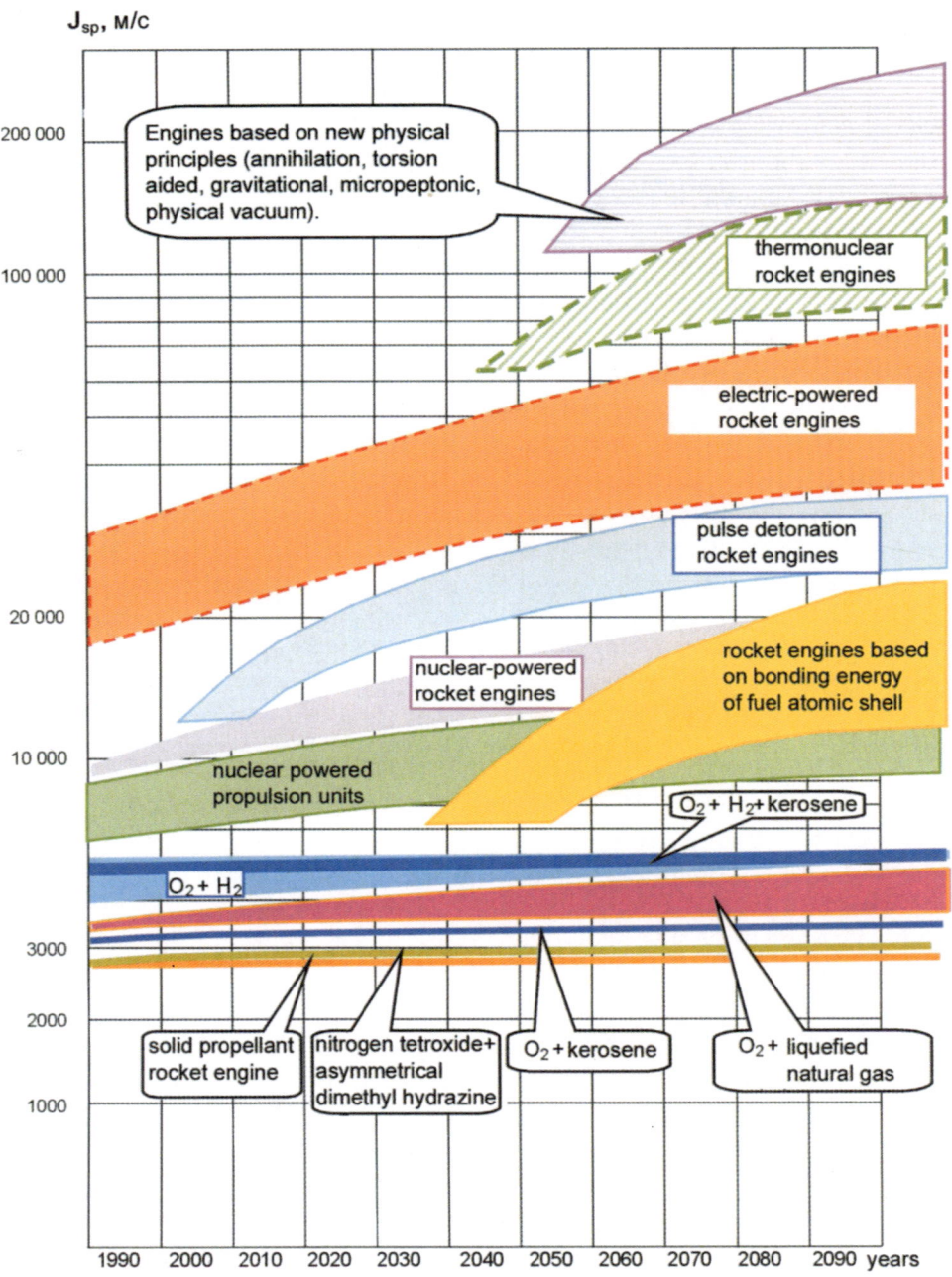

**Fig.56.** Prospects of development of rocket engines in the 21st century

Jsp – specific pulse

In chemical rocket engines the body energy is provided by various chemical reactions. Depending on the chemical structure of the fuel, engines are divided into two large classes: liquid propellant rocket engines (LPRE) and solid propellant rocket engines (SPRE). Standard LPRE are shown in Fig. 57, 58. An advanced LPRE is shown in Fig. 59.

*a)*                                                        *b)*

**Fig. 57.** Standard liquid propellant rocket engines

    a)    RD-253 engine (1st stage of Proton-M carrier rocket;

    b)    RD-0210 (2nd stage of Proton-M carrier rocket)

In LPRE the liquid fuel (fuel and oxidizer) is fed by the fuel supply pumps to the combustion chamber. The chemical reaction of burning raises the fuel temperature which makes the gaseous combustion products flow out at a high speed via the shaped nozzle, thus creating thrust. The speed with which gas flows out of the nozzle (hence, the specific thrust) depends on the temperature and molecular weight. The higher is the temperature, the greater is the speed. Conversely, the molecular weight of the combustion product should be as low as possible since the speed of outflow rises with its reduction. Such kinds of fuel as liquid oxygen and kerosene or nitric acid and dimethylhydrazine allow to obtain the specific thrust in the region of 300 s, i.e. the nozzle discharge velocity of about 3 km/s.

Hydrogen, as a fuel with a small molecular weight, offers a number of advantages. It has a high calorific value that ensures a high temperature of combustion products with the lowest molecular weight out of all substances known on Earth. Coupled with liquid oxygen, the liquefied hydrogen yields the specific thrust of around 450 s (the discharge velocity is around 4.5 km/s). In addition, the combustion products are absolutely non-toxic, since the combustion ends in generation of water vapor.

**Fig. 58.** B.I.Katorgin, NPO Energomash general designer
at the RD-191 liquid propellant rocket engine for the Angara carrier rocket

Due to its essential advantages as a fuel, hydrogen was first used in Saturn-5 and Energia super-heavy carrier rockets. Being ecology-friendly, the hydrogen-oxygen fuel is a dangerous explosive which forms in case of uncontrolled mixture the so-called "rattle gas". Therefore, rigid techniques have been elaborated for handling hydrogen., the essence of which lies in clearing all containers and thoroughfares of hydrogen, air and oxygen by thorough blowing with the help of nitrogen, by burning the evaporations of liquid hydrogen flowing out of drain mains and by automatic checking for presence of hydrogen. The disaster protection system gathers information as to condition and operation of all units and subsystems of the entire space rocket complex in order to prevent the systems malfunction and breakdown.

The take-off weight of the Energia – Buran space rocket system is 2,400 tons, of which 1,890 tons are consumable fuel. The near-Earth total thrust of four blocks of the first stage (740 ton force each) and of the central block (600 ton force) amounts to 3,560 ton force. The tanks of rocket blocks of the first stage empty in 165 seconds at the consumption of 2.4 tons per second. After 486 seconds at an altitude of 160 km the second stage too runs out of fuel. To accomplish the delivery to the target orbit the speed of 30-40 m/s is wanted. This is provided after separation from the Energia by the combined propulsion unit (CPU) of the Buran orbital spaceship. The activation of the CPU twice for 67 and 42 seconds forms an almost circular orbit with an altitude in apogee 256 km and in 252 km in perigee.

**Fig. 59**. Advanced three component rocket engine
(standing in the center V.S.Rachuk and A.A.Medvedev)

The supplied example shows how wasteful are the engines running on chemical fuel. Their fuel consumption and the power obtained are stunning. It's easy, for instance, to establish that a liquid propellant engine with a thrust of 100 tf generates during injection of SC into orbit the power well in excess of that supplied by the Bratsk hydroelectric power plant.

# Joint stock company "NPO ENERGOMASH named acad. V. Glushko"
## The company had based in 1929

Joint stock company "NPO ENERGOMASH named acad. V. Glushko" - conducting Russian enterprise for development of powerful liquid rocket engines. Engines of development by Joint stock company "NPO ENERGOMASH" are used in the basic space rockets: Souz, Proton, Zenit, Kosmos and etc.

Joint stock company "NPO ENERGOMASH" carries out a full cycle of engines development, including their designing, manufacturing and tests.

The basic activity directions:
· theoretical researches in the field of fluid jet engine creation;
· development and manufacture powerful liquid-propellant engine creation for first and second LV steps on low and high-boiling components of fuel;
· experimental improvement of designs and examination of liquid-propellant engine tests results, their units and aggregates;
· flight tests support;
· development of continuous chemical lasers of various capacity on working molecules HF and DF;
· development of the high technology engineering under commercial contracts.

The high technological level and quality of development have provided with Joint stock company "NPO ENERGOMASH" a victory in competition on development of the engine for the first step modernized space LV Atlas III (Lokhid Martin, USA). Successes of Joint stock company "NPO ENERGOMASH" are regarded by world press as economic break on the international market of high technologies.

Now Joint stock company "NPO ENERGOMASH" participates in realization of the international program "Sea launch", basing on use LV "Zenit 3SL" with engines RD-171 and RD-120 development of the enterprise on the first and second missile stages.

Also works on development single-chamber oxygen-kerosene engine RD-191 for new family Russian LV "Angara", developed by Khrunichev State Research and Production Space Center are conducted.

The general director and the general designer, corresponding member of the Russian Academy of sciences Katorgin Boris Ivanovich heads Joint stock company "NPO ENERGOMASH named acad. V. Glushko".

Prospects of the international communications development in the field of creation and operation of space engineering allow us to look with optimism in the future and to hope, that in XXI century Russia will keep the positions in the space industry.

Burdenko st., 1, Khimky, Moscow region, Russia, 141400
Joint stock company «NPO ENERGOMASH named acad. V. Glushko
el. (095) 777-2727, Fax (095) 251-7504, E-mail: energo@online.ru

Theoretical investigations show that with any combination of fuels and oxidizers, no matter how propitious, their specific thrust does not exceed 450 s, while in most cases one has to make do with 300 s.

This data gives rise to not unfounded doubts as to the practicality of expeditions to the planets of the Solar system with the help of chemical engines and launches conducted from the Earth surface. Such flights imply numerous dockings of modules with refueling in space or more powerful sources of energy, for example nuclear power installations. The nuclear rocket engine (NRE) is similar in many ways to LPRE. The principal difference lies in that the jet mass prior to its exit from the nozzle is heated not by burning but by warmth generated in the nuclear reactor. The nuclei division energy exceeds by 10 million times that of chemical reactions, the reason why the consumption of compound subjected to fissioning in the nuclear reactor is negligent compared to the fuel consumption in the LPRE. The NRE expends practically nothing but the actuating fluid. In this case, however, one must be always mindful of the reactor mass and the radiation protection.

Nonetheless there are projects (for example the MG-19 project developed under the guidance of V.M.Myasishchev and O.V.Gurko) in which a combined multi-mode engine is proposed for economy of the actuating fluid (the self-same liquid hydrogen). In the dense layers of the atmosphere (at altitudes 40 to 50 km) such an engine uses the atmospheric air which in passing through a special radiator is heated to a high temperature and without burning is flowing at a high speed out of the ramjet engine. As this occurs, the flight proceeds with engines consuming the minimum of fuel and the craft slowly climbing while the speed pick-up is at its highest (up to 10-12 M). After gaining an altitude at which the rarefied atmosphere does not significantly affect operation of the engine consuming the supplies of hydrogen, the gas jet discharge velocity can reach 25 km/s.

Calculations show that in the foreseeable future (with a 20 to 30% reduction in weight of structural materials) it will become possible to create single stage reusable horizontal take-off and landing systems for ventures into space.

In large thrust engines the actuating fluid is first heated to high temperature as a result of chemical or nuclear reaction, after which it is ejected outside as a gas jet through a shaped nozzle. In such systems the weight of the engine proper normally accounts for only a fraction of the weight of the space rocket The bulk of the weight is the actuating fluid (fuel). In modern carrier rockets the weight of fuel accounts for 90% of the take-off weight of the carrier rocket. Such draft systems create a huge thrust, but operate briefly. After a relatively short active portion the SC goes into a passive flight governed only by exterior forces.

It is well known that the upper atmosphere, solar wind and light pressure have an effect, small as it is, on the flight of the SC. They cause disturbing and braking effects which accumulate in the course of an extended flight and become quite significant. In operating the space systems it becomes necessary to introduce corrections (compensations) by way of counterbalance to the accrued effects of space. Thus the necessity arose to use small thrust engines.

The small thrust engines' prime function is to boost the operation of actuating fluid by using electric power. In such systems the weight of the engine is roughly equal to that of the actuating fluid. Electromagnetic rocket engines (ERE) are providing a small boost but for a lengthy period of time. There are various models of ERE. However, in all cases electric or magnetic fields or their combinations are used to boost the actuating fluid in the engine. Depending on the type of the ERE boosting system they are divided into ionic and plasma-operated.

In ionic ERE the jet is a flow of ions. Acting as a jet nozzle is the electrostatic field in which ions can be boosted to great speeds. Chosen as an actuating fluid is a readily ionizable compound, e.g. cesium. The actuating fluid is fed to the ionizer. The formed ions are boosted in the booster's electrostatic field. Ionic engines allow to obtain the specific thrust up to 20,000 s (discharge velocity is around 200 km/h).

Plasma-aided ERE use plasma as actuating fluid. The actuating fluid in them is boosted by the so-called Lorenz force which results from interaction of the Earth's magnetic field and the electric current in plasma. Launched in late 1964 into flight to Mars, the Soviet automatic station Zond-2 (Probe-2) had six small plasma engines that were used to orient the station. In 1988 a special space experiment was carried out, dubbed Plasma, which experimentally checked the effectiveness of using plasma engines on satellites and tested the resistance of SC communication to the interference of the operating plasma engine.

## 2.4.1 Prospective trends in improvement of power plants and propulsion units

Power plants (PP) and propulsion units (PU) of space rocket systems are labor-consuming products, complex to fabricate and optimize. The energy performance, weight characteristics, service life duration, reliability, cost effectiveness of PP and PU determine in many ways operational capabilities and cost effectiveness of the space rocket system as a whole. The duration of development and optimization of new examples of PP and PU is rather long varying between 5 and 7 years provided that sufficient experience and expertise are available in research and technology. This is precisely why these aspects should be treated as top priority issues. Only under such conditions is it possible to timely develop and build complexes and systems not inferior to their foreign counterparts in main characteristics and capable of competing against them on the world's market.

Fig. 60 shows results of predictions of power consumption in solution of prospective space tasks.

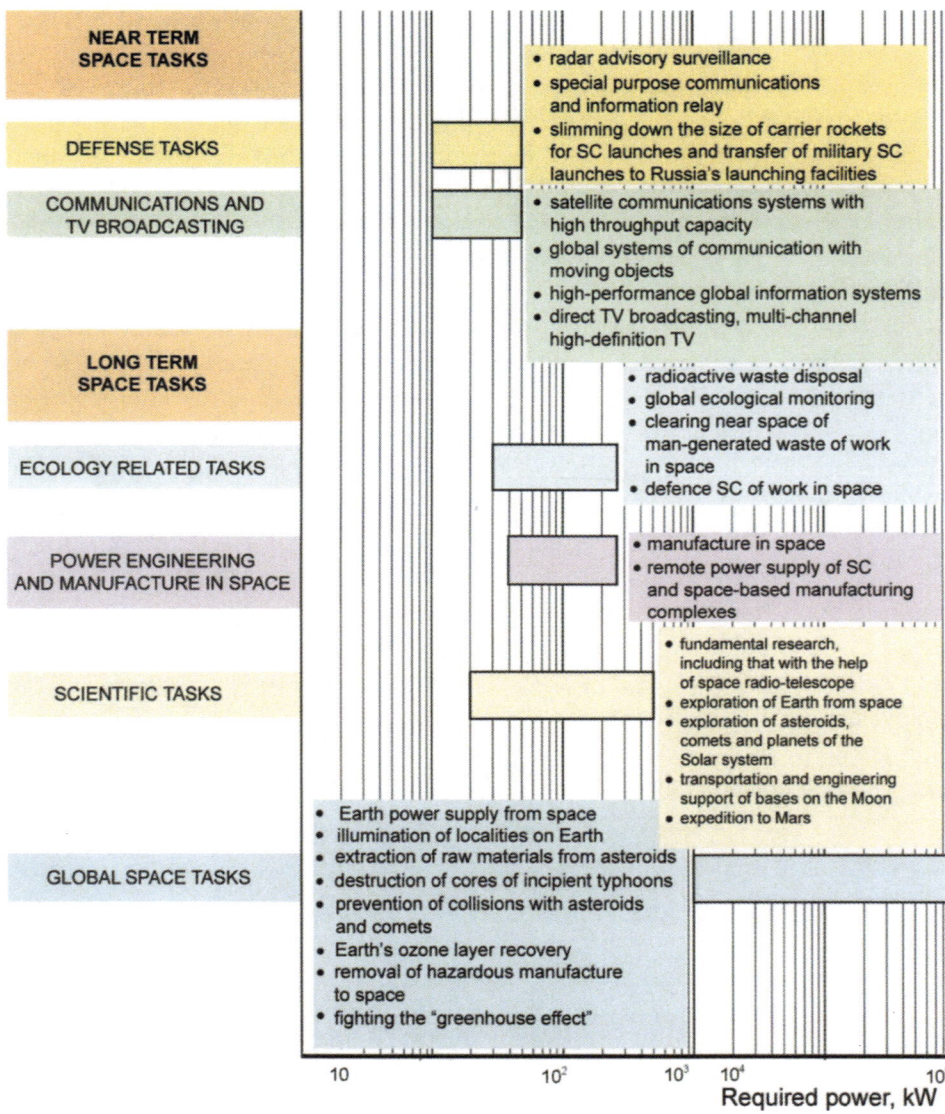

**Fig. 60.** Prospective space tasks requiring increased power consumption

In the 21st century space rocket engineering will become one of the main factors in development of the productive forces in the main areas of production that determine the welfare of the society: in power engineering, in materials and foods production, extraction of mineral resources etc. The results of investigations show that in addition to the effective utilization in today's businesses (global communications, information and navigation systems, space complexes for investigation of natural resources, ecology monitoring, etc) the 21st century may see the wide use of space rocketry for addressing such global interrelated issues as energy supply and environmental protection. Carrying out future-oriented research and development in the creation of means and systems of onboard power supply equipment (based on solar, chemical and nuclear energy) and utilization of the obtained expertise in the development of power supply systems and propulsion units will make it possible to increase the efficiency of SC and will provide the required base for implementation of space programs in the 21st century (Fig. 61).

Modern fuel engineering rests on two types of fuel – nuclear and chemical. Nuclear power engineering is based on the emission of part of bonding energy of nucleons in fission or nucleus synthesis reactions. The chemical power engineering is based on liberation of chemical bonding energy of fuel atoms and molecules. Also, there is an intermediate energy source. This is the liberation of bonding energy of atoms' electrons. The theoretical possibility of using this kind of energy is based on the physics of interaction of the powerful short pulse laser remission with the substance.

Liberation of bonding energy of atomic shells can occur in less than $10^{-17}$ s with ionizing reconstruction of atomic electronic shells in super-strong electromagnetic fields exceeding in intensity those within atoms without loss of energy for ionization and loss of heat. Today's level of laser tecnology makes it possible to effectuate the above said reconstruction by subjecting the substance to laser emission having the intensity of t above $10^{17}$ W/m$^2$ and duration of pulse impact t less than $10^{-13}$ s. Table 2. shows estimations of density of energy contained in above listed sources.

**Table 2.** Characteristics of energy sources

| Energy sources | Density of stocked energy, J/cm$^3$ | Source of energy emission |
|---|---|---|
| Nuclear fuel | ~$10^{11}$ | Mass defect |
| Fuel that provides liberation of bonding energy of atomic shells | ~$10^7$ | Electronic structure defect |
| Chemical fuel | Above $10^3$ | Change of chemical bonds of atoms and molecules |

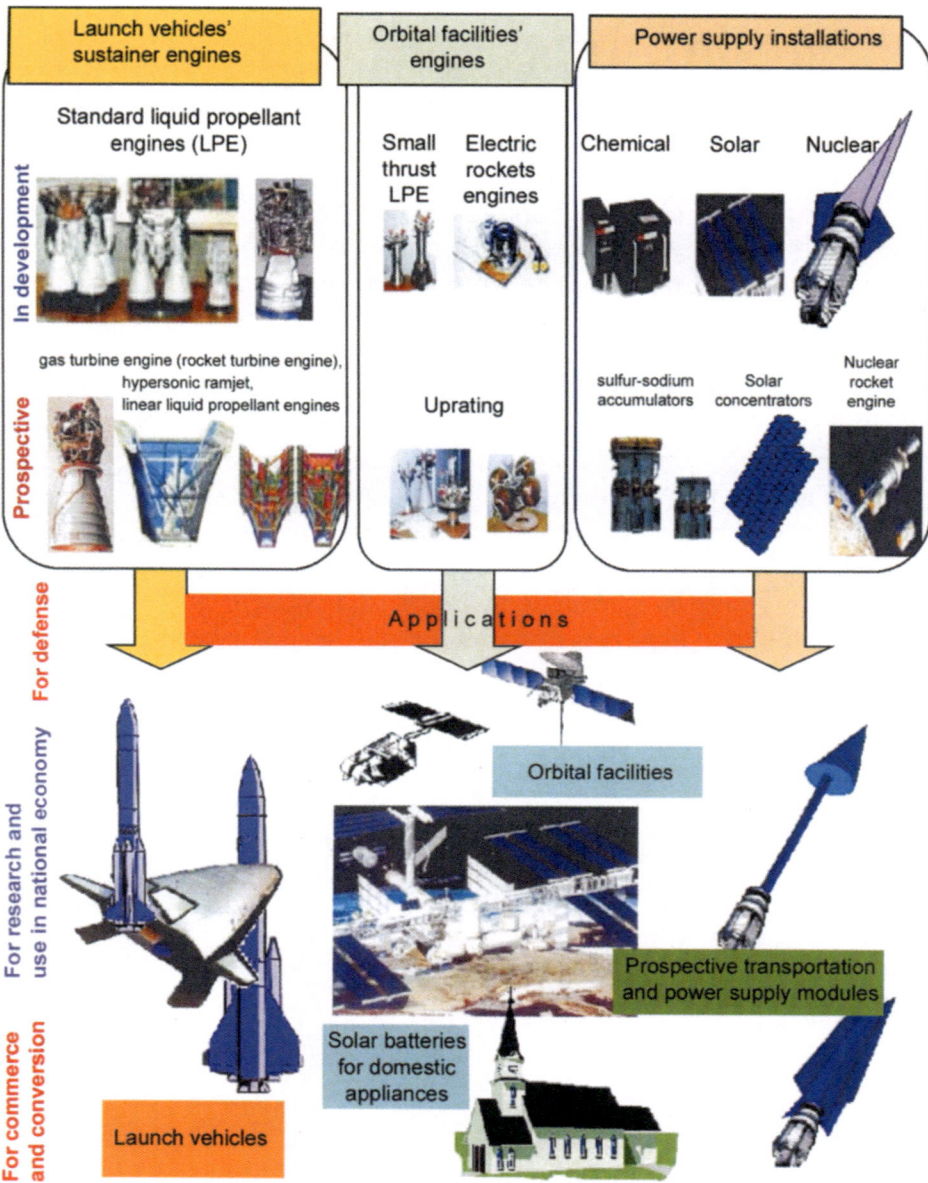

**Fig. 61.** Priority trends in future-oriented research and development of power equipment for space exploration

The interaction of high intensity ($t < 10^{17}$ W/cm$^2$) laser emission of ultra-short duration ($t < 10^{-13}$ s) with a substance can be by convention divided into several stages schematically presented in Fig. 62.

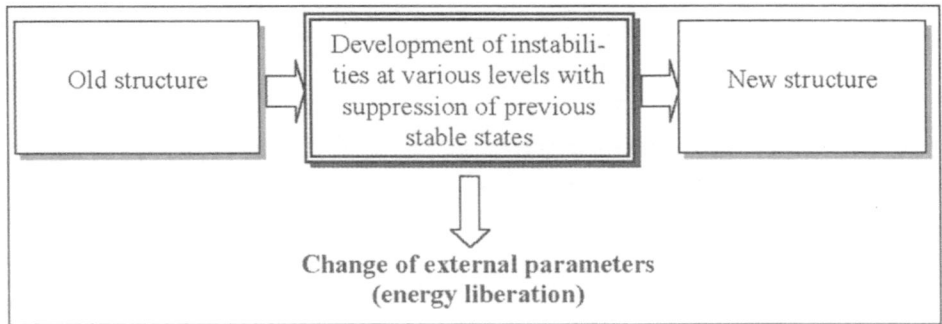

**Fig. 62.** The stages of interaction of high intensity short duration laser emission with a substance

1. The initial stage of impact is in essence the induction in the skin-layer (thickness in the order of $10^{-5}$ cm) of large scale vortical electronic structure having an electric field whose strength exceeds the atomic strength and the flow of free electrons featuring a concentration in the region of $10^{21}$cm$^{-3}$. Under such conditions proceed anisotropy of atomic ionization and anisotropy of electrons' heating in a plane perpendicular to the laser beam. The frequencies of the process are roughly equal to those of the plasma ($10^{15}...10^{16}$ Hz). Thanks to the development of those processes in the non-linear stage of the Weissbell instability the density of vortex energy and the density of anisotropy energy equalize. At this, the amplitude of the fields' magnetic induction of those fields attains the saturation in the region of 10 MGs. This stage of impact, dubbed induction, is characterized by accumulation of energy in the electronic component. As this occurs, the spontaneously generated magnetic fields are maintained because of the development of the Weissbell instabilities at the quasi-stationary level all through the time of operation of the laser pulse.

2. The development of high frequency potential and vortical instabilities of the forefront of laser pulse brings about the collective boosting of a small group of electrons with a high rate of boosting, more than 10 MeV/ps.

Electrostatic ionizing instabilities contribute to formation in the plasma trace of a considerable concentration of energy (more than $10^7$ J/cm$^2$) of potential oscillation as a result of the accumulated negative volumetric charge during the short periods of ionization.

The development of collective processes on the forefront of the laser pulse forms a fine structure of the front $\sim 10^{-7}...10^{-8}$ cm in size with an amplitude of intensities of magnetic and electric fields exceeding the threshold of the atomic stability. As this occurs laser emission converts to various types of energy.

3. The impact given to the atom by induced laser emission of vortical electromagnetic fields with intensity exceeding that of atoms results in spontaneous decomposition of upper shells of the atom (ionization explosion) during tunnel ionization $\sim 10^{-17}$ s in consequence of the reduction of the poten-

tial barrier caused by exposure to high frequency fields. The ionization explosion is accompanied by liberation of a flow of energy with intensity above $10^{17}$ W/cm$^2$.

The spin mechanism of the electron separation as per energy condition compacts the low lying shells down to K-shell. The induction of the magnetic field in an atom occurs spontaneously through the restructuring of its electronic shells. The restructuring starts with a threshold value of the magnetic field induction in the region of 10 MGs.

4. The compaction of low-lying electronic shells of the atom results in their deformation with rising intensity of electric and magnetic fields exceeding the stability reserve of those shells. As this happens, the probability increases for the nucleus to perform the K-capture of the electron. The duration of the K-capture dramatically decreases to the time of ionization of upper shells ($\sim 10^{-17}$s). The restructuring of upper and lower electronic shells proceeds in the single time scale of tunnel ionization ($\sim 10^{-17}$s).

Pattern of K-capture:

$$e^- + {}_Z X^A \longrightarrow {}_{Z-1} X^A + \upsilon$$

For example,

$$e^- + {}_4 Be^7 \longrightarrow {}_3 Li^7 + \upsilon + 0,864 \text{ MM} \ni$$

Contrary to the spontaneous capture, that of the K type is stimulated, has a greater probability rate and excites the inner restructuring of the nucleus. The energy liberated from the nucleus is consumed in regeneration of $\gamma$-, $\beta$-, $X$-radiations and formation of fast conversion Auger electrons. Further nuclear decay reaction is possible with emission of neutrons.

Such is the hypothesis of liberation of the atomic shells bonding energy.

## 2.4.2 Remote power supply system

The conversion of solar energy is probably the most promising means of obtaining power in the future since it requires no fuel and is ecologically clean. The preservation of biotope, including near Earth space, must also be one of the priorities in space work. Very important are the development and implementation of a number of programs aimed to take the edge off man's adverse impact on the environment by using space rockets and systems. The targets of such programs are as follows:

• preservation and recovery of the ozone layer;
• removal of especially dangerous industrial and power production waste to space;
  • clearing near-Earth space of man-caused debris and fragments.

Implementation of power generation and environmental protection programs will ask for dramatic improvement of performance and cost-effectiveness of space rocketry. This is needed for addressing a wide variety of complex scientific, technological, industrial and organizational issues.

A concept is now in development as to how to supply Earth with power from space. It rests on the conversion of solar energy and its conveyance to Earth in the form of radiation in microwave or optic range . The concept envisions gradual relocation of the bulk of power generation to space in which case space rockets will gain the major role in electric power generation.

The development and introduction of contact-free power transmission by means of electromagnetic radiation has been for quite some time one of the critical issues that determine the development of many industries. With reference to the needs of space work those problems must be resolved in view of the rising level of power consumption by SC being created and their growing service life. The remote power supply system (RPSS) is an effective alternative to autonomous onboard power plants since it improves mass and size characteristics of SC, reduces windage of objects, diminishes the orbit correction frequency and, hence, the amount of pulse required by correcting power plants. The likely users of the remote power supply system could be:
• orbital constellations of SC for various applications;
• single SC with a high level of power consumption
(also, in the longer term, piloted SC of the ISS type);
  • planetary (based on the Moon, Mars) bases and self-propelled robots;
• booster propulsion units of SC on the portion where transition is made from low to high operating orbits and sustainer propulsion units of SC for investigation of deep space on the boosting portion in near Earth space.

In addition, the energy beam can be used to recover the ozone layer of the Earth's atmosphere, to eliminate meteorites and debris of decommissioned SC as well as for other purposes.

The TsNIIMash (Central Research Institute of Machine Building) has designed in general terms a versatile (suitable for all users) and multi-purpose (suitable for operation with several users at a time) space power station of RPSS. In the case of using microwave radiation for power transmission the station mass will reach around 20 tons, which will enable it to be injected with a single launch of the advanced heavy carrier rocket, Angara. When deployed, the power station will measure along the axis of the reinforcement frame a maximum of 52 m. The wingspan of solar batteries (SB) will reach at the most 517 m. The advanced solar battery is shown in Fig. 63. Each of the two wings of SB measures 16 · 256 m. The power of the solar radiation they gather amounts to 11,206 MW while the electric power on the SB output at the efficiency of 12-28% will amount to 1,344-3,138 MW. The power of microwave radiation with the use of modern SB will reach 1.1 MW. For advanced ones it will be 2.7 MW. The specific mass of solar space power station with production capacity above 1MW, including combined film-coated solar batteries with ultra-high frequency radiators will amount, according to specialists, to 5 kg/kW.

**Fig. 63.**
Advanced solar battery

## Main characteristics of power station

The power of microwave radiation on the transmitting antenna, kW.................................... 1,162–2,712
Efficiency rate of transformation of solar radiation into microwave radiation, %.......................... 10–20
Total area of transmitting antenna, m$^2$ ..................................................................................... 10.4–24.4
Number of power transmission channels............................................................................................ 1–6

The mass related characteristics of power station are shown in table 3.

**Table 3.** Mass related characteristics of the main elements of space-based electric power station

| Elements of power station | Mass, kg |
|---|---|
| Solar batteries (SB) panels | 8,356 |
| SB frame and ribbing | 4,035 |
| SB deployment mechanisms | 1,024 |
| Active phased antenna array (APAA) | 461.4...1,077.6 |
| System of energy transformation and transmission from SB to APAA | 192.9...450.1 |
| Power frame | 2,000 |
| Control moment stabilizing gyroscopes | 1,000 |
| Integrated propulsion unit of control, orientation and stabilization system | 2,000 |
| Total mass of power station | ~20,000 |

The power station is designed for single use. After its service life has been depleted the station is to be removed from the operating orbit and destructed. Its servicing implies periodic refilling with fuel components for the integrated propulsion unit of the control, orientation and stabilization system. However, in case certain elements fail during operating life of the power station, e.g. the system of converting the solar energy into microwave radiation or the transmitting antenna assemblies, the failed units that can't be rectified by automatic servicing systems will be replaced with new ones. The replacement of SB in case of their damage is economically unfeasible since their mass accounts for up to 70% of the total mass of the station and their delivery to it practically amounts to deploying a new station into the orbit.

The structural arrangement and layout of the power station is based on the payload container of the Proton-M carrier with a fairing diameter measuring 5 m in diameter and 20 m in height. The basic idea is that the structural parts of the container are used as load-carrying structures of the power station. The side ferrule is used as the carrying base of the transmitting antenna which is an APAA for the microwave beam. Housed in the nose fairing are the integrated propulsion unit for control, orientation and stabilization and the electric power station's service systems.

The component package contains:
- power frame;
- solar energy pick-up system;
- the system of conversion and transmission of energy from SB to transmitting antenna;
- energy beam formation and transmission system;
- orientation and stabilization control system;
- system for monitoring the position of SC users and maintaining communication with them via pilot signal;
- system of power station's communication with ground-based (on-the-planet) services.

The calculations done for a laser-aided power transmission plant show that it can't be placed in orbit with just one launch of the Proton-M carrier rocket as its minimum total mass (at the radiated power of a mere 185 kW) exceeds by practically 30% the mass of the carrier's payload, in which case more than 50% of the station's mass is contributed by laser beam generators, their cooling system and the radiating antenna.

The design of the RPSS can also be regarded as a base for building the energy module of contact-aided power supply module for advanced power-consuming space systems of the ISS type whose total power rating is around 400 kW. A module like this is simple and fairly cheap to build. It is different from RPSS power station mainly in that it has no microwave radiation generating system and no transmitting antenna, but has a contact cable deployment system. The creation of the energy module will, on the one hand, provide a more solid base for designing and optimization of the remote power supply station while, on the other, its use will significantly diminish the windage of ISS and the associated vibration stress. The ruggedness and dynamic performance of the ISS structure also will generally improve. The energy module and the ISS are large size heavy objects fastened together not rigidly and forming a cable-connected orbital assembly The distance between them is determined on the grounds of safety and the capabilities of the conductor-and-support cable that automatically connects to the ISS (with the help of a smaller servicing SC). Given that the US is working now on an energy module for rigid docking with the Shuttle and the advanced piloted station is being built under international cooperation program, such a module could be of interest for all partner states and work on it can start shortly at any time. Later on, as the design of the station and the technology of beam-aided power transmission is refined, the piloted stations, including the ISS, may become users of RPSS.

One of the difficulties directly associated with power supply for propulsion is the organization of large (exceeding the current level by an order or more ) freight flows from Earth to the orbits of its artificial satellites and, further on, in the event of full scale deployment of Earth's power supply from space, to near Moon orbits and the lunar surface. Evidently, the solution of this problem also calls for the dramatic improvement in performance and cost effectiveness of launch vehicles sent into space, their interorbital transportation and maintenance in compliance with regulations of environmental protection which, in turn, necessitates the development of highly reliable, cost effective and ecologically safe carrier rockets, booster units, interorbital towing vehicles and other space rocket hardware as well as the required propulsion units and electrical installations.

### 2.4.3 Onboard solar power installations

The development of orbital equipment for various applications is characterized by its general growing level of power consumption and, hence, by its power supply per operating unit. This, naturally, extends its active service life. Specifically, the level of power consumption of onboard relay and service systems of the new generation communications geostationary SC will reach 5-10 kW at the level of power supply per unit varying between 1.4 and 2.0 W/kg which is about 2 to 3 times the respective performance value of the Russian-built equipment being currently in operation. The specific power of electric power supply systems (EPSS) which is the summarized indicator of their excellence has grown over the recent 10 years by approximately 1.5 times (from 4-5 to 6-9 W/kg). There is a theoretical possibility to further increase the specific power of EPSS by 2 to 3 times by improving the system of power generation, accumulation and conversion.

The important trend in raising the effectiveness of the use of space complexes and systems as well in extending their service life is employment of power plants and propulsion units fueled by solar or nuclear power installations or obtaining power from electrical rocket engines. Specifically, the use of a towing vehicle with an electric rocket engine (ERE) helps to increase by 2.5-2 times the weight of SC (target module) in GEO. Such advanced means of inter-orbital transportation are characterized by significant levels of power consumption (40 to 100 kW) and the high level of perfection of onboard power installations.

To comply with the ever stiffening demands made of power supply systems it is necessary to:

• improve the guaranteed specific characteristics of the elements of the EPSS at the terminal phase of SC active service life;

• transition to new types of elements;

• lowering the SB degradation level (initial reserve of its power);

• creation of adaptive systems for electric power conversion, for EPSS control and supervision of its functioning aimed to fully utilize the capabilities of this system's elements

• raising the resources and resistance of EPSS in case of exposure to open space.

The guaranteed specific values of mass and power supply onboard systems using the energy provided by solar batteries and chemical sources of energy can be improved along with the simultaneous increase of their resource by the following means:

1. progressive improvement of the traditional EPSS and their elements in order to achieve high specific characteristics of photoelectrical converters (PEC) based on Si ($170 \text{ W/m}^2$; $100 \text{ W/kg}$) and GaAs ($220 \text{ W/m}^2$; $60 \text{ W/kg}$), solar batteries ($100\text{-}110 \text{ W/m}^2$; $1.5\text{-}2.5 \text{ kg/m}^2$ towards the end of service life equal to 10-15 years in high orbits and $80\text{-}90 \text{ W/m}^2$ after 5 to 7 years of operation in low orbits) as well as improvement of the existing nickel-cadmium accumulators till they reach the level of specific energy $40\text{-}50 \text{ W} \cdot \text{h/kg}$ with service life up to 5 years in low Earth orbit and 10 to 15 years in GEO;

2. development of SB based on high performance PEC with increased efficiency rate and radiation resistance (specifically, in case of PEC based on InP the efficiency rate is >20%) and SB panels with solar energy concentrators for use on orbits especially dangerous in terms of radiation. The expected specific values of the SB panel with mirror concentrators is $120\text{-}140 \text{ W/m}^2$; $2\text{-}5 \text{ kg/m}^2$ in case of degradation of PEC based on GaAs by 2-3% over 5 years and, if based on Si, 15-20% over 5 years at the service life of 10 years.

For operation on high and less dangerous orbits SB can be used (now in development) that have lens-aided concentrators (the expected characteristics of the module with heterogeneous arsenide-gallium PEC ($200 \text{ W/m}^2$; $4 \text{ kg/m}^2$);

3. creation of SB based on ultra-fine PEC, flexible frame-free panels (achieving specific mass of SB panel between 1.5 and $2.5 \text{ kg/m}^2$);

4. development of new types of energy accumulators based on electrochemical current sources: nickel-hydrogen accumulator batteries (AB) with a common gas collecting main in order to achieve their specific energy between 45 and $60 W \cdot h/kg$ and the service life of 25,000 cycles on low orbits and 10 years on GEO; sulfur sodium accumulators AB with specific energy 110-140 $W \cdot h/kg$, specific power 120-180 $W/kg$ and service life of 4,000 cycles; regenerating power accumulators based on hydrogen-oxygen electrochemical system for operation as part of EPSS and energy propulsion systems with high specific characteristics (120 $W \cdot h/kg$) and extended service life (10 years) as well as development of EPSS based on electromechanical accumulators of energy with specific energy 30 $W \cdot h/kg$ for SC with extended cyclic life (10 years) and high current stresses;

5 development of adaptive complexes of electric power conversion equipment, including the system to convert electric power and to control electric rocket engines (the wide use of which is supposed to raise the effectiveness of interorbital means of transportation of SC and increase their service life) with specific power up to 100 $W/kg$ and service life between 10 and 15 years.

One of the promising trends in development of onboard power supply systems is the creation of solar gas turbine installations (SGTI) with electrical power of 10 kW and above. They offer the following advantages:
- higher efficiency of conversion of solar energy into electrical (25% with prospective 50%);
- smaller size of SGTI as compared to the size of an electrical installation based on SB;
- capability to generate higher voltage alternating current;
- extended service life (due to use of gas-lubricated bearings in the supports of turbo compressor);
- lesser cost of power installation;
- the promising nature of gas turbine energy converters that can be used together with nuclear and chemical sources of energy;
- possibility to use a thermal energy accumulator.

Currently, the Keldysh Research Center is building jointly with other companies a first generation SGTI-10 installation (a 10 kW module) for Mir type of orbital stations and ISS. Such modules can generate up to 100 kW power. The power levels in the region of hundreds and thousands of kilowatts (for functioning of tow vehicles powered by electric rocket engines, for carrying out Lunar and Martian programs) can be reached with the use of SGTI of the second generation (based on high temperature light receivers and turbine as well as drip-feed refrigerator-emitters). The projected specific characteristics of such SGTI are as follows: power 250-2,500 kW; specific power 100-250 $W/kg$.

At high levels of power consumption very promising is the use of nuclear power plants.

## 2.4.4 Nuclear power plants and electrical propulsion units

Over twenty five years ago in Semipalatinsk, the first nuclear reactor, IVG-1, was put in operation. It helped to start optimization of the rocket nuclear engine. Back then it was thought that such an engine would be needed for man's flights to Mars. As time went by, the insufficient funding of research slowed down the work. However, the scheduled for 2017 expedition to Mars stirred up the interest in the nuclear engine.

The nuclear engine is a reactor in which a flow of gas (hydrogen) passes along heat emitting elements with nuclear fuel. It cools the elements and is itself heated and quickly discharged from the nozzle, thus creating the thrust. As this occurs, a pulse is generated that pushes the rocket forward. The gas temperature at the exit must be very high – no less than 3,000 °C, the specific thrust must be 950 s. Only under such circumstances is the nuclear engine more effective than the standard one running on liquid fuel.

Nowadays, in spite of the half-frozen work in the area of nuclear rocket engines, Russia is here 15 to 20 years ahead of the USA. The work on nuclear power plants and nuclear powered propulsion units seeks to obtain advanced scientific and technological expertise in creation of base standardized elements, assemblies and units of nuclear power plants (NPPs) and propulsion systems. The priority trends in investigations that can demonstrate advantages of nuclear power supply over other variants could be the following:

• the development of technologies that ensure the creation of nuclear power station (NPS) with capacity varying between tens and hundreds of kilowatts (with a prospect of its further buildup);

• bringing the guaranteed service life of NPP to the level not lower than expected from solar energy systems (including up to 10 years and more on GEO);

• the development of technologies that ensure creation of bimodal nuclear electric propulsion units (running both in the mode of nuclear rocket engines on hydrogen and in electrogenerating mode for feeding the service equipment of SC or ERE);

• the assuring of nuclear and radiation security in the development and operation of nuclear power plants and nuclear powered propulsion units.

Investigations conducted by Russia's dedicated organizations show that at the power of 50-100 kW preference could be given to nuclear power plants due to their indisputable advantage over the conventional solar power installations in terms of mass, size, performance and economy. It's noteworthy that in the said range of power essential advantages are offered by heat emission NPP of the second generation based on further development of the technology created under the Topaz program whose important element was the successful test flights and engineering trials of the NPP Topaz-1 conducted in 1987-1988. It was precisely this circumstance, i.e. the use of the nuclear power plant, that makes quite specific the practice of designing SC since the layout scheme of the latter is becoming more dependent on the specific features of the power plant than on the characteristics and parameters of the target equipment.

It is also important that the NPP is used both as a source of electric power for onboard equipment and jointly with electric rocket engines for removing the SC from the radiation free orbit to the operating one.

Investigations carried out for determination of application areas of various types of energy used to power SC show that starting as early as 300 kW at the service life of SC more than 1 year the use of nuclear energy looks a better option. The results of theoretical studies show that a nuclear power plant can be built whose rated power of thermal emission conversion of energy is 7.5 MW and the specific mass characteristics are 6 kg/kW.

NPP with turbo-machine energy conversion (TMEC) can have an edge over heat emissive and thermo-electrical variants because of:
- significantly lower mass of the reactor installation with equal electrical power;
- higher efficiency rate;
- greater adaptability due to the much lower temperature of the actuating fluid;
- the capability to optimize the energy circuit separately from the reactor;
- higher reliability of TMEC thanks to lack of restrictions in duplicating the elements outside the reactor.

Therefore it makes sense to consider the NPP concept jointly with TMEC.

Also worth of note is the broad experience gained in development of nuclear powered engines, the availability of bench rigging, facilities and skilled personnel in Russia, as well as advanced expertise in science and technology obtained by the US in work under Nerva program. With such chosen electric power level (2 MW) the design of the reactor and its radiation safety are close to optimum in terms of specific mass, configuration, and refueling. In the meantime the specific mass of TMEC equipment reduces to 2-4 kg/kW.

This analysis of ballistic performance and design features of the energy transportation spacecraft (ETSC) determined the required parameters of the electric power as well as characteristics of the electric rocket engines. The main restrictions accepted in calculations are the following:

1. the mass and size of the installation must not exceed the capabilities of the Angara carrier rocket;

2. the radiation dose accumulated by the payload while crossing the Earth's radiation zones must not exceed $5 \cdot 104$ rad;

3. radiation secure is considered the circular orbit at an altitude of 600-800 km;

4. the service life of onboard ETSC must be 1 to 2 years in the initial phase and extended to 5-7 years during the follow-on optimization;

5. number of flights of ETSC throughout its service life is up to 10;

6. the total dose of radioactive exposure received in the instrumentation compartment from the work of the reactor and from Earth's radiation zones is as follows: gamma radiation – not more than $10^6$ rad; fast neutron fluence – not more than $10^{13}$ n/cm$^2$.

The Kurchatov Research Center has developed a draft NPP with turbo-machine energy conversion designed to meet the following requirements:

- heat rating – up to 10 MW;
- electric power – around 2 MW;
- energy conversion system – turbo-machine (Brighton cycle);
- total operating time – not less than $10^4$ h;
- number of activations during the service life – up to 30;
- maximum temperature of actuating fluid – up to 1,500 K.
- The results of the work done outlined the following characteristics of NPP:
- mass of the gas-cooled reactor – 1,000 kg;
- fuel – UC (U, Zr)C, UN with 90 % enrichment in U235, the fuel shell – Zr , $W^{184}$, reflector – Be
- radiation protection mass (LiH, W, $B_4C$) – 1,000 kg;
- energy converter mass (turbine, compressor and unipolar generator) – 3,500kg;
- actuating fluid – helium-xenon mixture (1-3% Xe);
- refrigerator-radiators – on heat pipes at the mean temperature of around 700 K, mass 3,000 kg;
- the area of refrigerator-radiator (effective) – about 300 $m^2$;
- the mass of the automatic control system, the power supply system – 1,000 kg;
- the mass of the NPP construction – 1,500 kg;
- the total mass of NPP – 11,000 kg;
- the specific mass – 5.5 kg/kW.

In terms of design, the ETSC whose component package includes the NPP consists of the power supply module with nuclear reactor and protection; TMEC placed in the cone of the radiation protection; refrigerator-radiator based on heat pipes and arranged in the carrying pattern; four opening planes of refrigerator-radiators shaped as semi-cylinders, and a drawer truss located inside the refrigerator-radiator. Placed on the drawer truss are :

- instrumentation compartment with a docking system, orientation, navigation and communication instruments and with an extra propulsion unit;
- sustainer electric rocket propulsion unit (specific pulse 4,600 s);
- xenon fuel tank.

The basic mass related characteristics of ETSC are: NPP – 11,000 kg; electric rocket propulsion unit (ERPU) – 5,000 kg; drawer truss, fuel tank – 1,000 kg, instrumentation compartment, docking system – 2,000 kg; extra propulsion unit, unaccounted elements – 1,000 kg; fuel (xenon) – 800 kg; total "dry" mass of ETSC – 20,000 kg.

ETSC provides for conduct of a wide variety of space research, creation of a lunar base and addressing of many other issues of national defense and economy.

The 21st century will bring forth the tasks that will consume a lot of power, such as the building of manufacturing complexes, investigation of comets, asteroids, etc. Their completion asks for more powerful propulsion systems. The demands made of the power of the propulsion unit are determined by the flight duration, the payload mass, the specific mass of the power plant (kg/kW), the specific pulse and efficiency rate of the engine. The estimated power required for conveyance of cargo to the Moon, for a 600 days' cargo flight to Mars with hundreds of tons of payload is in the region of 1-10 MW. The manned flights to Mars require energy sources with rated power of several tens of MW. Russia's and foreign experience allows to consider the concept of ETSC with nuclear electric rocket engine based on energy supply installation having the power of several MW.

**Nuclear power plant with electric power of 2 MW for energy transportation spacecraft.** The energy transportation spacecraft with nuclear power plant featuring around 2 MW and with electric rocket engines can considerably contribute to the investigation of the planets of the Solar system, to creation of a lunar base, to carrying out some purely scientific high energy experiments in space and, finally, their use can reduce by several times the cost of delivery of 1 kg of payload to geostationary and other high orbits. The Kurchatov Research Center has elaborated a concept of creating NPP with electric power of 2 MW for ETSC (Table 4).

**Table 4.** ETSC transportation capabilities

| Characteristics | Proton carrier booster unit | Angara carrier + booster unit |
|---|---|---|
| Orbit parameters for launching payloads (active duty orbit) | Hcircle = 600...800 km, i = 51° | |
| Mass of cargo launched by carrier into active duty orbit, ton | 14 | 17 |
| Mass of cargo delivered by ETSC to GEO, ton | 8 | 9 |
| Duration of direct flight in GEO, days | 35 | |
| Mass of cargo delivered by ETSC to lunar orbit, ton | 9 | 10 |
| Duration of direct flight to Moon, days | 45 | |
| Duration of return trip, days | 25 | |
| Mass of cargo delivered by ETSC to Mars's orbit, ton | 6 | 7 |
| Duration of direct flight to Mars, days | 320 | |
| Mass of cargo delivered by ETSC to Jupiter's orbit, ton | 1 | |
| Duration of direct flight to Jupiter, days | 500 | |

ETSC is space shuttle (interorbital towing vehicle). The ETSC is launched into low orbit by the Angara carrier rocket.

Obviously, the investigation of distant planets, the creation of a lunar base, the piloted flights to Mars and, finally, the global space-aided telephony require dramatic uprating of transporting capabilities of space hardware which predetermines the soaring power supply of SC.

**Nuclear electric rocket propulsion units with electric power of 2-10 MW.** The preliminary analysis of ballistic performance and design features conducted with reference to NPP shows that the optimum performance is obtained at the level of electric power in the region of 3 MW, which assures the following:

• the maximum mass of payload launched into GEO with the help of a nuclear electric rocket engine is placed during ejection from Earth in a payload container of the Energia carrier rocket;

• the time of transportation of the load to GEO does not exceed 100 days (in the event of not exceeding the admissible radiation dose while passing through the Earth's radiation zones);

• the specific pulse of the electric rocket engine (ERE) amounts to 5,000 s;

• the selected level of power is applicable to fulfilling a number of other tasks (transportation of cargoes to Mars, Moon, Venus, changing the orbital inclination of large space objects like research stations, conduct of research experiments and organization of industrial production on orbit).

Among the powerful electric rocket engines the best optimized both in terms of flight and in the development of subsystems are magnetic plasma and ionic electric power driven rocket engines. Studied now is the possibility of building a magnetoplasmadynamic (MPD) engine with power of 2.5 MW and with an external field. Such an engine operates with a discharge current of 10 kA and voltage of 250 V. The engine service life required for most space expeditions is accepted as equal to 10,000 hours. Therefore the developments aim mainly to increase the operating life of a separate engine. It has been demonstrated that MPD engines with power up to 40 MW can operate in the quasi-stationary mode. The plasma flowing performance is adequately described by equations of ideal magnetic hydrodynamics. The main parameters of some MPD engines are shown in Table 5.

The use of powerful MPD engines in experiments carried out over the last decades was ignored because of the low level of onboard power supply systems of the existing SC. The functioning of the installation at low power levels is disadvantageous for two reasons; first, under such circumstances the effectiveness of converting electric power into thrust comes down below the admissible limit. Second, high effectiveness at low levels of medium power can be obtained only in pulse operating mode of the propulsion unit. To ensure the pulse operating mode an energy converter is required that has auxiliary devices and whose mass is quite considerable. That is why small power propulsion units with pulse MPD engines cannot compete against other electric rocket engines.

**Table 5.** Parameters of some MPD engines

| Operating mode | Power, MW | Specific pulse, s | Efficiency rate |
|---|---|---|---|
| Stationary | 0.1 | 3000 | 0.25 |
| | 10 | 4000 | 0.38 |
| | 2.5 | 5000 | 0.5 |
| Pulse | 1000 | 10 000 | 0.7 |
| | 0.005 | 2000 | 0.13 |
| Quasi-stationary | 1000 | 10 000 | – |
| | 2.6 | 1300 | 0.2 |

The ballistics calculations have shown that the MPD engine is especially promising when used in a sustainer propulsion unit for interorbital flights if the SC component package contains an onboard one megawatt source of energy with which stationary MPD engines reach the required propulsion characteristics. Transportation of a large size energy source from a low orbit of artificial Earth satellite to GEO with the help of an engine running on chemical fuel requires a great deal of the latter which exceeds by 10 times the mass of payload. In case of using MPD engine the mass of the actuating fluid reduces by 5 to 10 times. Considering that the mass of MPD engine is of the same order as that of the engine running on chemical fuel, the gain in the initial mass of SC in the low Earth orbit proves quite significant. What is required for fulfilling such tasks is the reliable design of the installation using MPD engine with power in the region of several megawatts.

The optimum for the SC with the chosen power level is the reactor installation using fast neutrons. The concept of active zone of such an installation is based on the use of the uranium-intensive high-temperature composites in the form of twisted rod-type fuel elements or loose filling of ball-like fuel elements with an axial section equal to that of the heat carrier. The choice of fast-fission reactor is determined by the following factors: minimum size and mass; lack of restrainer which eliminates the problem of its resistance and cooling; virtual absence of reactivity effects caused by burning out and slagging; small initial supply and negative temperature effect of reactivity.

Nuclear safety in all phases of the SC service life in normal conditions and emergencies is safeguarded by active and passive means, including the following elements:
- regulating drums in the side reflector;
- withdrawable absorbent rods;

• resonance absorbers located in the active zone; programmable change of the wing's geometry in case of emergency.

The radiation protection of payload and control system is of the shadow type in the form of a truncated cone. It is determined by the maximum permissible level of radiation. Considered for use as the main components of protection are zirconium hydride, activated boron and lithium hydride.

The choice of turbo-machine technique of conversion by the Brighton cycle is conditioned by the small specific mass of the conversion system – less than 10 kg/kW which is much lower than its value for other means of conversion (30 kg/kW); the high degree of technological readiness; optimization of the main systems of the gas segment; the possibility to obtain compliance of output parameters of the electric generator and the required load; high efficiency rate of energy conversion (around 30%). The Brighton cycle of energy conversion is different from other dynamics methods in that it provides a simple launch, chemical inertness and non-activation of the actuating fluid under the influence of radiation.

The proposed power plant uses Brighton's direct regeneration closed cycle. Its principal points are turbo-compressor generator, recuperative heat exchanger and refrigerator-radiator (RR). The temperature of the cycle is 1,500 K which is not out of the ordinary with the use of modern ceramic-based materials for manufacture of turbine disks and heat-resistant alloys for hull parts and inlet branches. Materials operating in such temperatures are, however, excessively brittle under lower temperatures which asks for elaboration of the turbines launch algorithm. The construction of the recuperative heat exchanger consisting of a number of stamped sheets provides high intensity heat exchange and thus creates a compact and lightweight heat exchanger.

## Main parameters of the system converting the energy of NPP as per Brighton thermodynamic cycle

Electrical power, MW . . . . . . . . . . . . . . . . . . . . . . . . . . . . . . . . . . . . . . . . . . . . . . . . . . . . . . . 3
Cycle type . . . . . . . . . . . . . . . . . . . . . . . . . . . . . . . . . . . . . . . . . . . . . . . . . . direct regenerative
Heat carrier type . . . . . . . . . . . . . . . . . . . . . . . . . . . . . . . . . . . . . . . . . mixture of He and Xe
Heat carrier pressure, . . . . . . . . . . . . . . . . . . . . . . . . . . . . . . . . . . . . . . . . . . . . . . MPa 2.8
Heat carrier maximum temperature, K . . . . . . . . . . . . . . . . . . . . . . . . . . . . . . . . . . . . . 1,500
Regeneration degree . . . . . . . . . . . . . . . . . . . . . . . . . . . . . . . . . . . . . . . . . . . . . . . . . . . . 0.9
Refrigerator-radiator area, m2 . . . . . . . . . . . . . . . . . . . . . . . . . . . . . . . . . . . . . . . . . . . . . 800
RR mean temperature, K . . . . . . . . . . . . . . . . . . . . . . . . . . . . . . . . . . . . . . . . . . . . . . . . . 700
Energy conversion efficiency rate, % . . . . . . . . . . . . . . . . . . . . . . . . . . . . . . . . . . . . . . . . 27.5

SC consists of a power supply module based on nuclear reactor, of a propulsion module, a booster and a payload compartment. The power supply module comprises the reactor, shadow-type radiation protection, energy conversion system (ECS), heat pipes-based refrigerator-radiators and a sliding truss. The propulsion module contains a package of sustainer electric rocket engines (EREs), a fuel tank, engines control system, spacecraft control system and a nuclear power plant control system. The refrigerator-radiators of the electrical rocket propulsion unit are placed on the surface of the propulsion module.

The booster is a jettisonable rocket stage consisting of an oxidizer (oxygen) tank, a fuel (kerosene) tank and two engines with a total thrust of around 1 tf placed on a jettisonable truss. The truss is fastened to the surface of the ECS power framework and is jettisoned together with tanks and engines on a circular orbit at an altitude of $H_{circle} \approx 800$ km.

The payload compartment has a total volume of around 800 m$^3$ and separates from SC on GEO along the plane of docking with the propulsion module.

During launch into a low orbit the SC is placed in the payload container of the Energia carrier rocket. The payload container is opened and jettisoned after launch by the carrier at an altitude $H_{circle} \approx 200$ km. After that booster motors go into action and as soon as the SC reaches the support orbit at an altitude of $H_{circle} \approx 600...800$ km the boosters are jettisoned.

On the support orbit the RR trusses are unfolded and opened pursuant to a command from Earth. Next, the reactor is put in operation and power supply system is adjusted to the required level. After testing the SC subsystems, it is put in the position of gravitational orientation. After that the sustainer electric rocket propulsion units go into action.

### Preliminary estimation of mass related characteristics of SC and its elements

```
Mass of SC payload with NPP on GEO, t . . . . . . . . . . . . . . . . . . . . . . . . . . . . . . . . . . . . . . . . . 35
including:
      nuclear reactor and radiation protection. . . . . . . . . . . . . . . . . . . . . . . . . . . . . . . . . . . . . . 3.5
      energy conversion system  . . . . . . . . . . . . . . . . . . . . . . . . . . . . . . . . . . . . . . . . . . . . . . . 14.5
      construction of unaccounted elements . . . . . . . . . . . . . . . . . . . . . . . . . . . . . . . . . . . . . . . . 3
Mass of propulsion module, t 16
including:
      ERE . . . . . . . . . . . . . . . . . . . . . . . . . . . . . . . . . . . . . . . . . . . . . . . . . . . . . . . . . . . . . . . . 3
      ERE, SC, NPP control systems. . . . . . . . . . . . . . . . . . . . . . . . . . . . . . . . . . . . . . . . . . . . . 2
      of fuel tank . . . . . . . . . . . . . . . . . . . . . . . . . . . . . . . . . . . . . . . . . . . . . . . . . . . . . . . . . . 1
      of fuel . . . . . . . . . . . . . . . . . . . . . . . . . . . . . . . . . . . . . . . . . . . . . . . . . . . . . . . . . . . . . . 8
      of structure . . . . . . . . . . . . . . . . . . . . . . . . . . . . . . . . . . . . . . . . . . . . . . . . . . . . . . . . . . 2
Booster mass, t. . . . . . . . . . . . . . . . . . . . . . . . . . . . . . . . . . . . . . . . . . . . . . . . . . . . . . . . . . . 12
SC mass on support orbit ($H_{circle} \approx 800$ km), t . . . . . . . . . . . . . . . . . . . . . . . . . . . . . . . . . 72
SC mass on launch orbit ($H_{circle} \approx 200$ km), t . . . . . . . . . . . . . . . . . . . . . . . . . . . . . . . . . . 84
SC take-off mass, t. . . . . . . . . . . . . . . . . . . . . . . . . . . . . . . . . . . . . . . . . . . . . . . . . . . . . . . 104
including the payload container. . . . . . . . . . . . . . . . . . . . . . . . . . . . . . . . . . . . . . . . . . . . . . . 20
```

The estimated time of launching SC with above described parameters into GEO amounts roughly to 60 days, in which case most of the time SC will be in radiation zones of various intensity. If the protection system of SC and payload is fabricated from aluminum, thus making its specific mass not higher than 1 g/cm$^2$, the total radiation dose will not exceed $2 \cdot 10^4$ rad. After injection into orbit, the payload separates from SC. The SC, if necessary, is transferred to geocentric orbit.

Thus the investigations show the following:

a) the use of Energia carrier rocket and 3MW nuclear power rocket engine (NPRE) with a turbo-machine conversion and MPD engine having the efficiency rate ~ 0,7 and specific pulse 5,000 s enables a 35 ton payload to be injected into GEO in 60 days.

b) compared to liquid propellant engines the use of NPRE doubles the mass and volume of payload injected into GEO;

c) nuclear safety of SC in all phases of its service life in standard conditions and emergencies can be assured by active and passive protection;

d) he feasibility of the proposed concept of the electric rocket engine is verified by a number of experiments, calculations and theoretical works carried out in Russia and other countries.

Currently, Russia is capable of fulfilling this task since it is in possession of such a powerful carrier rocket as Energia and vast technical and scientific expertise in nuclear power plants and propulsion units for use in space.

In addition to NPP that feature heightened radiation hazard, conventional rocket engines too will continue to develop.

### 2.4.5 Prospective trends in improvement of chemical rocket engine

The current situation in the development of space transport vehicles is such that the capability of further improvement of chemical rocket engines of conventional types (based on stationary or slow running processes) are practically exhausted and are confined to minor amendments to energy and mass related characteristics achieved as a rule at the sacrifice of reliability, safety and ecological security. The leap in development of space transport vehicles can be obtained through elaboration and introduction of radically new engines using fast proceeding (explosive) processes, for example, pulse detonation engines (PDEs). The PDEs are characterized by the detonation mechanism of energy conversion, (up to ultrasonic level) frequency of operating cycles, lack of mechanical valve grating, capability to work both in rocket and aero-rocket modes.

The potentially pulsing detonation engines, as compared to conventional ones, offer a number of advantages. These are the high specific pulse (in the order of $10^3$ s); low specific mass (in the order of $10^{-3}$ kg/kgs); small specific size (in the order of 10...3 $m^3$/tf); capability to use in a wide range of speeds; jet-free mechanism of thrust generation (the zone of power and temperature manifestation of exhaust gas does not exceed 100 mm from the nozzle edge); simplicity, reliability and adaptability to manufacture (the main structure of the engine contains 5 to 7 rigidly fastened components without moving parts); low toxicity of exhaust gases.

The PDEs are often mistaken for pulse engines that operate on just one frequency because burning proceeds in an acoustically adapted chamber. Normally, the pulse engine has a low specific fuel consumption but is at that inferior because of its low specific pulse conditioned by the low compression of fuel and air mixture before combustion. By contrast, the PDE has a high specific pulse and is a mechanically simple device.

By the late 1980s the International Corporation of Applied Sciences (SAIC) sponsored by Defense Advanced Research Projects Agency (DARPA) has carried out a number of computer-aided studies of PDE. The results of the study had been reported over recent years at conferences of the American Institute of Aeronautics and Astronautics which testifies to the growing interest in this technology.

DARPA got interested in the capabilities of PDE for use as a propulsion unit in expendable craft, such as small unmanned flying vehicles and auxiliary penetration rockets.

There is a convincing proof that the Pratt & Whitney company and other developers of engines are or have been actively working on strictly classified projects aimed to create propulsion units based on the "unstable burning" effect. The officials of NASA and the Air Force had been informed about flight tests of the craft using propulsion units with PDE. The details, however, were carefully concealed. Those could be tests of both manned and unmanned flying vehicles.

The investigations conducted by Russia's research centers and design offices have shown that liquid propellant engines will remain the main rocke engines used by prospective launch vehicles in the coming 20 to 25 years. The creation of a propulsion unit using atmospheric air as an oxidizer, for example, a hypersonic ramjet engine which would considerably reduce the take-off weight of a launch vehicle calls for solution of many complex problems. They are associated mostly with the development of a propulsion unit and a flying vehicle as whole that operate in conditions of high speeds, pressures and aero-dynamic heating (1,500 K and higher). These problems put off the creation of the hypersonic ramjet to a more distant future, the fact confirmed by results of work under NASP program carried out in the US. The appearance of such an engine is shown in Fig. 64.

The progress of astronautics and the solution of problems that are crucially important for stable development of the civilization are hampered by the inadequacy of the existing launch vehicles as well as those being developed. This refers primarily to the high cost of launching payloads (from US$ 5,000 to 10,000 per kg of payload). Among the critical elements of modern launch vehicles are sustainer motors. The high cost, the risk of failure with disastrous consequences – such are the inherent shortfalls of modern expendable engine. These drawbacks can only be overcome based on radically new decisions in design, engineering and layout.

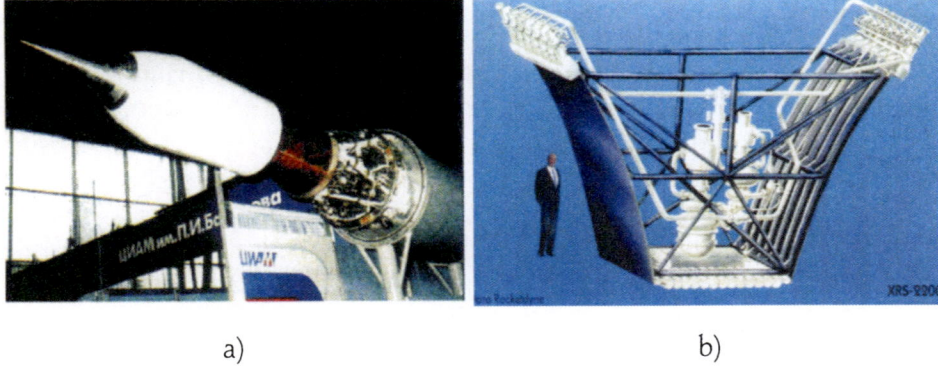

a)                                                        b)

**Fig.64.** SCRAMJET (a) and "revolutionary" linear liquid propellant engine Aerospike (b)

The development of a new concept of sustainer motors is necessitated by the fact that the already available rocket engines do not meet the main requirements imposed on advanced STS, specifically the following:

• the highest possible (close to absolute) safety, prevention of catastrophic consequences of possible failures of some elements and subsystems, primarily, engines;

• substantial, at least by an order, reduction of operating costs and the specific cost of performance of transport operations;

• reduction to the minimum of ecological damage caused by operation and the possibility to build up the Earth – to – orbit transportation;

• versatility and adaptability that provide for operation with the use of various launch complexes, reduction to the minimum of restrictions imposed on types of payloads allowed for transportation.

Currently, researchers and experts of the Keldysh Research Center jointly with the Central Research Institute of Machine Building have preliminarily estimated the optimum specifications of prospective launch vehicles and their propulsion units that meet the above requirements; acceptable solutions have been found that are worth more detailed study. Note worthy among such decisions and technologies are:

a) the use of liquefied natural gases (LNG) as an all-purpose ecologically clean fuel;

b) the use of new types of engine, specifically, with after-burning of reducing generator gas and effective systems of cooling the combustion chambers;

c) hot reservation of engines in propulsion units based on diagnosis and prediction of their likely failures and malfunctions;

d) the use of highly effective systems to control the quality and reliability of engines during their manufacture;

e) achievement of high reliability and adaptability of engines to manufacture, low labor consumption and low cost of manufacture;

f) the use of new optimum layout of three-component LPREs.

The above described requirements are best of all satisfied by the following three types of prospective LPREs: first, extra-reliable and cheap to operate engine intended for multiple use as part of the first stages of launch vehicles and designed to operate in the so-called "sweet" fashion (i.e. with excess of fuel in the generator gas which precludes the formation of soot in case of using liquefied methane as a fuel and liquid oxygen as an oxidizer); second, a three-component engine running on the $O_2 + H_2$ + hydrocarbon used as a sustainer motor in prospective single stage launch vehicles; third, an engine of the first stages of reusable launch vehicles running on accumulated and liquefied in flight atmospheric air, the so-called liquid air cycle engine (LACE).

Let us briefly consider the features of those types.

**Oxygen-methane LPRE.** The creation of such an engine can be regarded as an integral part of Russia's program aimed at expanding the use and raising the efficiency of using liquefied natural gases in automobile, railroad, aviation and space transport systems. The implementation of such a program is necessitated by shortage of fuel produced from oil, by large stocks of easily available and relatively cheap natural gas, as well as by ecological advantages of using it.

The wide investigations of various types of fuel, including synthetic, have shown that LNG consisting of methane to the extent of 95-98% can be regarded as the prospective fuel for space rocketry, the fuel that will meet the main requirements related to cost, reliability and ecological safety of the new generation sustainer LPREs without detriment to energy and size characteristics. The use of the oxygen / methane fuel with consideration for reduced losses in its suspended cooling raises the thrust's specific pulse as compared to oxygen/kerosene engine by 200-250 m/s which makes up for the 15% loss of fuel components density. The increase of the thrust's specific pulse results from higher thermodynamic characteristics and the cooling capability of methane.

The experiments carried out by the Thermal Processes Research Institute have shown that it is possible to develop an engine fueled by oxygen and methane with the use of reducing generator gas, i.e. realization of the idea which provides plenty of capabilities to improve the reliability of LPRE and build a reserved propulsion unit. Compared to liquid hydrocarbon fuel of RG-1 type the liquid methane is cheaper and its raw materials base is practically unlimited. The problem of building a ground infrastructure (installation for liquefaction of natural gas and its transportation, as well as the refueling equipment) has been in the main resolved. The transition to methane allows to settle fairly easily many problems of turnaround servicing of the engine since after work it remains clean and without deposits of resin and soot, the phenomenon typical of the RG-1. Thanks to this, the time and cost of servicing are reduced to the minimum. It could be expected that the ecology related problems would also be resolved due to the plummeting content of obnoxious compounds in fuel combustion products (especially $CO$).

The building of oxygen/methane LPE can be based on the latest achievements of Russia's space rocket engineering, including those obtained in development of oxygen/kerosene (RD-170) and oxygen/hydrogen (RD-0120) engines. This diminishes the technological risk and the loss of time and funds. The most appropriate applications of such LPREs are multi-engine reserved installations of the first stages of two-stage new generation launch systems.

The design of LPRE must be developed based on requirements of its repairability and maximal adaptability to manufacture. One of the variants of the proposed engine can operate in an open circuit with a pass-by of generator gas into the supercritical part of the nozzle. The use of the unclosed circuit of engine operation (or circuit with reducing generator gas) makes it possible to optimize the engine's performance on a piece-by-piece basis and to considerably reduce the refinement period and its cost.

**Three-component engine.** To develop in the future more effective single stage launch systems it is necessary to create a new generation of LPRE operating with the use of two fuels (hydrogen and hydrocarbon fuel) with liquid oxygen.

The main advantage of three-component LPRE as compared to two-component oxygen/hydrogen engines is the reduced by 1.5 to 2 times supply of hydrogen. This will cut down the cost of launching the payload. Also, this will reduce the "dry" mass of the carrier structure.

The investigations have shown the competitiveness and substantial effectiveness of LPERs running on three-component fuel (liquid oxygen – hydrocarbon fuel – liquid hydrogen).

**Liquid air cycle engine (LACE).** The development of engines operating with the use of atmospheric air entails the solution of many new scientific and technological problems which, as analysis shows, puts off the creation of the liquid air cycle engine (LACE) until more distant future.

LACE should be regarded primarily as a prospective engine for single stage reusable aerial space systems. Work on this type of engine has been going on in Russia at the Thermal Processes Research Institute since the early 1960s.

LACE is a combined type of engine which, depending on the flight speed of the aerospace system, functions with the use of an oxidizer represented by either compressed atmospheric air or liquid oxygen from the system's fuel tanks. As distinct from other engines working on this principle, it operates in the speed range from zero to orbital. Therefore the propulsion unit of the aerospace system can be formed solely of LACE without resort to other engine types.

Estimates show that in the air mode LACE will feature the specific thrust pulse in the region between 23,500 and 14,000 m/s and the specific thrust in the air varying between 2,300 and 1,550 m/s. In the rocket mode the specific pulse will be no less than 4,600 m/s. The engine specific mass in this case will vary between 65 and 45 kg/tf.

The main potential advantages of LACE are:
- the capability to be used for propelling flying vehicles equipped with single type engines;
- the high value of the specific thrust in the air as compared to similar characteristics of air jet engines of other types (smaller mass of the air intake and the engine gondola, as well as smaller aerodynamic resistance of the flying vehicle);
- the capability to optimize the engine on the ground with the use of the mainly available bench rigging and test facilities used in work with oxygen/hydrogen LPRE;
- the reduction of take-off mass of the launch system (compared to the mass of launch vehicles with LPRE) by 1.5 to 2.0 times;
- fairly mild temperature operating modes of the carrier structure in the delivery of payload as compared to the variants of the aerospace system of NASP type with SCRAMJET engine;
- the capability to utilize aeroplane's pattern of horizontal take-off of the flying vehicle equipped with LACE, to land it on an aerodrome and independently relocate.

The major problem in development of LACE is the creation of a high-performance, low weight heat exchanger for liquefaction of atmospheric air during flight of the aerospace system at $M \approx 0...5$.

In countries other than Russia attempts to introduce the LACE (RB.5A5) have been made in Great Britain under HOTOL program. In Japan, research on LACE is being conducted by Mitsubishi: air liquefaction system has been tested and demonstration trials are now being prepared.

In Russia, engineering, design and experimental work on the creation of the bench-based demonstrational prototype of LACE together with elaboration of make-up and specifications of the engine's live prototype is being carried out by the Keldysh Research Center in cooperation with a number of Russian companies. Specifically, preliminary study has been done of the make-up of the engine's live prototype, assembly units of the experimental heat exchanger are being fabricated, air liquefaction that takes place in it is being studied and means of fighting the ice build-up are being investigated.

The considered types of engines united by the single concept of the engine of the 21st century will assure, as results of the systematic analysis show, the successful realization of prospective space programs in various scenarios of development of Russia's and the world's astronautics. The work on this concept will preserve and assure the effective use of the available high potential of Russia's rocket motor building. The future-oriented work done in the initial phases of the endeavor will, the authors believe, serve as a solid basis for the progress of mutually beneficial international cooperation.

**Low thrust liquid propellant rocket engines** (LTRs) are the main type of executive organs in spacecraft's systems of correction, stabilization and orientation. More than fifty types of LTR developed by the industry are successfully used in component packages of SC of various applications. A number of engine types are being experimentally optimized. By now the following four main types of LTR have emerged that are called upon to fulfill the tasks of the current situation:

1. engines running on long storable self-inflammable fuel components (e.g. nitrogen tetroxide + asymmetrical dimethyl hydrazine) intended for SC (mainly automatic ones) with a long service life (up to 15 years and longer). Those engines must have the firing endurance in the region of hours, high reliability and adequate characteristics in terms of energy, mass and dynamics. In addition, they must have a multi-role capability and an ability to be standardized for use on other spacecraft;

2. engines functioning with the use of both conventional self-inflammable fuel components (e.g. nitrogen tetroxide + asymmetrical dimethyl hydrazine) and high energy performance prospective types of fuel intended for highly maneuverable small size special purpose SC (e.g. ABM facilities). The said engines must have very high dynamics (less than 5 ms), the smallest possible mass (thrust to mass ratio is 1,000 kgf/kg and higher), high specific thrust pulse and relatively small firing endurance (10 to 50 s, or 200 s in individual cases).

3. engines running on non-toxic and, as a rule, self-noninflammable fuel components and intended for prospective piloted SC. The main characteristics of those engines are similar to those of LTR of the first type;

4. engines running, as a rule, on single component fuel (hydrazine) and providing precision orientation and stabilization of SC. Such ultra low thrust rocket (less than 0.5 kgf) must have a firing endurance in the region of tens of hours, high reliability, stable thrust and stable pulses; also they must have the capability of being standardized;

The main trends in the radical improvement of LTR must be the development of:

• combustion chambers made from advanced heat resistant metal alloys without protective coatings, from non-metallic and composite materials; it's noteworthy that the combustion chambers of engines of the first and third types could be best made from alloys without coatings and metal ceramics since they are able to operate throughout extended firing life, while the engines of the second type – mainly from carbon-carbon composite materials having high heat resistance and low density which provides for mass reduction of the chamber to the lowest possible minimum at the high (up to 2,000 K) temperature of its wall and highly economical performance of the engine;

• in case of the fourth type those should be materials resistant to hydrazine and its decomposition products throughout the required firing life;

• highly effective and reliable mixing heads whose hydraulics operates in a stable fashion throughout extended service life, heads that prevent obliteration of nozzles' throats of the second, third and fourth types of engines as well as high completeness of fuel combustion despite their small size and very small length of the combustion chamber of the second type engines;

  • reliable high speed valves remarkable in that while for second, third and fourth types of engines the crucial points (already achieved in many respects) are their endurance of the number of activations and reliability, critical for LTR of the second type are the dynamics of the valve's openings and closings as well as its mass;

  • new basic components for advanced propulsion units with LTR and techniques of optimizing its parameters.

## 2.4.6 Electric rocket engines

The practical work on electric rocket engines began in 1970s and was urged by the limited capabilities of onboard power supply systems of SC whose maximum electric power was around 1 kW. Such a level of power and requirements imposed on electric rocket engines (ERE) made it feasible to develop stationary plasma engines (SPE) based on the Hall plasma accelerator with a closed electron drift. The first examples of SPE were made by the Kurchatov Atomic Energy Research Institute, and later by Fakel (torch) Design Bureau. Those engines feature fairly high performance at the said power level and specific pulse roughly equal to 20 km/s; the actuating fluid is xenon.

In the coming two decades the ERE will be used to correct spacecraft's orbits in order to rectify mistakes made while launching or to maintain (or change) them during SC's active service life, also to compensate for disturbances acting on spacecraft as they function on both geostationary and low Earth orbits; to ensure high precision orientation and stabilization of SC; to deliver SC from support near Earth orbit to a higher one, e.g. geostationary.

Afterwards, the ERE must find wide use in sending spacecraft to the Moon and planets of the Solar system. Currently, the ERE are especially effective and economically justifiable in use as part of geostationary communications satellites. Estimates show that the replacement of an orbit corrective LTR by an ERE will reduce the mass of the type satellite by 15% with its active service life (ASL) being 5 years and by 40% at ASL being 15 years. In case of constant mass of the satellite the number of channels and the amount of information transmitted will be increased respectively by 30 and more than 50%.

The analysis of tasks of space rocketry in the near term shows that equipping the bulk of SC will require ERE with electric power from 1 to 10 kW (subsequently – up to 100 kW) and with specific thrust pulse from 15 to 30 km/s while the operating life will be in the region of 10,000 h. These parameters of the ERE are optimal for correcting and orienting spacecraft of various applications, for interorbital flights, for maintaining the orbits of low-flying satellites and so on.

Among the prospective trends in developments of the near future are SPE operating with the use of more readily available compounds than xenon since the supply of actuating fluid required for instance for means of interorbital towing runs into tens of tons.

In the longer term, as challenges of global energy supply, ecology related issues and problems of growing scale of interorbital transport operations are successfully settled, the economic effect of using ERE will soar up. The future piloted interplanetary expeditions and large scale development of the Moon are hardly possible without such engines. Therefore, perfecting the ERE that consume tens and hundreds of megawatt of energy, have the specific thrust pulse in the region of 15 to 80 km/s, the service life up to 5 years and the capability to reliably operate on such readily available compounds as argon, krypton, lithium, sodium, seems to be one of the most promising and imperative trends in the evolution of space rocket motor building.

## 2.4.7 Unconventional rocket engines and techniques of space travel

The creation of unconventional rocket engines is based primarily on the use of unconventional power sources.

Some authors (A.Ye. Akimov, A.F. Okharin, G.I. Shipov and others) propound plans of using ways and means of space travel based on radically new physical fields (torsion, microleptonic, mesic) and the energy of physical vacuum. The experimental results confirming the feasibility of such techniques and providing support for building such space vehicles are not yet available.

Dedicated literature draws attention to the possibility of building the so-called "flying saucers" whose lift is obtained by rotation of disk-like bodies. The idea is allegedly borrowed from the works of Sarle in Germany. Unfortunately, the experimental verification of the feasibility of building the "flying saucers" based on Sarle's design and using the drawings supplied in the literature yielded no positive results.

Nonetheless, the disks rotation was used by S.M. Polyakov in the model of a "flying saucer" that created a certain lift. The author calls the principle of the model's movement gravitational and substantiates it mathematically, suggesting at the same time the hypothesis of interdependence between gravitation, magnetism and rotation. Here, based on experimental results, an explanation is supplied as to the nature of magnetostriction as a secondary gravitational effect.

The Russian inventor Ye.Podkletnov has conducted experiments with a fast rotating disk made of high-temperature ceramics. The disk was placed in a cryostat with liquid nitrogen. The effects observed during this procedure can be regarded, according to experts, as a proof of anti-gravitation at work. However, even though other works of other authors on the same subject are available, the writers of this book believe that the proof of anti-gravitation is absolutely insufficient.

Anyway, major space rocket companies of many countries are now ever more interested in this issue. Among them are such state-supported organizations as the NASA and British Aerospace Association. Also, work is underway in Canada, Finland, Greece and other countries.

Quite interesting is the development of a radically new type of fuel enabling the use of liberated energy of atomic shell bonds. In terms of stored energy (around 10 MJ/cm$^3$) such fuel exceeds by roughly four orders the chemical fuel and is inferior only to the nuclear type. The experimental results do not preclude the possibility of creating such kind of fuel. The principle of operation of the appropriate rocket engine is based on the emission of energy of atomic shells bonds under the exposure of the working medium to laser pulses of a femtosecond's duration with power density to the excess of 1,017 W/cm$^2$. The estimated specific pulse of the rocket engine running on such fuel is several hundreds of kilometers per second. The design of the projected engine has much in common with laser rocket engines, the possibility of whose building was studied in the previous decades.

Also, laser rocket engines have been developed the energy to which must be fed from Earth-based laser installations. Considering the complete set of properties, the laser method may prove quite promising as long as efficiency rate is substantially raised all through the energy conversion and transmission as well as in case the problem is resolved how to home the laser beam precisely while it is subjected to fluctuating atmospheric impacts.

Also worth of note is the energy generation in an inertial thermonuclear reactor. This technique is based on the existing knowledge of physics and the magnetostriction phenomenon. This method is ecologically clean, since water, including deuterium, is used here as fuel, helium and radiation passive elements being the end products. The energy liberated in the course of controlled thermonuclear synthesis is converted into electrical power. Used for initiation energy here is the blast wave energy generated by the magnetostriction generator of solitons. Used as a target is the cavity formed by ultrasound. The reaction proceeds in the form of a micro-explosion in a chamber filled with current-conducting liquid. The conversion of energy occurs in the magnetic hydrodynamic generator. This method calls for further detailed theoretical and experimental refinement. It serves as a basis for the magnetic plasma technique of boosting payloads proposed by V.A. Zolotukhin. Essentially, the magnetic plasma technique of boosting payloads lies in the following. The boosted object (capsule) shaped as an ellipsoid of revolution is surrounded by a poloidal magnetic field. The boosted object-capsule is exposed to a plasma jet having the ions' speed in the region of 25 km/s. Under the influence of fairly powerful jet of plasma the object starts accelerating. The acceleration is regulated by the capsule's magnetic field and the power of the plasma jet. The introduction of this technique calls for solution of a number of complex scientific and engineering problems.

Also worth of note is the possibility of a significant reduction of aerodynamic resistance of the environment by forming its specific movement around the moving object.

Quite interesting is the use of cable systems for conveying payloads from one orbit to another. This technique is particularly effective in the case of large and equal in size counter flows of cargoes (moving upwards and downwards) as in this situation the power consumption in towing cargoes will be minimal. Some authors ponder the possibility of transporting cargoes by "the space elevator", which is a cable stretched from the Earth's equator over to the geostationary orbit. Cargoes are supposed to move along such a cable.

Thus the following conclusions can be drawn.

1. Further improvement of space power plants and propulsion units is the major trend in evolution of astronautics in the 21st century. It implies both perfecting the conventional propulsion units (running on chemical and nuclear fuel) and creating radically new technical systems.

2. The development of chemical engines in the 21st century is envisioned both as improvement on chemical components of rocket fuel and development of new patterns of chemical reactions. This could be exemplified by the theory of pulse detonation engines, the theory now being actively developed.

3. The development of nuclear space power engineering in the 21st century will proceed mostly through increasing the power of electric installations to tens of megawatts by using new techniques, particularly those based on magnetic plasma processes.

4. The 21st century may bring about unconventional techniques and methods of power production based on torsion and microleptonic processes, on extraction of physical vacuum energy and on a number of other processes.

5. One of the biggest problems in power installations is the search of more effective unconventional means of using the Solar energy not only in space research but also in solution of mankind's global problems in the 21st century, such as power supply of Earth from Space

## 2.5 The necessity of dual use technologies in space

Currently, ever more states come to realize the importance of using space to address the issues of national security and socio-economic progress. That is why they intensify their space efforts. In case of Russia, in spite of being, thanks to its high potential in space engineering, on a par with the leading space powers, it faces now the danger of falling behind others in the space race. This is attributed to the specifics of today's development of astronautics in Russia which is characterized by the drastic reduction of the state's funding for the space industry, Russia's uncertain position in the international space cooperation, deformation of the national military industrial complex which had been over decades the source of technologically advanced hardware for Russia's space effort whose military streak had always been distinctly pronounced. Astronautics in Russia, as well as in other countries, has developed in the conditions in which its achievements were primarily utilized for national defense.

The subsequent perfection of space systems, the cost reduction of their manufacture and operation, and the dramatic improvement of their performance have raised the cost effectiveness of using them. This, in turn, triggered the rapid development of the civilian types of space hardware which in many aspects became not only self-repaying but quite profitable when used on a commercial basis.

In this context it should be pointed out that until the early 1990s the civil segment of space work used to get far less attention both in terms of funding and organization of effective interaction with the military segment in order to make the best use of the combined space potential for fulfilling various tasks. The latter circumstance is caused by a number of factors, the most important of which are as follows:

• the amount of funding that enables to create in peacetime redundantly equipped orbital constellations for military uses made it as a rule unnecessary to conduct a feasibility study of using civil systems for addressing the issues of national defense and security;

• the principle of permanent operational readiness of military space facilities precluded the effective utilization of the redundant component packages of military space systems in peacetime for fulfilling civil tasks which in turn resulted in unasked for space hardware (according to some estimates it varied between 30 and 70%).

The independent development of military and civil systems is largely a matter of convention since the main demand which is made of them and which determines their overall performance is the adaptability to operation in space. It is only now that the necessity of building dual use space systems became obvious and undeniable. The dual use implies projecting the system with due regard to its application both in civil and military roles. Understandably, the cost of one SC capable of performing in both capacities will generally be higher than that of a purely military or civil one, but indisputably lower than the cost of two SC taken together.

The problem of "the gap" between military and civil space technologies employed to tackle a variety of tasks, is not exclusively Russian. In the US, for instance, it resulted in an unjustifiable rise of the cost of military developments because of the reduced access of the military sector to some commercial technologies rapidly developing in some areas and due to difficulties of using in the civil sector the results of massive federal investments in the military sphere. In addition, a tendency has emerged for relaxation of the competition in the area of military research and development because of reduced number of potential customers in consequence of mergers of companies whose activity was associated with defense orders (compared to 1985, in the area of carrier rockets production their number slid by 25%, in spacecraft production – by 25%, power plants – by 38%, combat missiles – by 57%, reconnaissance facilities – by 40%).

The danger of such a gap was understood in the US in due time which generated the mechanism for handing over the information received from the military space systems to civil agencies. Also, it gave birth to mechanisms of drawing civil and commercial space systems into tackling the military tasks. Thus the civil systems are widely used by the military agencies, primarily through renting the channels of commercial communications satellites. The US DoD also receives a large amount of information from civil natural resources surveillance satellites, from geodesic and meteorological survey satellites. It uses more than 20% of information provided by the American Landsat system and complements it with information supplied by ERS satallite, Spot (France) and Mos (Japan). The US Cartography agency of the DoD is the second largest, after the Department of Agriculture, buyer of images obtained from the natural resources surveillance satellites. Also, interaction was organized of the leading coordinators of development of new technologies of military and civil agencies (DARPA, NASA and others) in the form of joint projects (TRP project) and bilateral agreements for coordination of work in the sphere of new technologies (agreement between NASA and the Air Force Space Command concluded in February 1997).

The US is leading in the use of military space systems for civil purposes and in the use of commercial satellites for military purposes, which substantially reduces the overall expenditure on space work. Other countries profit mostly by wide use of commercial satellites in the interests of armed forces.

In Russia the problem of integration of the military and civil space efforts increasingly sharpens from day to day. In the US this problem and its causes are well understood and get due attention. By contrast, in Russia, this problem is largely the struggle for survival among manufacturers of space hardware driven to sell (often without supervision) high-tech space technologies or to provide space work related services. The former often results in drain of high-tech space technologies abroad while the latter creates a situation in which a number of Russia's space industry companies (such as Khrunichev State Space Research and Production Center) have by now accumulated positive experience of using space systems previously developed to the orders of the MoD for civil applications as well as for proposals in the international market of space services on a commercial basis. However, the truly effective integration of the military and civil space efforts for the development of Russia's space potential is impossible without clear-cut delineation of the problems of dual-use space systems and without finding on this basis the means of resolving them. But, importantly in this case, the very interpretation of the term "dual use" must imply not only the use of space systems (and technologies) for solution of civil tasks but also the use of civil space systems (appropriate technologies) for fulfilling the tasks of defense and security. As noted previously, the overall performance of any space system is primarily determined by operating conditions in space, not by its planned function in tackling the military or civil tasks. It means that in principle there is a possibility to expand the application area for the already existing space systems.

The predominant in early 21st century view of space systems' application areas allows to start immediately to optimize the product line of the Russian-made space systems and complexes, primarily by slimming them down in size. For optimization of space systems (complexes) the Federal Space Program envisions the state's support and participation financing of dual purpose and dual use space systems and hardware that will surely continue to be used in the 21st century. What is meant here by dual purpose hardware are space systems (complexes) and the hardware of ground-based facilities of the space infrastructure, which systems are specifically created for fulfilling both military and civil tasks. Meant by dual use hardware are space systems (complexes) and the equipment of ground-based facilities of the space infrastructure, which systems are created specifically for tackling only military or only civil tasks and used (those that can be used) for fulfilling the tasks in the adjoining (relatively, civil or military) area. By the beginning of the 21st century it has become standard practice to use in dual application the domestically manufactured space systems and hardware. So the Altair space communications system which is being created for defense roles was originally designed also to maintain continuous communication of the Mir complex with the flight control system. Such utilization of the space system for dual purpose was conceived in the late 1970s – early 1980s. By now, the dual purpose of space systems has become legalized in the form of an agreement between two agencies (the MoD and Rosaviakosmos) whose main provisions stipulate joint financing of the development and operation of the following systems (efforts):
• Luch (ray) – space complex of global space relay system with a spacecraft in geostationary orbit;
• Resurs-DK (resource) – space optoelectronic observation system;

- Kometa (comet), Bars (panther) – space cartography systems;
- GLONASS – space navigation system;
- Nadezhda-M (hope) – space navigation and distress detection system.

One of the realistic ways to preserve the possibility of using orbital facilities for fulfilling military tasks is the maximal involvement of the orbital civil use hardware. The use of civil SC intended for communication, observation, time-coordination, hydrometeorological survey and other roles in military missions is the major trend in the global military space effort. The users of civil space communications systems from defense and law enforcement agencies can now  obtain the channels of the main and zonal communication (subscription on a permanent or temporary basis) via communications SC. Under such circumstances it is possible to organize the so-called issued (agency controlled) network based on VSAT technology (network using ground-based stations with a small antenna aperture)  in which via the relay channel a two-way radial communication is provided between the junction center and a group of subscribers.

The information support in hard-to-reach areas is accomplished by satellite communications transportable subscriber stations.

Already now the TV broadcasting channels can be used to transmit via SC of Gorizont (horizon) or Express type to the Moskva (Moscow) type of station (antenna diameter ~2,5 m) topographic and electronic-digital locality maps, synoptical maps and other image-based documents. In the 21st century these capabilities will substantially expand.

In the longer term, the advanced data transmission SC of the Zerkalo (mirror) type could be used to organize the transmission of large amounts of information either directly from SC or via relay from ground-based information reception posts. The Express-M platform can be used for installation of appropriate instruments right now (the active service life is around 10 years, power supply in the region of 5.5 kW). It has been developed in the course of implementation of civil projects, including those proposed by foreign customers. Later on it can be replaced by its modifications.

Protected communications channels in civil space systems can be obtained by coding information at users' level. Such coding protects information at the required level for a group of network users. The protection is inaccessible to other subscribers. In this case, within the assigned frequency band the military users of communication services can carry out "an exchange" of throughput capacity for a degree of coding. Currently, such tasks can be tackled based on communications SC deployed in geostationary orbit (of Gals and Ekran types) with the use of mobile transceiver  stations with antennas measuring 1.5 to 4 m in diameter, with a transmitter having an output of 200-500 W and with the aid of individual receivers with antennas measuring from 0.45 to 1 m. The use of NTV SC provides a fixation of observation data to digital locality maps.

For serving customers from law enforcement departments, the hydro-meteorological service traditionally uses medium altitude SC of the Meteor-2(3) type and Elektro SC on geostationary orbit. These craft provide for continuous global observation at relatively low (worse than 0.5 km) resolution of measuring instruments.

The data can be discharged at autonomous information reception posts. At work now are around 50 autonomous information reception posts (AIRP) receiving data from those craft within Rosgidromet (Russia's national weather forecast network) and around 80 AIRPs within the network of the Defense Ministry.

The technologies utilized by the Elektro SC will make it possible to observe nebulosity, determine direction and speed of near-Earth wind, vertical temperature and humidity profiles for the routine needs of armed forces and for use of high precision weapons. The capabilities of the onboard relay complex of the Elektro type SC allow to organize information reception directly from SC or from information processing posts on to small size mobile terminals with an antenna measuring ~1,5 m in diameter, or smaller in the future.

In a way, all Earth remote sounding craft can be utilized for military observation purposes. The operational observation can be conducted by optoelectronic surveillance SC (of Resurs-01 type providing the resolution of onboard equipment in the visibility range of ~25 x 35 m, and others). The tasks of operations-free observation can be assigned to photography surveillance SC (of Resurs-F1M type with resolution of 5 to 8 m) and to Resurs-DK SC being now in development and featuring the resolution between 2 and 3 m in the visible and IR ranges.

Useful for Russia's MoD could also be the following types of civil orbital equipment being developed under Russia's space program:
• The Express type of SC for organization of fixed communication and data transmission;
• Yamal-200, -300 types of communications SC for organization of fixed and mobile communications;
• Triada type of communications SC for organization of mobile satellite communication with the use of SC on geostationary, high elliptical and medium altitude orbits;
• Gonets (messenger) type of communications SC for organization of personal satellite communications and for information transmission in the interests of various official entities;
• Ocean-O type of observation SC for organization of operational observation of the world's ocean and the glacier situation;
• Resurs-01 type of observation SC for organization of all-weather observation of sea surface, glacier situation, and for hydro-meteorological information support;
• Meteor-3M type of meteorological survey SC for organization of timely

receipt of global hydro-meteorological information for weather forecast, for monitoring the ozone layer and radiation situation in near-earth space;

• Elektro type of meteorological survey SC – for receipt of images of nebulosity and the Earth's underlying layers in equatorial and medium latitudes, for carrying out helio-physical measurements, for gathering and relaying hydro-meteorological and service information.

What, in this context, could be the crucial problems related to the area under consideration? Not a rhetoric question since these problems, if ignored, may hamper the progress of astronautics in the 21st century.

Dividing the entire spectrum of problems for convenience sake into organizational and technical, let us dwell first on the principal problems of organization. Among them, the following issues should be singled out as critical.

Lack of an effective mechanism for formulating the state's single policy in using space, the policy based on the comprehensive consideration of the state's interests in various spheres (defense, economy, public welfare, etc) and manifest in the single space program that provides for the most effective utilization of funds made available. To resolve the problem it is most expedient to establish an agency similar to the American national council for space which coordinates space work of civil and military organizations on the basis of appropriate legislature and the Presidential instructions. This entity is also called upon to support on behalf of the state the enterprises and organizations involved in development and creation of dual application space systems, including special taxation rates and preferred crediting treatment. One of the main results of such entity's work could be the Program of Building Dual Application Systems covering a wide range of issues.

The organization of effective interaction between performers of space work is, in turn, impossible without elaboration of the legal regulatory basis that fully regulates the issues of building the dual application space hardware and technologies. The legal regulatory basis must rest on legal acts elaborated within the framework of Russia's Law on Space Work. The leading role in those documents must belong to the document that regulates the interrelationship between Rosaviakosmos, the Defense Ministry and other interested departments in the phase of development and creation of the dual application space hardware (Fig. 65). It must unfailingly specify such key aspects of the interaction as the funding of work (including participation financing) for creation of dual application space systems with due regard to their possible future uses, establishment of a well-defined procedure for settlement of financial disagreements between the subjects of space work; the procedure of admission of various organizations (including commercial ones) to work on creation of dual application hardware, primarily for prevention of information leakage and advanced technologies drain; the procedure of deployment and commissioning of space systems, complexes and equipment.

Also, quite a big job is the development and realization of mechanisms of using dual application space systems in various conditions that assure the effective distribution of space systems' resources in case of their restriction (as in case of space systems for observation, communication and other uses); planning of space systems' applications; reception of information from space systems, its processing and storage; providing of space related information (services) to users and its distribution on the market (including expert analysis of information obtained from space observation systems in order to extract from it any intelligence concerning classified objects); financial and economic activity in course of operation of space systems (settlements with users, revenues distribution, offsetting operating costs, etc.)

**Fig. 65.**
Discussing the prospects of dual use of space in the 21st century (academician V.F Utkin and V.A.Menshikov)

One of the principal problems in building dual application space systems is the necessity to radically improve the ground-based infrastructure both in terms of its structure and technical equipment of facilities. Considering the existing state of the ground-based infrastructure it would be proper to regard as its basis the territorially distributed system of facilities controlled by the Defense Ministry. However, its structure now does not fully meet the requirements of effective use of the entire range of dual application space systems, while the relevant hardware components that make up the base of computer systems for information processing, storage and transmission are obsolescent and need in-depth modernization which fact must be borne in mind while creating dual application systems.

The fastest solution of the above described problem calls for perfection of the ground-based space infrastructure with due regard to the tasks of dual application of space systems based on the development of its optimal adaptive structure and proposals for step-by-step development; creation of a ground-based infrastructure of highly effective means of launching SC;

determination of composition and specifications of equipment used to receive, process and distribute information, development of such equipment and its commissioning, including systems like mobile and small size terminals for information reception and processing, SC mobile control systems; refinement and introduction of methods and equipment of onboard and ground-based information processing based on modern information technologies; introduction of advanced technologies for controlling SC with the aid of space relay and navigation systems.

In issues under consideration particular attention is being given to the problem of developing multi-role SC of dual application space systems. The essence of the problem lies in the different (often very much so) specifications requirements imposed on the SC onboard equipment used to fulfill various tasks. At the same time, for obvious reasons it is economically unfeasible to install on board SC the equipment used to address the entire spectrum of space system's dual application tasks because of the various required number of SC in the structure of the system for resolving various groups of tasks. The problem can probably be resolved by developing and creating SC based on standardized orbital platforms outfitted with modular onboard equipment whose performance assures its high operational readiness for launch.

The next problem is associated with the development of optimal structures of dual application space systems. The crux of the problem lies in the difference of optimal structures of space systems synthesized for monopolistic solution of each individual group of tasks being resolved by the dual application space system. Considering this, there are two ways of resolving the problem. The first way envisions synthesizing the structure of a dual application system based on the solution of a multi-criterion task eventually resulting in a structure which is optimal "on the average" for resolving the entire spectrum of tasks with due regard to their significance. The second way envisions creation of a space system having an adaptive structure and providing the most effective fulfillment of urgent tasks during the time period in question as well as a mechanism for adapting the structure to changing conditions. The authors believe that the second way is preferable. Its realization however entails a number of extra technical problems that have been described above (those problems are primarily associated with the creation of mobile systems for information reception and processing, with creation of SC control and high-performance launch systems, with the reduction of time required to prepare SC for operation).

# PART 3
# The Ground-Based Infrastructure

## 3.1 Russia's spaceports, state and prospects

### 3.1.1 The history of spaceports' construction

A spaceport is a territory provided with engineer works on which functionally interconnected structures and technical facilities are located that accept space rocketry components from manufacturers, prepare launch vehicles and spacecraft for take-off and launch them into space. In case of using reusable launch vehicles, the spaceport can be supplied with repair and maintenance facilities for giving that hardware an after-flight service.

Spaceports' main technological facilities are: take-off and engineering complexes, refuel and neutralization stations, storage facilities for various applications, landing complexes for reusable launch vehicles, areas allotted to jettisoned hardware components, command and instrumentation complexes. In addition, spaceports have a number of ancillary facilities, such as rocket fuel components production plants, aerodromes, rail and motor road approaches, engineer communication networks, computing center, residential areas with consumer establishments and every day care centers. The typical diagram of a spaceport is shown in Fig. 1.

The history of spaceports goes back to the creation of small take-off platforms for launching experimental rockets.

The building of spaceports in the USSR started with a very simple launch facility at a small test site near the village of Nakhabino in the neighborhood of Moscow. Here, on August 17, 1933, Russia's first liquid propellant rocket, GIRD-09, designed by Mikhail Klavdiyevich Tikhonravov, was launched under the supervision of Sergei Pavlovich Korolev.

In the US the history of spaceports began with developments of professor Robert Goddard. About 10 years after his book "A Method of Reaching Extreme Altitudes" had been published (1919) Goddard launched several liquid propellant rockets. Among them was the Nell rocket with take-off mass above 38 kg that rose up over the New Mexico desert to an altitude of 2,300 m at near sonic speed. Contrary to Soviet and German researchers, Goddard worked all on his own. He had no opportunity to gather a team of gifted people who could later make up the pivot of further space research. During his lifetime (R. Goddard died in 1945) the military turned to his talent only once when they needed to develop rocket boosters for aeroplanes.

In Germany of the very early 1930s the situation in space rocket engineering was about the same as in the USSR. In 1930, the "The Space Bible" of Oberth was seven years old. That work contributed largely to the establishment in 1927 in Breslau of the German Interplanetary Flights Society. The rocket motors had advanced so far a rocket driven cart was tested in Russelheim.

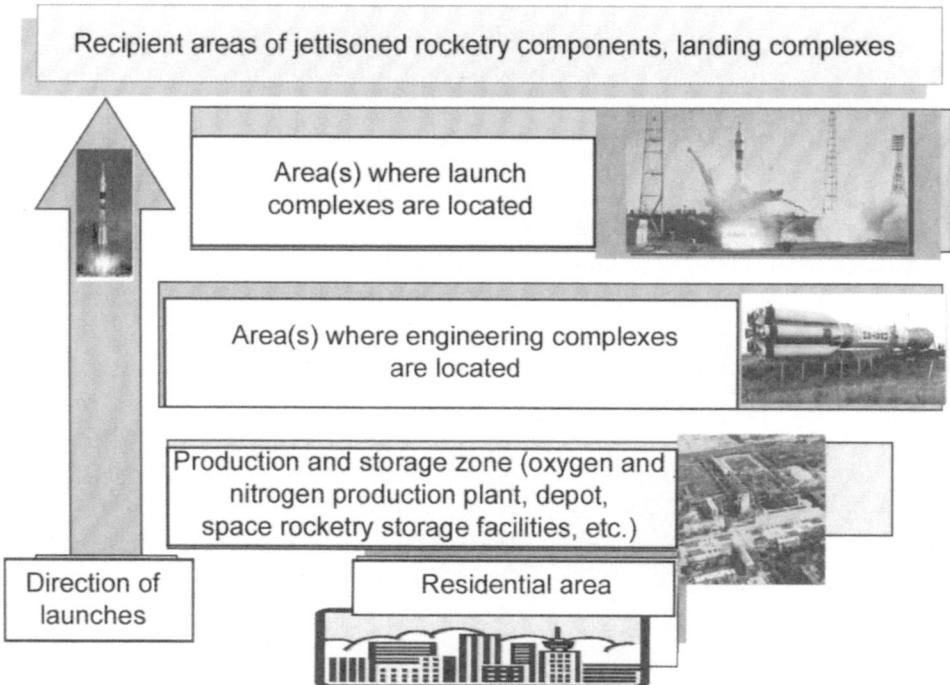

Recipient areas of jettisoned rocketry components, landing complexes

Area(s) where launch complexes are located

Area(s) where engineering complexes are located

Production and storage zone (oxygen and nitrogen production plant, depot, space rocketry storage facilities, etc.)

Direction of launches

Residential area

**Fig. 1.** Typical spaceport diagram

In 1930, the Society of Interplanetary Flights conducted tests at the airdrome of Kummersdorf near Berlin. Among those attending the event was Werner von Braun, a student of the Berlin University, and captain Walther Dornberger who had assured the German Army Command to donate 5,000 marks (around US$ 1,200) for carrying out further research and investigations. The German team of rocketry enthusiasts had financial difficulties, but was well organized and, most importantly, was well versed in testing different types of motors. As Nazis came to power in Germany, information about the work of this team began to disappear from mass media. The several take-off complexes of different types called for allotment of considerable areas to test sites and enlistment of numerous personnel to service launch complexes, engineer communications and a number of other facilities on test sites. Starting 1935, the German rocketry engineers had been working under the auspices of Wehrmacht and their work became strictly classified.

In early August, 1936, German Air Force general Kesselring and Army colonel Berker ordered to allocate an allotment on the Usedom Island near the Baltic Sea coast for building on it Penemuende military research testing site. Here, in a small town on a sea coast, missiles were fabricated that the third Reich's propaganda minister christened Vergeltungswaffe, or "retaliation weapon".

Just as with the Manhattan project that aimed to build an atomic bomb in the US and which was supervised by general Groves and research physicist Oppenheimer, the Penemuende project was also supervised by two officers: military in the person of general Walther Dornberger and scientific in the person of doctor Wernher von Braun. The Penemuende project was rigidly controlled by the Reich's top officers. "I was amazed by those people of technology, by their fantasy and mathematical romanticism. Paradoxically, those callow young men were allowed to realize ideas that seemed absolutely utopian", recalled later the third Reich's armaments minister Speer.

The first long-range ballistic liquid propellant rocket, A-4, was designed, constructed and successfully tested in the static mode on 18 April 1938, seventeen months before World War Two broke out. On 7 July 1943 the first successful trials of this rocket took place. The launches were filmed and shown to Hitler at his quarters in Eastern Prussia. Hitler responded instantly "This is not a missile. This is a means of winning the war. What an encouragement to the front it will be once we strike a missile blow against England!" The visit of Penemuende top managers to Hitler had two aftereffects: baron von Braun was awarded the title of professor and the rank of SS obersturmbannfuehrer; Dora-Mittelbau concentration camp was set up in the town of Nordhausen. Forty thousand POW at Dora were building first an underground facility near Nordhausen, and then worked in its underground sections.

The V-2 rocket was the largest of rockets known by then. Technically it could have been made still larger, but the designers deliberately confined it to the length of 14 m and diameter of 1.65 m thus making it transportable by a trailer over highways and dirt roads of Western Europe. Chosen for fuel components were 75% ethyl alcohol and liquid oxygen. At sea level the rocket engine had the thrust of 25.4 t f (take-off mass of the rocket was 13 tons). After the rocket reached at the $66^{th}$ second of its flight the speed of 1,600 m/s, its engine went out of action and the rocket fitted with a warhead weighing around 1,000 kg reached the maximum altitude of around 90 km and hit a target at a range of 270 km. Even though the first examples of V-2 rockets were not very reliable, eventually they were produced in large numbers. According to W.von Braun, one plant in Nordhausen alone turned out up to 300 rockets per month.

Rockets were launched as part of development flight tests both from the test sites in Penemuende and in Debice (Poland).

Beginning in 1950s the highly industrialized countries like USSR and USA started building spaceports for space research and practical uses. A few years subsequently the world witnessed the greatest of achievements – the flight of the first artificial Earth satellite followed by flights of spacecraft to other planets and then the first flight of man into space and landing on the Moon's surface.

The first spaceport in the USSR was the test site at Kapustin Yar.

For carrying out flight tests of long range ballistic missiles a decision was taken to build the State Central Test Site (SCTS) under control of the USSR Defense Ministry. In choosing the locality for the SCTS a number of factors were taken into account: the availability of transport approach routes to deliver rockets, equipment and fuel components for tests; fairly close location to an industrial center for using its industrial potential, availability of free areas for locating on them the test site proper, engineering and launch facilities, recipient territories for falling rockets or their jettisoned parts. On the basis of these basic requirements the State Central Test Site (SCTS) of the USSR Defense Ministry was located about 100 km south-east of Stalingrad, near the village of Kapustin Yar, Astrakhan region. The commissioning of the Site was scheduled for August 1947. Lieutenant-general V.I.Voznyuk was appointed its first chief.

The SCTS was built gradually. First a rocket firing test bench was built along with engineering and launch complexes, a mess, approach roads, power and gas supply lines. The bench was built in order to conduct rocket motor firing tests and to timely prepare rockets for launches. The first firing test of the A-4 rocket on the bench was conducted October 16, 1947.

Initially, the engineering complex was just a few wooden barracks in which horizontal tests of rockets were conducted. The massive permanent building for assembly and testing with all its laboratories and services was erected later.

The launch complex was built at a distance of 4-5 km from the engineering complex and was a concrete platform fitted with support equipment and facilities. Installed on it was a take-off pad with a weigh-in device, banked-off structures and caponiers in which stood various test equipment mounted on truck chassis. Rocket maintenance and fuel components refueling equipment were in the immediate vicinity of the launch channel and were removed to a safe place before launch.

As time went by, residential buildings, hotels, canteens, shops and other buildings were erected at the Test Site. The newly built launch complex was moved 30 km away from the engineering complex, a bunker was built from which control was exercised of pre-launch tests and rocket launches proper. Highways approach branches were built.

Located in the vicinity of the launch complex were structures accommodating cine-theodolitic measuring instrumentation, communications center, single time service, computing bureau and others. Placed in the near-the-launch area were ground-based stations for receipt of telemetric information from the rocket.

The first launch of the A-4 rocket in the USSR was performed October 18, 1947, at 10:47 Moscow time.

The test site allowed to successfully conduct development flights of carrier rockets until their range was within 1,000-1,500 km. The debris recipient areas in this case were uninhabited deserts. The territories on which rocket heads were falling also met this requirement (they moved gradually eastward till they reached lake Balkhash). The operation of the flight lateral control radio systems was not impeded by any formations on Earth's surface (mountains, hills, etc.); ground-based instrumentation posts functioned reliably throughout the entire flight route receiving telemetric information about the state of on-board equipment of rockets flying along the trajectory. In case of rockets' abortive launches the damage was usually confined to a single launch site. The size of the entire test complex allowed to simultaneously conduct launches of several types of ballistic missiles not only for purposes of the Defense Ministry but also for academic studies.

The situation changed drastically once work began on a radically new multi-stage ICBM, R-7. The old testing base of the site appeared too small for the new rocket. Its range exceeded 8,000 km, the flight trajectory passed over nearly all Asian part of the USSR in the eastward direction. The flight tests of the R-7 asked for allocation of new territories to receive the jettisoned rocket stages, for creation of new instrumentation posts, selection of suitable areas for ground-based rocket flight radio control centers, equipment of areas in the eastern parts of the country (Kamchatka peninsula and areas of the Pacific Ocean) upon which rocket forebodies could fall. Also, it was necessary to develop a transportation system to haul the R-7 rockets  some of whose component units were noted for their large size, to the launch site  (suffice it to say that each of the four side units forming the first stage of the R-7 was comparable in size with the previously created first generation rockets while the central block considerably surpassed them). A new test site was needed to conduct development test flights of the new generation rockets. To resolve this problem, a special commission was set up in the early 1950s, basic requirements were formulated with reference to the new test site, its desired characteristics were specified and the preferred areas in the country where it could be built were pointed out.

The commission started to reconnoiter on locality and to study in depth the possible variants of placing ("sitting" as per wording of the commission members) of the new test site. Three variants had been considered with particular care.

The first variant – in Mordovia where during the war years large wooded areas had been cleared, the population was rather scarce and approach roads in fairly good repair. The detailed study, however, showed that that variant failed to meet quite a few demands made of the would-be test site.

The commission looked then to the second variant. That was the Caspian Sea coast (the territory of Astrakhan region and Dagestan). It appeared, how-ever, that in case of locating ICBM launch complexes in that area insurmountable difficulties would emerge in placing there R-7 rocket flight radio control centers. The local hills and mountains prevent the radio beam of the ground-based control center from reaching the missile in some portions of its flight (primarily in the critical portion, i.e. the first tens of seconds after the missile clears the take-off pad).

If in choosing the variant it were known that missile flight control was just a temporary arrangement and subsequently all ballistic missiles would be using only autonomous control systems that have no need for ground-based radio control centers, that variant could have become quite attractive for the following reasons: the available transportation thoroughfares (railroads, maritime and aerial routes), relatively lenient climate not debarring people's work and habitation, the proximity of the Volga river, a nearly unlimited source of industrial and drinking water, the flight path passing over deserts and semi-deserts creating no problem with the falling rocket forebodies and combat parts. Those reasons being unknown back then, the above variant had also to be rejected.

The third variant, Kazakhstan with its area stretching from the Aral Sea to the town of Kzyl-Orda, proved suitable for fixating the test site. It was taken into consideration that near the station of Tyuratam a narrow gauge rail track had been preserved that led to a small quarry located 30 km away from the station in the steppe. The big merit of the variant was that the Moscow – Tashkent railroad was passing through Tyuratam, while the neighboring Syrdar'ya river offered a source of water for construction, for technological processes in rockets tests and launches, for needs of the would-be residential areas. In addition, that territory was practically free from any structures or settlements. It was nearest to the equator which was an advantage in launching rockets towards the east (maximum utilization of the Earth's natural revolution for boosting the launched rocket which is very advantageous in terms of energy saving). Nor were there any difficulties with areas which received jettisoned rocket stages or on which centers were located to maintain radio control, to receive telemetric information, to control the rocket flight trajectory with the aid of electronic equipment. The package of these factors determined the final choice. On 12 February, 1955 a decision was taken by the USSR Council of Ministers authorizing the construction of a new site for development flight tests and launches of artificial earth satellites in the neighborhood of Tyuratam railroad station, Kzyl-Orda region of the Kazakh Republic of the Soviet Union.

Artillery lieutenant-general A.I. Nesterenko was appointed Chief of the test site, colonel A.S. Butsky – Chief-of-Staff. The core management were A.A. Vasil'yev, A.I. Nosov, A.P. Metelkin and about 50 other officers with experience of test work at Kapustin Yar test site.

The designers, rocket engineers and construction workers envisioned the test site as a sophisticated complex of interrelated facilities designed to prepare and launch rockets with spacecraft. The original task was to launch a R-7 rocket with a thermonuclear warhead. Its size and weight were determined by A.A. Ilyushin, corresponding member of the USSR Academy of

Sciences. The initial development of the combat intercontinental missile, R-7, was based on those parameters.

The test site needed a ground-based launch facility with an underground command post and auxiliary services for launching the R-7 type of rockets. This implied removing more than a million cubic meters of earth. Close to it, a rocket motor assembly and test complex was to be built resembling a huge factory shop for putting together rockets and testing them in a horizontal mode. A specially clean building was required for pre-launch preparation of space hardware – spacecraft, spaceships and interplanetary stations. Also, structures were needed for refueling spacecraft. It was necessary to build assembly and test buildings for preparation of spacecraft, fuelling stations for refueling spacecraft and booster units, storage facilities for rocket fuel components, oxygen production plants for production of liquid oxygen (one of the rocket fuel's main components), nitrogen, helium (subsequently, liquid hydrogen production plants also appeared), railroad branch lines to deliver rockets, cargoes, fueling containers to the launch site and much else which included logistic support and maintenance of a military facility.

Thus the building of the test site was growing into a package of problems that called for work of hundreds of various specialists, a strenuous effort on the part of construction workers, fitters of technological systems, geodesists, test engineers, military and civil experts.

The job to be done was to erect the structures of the site's command and instrumentation posts, to outfit them with equipment that controls rockets and spacecraft flights, monitors the operation of onboard systems installed on them. All these structures and the equipment installed in them needed a power supply system, electric power transmission and engineering networks, wire- and radio-aided communication systems and much more.

A powerful computing center couldn't be dispensed with, either. And, sure enough, a test site was unthinkable without a modern town for thousands of people to live in, a town with a developed network of motor roads, up-to-date airdrome and much else which is required for normal life and work of a large group of specialists.

The construction of the Defense Ministry's Scientific Research Site number 5 (the Russian acronym NIIP-5 MO) began in 1955 in uninhabited sandy terrains featuring an adverse climate (temperature in summer reaches 40°C in the shade with frequent sandstorms and dry winds, dropping occasionally to –30°C in winter with heavy snowstorms). In 1957 the site ushered in the space era.

The military builders of the site worked under the supervision of general G.M. Shubnikov, an experienced specialist and war veteran. The first detachment of builders headed by lieutenant I.N. Denezhkin arrived at Tyuratam station to prepare the area for receipt of groups of military construction workers as early back as January 12, 1955. In June 1955, a directive of the

General Staff specified the personnel arrangements and organizational structure of the site; in August 1960 by the order of the Defense Minister of the USSR the 2$^{nd}$ of June 1955 was defined as the site's foundation date which is marked annually as a holiday of troops and units serving the NIIP-5 MO.

It won't be too much to say that the builders of the site and its specialists had committed acts of heroism. As early as the beginning of 1957 the basic structure had been built which enabled to start flight test of the world's first ICBM, R-7, that became later a peaceful carrier which largely contributed to the success story of Russia's astronautics, namely, the launch of the world's first artificial satellite on October 4, 1957, the flight of the first man, Soviet citizen Yu.A.Gagarin, in space round the Earth on 12 April, 1961, launches of the first interplanetary stations and much more that made a proud page in the history of the Soviet and the world's astronautics.

Speaking in 1961 at a meeting of Baikonur construction workers soon after Yu. A.Gagarin's triumph, S.P. Korolev said: "I was sure the military builders would not let us down. But I dared not hope that so much could be built so fast and so well. Many thanks to you, dear comrades!" Those words express the appreciation of work of builders and those who accepted the site for operation.

Later on, in addition to the R-7 rocket the test site tested other types of S.P. Korolev's rockets, N-1 lunar carrier rocket, as well as rockets built by other design offices with other characteristics: combat ICBM by M.K.Yangel, the Proton heavy carrier by V.N.Chelomei, combat silo-based capsulated ICBM by the same author, Energia – Buran reusable space transportation system that emerged ahead of its time, and many others. Ground- and silo-based launch complexes had been built for conducting launches of various types and designs, the capabilities of the test base were permanently improved etc.

The rapid development of astronautics in the 1960s essentially increased the launching rate of carrier rockets. In 1963 a decision was made to build a spaceport in the North of the country near the settlement of Plesetsk, Arkhangelsk region. In the end of 1967 the first carrier rocket was launched from that spaceport.

To obtain an independent access to space after the breakup of the Soviet Union, Russia built the spaceport of Svobodny in the Amur region. Currently, it is used to launch space rockets being created on the basis of interplanetary ballistic missiles.

Since the 1950s the USA too has been actively engaged in work on the building of spaceports for purposes of the DoD and for scientific research. Eastern test site and J. Kennedy Space Center are located actually on the same territory, but have independent technological complexes for fulfilling tasks for the Air Force and NASA. In the state of California, on the Pacific coast, the US Western test site is located which launched the first space rocket back in February 1959.

France became third among the leaders in space research. It had built by November 1965 its Hammagir spaceport on the territory of Algeria. China, Japan, India, Israel too have spaceports of their own. Separate rocket launches from test sites can be attempted by Brazil, North Korea and some other countries.

- Spaceports were built with consideration for the following requirements:
- provision of the required set of trajectories for launching space rockets;
- provision of recipient areas for jettisoned hardware components;
- remoteness from densely populated areas;
- lenient climatic and seismological conditions;
- non-passage of active portions of space rocketry trajectory over densely populated areas and other countries' territories;
- possibility of using the existing transport communication lines;
- relative proximity to space industry facilities.

As it happens, all the world's spaceports are located on oceanic coasts. Only China and Russia have built spaceports inside the continent.

### 3.1.2 Russia's spaceports' ground infrastructure today

Baikonur spaceport was built in 1955 as a test site of space rocketry. The spaceport is used to prepare and launch carrier rockets of light, medium and heavy classes. Russia's international cooperation in space was based largely on the use of capabilities afforded by Baikonur spaceport. Baikonur accounts for more than 50% of SC launches, including all launches into GEO and launches of heavy craft. Only Baikonur has two launch complexes for the Proton heavy class carrier rockets.

The total area of the basic and auxiliary facilities of the spaceport was 6,717 km². According to an agreement between Russia and Kazakhstan the space port continues to be currently used on a leasing basis. The spaceport borders have been revised, the recipient territory for jettisoned hardware components has been reduced.

The location of basic facilities at the Baikonur spaceport is shown in Fig. 2. The spaceport comprises the centers for testing and studying the application and control of space hardware. Those include engineering and testing facilities, instrumentation posts and maintenance and support services. All in all, the spaceport has:

- 11 assembly and test shops housing engineering complexes for pre-launch preparation of carrier rockets, booster units and spacecraft as well as refueling and neutralization stations for refueling booster units and spacecraft, which stations are functionally integrated into the engineering complex;
- 9 launch complexes (15 launch installations) for launching Soyuz, Cyclone-2, Proton, Energia, Zenit, Rockot carrier rockets;
- instrumentation complex to monitor and control carrier rockets in flight;
- launch installations for testing ICBM.

In building the instrumentation complexes of the spaceport taken as a basis was the principle of centralized collection and processing of instrumentation related information, in which case all functions of processing telemetric information are concentrated in one point, the site's computing center. The efficiency of such a concept of the site's instrumentation complex was proved while drawing up the preliminary design of the Buran reusable space system. The concept was realized in preparation of the instrumentation complex for flight tests.

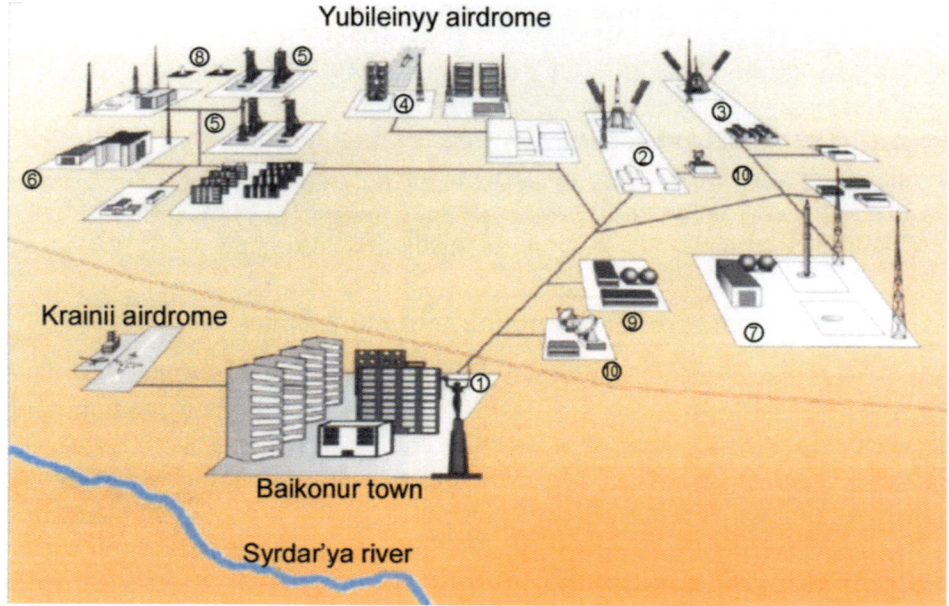

**Fig. 2.** Location of principal facilities at the Baikonur spaceport

    1 – residential area;
    2 – Soyuz carrier engineering and launch complex;
    3 – Soyuz and Molniya carriers' engineering and launch complex;
    4 – Energia and Buran reusable space system's engineering and launch complex;
5 ... 8 – engineering and launch complexes of Proton, Cyclone-2, Zenit and Rockot carriers respectively;
    9 – oxygen and nitrogen production plant;
    10 – instrumentation post

In course of that work the trajectory instrumentation posts were outfitted with the Svyaznik (communicator) satellite communication stations and operative selection equipment registered by radio electronic means of telemetric information which made it possible to abandon using on them data processing complexes with relevant logistic support and extra personnel. This simplified the technical structure of the trajectory instrumentation posts, equipping them with only means of instrumentation data reception, registration and transmission.

The analysis of the state of ground complexes of the Baikonur spaceport shows that by now they have nearly exhausted their service life.

The service life of the launch complex of the Cyclone-2 carrier put in operation in 1967-1968 is now practically expended. The general overhaul of the launch complex for launching R-7A type of carriers was scheduled for 2000-2005.

The ground complex of the Zenit carrier, remarkable for its high automation and short carrier launch preparation time went into operation in 1988. One launch installation of this launch complex, installation damaged during the accident while launching the Zenit carrier on 4 October 1990, requires a major repair.

The Baikonur spaceport has two launch complexes for the Proton carrier, two launch installations on each complex. The ground complexes of the Proton carrier have expended their guaranteed service life. Actually, only one launch installation is fully operable now. Others are either in repair or in preparation for it.

For preparation of the Proton carrier booster units the Baikonur spaceport uses two engineering complexes that are in operable condition and prepare the existing booster units for the Proton carrier. Currently, Energia space rocket corporation prepares booster units on the engineering complex of the Buran reusable space system.

Only two refueling stations can be used at the Baikonur spaceport to refuel booster units and spacecraft. They are intended for receipt, storage, preparation and refueling (drain) of fuel components and compressed gasses.

The infrastructure, the social facilities and communal service establishments of the spaceport are also rather worn.

The construction of **the Plesetsk spaceport** began in 1963 as a space rocketry test site (the official status of the spaceport was granted in November 1994). The facilities of the Plesetsk spaceport can be divided as per their application as follows: launch and engineering complexes, refueling and neutralization stations, oxygen and nitrogen production plant, other technological facilities, including site instrumentation and communication complex plus the town of Mirnyy, engineering and storage support facilities, airdrome, construction workers' base.

The placement of the principal facilities at the Plesetsk spaceport is shown in Fig. 3. The Plesetsk spaceport has considerably contributed to the realization of the space program. Around 30% of spacecraft fulfilling tasks of the Federal Space Program are launched from here. Launched from the Plesetsk spaceport are carriers of light and medium classes. The spaceport launches SC for military, research and business purposes as well as part of an international space effort. Eight launch complexes are now located at the spaceport. Most of ground complexes are in operating mode and function as specified by the launch program. The spaceport's ground facilities are used to prepare launches and to launch four types of carriers: Cosmos, Cyclone, Soyuz (union), Molniya (lightening).

The Cosmos carrier launch complex with two take-off pads was built in 1967 under the project of the Design Bureau of Transport Machinery.

One launch installation is now mothballed. This is explained by the fact that insufficient funding has precluded major overhaul of the launch installation, technical systems and building structures of the entire launch complex.

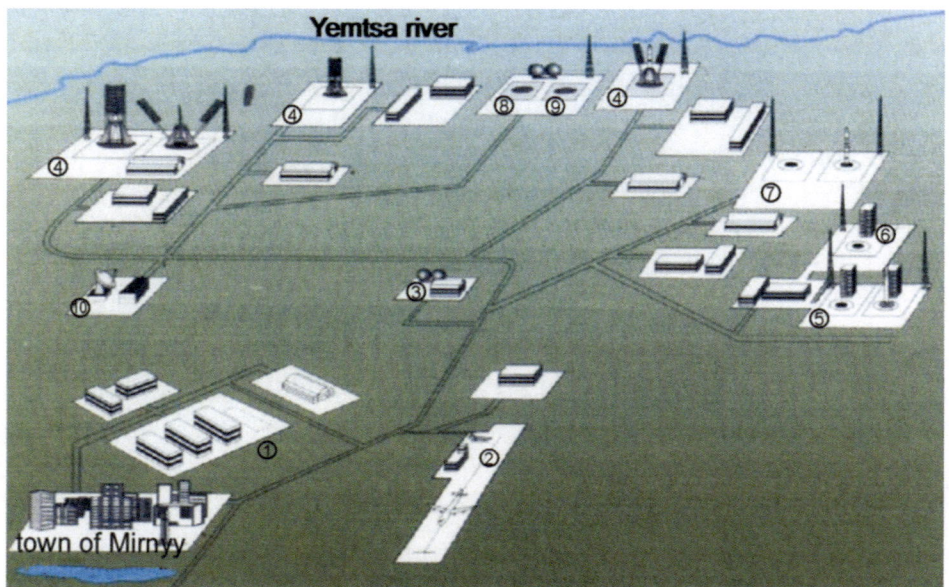

**Fig. 3.** Location of the principal facilities at the Plesetsk spaceport

1 – administration and utilities management sector, Cosmos carrier engineering complex;
2 – Plesetsk airdrome;
3 – oxygen and nitrogen production plant;
4 – Soyuz-U and Molniya-M engineering and launch complex;
5 – Cosmos-3M carrier launch complex;
6 – Rockot carrier launch complex;
7 – Cyclone-3 carrier's engineering and launch complex;
8,9 – Angara carrier multi-purpose launch and engineering complex;
10 – instrumentation post.

The engineering complex of the Cosmos carrier was built in 1967 simultaneously with the launch complex. Until now no repair has been done on any of the building structures or engineering systems. They are in an unsatisfactory condition from overuse and need a major overhaul.

The Cyclone carrier's launch complex has been developed by the Design Bureau of Transport Machinery on the principle of "no-people launch". The launch complex has organized for the first time the carrier preparation and launch procedure control under a single program in a fully automatic mode, starting from the delivery of carrier to the take-off pad. The automation of the carrier preparation procedure reaches 100%, being no less than 80% for the launch complex as a whole. In terms of performance this complex surpasses all its foreign counterparts known to date.

The Cyclone carrier launch complex contains two launch installations. Its engineering complex has two work areas to prepare carriers.

The engineering complex comprises installation and test shop with two work areas and the systems servicing shop. Detected now on the engineering complex are the same faults as on the launch complex, equipment wear, depletion of kits of spares and accessories, structural defects in buildings, partial disablement of the ventilation, heating, water supply systems, personnel's work conditions defy sanitary standards. In this context it has been proposed to carry out repair and take preventive measures on the engineering complex in order to ensure the operability of the space rocket complex as a whole.

Over the time since the launch and engineering complexes began to operate, the technological equipment, building structures and technical systems have expended their designed guaranteed dates and life cycles. Now in operation is only one launch installation, the other one being mothballed.

Considering the actual condition of the equipment and ground-based technological systems, buildings, structures and engineering complexes of the Cyclone carrier it can be concluded that a major overhaul is needed in order to prolong the service life of hardware and ensure its further operation.

The launch complexes of the Soyuz and Molniya carriers were originally built for launching R-7A combat missiles. Later on they were re-equipped to launch space rockets. While in use, those launch complexes had undergone periodic overhauls, reconstructions and repairs after accidents and major failures.

This work as well as a package of restorative and preventive measures plus maintenance of those launch complexes prolonged their service life and ensured their further operation.

The launch complex of the Soyuz and Molniya carriers launches now space rockets for research, business and national defense.

The service life of one of the launchers that had taken an overhaul expired in 1983, after which it was periodically repaired and given preventive maintenance. This extended its service life. Beginning 1989 the launch complex has been used for training and instruction.

In 1986, construction of a ground complex for the Zenit carrier rocket began at the Plesetsk spaceport. The commissioning date of the first phase of launch complex and technical station was repeatedly adjourned. The launch complex of the Zenit carrier was built as a new facility. Work is now under way aimed to build on the basis of the Zenit ground complex a new multipurpose ground complex for launching the Angara carrier.

The multi-purpose launch complex (MPLC) is intended to receive, to prepare for and send into flight space rockets of the Angara family (Fig. 4). It is a
group of equipped structures of the unfinished construction of the Zenit space
rocket complex at the Plesetsk spaceport combined with launch complexes
being built anew. Plans are made to build a second launch installation in the
framework of the MPLC.

The MPLC comprises:
• building structures (launch installation, command post, nuclear fuel components storage facility, compressor and pump stations, fabricating facilities, industrial effluent collection facilities, neutralization station, thermostating station etc.);
• technological equipment;
• engineering systems;
• auxiliary equipment;
• automated MPLC control system;
• set of checking equipment for carrier rockets, booster units, spacecraft.

**Fig. 4.** Construction of the multi-purpose launch complex for the Angara space systems

For the first time in the world's and national spaceports construction practice a multi-purpose launch complex is being built that provides for launching
light, medium and heavy class carrier.

The Svobodny spaceport was built in 1993. The official status of the
spaceport was granted in 1996. The area assigned to the spaceport's principal
facilities lies in Svobodnenskii district of the Amur region, Khabarovsk

Administrative Territory. The existing infrastructure has been built for operations of a division of the Strategic Rocket Forces. It includes:
- silo-based launch installations and command posts of rocket regiments;
- technical rocket base with its building structures and specialist equipment;
- repair and maintenance base with building structures, general and special purpose equipment;
- division head quarters, main and stand-by division command posts, communication centers with communication and combat control equipment, with auxiliary technological equipment;
- residential area with housing complexes, barracks, utilities management offices, storage facilities and communal services establishments.
- motor vehicle pool, railroad depot and helicopter landing facility with automotive vehicles, locomotives, car fleet, helicopters, approach roads, structures, helicopter support services, auxiliary equipment and systems;
- networks, systems and objects of heat supply and water supply, sewage and electric power supply (including transformer substations, electric trains, diesel power plants) located in the neighborhood of the division.

The spaceport is located on a well developed area with railroad and motorway networks and engineer communication lines. The location of the principal facilities at the Svobodny spaceport is shown in Fig. 5.

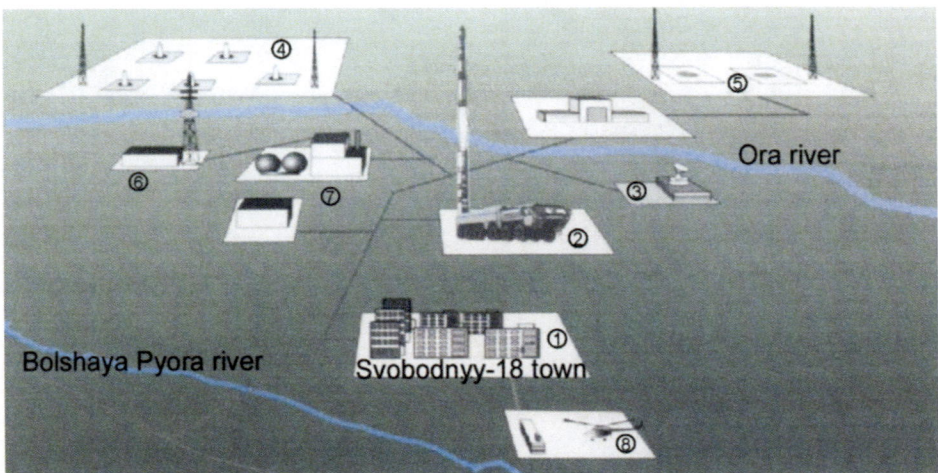

**Fig. 5.** Location of the principal facilities at the Svobodny spaceport

    1 – residential area,
    2 – Start and Start-1 engineering and launch complex;
    3 – instrumentation post;
    4 – Rockot carrier launch complex;
    5 – Angara carrier multi-purpose launch and engineering complexes (project);
    6 – radio transmission center;
    7 – oxygen and nitrogen production plant;
    8 – helicopter landing site

The existing town of 5,000 people is built up with permanent housing complexes having all modern conveniences.

The ground infrastructure of a rocket force division is expected to be used for building space rocket systems derived from rockets reworked to launch light class spacecraft. Silo-based launch installations are intended to be used for this purpose. The area began to be used in 1974 with a guaranteed period of 15 years. In 1988, after revision and renovation work the guarantee was extended for another 5 years. The condition of the silo installations, structures and engineer systems is satisfactory.

The examination of the system of internal and external power supply of launch positions has shown that it is in working order, though some of its components need to be replced and re-equipped with new systems of automatic re-switching and protection. Underground high voltage cable lines also need to be replaced.

The engineering structures built in the 1960s and 1970s are now in a satisfactory condition. A revision carried out found it appropriate and feasible to place the rocket engineering base of the complex on the available platforms.

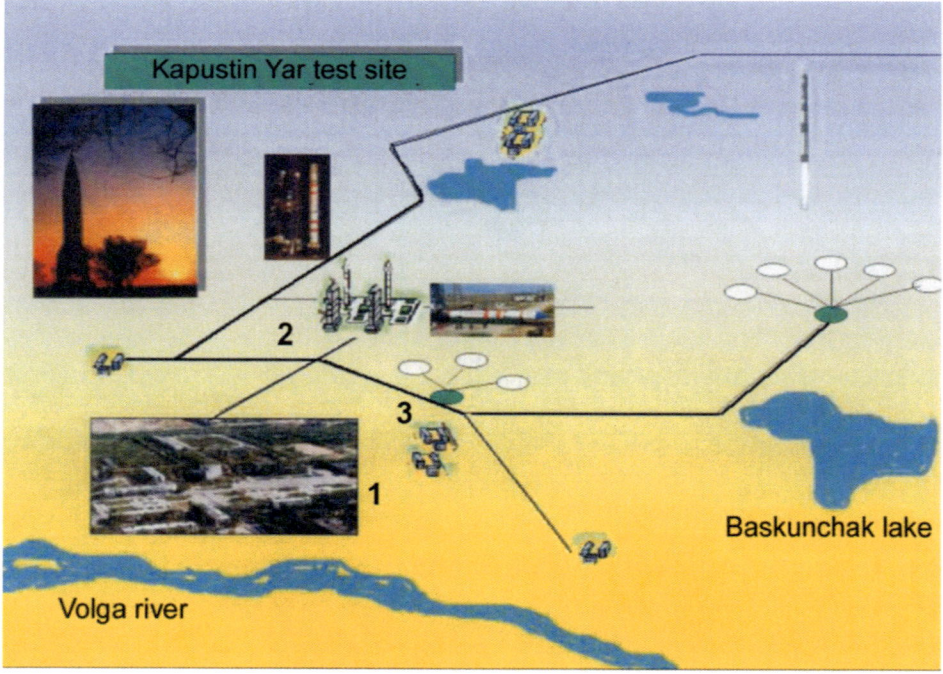

**Fig. 6.** Location of the principal facilities at the Kapustin Yar test site:

    1 – residential area;
    2 – Cosmos carrier launch and engineering complexes;
    3 – instrumentation post

The 4th State Central Test Site of the Russian Federation Defense Ministry (Kapustin Yar) was built in the neighborhood of Kapustin Yar village of Astrakhan region. From the first days of its existence the site has been the base for carrying out flight and on-the-ground tests of Russia's first ballistic missiles of all types. Also, it is used to train crews and has become a powerful scientific research and test center. The location of principal facilities at the Kapustin Yar site is shown in Fig. 6.

The first successful launch of a guided missile, R-1, manufactured by Russia was performed at the Kapustin Yar site on 10 October 1948. Later on, flight tests of the second series of R-1 rocket, R-2, R-11, R-11M were carried out. Their acceptance for service played an important role in arming the Armed Forces with a new type of nuclear weapon.

In addition to their principal application the ballistic missiles were also widely used for research purposes: execution of high altitude experimental launches with animals on board and their return to earth, upper-air sounding, launching of aerodynamic models and research equipment, conduct of other scientific experiments.

Later on, launch complexes for Cosmos carrier rockets were created at the Kapustin Yar site. Those were supposed to launch Cosmos and InterCosmos type of spacecraft as well as vertical take-off geophysical research rockets. The spaceport has been used to launch Vertikal geophysical rockets, satellites of the Cosmos and InterCosmos series, the Indian Aryabhata and Bhaskara, French Sneg-3 satellites.

### 3.1.3 The main trends in development of Russia's spaceports

The existing launch system comprises space rockets of light, medium and heavy classes based at Russia's spaceport Plesetsk and the spaceport of Baikonur located on the territory of Kazakhstan. The locations of Russia's spaceports are shown in Fig. 7.

The bringing of space infrastructure facilities under the jurisdiction of the former Soviet Union's republics raised a number of problems before Russia:
- ensuring independence in space work, primarily in military applications;
- adequate distribution of supplies and means among the existing facilities of the space infrastructure for sustaining their operability;
- redistribution of the right of ownership of the space infrastructure facilities;
- defining of rational ways to develop spaceports (the main means of solving this problem in the conditions of the transitory period is to politically and juridicially provide Russia with a possibility to use on a negotiated basis the space facilities located on the territories of the former Soviet Union's republics);
- creation in the longer term of Russia's spaceport infrastructure able to satisfy fully all Russia's needs.

Fulfilling such complex, costly and labor-intensive tasks under the pressure of necessity to implement programs of high energy orbit launches that can only be performed from the Baikonur spaceport is unthinkable without a transition period that envisions Russia's participation in operating the Baikonur spaceport and simultaneous development of the Russian test site base. The transfer of research, business and commercial SC launches from Baikonur to Russia's spaceports must be primarily determined by economic feasibility. In doing so, account must be taken of the expenses related to the development of Baikonur's infrastructure, including the cost of repair and modernization of the Proton (Proton-M) complex, the development of required booster units as well as restrictions concerning the number of carrier rockets being launched.

**Fig. 7.** Location of Russia's spaceport

Most of space rocket complexes now in use have been built around combat ICBM complexes. Those are complexes of the Cosmos, Cyclone-2, Cyclone-3, Soyuz, Molniya and Proton carrier rockets. Such carriers, except Soyuz and Molniya, use high toxic fuels and hardware component elements that are currently not easy to manufacture because they are obsolete and their production facilities are located outside Russia. The ground complexes of most carriers have exhausted their service life and have been more than once overhauled.

In the mid 1970s a decision was taken to develop a standardized series of advanced carrier rockets including light, medium (Zenit), heavy and super-heavy class carriers. Before disintegration of the Soviet Union, most of them used to be launched from the Baikonur spaceport.

In the 1980s, this series of carriers was used to build a Zenit carrier rocket which went into service and was launched from the Baikonur spaceport. Also, work began on testing an Energia carrier rocket.

The geopolitical and economic changes that took place in Russia in the 1980s and '90s made it necessary to revise the previously formulated concept of development of launch systems.

Based on this, prospective work has been mapped out aimed to develop launch systems and funds have been made available for the purpose on a priority basis; the work to be done implies:

• transfer of payload launches from Baikonur to Plesetsk spaceport with expansion of its capability to launch SC in a wide range of inclinations and operating orbit altitudes, including geostationary;

• modernization of R-7A type of carriers that in addition to extending the service life of space rocket complexes enhance energy performance of carrier rockets and improve their ecological characteristics;

• completion of construction of the first phase of the engineering complex and launch complex for the Zenit carrier at the Plesetsk spaceport;

• modernization of space rocket complex (SRC) with the Proton carrier at the Baikonur spaceport as the only heavy class SRC that ensures delivery of SC to GEO;

• building heavy class SRC at the Plesetsk spaceport using Russia's industrial base.

Launches of SC are intended to be conducted at the Plesetsk spaceport as it is now Russia's only facility of its kind with a developed infrastructure.

After the above described measures aimed to develop launch systems are realized and the bulk of work under the space program is shifted as planned to Russia's territory, the Plesetsk spaceport may assume most of jobs for the Defense Ministry and a good share of them for research and business purposes. In this context the development of the Plesetsk spaceport is particularly important.

The development of the Plesetsk spaceport envisions:

• creation of a multi-purpose ground complex for launching the Angara carrier (Fig. 8) based on Zenit carrier rocket ground complex;

• creation of a work station for preparation and maintenance of SC transferred from Baikonur spaceport along with prospective ones;

• rework of the R-7A carrier's ground complexes to suit the Soyuz-2 carrier;

• completion of construction of oxygen and nitrogen production plant;

• reconstruction of refueling and neutralization stations;

• putting in operation the rocket fuels preparation and storage facility;

• additional equipment of the spaceport's instrumentation complex with systems required to launch carrier rockets and control spacecraft in orbit;

• reconstruction of the spaceport's communication facilities;

• extra equipment of launch trajectories and areas upon which jettisoned rocket hardware is dropped, expert examination of the ecological situation, certification and cleaning of spaceports and debris recipient areas;

• reconstruction of power-, heat- and water supply systems, environmental protection, protection of motor- and railroads and airdrome.

**Fig. 8.** Plesetsk spaceport. At issue is the building of the Angara launch complex (seen in picture A.N.Perminov, G.P.Biryukov, A.A.Medvedev)

The trends in development of the infrastructure used to prepare SC launches are formulated on the basis of tasks and targets of using space systems. The principal features in development and modernization of the ground infrastructure used to prepare SC for launches are the wide variety of space equipment used and uncertain dates of starting tests of new SC. In addition, it should be borne in mind that the deployment of new space complexes will proceed simultaneously with operation of SC built previously, i.e. the development of new ground-based equipment for preparing SC will go on not only via creation of new technical means, but also through modernization and reconstruction of the existing equipment of ground complexes.

In considering the versions of development and modernization of ground-based equipment for preparation of SC launches the following should be taken into account:

• the possibility of using the existing buildings and structures for deploying in them new engineering complexes of SC;

• the necessity to build permanent structures for new engineering complexes;

• the necessity to modernize, reconstruct or build new elements of ground infrastructure for preparation of SC;

• uncertainty as to location of some elements of ground infrastructure used to prepare SC in the spaceport's positional district;

- the feasibility of categorization of engineering complexes as per application (military, national economy, research) and (or) as per business profiles of the developer enterprises;
- the duration of operation and the technical condition of elements of infrastructure for preparation of SC;
- performance indicators showing the workload of workplaces and personnel employed in preparation of SC and carrier rockets.

The creation and development of infrastructure facilities of Russia's new spaceport Svobodny will cut down the workload of the Plesetsk spaceport. The spaceport of Svobodny is expected to conduct launches using SC carriers built around ICBM and the Angara carrier. However, the operation of the Angara heavy class SRC at the spaceport of Svobodny cannot begin before 2010. The ground complex of the Angara carrier will be created so that to fully utilize engineering decisions regarding the multi-purpose ground complex being created at the Plesetsk spaceport. The progress of work at the Svobodny spaceport will expand the residential area and increase the population by 25 to 30 thousand people. This in turn will intensify the construction of social and entertainment facilities and communal services establishments. The duration of building a multi-purpose ground complex at Svobodny spaceport is 7 to 10 years.

To refuel spacecraft and booster units for various applications, plans are made to install at the spaceport a multi-purpose fuelling and neutralization station developed by the Design Bureau of Transportation and Chemical Engineering. Such a station would offer a wide variety of space rocket fuels and compressed gasses as well as a possibility to prepare space fuels (thermo-stating, saturation, degassing) during refueling.

Due to the fact that the prospective carrier rockets use liquid oxygen as an oxidizer, it is necessary to build at Svobodny spaceport an oxygen and nitrogen production plant. The plant must be located as close as possible to the Angara multi-purpose ground complex, the principal consumer of its products. The duration of construction is 3 to 5 years.

In the first phase of building Svobodny spaceport it is possible to use part of the existing silo-based launch installations, work station structures, control and communication systems to ensure the functioning of the Rockot complex with minimum optimization. In the second phase a construction is expected of the Angara carrier ground-based complex. The experience of building spaceports and service facilities for the reusable space system Buran at Baikonur, in which case creation of the system was one of the state's top priority tasks makes it possible to conclude that with proper organization of construction, adequate supply of materials and sufficient funding the construction of a spaceport can be completed in 10 to 13 years.

The spaceports of Plesetsk and Svobodny allow to launch practically all types of payloads from Russia's territory and to put enough launch equipment in stand-by mode.

The spaceport of Baikonur has a high potential for optimum use of the ground complex for the Buran reusable space system. That complex can be regarded as a base complex for optimization and operation of the Angara and Zenit carriers and aerospace complexes.

The spaceport of Kapustin Yar can be regarded as Russia's prospective spaceport should there appear space rockets with reusable elements (first cruising stage and discard of disposable elements into the ocean). The proximity to space rocketry manufacturing zone and the developed transportation infrastructure will speed up the creation of space rocket complexes with reusable elements.

In building sea-based space rocket complexes the Far East regions look particularly promising. The network of seaports, the latitude at which the ports are located as well as the availability of first class airdromes provide for rapid delivery of space rocketry hardware to the space rocket launch site.

Thus the following phases in the development of spaceport structures are proposed.

1. **In lightweight class:** operation of the existing ground complexes of Cosmos, Cyclone-2 and Cyclone-3 carriers (till depletion of carrier reserve), re-equipping them to enable launches of Rockot, Strela, Start (Start-1) space carrier rockets and conduct of research and development for a prospective light class carrier.

**In medium class:** operation of the existing ground complexes of Soyuz, Molniya and Zenit carriers, beginning of modernization of type R-7A complexes to suit Soyuz-2 carriers, construction of the first phase of engineering and launch complexes of the Zenit carrier at Plesetsk spaceport for launching light class Angara carriers.

**In heavy class:** operation of the existing ground complexes of the Proton carrier (Proton with booster unit), completion of its modernization, beginning of creation of multi-purpose ground complex for launching the Angara series carriers of the heavy class based at Plesetsk spaceport (the spaceport of Svobodny is also planned to be used in the longer term).

2. **In light class:** operation of the Rockot carrier (Strela), Start (Start-1) at the spaceports of Plesetsk and Svobodny. The Cosmos carrier may continue to be operated in case it gets an upgrade. The commissioning of the first phase of the multi-purpose launch complex at Plesetsk spaceport.

**In medium class:** operation of the modernized Soyuz-2 carrier (Soyuz-2 with booster unit), optimization of the multi-purpose ground complex for launching Angara medium class carrier (Zenit carrier with booster unit so long as purchases of carrier from Ukraine continue).

**In heavy class:** operation of the modernized Proton-M carrier (Proton-M with booster unit), construction of the second phase of multi-purpose ground complex for heavy class carriers (Angara with booster unit) at the spaceport of Plesetsk.

3. **In lightweight:** class: completion of operation of the Rockot, Strela, Start (Start-1) carriers at the spaceports of Plesetsk and Svobodny. Completion of creation and beginning of operation of the prospective light class carrier at the spaceports of Plesetsk and Svobodny.

**In medium class:** operation of the modernized Soyuz-2 carrier (Soyuz-2 with booster unit), Zenit carrier (Zenit with booster unit so long as purchases of carrier rockets from Ukraine continue).

**In heavy class:** operation of the Angara heavy class carrier (Angara with booster unit) at Plesetsk spaceport, operation of a modernized Proton-M carrier (Proton-M with booster unit) before moving payloads to the Angara carrier rocket.

Creation of a ground complex for the Angara carrier rocket at Svobodny spaceport. The important task for Svobodny spaceport now is the use of the advanced Angara carrier with a reusable first stage. In case no areas are required to receive carriers' jettisoned components, there is much promise in building a ground complex for space rocket systems at the test site of Kapustin Yar.

The transition to carrier rockets with reusable elements and the employment of reusable carriers will necessitate the incorporation of a network of airdromes into the spaceport infrastructure.

The development of Russia's spaceport infrastructure will allow to:
• provide, regardless of relationship with countries of the "near abroad", for unfailing deployment and maintenance of orbital constellations of space complexes and systems for military, scientific and economic applications within planned schedules;
• transfer the launches of military SC to Russia's test sites;
• preserve the country's accumulated scientific and manufacturing potential of space industry in modernization of existing and creation of prospective multi-purpose ground complexes and reusable space systems;
• raise energy performance of carriers and expand the range of orbits that can be reached in altitude and inclination;
• reduce the product line of space rocket systems in operation;
• reduce the areas allotted to receipt of rockets' jettisoned components;
• raise ecological and operational safety of space rocket systems.

### 3.1.4 Comparative estimation of variants of launch complexes

A launch complex is an integral part of a space complex designed to conduct a pre-launch preparation of launch vehicles and space objects and their launch. Launch complexes are categorized as per the following features:
• carrier rocket class: light; medium; heavy; super-heavy;
• assembly and transportation methods: horizontal; vertical;
• method of preparing the space rocket: fixed; mobile; mixed;
• location: continental (on the ground, dug-in and underground); maritime

(ship-, barge-, submarine-, platform-based); aerial (with the use of various flying vehicles);
• possibility of travel: mobile (moving); stationary;
• number of launch installations;
• multi-role capability: special purpose (for launching a certain space rocket); multi-purpose (for launching space rockets of various applications).

The choice of the variant of launch complex is determined by targets set before the space complex and the technical decisions taken while creating it. The determining factor in formulating the overall specifications of a launch complex are the characteristics of a space rocket launched from it. Such main characteristics are as follows:
• mass of the launched space rocket;
• rocket fuel components chosen for the space rocket;
• space rocket transportation conditions;
• space rocket launch requirements;
• preparation of space rocket for launch and launching it;
• operative characteristics of the space rocket while in certain conditions of readiness for launch;
• launching method;
• safety requirements.

Categorization of launch complexes as per carrier classes is shown in Fig. 9.

Known to date are two main methods of preparing carrier rockets for launches. These are fixed and mobile methods. The essence of the methods lies in the following.

In the mobile method of preparation the carrier rocket is assembled and gets the required checks at the assembly and test shop of the engineering complex and is transported together with the space nose tip to the launch complex where it is further prepared and launched. This method of preparation implies the availability of two territorially separated complexes for preparation of the carrier for launch: engineering complex (EC) and launch complex (LC). The EC and LC are interconnected by transport communications for delivery of the carrier assembled at the assembly and test shop of the EC to the launch installation of the launch complex. The mobile method of preparation of the carrier rocket for launch provides fairly high values of such characteristics of space rocket systems as output rate and the launching rate. In using this method of preparation the assembly and testing of carrier rockets take place in the rooms of installation and assembly shop in conditions which preclude adverse effects of environment on the personnel and components of carrier rockets. This enhances the convenience of the manufacturing process, the quality of goods being manufactured and decreases their dependence on meteorological conditions. However, with the growth of size and mass of the rocket it becomes necessary to build larger installation and test shops as well as special transport communication lines between EC and LC for transportation of assembled space rockets.

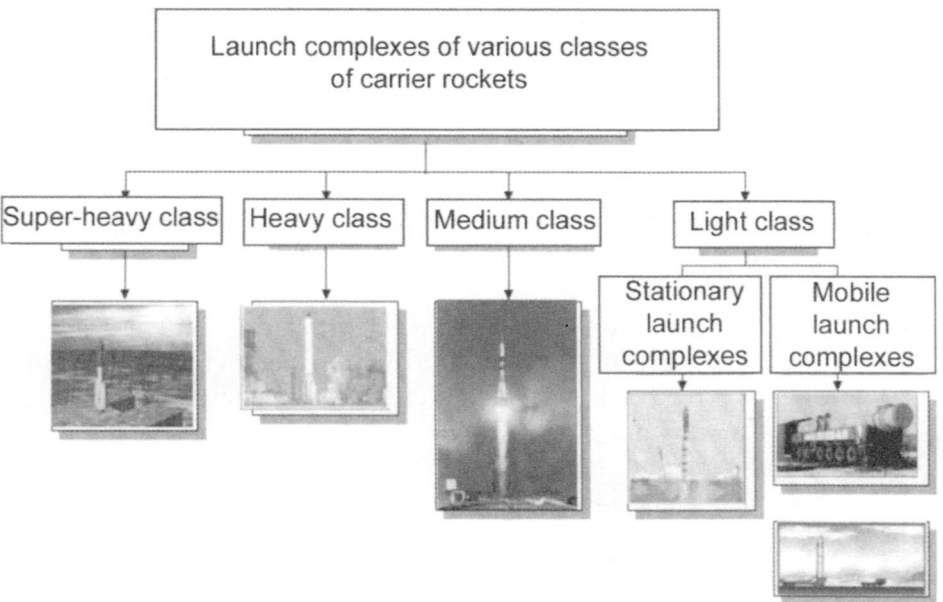

**Fig. 9.** Categorization of launch complexes as per carrier classes

Transporting a heavy weight space rocket entails substantial financial expenditures associated with creation of both transport communication lines and means of transportation. One of the main factors determining the overall performance of the engineering complex in case of mobile preparation of the carrier for launch is the method in which the carrier is assembled and transported along with its nose part from the EC to the LC. There are two methods of assembling and transporting a carrier rocket with a nose part from the EC to the LC: horizontal and vertical. The choice of the assembling technique influences the component package and the design features of the technological equipment of the EC and LC, the time required to prepare the carrier and the cost of building the space rocket system. Apart from the horizontal method of assembly and transportation of carriers with nose parts from the EC to the LC, the vertical mode, too, has gained wide acceptance. The Russian space industry has been using to date only horizontal method of assembling and transporting carrier rockets with a nose part from the engineering to the launch complex.

In the case of the fixed preparation of the carrier rocket at the assembly and test shop of the engineering complex, only separate units of the carrier undergo the required preparation. The assembling and testing of the carrier are accomplished on the lift-off installation of the launch complex. In this case the engineering complex becomes practically unnecessary.

The advantages of this method are the substantial reduction of size and cost of the assembly and test shop, diminished expenditure on building transport communication lines between the EC and LC and elimination of necessity to use special transportation means to haul the assembled carriers with nose parts from the engineering complex to the launch site.

Among the shortfalls of the fixed method are the low output capacity and low rate of launches from one launch installation as well as equipment sophistication due to additional aggregates needed to assemble carrier rockets. Furthermore, the necessity to preclude adverse exposure considerably complicates the relevant units of the launch complex.

Till now Russia has been using only mobile method of preparing carrier rockets. However, the analysis of information available with reference to foreign space rocket systems as well as research carried out in Russia show that with the growth of take-off mass of space rocket systems the mobile method of preparation and the horizontal method of assembling and transporting space rockets from engineering complex to the launch complex lose their decisive indisputable advantages.

The main means of transportation for mobile launch complexes can be motor-road, railroad and air-borne transport vehicles.

In creation and application of mobile launch complexes the space rockets can be transported both in a ready-to-launch state and in condition requiring pre-launch work, for example refueling of the carrier rocket.

The specific features of the space rocket as an object of transportation are its large mass (especially in the case of carrier rockets with solid propellant engines) and considerable size, the rocket hull's limited capability to accept exterior (static and impact) stresses and bending moments, presence of onboard equipment susceptible to impacts. It is desirable to transport liquid propellant engine rockets unfuelled. In doing so it is necessary to include refueling equipment and fuel transportation gear into the component package of the mobile launch complex.

Transportation of rockets over long distances by the existing highway network implies compliance with mass and size requirements imposed on an object being transported (normally, the mass should not be above 45 tons, the cross section should not exceed 4 m).

Carrier rockets are transported by railroad using standard or special purpose cars, containers, platforms, trucks, handling gear, etc. During such transportation of carriers, especially in case of long distance haulage, the carrier rocket may be disabled by prolonged impacts, thrusts and vibration. To protect the carrier rocket against these adverse effects shock absorbers of various constructions are used and the cars' coupling mechanisms are fitted with special shock absorbing devices.

Over recent years at issue are the creation and use of unconventional means of transporting carrier rockets.

Carrier rockets and launch containers with space rockets in them can be transported by transport aircraft.

In using aerostatic means as aerial launch complexes the main problem is separation of the multi-ton space rocket from aerostatic equipment.

Transportation of carrier rockets by water is accomplished by barges and ships. The main advantage of transportation by water is the smallest restrictions with regard to mass and size of rockets being transported. The limiting factor is insufficient lifting capacity of ports' loading and handling equipment. Transported by water are carrier rocket stages weighing up to 450 tons, measuring more than 10 m in hull diameter and over 50 m in length. The stages are unloaded by rolling with the help of a prime-mover or electrically driven winches.

The aerial transportation of launch complexes and space rockets can be accomplished by means of:
- transport aircraft;
- lead aircraft;
- escort aircraft;
- refueling aircraft;
- ground-based servicing equipment at cargo dispatch and receipt airdromes (ground-based lifting and handling gear and transportation equipment);
- air traffic control equipment.

The promising trend for heavy and super-heavy class carriers is the use of vertical assembly, including assembly at the launch complex.

The launch complexes have been created at Russia's spaceports for launching one type of carriers. The amount of launch complexes and the variety of their types called for allotment of considerable territories to test sites and employment of numerous personnel to service the launch complex, engineer communication lines and other facilities of the spaceport. However, under some programs and with certain rates of launching various classes of space rockets it might appear feasible to build multi-purpose launch complexes.

The creation of high-performance MPLCs for launching various classes and types of space rockets allows to adapt quickly to the new generation of launch vehicles.

The MPLC as compared to a system of special-purpose launch complexes offers the following advantages:
- lower cost of creation and operation;
- simpler general infrastructure of the spaceport;
- smaller total number of combat crews and servicing personnel;
- higher rate of launch;
- excellent upgrading capability of the launch complex.

At the same time, the MPLC has a number of shortfalls:
- the total throughput capacity and launching rate is lower than those of dedicated launch complexes;
- more time is required to prepare the follow-on launch of a space rocket of a different type because of re-adjustment of the launch complex;
- more stringent reliability and upgrading capability requirements.

In choosing the location for a launch complex on a continental part of Earth, account is taken of restrictions imposed by reasons of safety for the launch complex and carrier rocket. Chosen for construction of a launch complex are lands least suitable for farming in sparsely populated areas of the country. The flight trajectories of space rockets and debris recipient areas must not be located in densely populated areas or places with industrial or any other facilities on them. Aerial and maritime launch complexes are adaptable to such restrictions best of all.

Experience has been accumulated in creation of various variants of sea (ocean) based launch complexes. Launches from the stationary sea-based platform of San-Marco located off the coast of Kenya had been conducted as far back as 1967-1988. Used now as a sea-based launch complex is the floating spaceport of Sea Launch (based on the Odyssey oceanic platform). The principal advantage of the project is the possibility to launch geostationary satellites from equatorial waters. The complex component package includes the Odyssey launch platform and an assembly and command vessel (ACV).

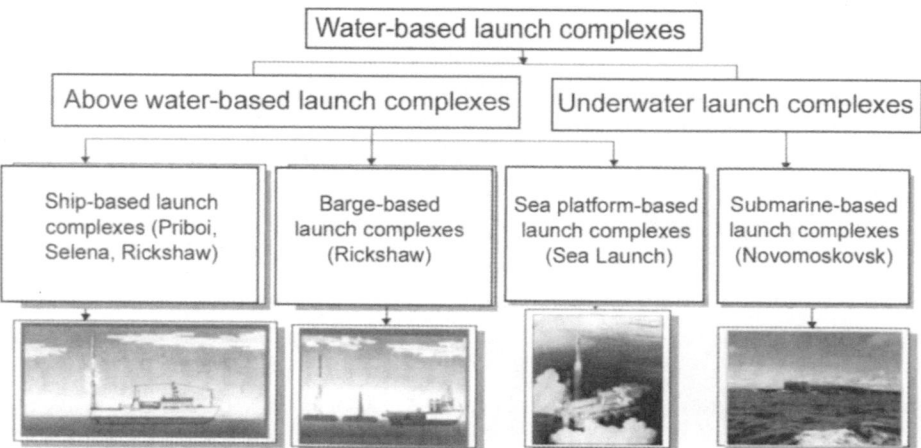

**Fig. 10.** Classification of water-based launch complexes

The assembly of a space rocket is accomplished on board ACV at Long Beach Haven, California. After that, a Zenit space rocket is reloaded onto the Odyssey and both vessels head for the Caiman Islands in the Pacific. Before placing the rocket on the take-off pad the servicing team leaves the platform and the ACV moves the personnel 5 km away to a place where operators control the preparation of the rocket, its fuelling and launch. All these operations are performed automatically.

Among sea-based launch complexes is the mobile underwater launch complex (Berlin Technical University and the V.P.Makeyev State Rocket Engineering Center). On July 7, 1998, the nuclear-powered K-407

Novomoskovsk submarine launched from submerged position a Shtil-1 carrier rocket with two satellites, Tubsat-N and Tubsat-N1, weighing 8.5 and 3 kg. During the launch the submarine was in the Barents Sea at 69.5° n.l., 43.2° e.l. Ground-based equipment of the command and instrumentation complex was used to launch SC.

Sea-based launch complexes mounted on ships, barges or floating platforms have home ports of their own. Sea- (ocean) based launch complexes do not provide a high yearly throughput because the time required to return to the home port for replenishment of space rocket supply can be rather long while the number of places to store the carriers, as existing plans show, is rather limited. The admissible storage dates of carrier rockets and spacecraft supply in sea conditions can be much shorter than in conditions of ground-based storage as they can be likened, with a great measure of certainty, to transportation conditions where the tolerance for transportation duration is known to be essentially lower than for storage dates.

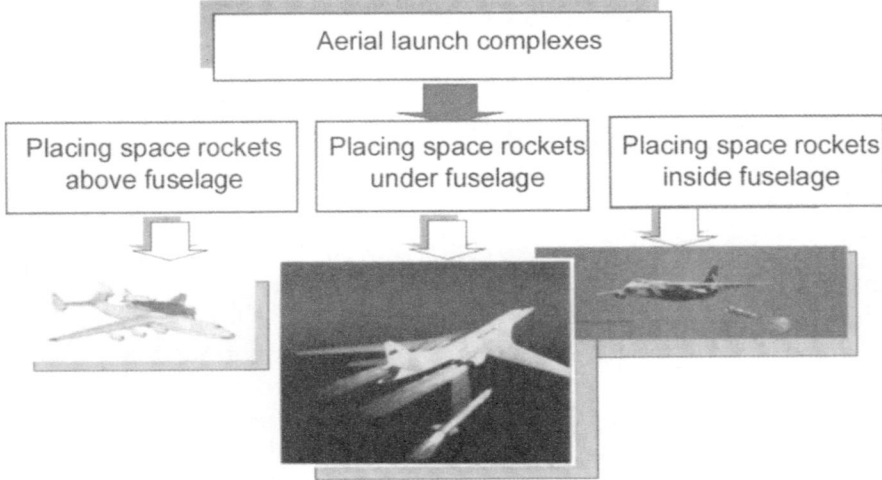

**Fig. 11.** Classification of aerial launch complexes as per method of placing space rockets

The possible classification of water-based launch complexes is shown in Fig. 10.

Aerial launch complexes can launch carrier rockets practically from any point of the aerial ocean both over land and sea (ocean) surface which makes it possible to choose a launch point that provides the maximum energy performance in launching spacecraft and optimal in terms of restrictions imposed by flight trajectories and areas receiving jettisoned stages of carrier rockets.

Aerial launch complexes must have a network of base airdromes, stand-by and auxiliary airdromes for inter-flight servicing and storage of supplies of expendable components, carriers, SC and fuel for basing lead aircraft, escort aircraft, tracking and control aircraft. The classification of aerial launch complexes as per methods of basing carrier rockets is shown in Fig. 11.

### 3.1.5 Problems to be addressed in building spaceports.
### Trends in development of spaceports

In building spaceports a variety of problems arise, which primarily relate to ecological issues. Such technological facilities of spaceports as launch complexes, refueling and neutralization stations, facilities for storage of rocket fuels, solid propellants and pyrotechnical materials are sources of extra hazard. In addition, launched space rockets pose a serious threat to people and environment during flight. Rocket fuel components are in the main highly toxic compounds that require special care in handling them. Therefore technological facilities of the spaceport must be located at a safe distance from residential areas while carriers' flight trajectories must pass over sparsely populated territories.

The important factor in building a spaceport is the delivery to it of space rocketry hardware components. The existing railroads impose certain restrictions on freight size which precludes transportation of large size pieces of equipment characteristic of space rockets complexes of heavy and super-heavy classes. The problem can partially be resolved by using earth road motor vehicles to haul materiel over short distances within spaceports. The use of water transport essentially enhances the transportability of space rocketry hardware components. This advantage is offered by spaceports situated on ocean coasts. Spaceports without waterways for communication with manufacturing works have to resolve this problem by reducing the size of space rocket components and by performing manufacturer's assembly operations (technology accepted at the spaceport for assembling the N-1 carrier rocket) or by creating special aviation means. So, in order to deliver tanks of the Energia carrier and the Buran orbital ship to the spaceport, special transport aircraft, 3M-T and An-225 (Mriya aircraft developed by Antonov Design Buireau as per task order and technical specifications of the NPO Energia (research and production association). Also, other means of delivery of space rocketry hardware from manufacturing works to the spaceport have been considered. Furthermore, attention has been given to the delivery of landed spacecraft and their parts and techniques have been elaborated allowing such delivery in various conditions. Best suited to these purposes is a package of transportation means consisting of vertical take-off and landing craft (helicopters, aerostatic equipment and combined helistats), air-cushioned craft, all-terrain vehicles and aircraft. Such a package can be used to serve several airdromes.

The very nature of spaceport's work has an adverse impact on environment. This impact stems from the work done at the spaceport, from the inherent characteristics of launch vehicles used, from the rocket fuel components.

Work is underway on issues aimed at reduction of rocket fuel impact on environment. This could be attained by the following ways:
- reduction of rocket fuel impact on the Earth's surface;
- reduction of the rocket fuel impact on the Earth's atmosphere.

The reduction of rocket fuel impact on the Earth's surface and atmosphere can be attained by using low toxicity fuels, new methods of neutralizing their remnants and discharges into atmosphere. Perfecting the techniques of controlling the flight of a space rocket and its jettisoned parts will considerably reduce the number of debris recipient areas and their size.

To reduce the impact of spaceports on environment it is necessary to replace in space systems and their units the chlorfluorocarbons that ruin ozone. Based on the UN Vienna's convention on protection of the Earth's ozone layer and the Montreal protocol on compounds that ruin the ozone layer, measures have been elaborated aimed at reduction of use and cessation of production of substances ruining ozone. According to an agreement between signatories of the Montreal protocol, the production of groups of compounds, including specific coolants, ruining the Earth's ozone layer, must be ceased completely. The use of a new alternative coolant can quickly resolve this problem. Modernization of the existing fleet of refrigerator equipment is anticipated with replacement of the banned chladone R12. Realization of the plans provides for modernization of the refrigeration equipment now in operation, development of a number of base line refrigerators that will meet today's demands of ecological safety and will standardize the refrigeration equipment of spaceports.

In order to reduce the environmental pollution the landing complexes are fitted with special processing equipment for collection and disposal of jettisoned hardware. Work is now under way to make possible space rockets' switch to pollution-free fuels and ultimate burning of the remaining fuel while rocket stages and jettisoned subunits fall to the ground. A step forward in the solution of this problem was the employment of reusable launch vehicles and their elements. Reusable rockets and their stages land on special landing sites and airdromes. After-flight servicing of reusable rockets and their stages will be performed on specially prepared work stations.

To date, twelve spaceports have been created and are intensely used across the globe. Each of them is a unique technological complex providing for launches of various types and classes of space rockets for various uses. Most spaceports are located on ocean coast because of necessity to allot territories to debris recipient areas and to exclude the flight of space rockets over inhabited localities. The proximity to the equator expands the capabilities of a spaceport in terms of the mass of payload launched into orbit. Recently, a tendency has emerged to build autonomous, mobile space rocket complexes for launching light and medium class space rockets. Those are maritime complexes, wheeled and tracked ground vehicles, airborne systems.

Further development of the world's spaceports is closely associated with the prospects of development of launch vehicles and features the following tendencies:

• use of expanses of seas and oceans as areas to launch spacem rockets and receive their jettisoned parts;

- duplicating spaceports by the leading nations;
- construction of spaceports on territories of other nations having advantageous geographical position near the equator;
- conservation of unused ground complexes or their modernization for use by prospective launch vehicles;
- creation of a new infrastructure for reusable launch vehicles (runways, repair and preventive maintenance stations, etc.)
- use of new techniques and devices for propelling launch vehicles.

The tendencies in development of spaceports and their shares in launches of SC are shown in Fig. 12.

The development of spaceports in Russia can proceed both as modernization of the existing infrastructure and as creation of new ground-based complexes.

The creation of high-performance multi-purpose ground complexes will provide for boosting various types and classes of launch vehicles which will cut down the cost of space rocket launches.

The use of aircraft transportation package will enhance the efficiency of transporting space rocket components from manufacturing works, arsenals and spaceports. Also, this will fulfill a number of target tasks (interception of descending craft and elements of space rockets, transportation of space rocket complexes, refueling of aerospace complexes, etc.)

The issue of proximity to the equator is settled by Russia through creation of sea-based complexes or construction of ground-based complexes on other countries' territories, for example, Australia.

The use of reusable elements of launch vehicles calls for modernization of the existing complexes and creation of new ground-based technological equipment for landing those elements and for carrying out preventive maintenance prior to their re-use.

The complexes that use the aerodynamic property of the wing will require improvement of the country's airdrome network. The extra equipping of air dromes must include improvement of runways and creation of ground-based equipment that puts space rocket complexes in various states of readiness.

The use of aerostatic launch complexes for launching space rockets from specific altitudes will allow to:
- assemble space rockets at manufacturing works proper;
- deliver the assembled space rockets to the launch area;
- launch by means of aerostatic launching device from specific altitudes and places.

Sea-based launch complexes can in the longer term be based on platforms mounted permanently on an off-the-coast shelf or other shallow places of the ocean. By using optimized technologies of installing sea drilling rigs for creation and operation of sea-based complexes, it is possible to change their placement and to gradually build up their capacity with regard to various layout arrangements.

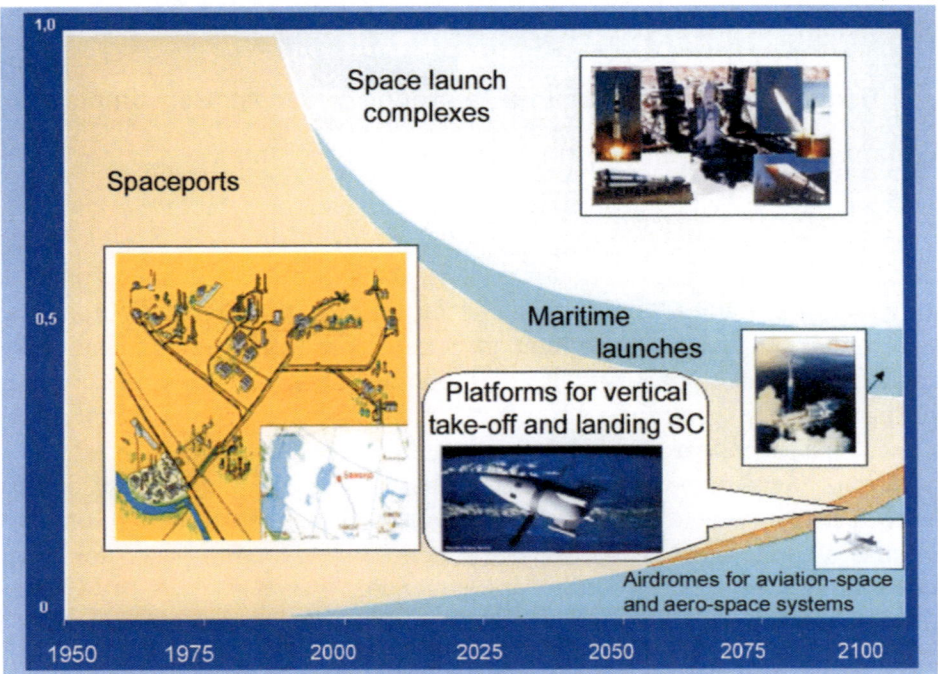

**Fig. 12.** Tendencies in development of spaceports and their involvement in launches of SC

The longer term prospects anticipate the use of various stationary ground-based accelerators. So a reusable magnetically suspended platform fabricated as a linear motor can create the required pulse for carrying SC into space (from a silo or with the use of terrain features).

Thus, the development of spaceports is mainly determined by the development of launch vehicles themselves and by the tasks they are supposed to takle. Once spacecraft and launch vehicles emerge that are able to take-off vertically from any unprepared place, spaceports may radically change both in their appearance and functions.

## 3.2 The state and prospects of development of spacecraft ground-based facilities and control systems

### 3.2.1 General principles of building a ground-based control complex

#### Aim and application of a ground-based control complex

The ground-based control complex (GBCC) means an aggregate of ground-based electronic and computing systems with associated software intended to control SC in flight and monitor the functioning of its onboard equipment. GBCC operates during SC's orbital flight starting from the moment it separates from the carrier rocket till the moment it ceases its active service. The main tasks of the GBCC can be formulated as follows: support of the space system's orbital structure, monitoring and control of performance of dynamic operations in space, onboard SC equipment control, control and diagnosis of its condition, interaction with complexes that provide for targeted use of SC as well as with complexes providing for launch, test site instrumentation and landing. In addition, in case of piloted SC the GBCC is tasked with providing telephone-, cable- and TV-aided communication between crews of SC and ground-based centers.

As per tasks tackled, the following functions can be singled out in SC control:
- command software support;
- navigation and ballistic functions support;
- information and telemetry support.
- The command software support includes:
- development of long- and short-term plans of SC and GBCC systems operation;
- formulation and transmission aboard SC of command software information (CSI) and receipt of acknowledgement as to its passage;
- control of execution of SC flight program based on results of analysis of telemetric and trajectories related data.

The basic data for planning the operation of SC is the long-term flight program and customers' requests for purpose-oriented information as well as results of the previous control of operation of onboard systems based on telemetric measurement data. The control of operation of SC onboard systems at a specified period of time is accomplished via a package of command software messages. The package includes both separate once-off commands executed on board the SC as they arrive and temporary programs recorded in the onboard memory device. Such programs ensure autonomous operation of SC beyond radio visibility zones by means of GBCC.

The navigation and ballistic support (NBS) of the flight includes the following main operations:
- measurement of current navigation parameters (MCNP) of SC;
- determination and prognostication of parameters of SC orbit based on results of processing MCNP data;

- computation of basic data for correcting the orbit in order to maintain the set parameters or maneuvers of SC for transition from one orbit to another;
- computation of ballistic data required for operation of GBCC systems and centers for receipt of information from SC: SC radio visibility intervals by ground-based means, target designations for guiding antenna systems etc.

Information and telemetry support includes the following operations:

- receipt of information from aboard SC, its primary processing and registration at ground-based centers of telemetric (TM) information;
- transmission of TM information to the flight control center (FCC) and processing centers;
- processing of TM information at FCC and its presentation in a format convenient for the subsequent analysis;
- analysis of TM data concerning the state and functioning of SC onboard systems and medical condition of the crew (in case of piloted SC).

The telemetric information and results of trajectory measurements are the basic data for evaluating the quality of performance of the prescribed technological cycle of SC control.

The spatially separated multi-level structure of the GBCC requires a strict organization and coordination in control of its technological systems.

Instrumentation posts' systems based on source technological information (operating frequencies of onboard equipment, command information code numbers, ballistics data regarding the orbit of SC, etc.) search and capture signals from SC as it enters the radio visibility zone and switch to the stable tracking mode to conduct information exchange and carry out trajectory measurements. As the communication session ends, instrumentation posts issue to FCC the results of executing the program of operation of instrumenta-tion post's systems and evaluation of the quality of information received from the SC.

## Demands made of the main characteristics of GBCC

Demands made of the GBCC are determined by the purpose of guided SC, by functions they perform in orbit and by the ballistic structure of SC orbital constellation.

Automatic SC of adjacent and medium space, which include SC for study of the Earth's natural resources, for geodesic investigations, for communication and TV broadcasting do not as a rule require extra-high control performance in case of normal operation. The SC onboard systems' operation software can be prepared in advance and transmitted aboard the SC no more than once in 24 hours or once a month for communications and TV broadcasting SC.

The accuracy of determination and prognostication of SC orbit parameters based on MCNP data by means of GBCC is normally supposed to meet the requirements of:

- target designation of GBCC radio electronic equipment for establishing communication with SC:
- planning the work of onboard equipment controlling the SC payload;
- time and geographical position-related fixation of scientific and special purpose information received aboard the SC.

The requirements imposed on accuracy of information concerning SC orbit parameters needed to provide communication for the GBCC systems are relatively low (essentially below the requirements imposed by SC target missions) and normally the accuracy of prognostication along the orbit of several tens of kilometers is sufficient.

The requirements concerning the accuracy of fixation of various instrumentation or research information received at SC is much more stringent and can be met with the aid of a posteriori methods of processing MCNP data if no special demands are made here of the speed of determining the orbit parameters.

The most stringent demands are made of navigation and ballistic support of SC control in cases when it is necessary to plan the use of SC dedicated equipment which must be guided to a certain point in space or must keep track of some route on a territory under the satellite.

Based on above, in determining the requirements imposed on NBS of the SC flight the following values of admissible errors in prognostication of SC movement can be taken for the basis:
- for SC with orbits H=200–300 km (for 24 hours): along the orbit – 50 km, in lateral direction – 2 km, in altitude – 1.5 km;
- for SC with orbits H = 700–1,000 km (for 7 days and nights): along the orbit – 5–7 km, in lateral direction – 1–2 km, in altitude – 1 km;
- for geostationary satellite (for 7 days and nights): along the orbit – 10 km, in lateral direction – 10 km, in altitude – 10 km.

The maximum accuracy of determination of orbit parameters are required for SC of space navigation systems since in this case their users will depend on them for navigation accuracy. Today's navigation spacecraft of Russia's GLONASS system or the USA's GPS, spacecraft placed in orbits at an altitude of 20,000 km must provide users on Earth's surface and in near Earth space with navigation accuracy in the order of tens of meters. Similar, or even higher, must be the accuracy of ephemerid support of the said systems.

## The main systems of GBCC

All systems of GBCC can be divided as per their functions into the following main groups:
- SC flight control center systems;
- systems of information exchange with SC;

- systems of communication and data exchange between FCC and other elements of GBCC, as well as between FCC and exterior interacting complexes;
- single time system devices.

The information exchange between GBCC and SC is accomplished via radio channels of command and instrumentation (CI) and telemetric (TM) systems, via radio lines of purpose-oriented information exchange with SC, as well as via radio channels of a space relay system. The ground-based systems of purpose-oriented information receipt may not be comprised by GBCC but be part of special systems of transmission, receipt and processing of purpose- oriented information.

The CI radio lines are used to transmit to spacecraft CSI, measurement of SC navigation parameters and summarized control of the state of onboard systems based on information transmitted to GBCC via CI return channel. Transmitted from SC to GBCC via TM radio lines is detailed TM information about the state and functioning of SC onboard systems.

In some cases the GBCC component package contains combined radio lines enabling simultaneous transfer of several types of information down to the single radio line for all types of information exchange between GBCC and SC.

The combined radio lines allow to minimize the component package of the required onboard and ground-based equipment.

However, the combining of various functions in one radio line complicates the transceiver equipment and precludes in a number of cases characteristics optimization of onboard and ground-based systems.

The principal element of the GBCC is FCC which maintains automatic round-the-clock flight control of SC. The analysis of tasks set before FCC shows that for all variety of specific features encountered in control of various SC the organizational structure of the FCC must contain the following elements:

- the command post of FCC coordinating the operation of all its segments, services and complexes;
- long and short term SC operations planning sector;
- sector of integrated analysis of condition and functioning of SC, of estimation and formulation of CSI;
- sector of navigation and ballistic support of flight control;
- sector of telemetric support of flight control;
- information presentation sector;
- sector of accompaniment of software support of the FCC automation systems;
- sector to support telephone, cable- and TV-aided communication with crews of piloted space complexes;
- communication organization and GBCC data transmission service.

The technical systems of FCC contain computing complexes with mathematics support for automatic solution of problems relating to SC control, formulation of CSI; communications station; systems of individual and collective representation, documenting and storage of information.

GBCC systems of communication and data exchange (SCDE) are intended for information exchange and communication by means of telephone, cable, TV or fax between elements of GBCC. The component package of SCDE contains automatic centers of commutation, modems, channel forming and subscriber equipment, cable, wire, optic fiber, radio relay and satellite communication channels, integration means of various communication subsystems.

The single time system (STS) is intended for time synchronization of operation of all elements of GBCC. It includes the STS central post and local posts located at FCC and all CI posts. The central post is the source of STS primary signals formed on the basis of highly stable timers. The signals of STS central post are transmitted with the help of communication means to local posts where they are used to synchronize local frequency standards. The signals of the latter are, in turn, transmitted to radio electronic systems of CSI and synchronize their operation.

Currently, the local GBCC STS posts use local frequency standards based on quantum standards with a relative variation of $10^{-12} - 10^{-13}$ within a 24 hour period.

## Command and instrumentation posts of GBCC

The ground-based equipment of radio electronic systems of GBCC intended for information exchange with SC is located in command and instrumentation posts (CIP). The GBCC component package normally contains several CIPs. They can be stationary or mobile (located on floating ships, trucks, aircraft, et.) The placement of CIP on Earth surface depends largely on the ballistic structure of the orbital constellation of a guided space system, on demands made of NBS accuracy and the global nature of information interaction of GBCC and SC.

The observation conditions of SC from CIP are primarily determined by the altitude of its orbit and the trajectory of the under-the-satellite point on Earth surface (routes of SC orbit) relative to CIP. In terms of control the re-quirements imposed on observation conditions of SC by means of GBCC are generated by the necessity to realize during communication sessions the technological cycle of SC control. Normally, in the course of a standard flight the continuity and global scale of interaction between SC and GBCC are unnecessary. However, in case of deviation from the flight program and in emergencies, the possibility to establish at any time a channel for communications with SC (in order to quickly eliminate the problem) is an im-portant measure in enhancing the reliability of flight control.

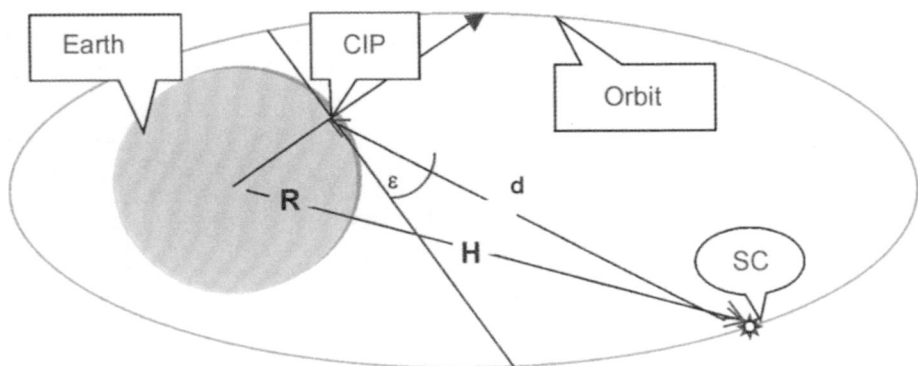

**Fig. 13. CIP radio visibility zone**

Most space radio lines operate in ultrahigh frequency ranges, due to which fact the communication sessions between SC and CIP appear possible only while SC remains in the zone of its direct visibility. The CIP radio visibility zone is all near Earth space visible from CIP at elevation angle above the local horizon to the excess of 7°. In this zone assured radio communication is provided between ground-based and onboard radio equipment. Also, the targeted accuracy of trajectory measurements is ensured. However, sometimes also considered as practical are the radio visibility zones at elevation angles less than 7°.

The geometrical constructions suggested in Fig. 13 show that the distanced from CIP to SC located on the border of radio visibility (sometimes this distance is called the radius of radio visibility zone) is determined by the expression:

$$d = \sqrt{(R + H)^2 - R^2 \cos^2 \varepsilon_{min}} - R \sin \varepsilon_{min} \quad (1)$$

where R – mean radius of the Earth, R=6,371 km; H - orbit altitude; $\varepsilon_{min}$ – elevation angle corresponding to the border of radio visibility zone. For $\varepsilon_{min}$ = 0 expression (1) is simplified:

$$d_0 = \sqrt{2RH\,(1 + H\,/\,(2R))}.$$

At H/(2R) << 1 a simple formula is obtained

$$d_0 = \sqrt{2RH}.$$

For the range of SC orbit altitudes from 200 to 1,000 km within which lie the orbits of most SC of the adjacent space, the radii of CIP radio visibility zones at εmin = 7° will be vary between 1,000 and 3,000 km.

The maximum duration of a communication session between CIP and low orbit SC reaches 4 to 13 minutes depending on orbit altitude so long as the flight trajectory passes through the point where CIP is located. The short duration of communication session between CIP and SC brings about serious difficulties in providing the global scale and continuity and of the information interaction both for flight control and for receipt and transmission of target information. One of the ways to resolve this problem is the multi-point dispersed structure of GBCC consisting of CIPs scattered over the Earth's surface.

Figure 14 shows the route of SC circular orbit flight with parameters of $H = 500$ km, $i = 63.0°$. Also it indicates radio visibility zones of five CIPs located in the neighborhood of Moscow, Ulan-Ude, Dzhusaly, Ussuriysk and Petropavlovsk Kamchatskiy.

The number of CIPs in the component package of GBCC depends on requirements imposed on accuracy, speed of determination of SC orbit parameters by means of GBCC and on requirements imposed on SC radio visibility intervals for CSI transmission, for control of operation of onboard systems, for receipt of purpose-oriented information and for radiotelephone and TV communication with crews of piloted space complexes.

Analysis of the expression (1) and Fig. 14 allow to formulate a number of general requirements in relation to placing CIPs on Earth's surface in a way that assures the best possible conditions for communication with low orbit SC having a limited component package of CIPs:

1) for communication with SC at the greatest number of orbits in a 24 hour period CIPs must be evenly spaced over Earth's surface at geographical longitude at maximum distances from each other;

2) for continuous communication with SC the maximum interval between two CIPs must not exceed the diameter of the CIP radio visibility zone;

3) for communication of each CIP with SC at the greatest number of orbits in a 24 hour period the geographical latitude of CIP's location must be somewhat less (by 5-10°) than the angle of inclination of SC orbit plane relative to the plane of equator.

Specifically, requirement 3 implies that each orbital constellation of SC characterized by a ballistic structure with an identical orbit inclination angle has its optimal (in terms of radio communication conditions) geographical latitude for placing CIPs.

For medium class SC in high elliptical orbits (HEO) with an apogee up to 100,000 km and for geostationary SC the interval of radio visibility from a ground-based CIP grows essentially. It increases to several hours for SC in HEO due to which fact radio visibility becomes available to SC in GEO at all times.

In view of this condition the number of CIPs used in the component pak-
kage of GBCC for controlling the said SC can be reduced to one or two in case
of their appropriate location on Earth's surface.

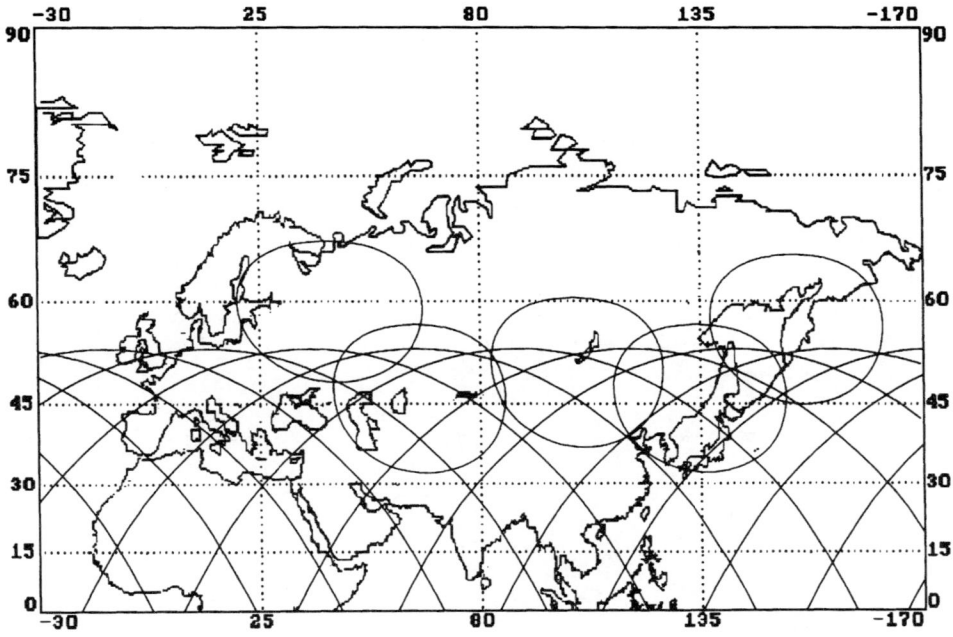

Fig. 14. SC flight route and CIPs' radio visibility zones

Specifically, requirement 3 implies that each orbital constellation of SC
characterized by a ballistic structure with an identical orbit inclination angle
has its optimal (in terms of radio communication conditions) geographical
latitude for placing CIPs.

For medium class SC in high elliptical orbits (HEO) with an apogee up
to 100,000 km and for geostationary SC the interval of radio visibility from
a ground-based CIP grows essentially. It increases to several hours for SC
in HEO due to which fact radio visibility becomes available to SC in GEO at
all times. In view of this condition the number of CIPs used in the component
package of GBCC for controlling the said SC can be reduced to one or two
in case of their appropriate location on Earth's surface.

## Information relay space system

The increased amount and complexity of missions performed by today's
SC together with the more stringent demands made of control reliability
necessitate the continuous and global information interaction between GBCC
and SC. The efficient means of achieving this goal is the use of space relay
systems (SRS) for information exchange with SC.

The technical structure of SRS comprises two or three relay satellites (RS) in GEO, one or two ground-based stations (GBS) of satellite communications and subscriber equipment installed on board service recipient SC.

Fig. 15 shows that three RS in GEO with geographical longitudes of under-satellite points spaced at 120° provide with overlapping the observation of practically all Earth's surface (and adjacent space) except near-pole regions up to ~80° of northern and southern latitudes. In order to simplify the structure the system can utilize two operating RS with insignificant deterioration in the global coverage of communication. The information interaction of GBCC with SC via RS is accomplished through the following scheme:

direct channel: FCC – GBS – RS – SC;
return channel: SC – RS – GBS – FCC.

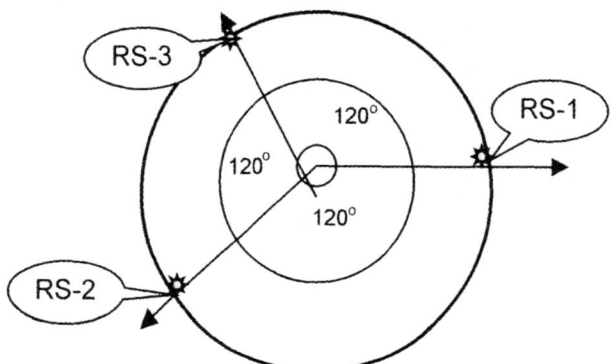

Fig. 15.
Arrangement of RS of space
relay system in GEO

The space relay system is exemplified by the American system TDRSS put in operation in the 1980s and designed for practically global servicing of currently operating and future low orbital SC, such as Space Shuttle system and Spacelab laboratory. The system utilizes two active duty and one stand-by geostationary relay satellite. The coordinates of subsatellite points of active duty RS (41° w.l. and 171° w.l.) are chosen so as to attain overlapping of their radio visibility zones on Earth's surface on the American continent in a way that provides radio visibility of both RS from the ground-based station at the White Sands base in the state of New Mexico, USA. To assure a simultaneous operation via three RS, installed at the station are three suites of receipt and registration equipment with antennas measuring 18 m in diameter. Techno-logical information and operation program needed to organize the SC control via RS are fed from NASA's and DoD's control centers through communication channels to the ground-based station. The information received from SC is conveyed in reverse direction.

The TDRSS system assures the global coverage of servicing low orbit (alti-tude ≤ 2,000 km) SC in the region of 0.9. To attain virtually full global coverage an extra RS is needed with an subsatellite point in the Eastern hemisphere. This, however, will necessitate a second ground-based station outside the territory of the USA or the establishing of inter-satellite communication between RS.

The similar space relay system was used in Russia in the control package of Mir manned orbital station and Buran reusable space system. The base of the system is two geostationary RS Luch (ray) with coordinates of subsatellite points 16° w.l. and 95° e.l. plus a ground-based satellite communications station near Moscow in the radio visibility zone of both RS. The station is outfitted with two suites of transceiver equipment having antenna systems. In the 1990s the Luch orbital constellation ceased to exist because of exhaustion of satellites' service life.

The SRS assure the exchange between SC and launch vehicles of all types of information needed to control the flight, perform telemetric operations and communication with the crew of manned complexes. Also, they provide for high-speed (up to hundreds of Mb/s) digital information transmission from SC mission-oriented equipment.

The drawback of using SRS is the need for extra reserves of mass which is required for installation of subscriber equipment and onboard antennas that provide for the required performance of SC – RS radio lines, especially in non-oriented flight of SC.

## Structural peculiarities of deep spacecraft GBCC

The structural arrangement and radio electronic systems of deep spacecraft GBCC have a number of essential differences on account of large distances and the nature of movement of lunar and interplanetary space stations.

At distances beyond hundreds of thousands of kilometers the visible movement of SC over the sky resembles the movement of planets: within a communication session the position of SC relative to stars remains for an observer on the ground practically unchanged. The angular data of SC relative to the ground-based station change mainly due to the Earth's rotation roughly by 15° per hour.

The change in SC radial speed within the communication session and hence the Doppler shift of frequency of a received radio signal is also determined mainly by the Earth's rotation in 24 hours and can be predicted with a fairly high degree of accuracy.

As SC moves farther away its zones of radio visibility from Earth expand, reaching at the maximum the size of the Earth's hemisphere while the duration of radio visibility intervals from one CIP can reach 12 hours. Thus to obtain a continuous round-the-clock communication with distant SC (DSC) flying in the proximity of the ecliptics plane it is sufficient to have three CIPs located in equatorial or medium latitudes and spaced over the Earth's surface at an angle of ~120° in geographical longitude.

The listed factors allow to simplify the structure of deep spacecraft GBCC as compared to the GBCC of near Earth SC, confining the number of CIPs to two or three units providing the global communication with DSC in the region of 0.7 to 1.

The main difficulties in creating and operating radio electronic systems of far-out space system from large interplanetary distances to SC and stringent limitations concerning the energy, mass and size of their onboard equipment. While for near space SC the maximum communication distances are restricted by radio visibility zones and do not exceed 3,000 km, for far-out space SC the characteristic communication distances are determined by the following values: 380,000 km – distance from the Earth to the Moon, ~40,000,000 – 260,000,000 km to Venus, ~80,000,000 – 380,000,000 km to Mars, ~6,000,000,000 km – the radius of the Solar system.

To assure the required communication range, it becomes necessary – in spite of the above described restrictions in mass and size and the appreciable sophistication of DSC construction – to use acutely directed rather large parabolic antennas. However, the major role in increase of energy performance of SC radio lines is played by ground-based systems which comprise very large antenna arrangements, powerful transmitters and highly sensitive receivers. The considerable effect is obtained by choosing the adequate range of radio waves for long distance space communication, the optimal means of information receipt, of jam-proof coding, etc.

The long duration of the flight (up to several years) and restricted capability to duplicate all elements of DSC onboard systems impose especially stringent requirements on their reliability and longevity.

## Choice of design characteristics of long distance space communication radio lines

The choice of design characteristics of space radio lines is a complex engineering task that requires consideration of a large number of various factors affecting the energy performance of radio lines and the quality of information transmission over the required distance.

Let us consider an expression that determines the dependence of maximal distance of communication D on the parameters of space radio line,

$$D = \sqrt{\frac{PS_\delta S_{\eta\Sigma}}{\lambda^2 \kappa T_{\ni\phi} \Delta f q_\Pi}} \qquad (2)$$

where $P$ – transmitter power; $S_\ni$, $S_\ni$ – effective areas of onboard and ground-based atennas; $\eta_\Sigma$ – ratio determined by total losses of signal energy in the elements antennas´and feeders´routes and while spreading in environment;

$\lambda$ – radio wave length; $\kappa = 1,38 \bullet 10^{-23}$ J/K – Boltzmann constant; Тэф – effective temperature of noise at the recipient system input; $\Delta f$ – receiver's width of pass band to detector; $q_\Pi$ – threshold ratio of signal power to noise power at recipient device input at which the intended quality of received message is assured.

Formula (2) holds true both for inquiry (Earth – board) and reply (board – Earth) radio lines. In the subsequent consideration we shall focus on the reply radio line as the busiest one in obtaining the required energy potential.

One of the main problems in designing long distance space communication radio systems is the choice of the optimum range of radio waves.

In formula (2) along with the evident dependence of $D$ on l, the effective noise temperature Тэф and components of ratio $\eta_\Sigma$ are functions of the wave length; both of which characterize the loss of signal energy in the Earth's and planets' atmosphere and in interplanetary plasma.

The losses of radio wave energy in ionosphere are ac-counted for by their reflection and dispersion as well as by rotation of polarization plane during distribution. Those effects weaken as the radio wave length decreases. At $\lambda < 3$ m the losses in ionosphere are insignificant while in the decimeter range they are negligent.

By contrast, the losses of radio wave energy in the Earth's troposphere grow as the wave length decreases and are quite small at $\lambda > 3$ cm. As the wave length decreases $\lambda < 3$ cm, the losses start growing because of the presence of resonance lines of absorption in the spectrum of molecular components of water steam and oxygen of the Earth's atmosphere.

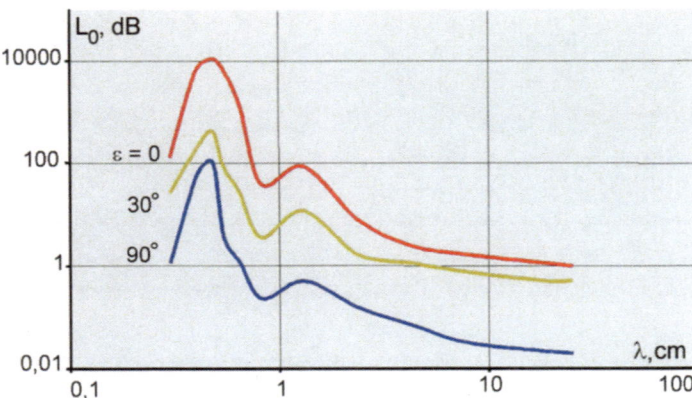

**Fig. 16.**
Absorption of radio waves in Earth's atmosphere

The graphically presented dependence of radio signal en-ergy loss in Earth's atmosphere on the wave length at the antenna's various elevation angles is shown in Fig. 16. The absorption of radio waves caused by precipitations, mainly rain and fog, increases with the decreasing wave length and depends on the intensity of precipitations. It becomes significant at $\lambda < 5$ cm.

Fig. 17 shows the curves of dependence of radio signal energy losses at a stretch of 1 km on the wave length l and intensity of the rain (in points).

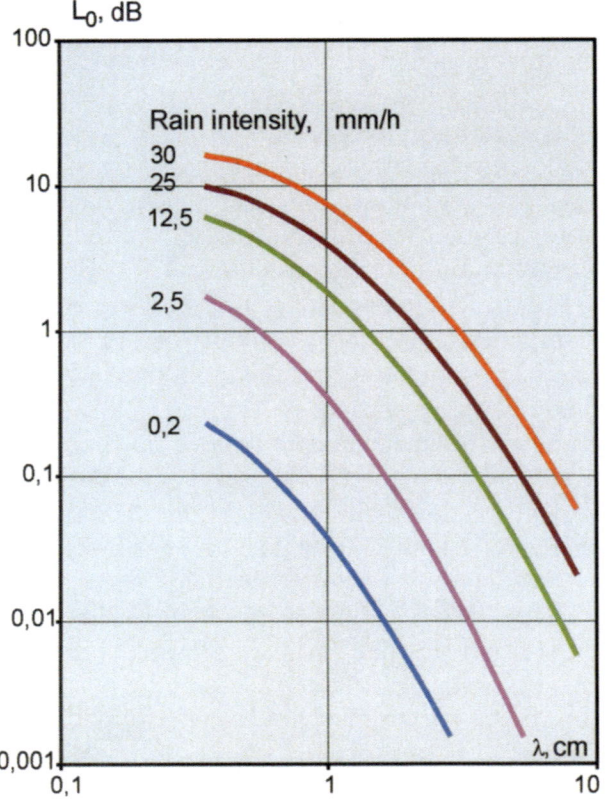

**Fig. 17.**
Absorption of radio waves caused by rain at a stretch of 1 km

It should be borne in mind that other planets of the Solar system also have atmosphere, which is why in choosing characteristics of channels of radio communication with SC located on the planet's surface, account must be taken of conditions in which radio waves spread in the planet's atmosphere.

The noise level at the receiving system input is deter-mined by own noises of the receiver input devices, by losses in the elements of the antenna-feeder device (AFD) and by exterior noises generated by the thermal radiation of the atmosphere, Earth surface, Galaxy and discrete space sources – the Sun and the Moon.

The atmospheric noises at high frequencies come about as effect of absorption of radio wave energy in the gases of the troposphere and in atmospheric formations (clouds, rain, fog). The intensity of those noises depends on the length of the path covered by the radio signal in the medium. Fig. 18 shows charts of dependence of brightness temperature of standard atmosphere on the length of radio wave l and the antenna elevation angle e. The charts have been calculated for average conditions of the European part of Russia. The same figure shows charts of dependence of the space radiation brightness temperature on the wave length corresponding to the galactic center ($T_{kmax}$) and the "cold" sky near Galaxy ($T_{kmin}$). Those charts can be used for approximate estimation of the contribution of atmospheric and galactic noise to the effective temperature of noise at the receiving device input.

It is evident from Fig. 18 that at wave lengths less than 10 cm the atmospheric noise dominates over galactic noise. At $\lambda = 1.35$ cm and $\lambda = 0.5$ cm the atmosphere brightness temperatures are at the maximum which is caused by the resonance radiation of water steam and oxygen molecules.

The Earth surface also emits radio noise with equivalent temperature around 290 K which can be received by ground-based antennas via side lobes of a directional diagram. For large antennas operating at large elevation angles the impact of thermal radiation of Earth surface on the effective temperature of the noise at the receiving device input will be insignificant.

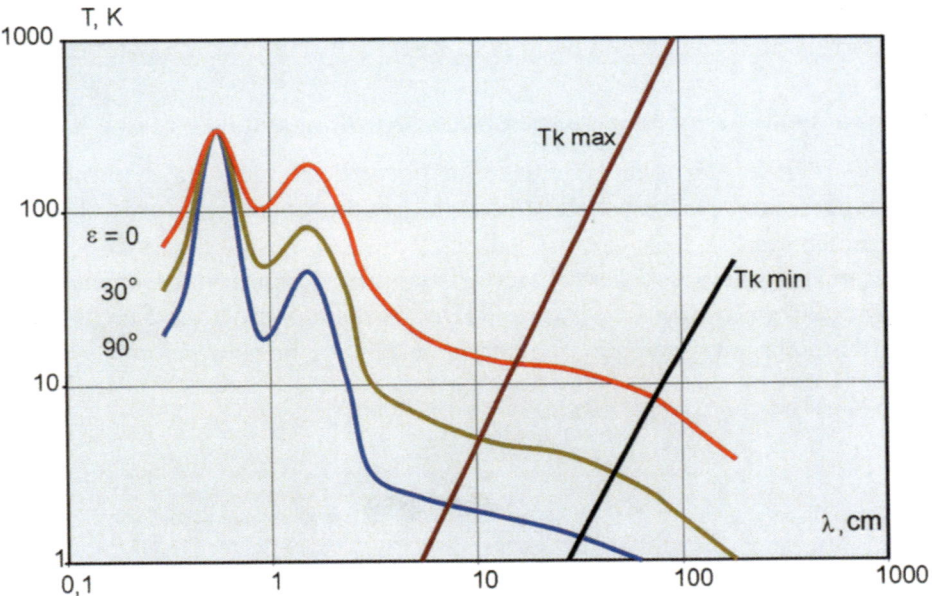

**Fig.18** Radiation temperature of standard atmosphere and galactic background

Thus the analysis of expression (2) and Fig. 16 – 18 shows that the optimum range of radio waves for obtaining the best energy performance of space radio channels and maximum distance of communication with SC by ground-based means corresponds to wave lengths from 1.5 to 30 cm. In this range the effective temperature of exterior noises brought in by radiation of the Earth's atmosphere and space will amount to 5-40 K, while integral absorption of radio wave energy in troposphere in the absence of precipitations will not exceed 0.1 dB at the antenna's mean elevation angles.

Apart from the described range, it is also possible to or-ganize space communication with DSC in the millimeter, sub-millimeter and optic wave ranges in "clarity windows" of the Earth's atmosphere.

To reduce the equivalent temperature of own noise at the input of ground-based station's recipient system, low-noise quantum – parametric or high frequency molecular amplifiers are used. They are cooled by cryogenic systems to the temperature of liquid helium. In the best amplifiers used in modern space electronic long distance communication complexes the noise temperature is reached in the region of 10-20 K.

The value of product $\Delta f q_{\Pi}$ in formula (2) depends on the type of signal modulation, coding method and signal extraction as well as the admissible error probability in information receipt. In terms of the best utilization of the radio line energy the optimum are multi-position or-thogonal digital signals and phase manipulation of carrier frequency.

In modern radio electronic systems the minimum value of product $\Delta f q_{\Pi}$ is 3 - 4 $F$ at error probability in signal receipt $10^{-3}$, where $F$ – speed of digital information transmission.

## Antenna systems for long range space communication

Fully rotable parabolic antennas and phased antenna arrays formed by several antennas are most widely used in ground-based complexes of space communication.

Antenna's most important characteristics that determine the energy performance of radio lines are the directive gain (DG) and the effective area $S_{\Im\phi}$ of antenna mirror which are interconnected by the expression

$$DG = 4\ \pi S_{\Im\phi}/\lambda^2\ ,\qquad\qquad(3)$$

where $S_{\Im\phi} = K_{N.\Pi} S$, $S$ – geometrical area of mirror aperture, $K_{N.\Pi}$ – area utilization ratio.

The area utilization ratio characterizes the quality of construction and antenna's workmanship and is determined by the amplitude and radio waves phase distribution pattern in the mirror aperture. The following approximate expression has been obtained for antenna DG depending on deviation of the mirror's shape from the designed one because of inaccuracy of its fabrication and in consequence of deformations caused by the antenna's own mass, wind stresses and uneven heating of its construction by the sun:

$$DG = K_0\ (\pi\varnothing/\lambda)^2\ [1 - (4\pi\sigma\varnothing/\lambda)^2\ ],\qquad\qquad(4)$$

where $\sigma = \delta/\varnothing$ – relative accuracy of mirror fabrication; $\delta$ – mean square deviation of the mirror's actual surface from the designed one; $\varnothing$ – antenna mirror diameter; $K0$ – antenna area utilization ratio in case there are no deviations of the mirror shape from the designed one.

The expression (4) has an extremum relative to l. The DG reaches the maximum value equal to $K0/(64\sigma)$ at $\lambda_{опт} = 4\pi \sqrt{2}\ \sigma\varnothing$. At the set values of $\delta$ and $\Delta$ the magnitude of $\lambda_{опт}$ is the minimum operating wave length for the antenna at $\lambda < \lambda_{опт}$ DG, and hence $S\ni\phi$ quickly decreases and the antenna ceases to operate efficiently.

The relative accuracy of fabrication of the best speci-mens of modern large size parabolic antennas is $(1,5...5) \cdot 10^{-5}$. Then, taking for optimum wave length used in deep space communication $\lambda = 3$ cm and the attained accuracy in mirror fabrication $\sigma = 1,5 \cdot 10^{-5}$, the maximum diameter of the parabolic antenna will be:

$$\varnothing_{max} = \frac{\lambda_{опт}}{17.6\sigma} = 114 \text{ m}$$

i.e. in given conditions it is unpractical to create fully ro-table parabolic antennas with a diameter larger than above described.

In real life conditions the mean square deviation of the mirror shape from the designed one due to deformations caused by gravitation, wind stresses and heating by the Sun, may be as high as tens of millimeters for antennas measuring 60 to 70 m in diameter which results in the corresponding increase of $\sigma$. These factors, too, are a stumbling block in an effort to increase the size of this type of antennas and must be taken into account in choosing the specific place to locate an antenna and in choice of mirror designs. To further increase the energy potential of deep space communication radio lines it is advisable to build antenna arrays based on several large size antennas and coherent addition of signals received by each antenna.

A number of unique fully rotable parabolic large diameter antennas function nowadays across the globe. They are used in deep space radio communication complexes and in radio astronomy. The largest of them has been built in Germany. This is a radio telescope of the M. Planck Institute of Astronomy near Bonn. Its mirror diameter is 100 m. The radio telescope has very high specifications compared to antennas of the same class. Table 1. shows characteristics of some large size parabolic antennas operated throughout the world.

The maximum size of the onboard antennas of DSC is limited by the carrier rocket's fairing diameter in case of rigid construction of the mirror and equals 3 to 4 m. Currently, much progress has been made in fabrication of unfolding parabolic antennas of the umbrella type. The required accuracy in the retention of the paraboloid shape for $\lambda = 3$ cm is obtained at the diameter equal up to 10 m and more.

**Table 1.** Large size antennas for long distance space communication

| Location,town (country) | Mirror diameter, m | Effective area, m2 | Operating wave range |
|---|---|---|---|
| Edelsdorf (Germany) | 100 | 5,500 | cm |
| Goldstone (USA) | 70 | 2,600 | cm |
| Madrid (Spain) | 70 | 2,600 | cm |
| Canberra (Australia) | 70 | 2,600 | cm |
| Parks (Australia) | 64 | 1,900 | cm |
| Usuda (Japan) | 64 | 1,900 | cm |
| Jodrell Bank (England) | 76 | 1,100 | dm, m |
| Yevpatoriya (Ukraine) | 70 | 2,600 | cm, dm |
| Ussuriysk (Russia) | 70 | 2,600 | cm, dm |
| Medvzh'i Ozera(Russia) | 64 | 1,900 | cm |
| Kalyazin (Russia) | 64 | 1,900 | cm |

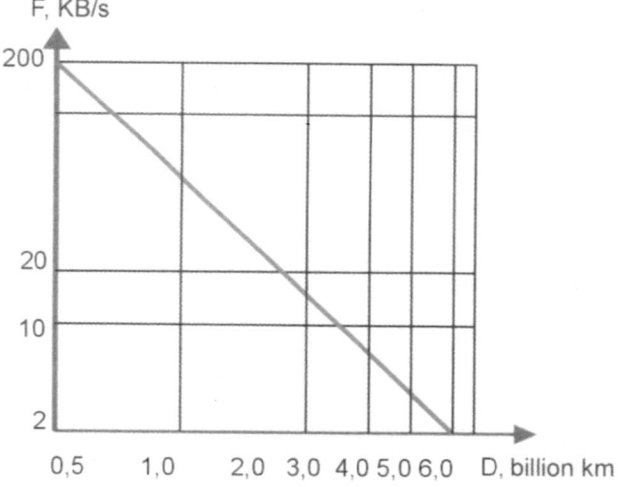

**Fig.19.**
Dependence of information transmission speed in radio line on communication range

Fig. 19 shows the dependence calculated on the basis of formula (2) of the maximum speed of information transmission via SC – Earth radio line on the communication range at the following values of the system parameters:

$\lambda = 3.6 \cdot 10^{-2}$ m; $P_\delta = 20$ W; $\varnothing_\delta = 3.7$ m; $K_{N.\Pi.\delta} = 0.5$; $\varnothing_\mathfrak{z} = 70$ m; $K_{N.\Pi.\delta} = 0.7$; $\eta_\Sigma = 0.7$; $T\mathfrak{z}\phi = 35$ $K$; $\Delta f q_\Pi = 4F$, where "$\delta$" index means board, "$\mathfrak{z}$" – mirror.

These parameter values are close to the limit reached in modern radio electronic complexes for far-out space applications. Both Russia's and NASA's (USA) DSC GBCC are built with due regard for the special features characteristic of the control systems for DSC.

The US NASA's DSC GBCC comprises:
- three long distance space communication centers (LDSCC) located near the towns of Goldstone (USA), Madrid (Spain) and Canberra (Australia);
- flight control center in the town of Pasadena (USA);
- means of communication and data transmission.

All ground-based LDSCC are united into Deep Space Network (DSN) system and are spaced over the Earth surface at approximately 120° in longitude and at 35–40° in latitude. Thus the DSN stations provide for continuous round-the-clock communication with interplanetary SC.

Each DSN station is equipped with three parabolic antennas: one measuring 70 m in diameter and two 34 m. The operating frequencies of Earth – SC radio lines are 2.1 GHz ($\lambda$ = 14 cm); SC – Earth – 2.2 GHz ($\lambda$ = 13 cm) and 8.4 GHz ($\lambda$ = 3.6 cm).

In the 1980s NASA modernized the ground-based network of DSN stations in order to enhance energy performance of its radio channels to ensure the flight of Voyager-2 SC launched in 1977 and which had reached by August 1981 the vicinity of Saturn, by January 1986 – Uranus, by August 1989 – Neptune. The effective aperture of ground-based recipient systems was increased by creating a synthetic lattice out of several antennas spaced apart and by increasing the diameter of the mirror of LDSCC main antennas from 64 to 70 m.

During 1981–1989 the Canberra-based antenna complex was joined by radio telescopes 64 m in diameter based at Parks, Australia, and Usuda, Japan. The complex at Goldstone used additionally a super-large antenna array comprised of 27 antennas measuring 25 m in diameter. The complex belongs to the radio astronomy observatory of New Mexico, USA.

As a result of work done the total amplification of the radio signal from the Voyager-2 SC at the stations in Canberra and Goldstone was increased by more than 5.5 dB.

The joining of the Australian and Japanese antennas to the DSN network also increases the reliability of signal receipt from SC in case of adverse weather conditions in an area of one of the antennas.

The aim of further development of the DSN network is joint use of the American antenna network that unites the complex at Goldstone and a super-large antenna array in New Mexico with the Australian-Japanese network comprising the complex at Canberra and radio telephones at Parks and Usuda.

The DSC GBCC of the Russian Federation comprises:
- two centers of long range space communication: western – at Medvezh'i Ozera facility in the Moscow region and eastern – near Ussuriysk;
- flight control center at S.A.Lavochkin NPO
(research and production association) (the town of Khimki, Moscow region);
- communication and data transmission means.

**Fig. 20.**
General view of long range
space radio telescope
(at Medvezh'i Ozera)

The western and eastern LDSCC are spaced apart at 100° in longitudes and provide for continuous communication with interplanetary SC for 18 hours in a 24 hour period. The LDSCC at Medvezh'i Ozera is outfitted with a parabolic antenna 64 m in diameter (Fig. 20). The LDSCC near Ussuriysk comprises three large size antennas 70, 32 and 25 m in diameter. The operating frequencies are:

• Earth – SC radio line – 5 GHz ($\lambda = 6$ cm) and 0.77 GHz ($\lambda = 39$ cm);
• SC – Earth radio line – 8.4 GHz ($\lambda = 3.6$ cm); 5.9 GHz ($\lambda = 5.1$ cm) and 0.94 GHz ($\lambda = 32$cm).

In order to improve the reliability of DSC control and to carry out high precision measurements of their angular coordinates by radio interferometric technique, in the package of GBCC used as extra means is the LDSCC located near Yevpatoriya and having in its suite high-performance antenna systems measuring 70, 32 and 25 m in diameter.

The energy performance of radio lines of DSC GBCC LDSCC allows to reliably control DSC, receive from them research information and conduct trajectory measurements at ranges exceeding the radius of the Solar system (6 billion km).

## 3.2.2 The state and the main trends in development of GBCC

## The experience of creating and operating Russia's ground-based SC control complexes

Before the early 1990s the sole prime contractor of space hardware and SC control systems was the MoD. The GBCC of all SC for research, business and military purposes were built within the framework of ground-based automated control complex (GBACC) of the MoD which implies the entire set of ground-based complexes for controlling various types of SC.

This made it possible to use the multi-point technology of controlling SC which has expanded the radio visibility zone of SC from the country's territory. Also, this ensured the reliable execution of technological control cycles (the probability of successful execution of communication session was 0.99). A possibility emerged to reserve communication session points and redistribute technical means in case of various malfunctions both on board SC and on the ground-based complex.

However, the availability of multi-point GBCC providing for a large number of communication sessions with SC in a 24 hour period in conditions when all expenses related to operation of GBACC were borne by the MoD deprived the developers of SC of economic drives to improve the control technology of SC, to reduce the number of measurements and control sessions and to transfer the supporting functions (navigation, condition diagnostics) aboard SC. As a result, the control technology had not changed over many years and was rather costly.

Moreover, the use of multi-point technologies with involvement of a large number of ground-based systems leads to considerable inertia in replacement of the already functioning controls by those of a new generation. This process is heavily behind times due to cuts in budgetary allocations to maintaining the technical readiness and to modernization of GBACC.

The SC control centers were also created at the GBACC facilities of the MoD. This is explained by a number of reasons:
• availability of sufficient number of skilled operators and managers at the MoD;
• necessity to employ costly and the then sophisticated computing hardware used in information and computation complexes for controlling SC;
• the commonality of means of management technology and software support proposed by chief designers for SC of both military and business and research applications. Developed by one and the same company, those SC enable one information and computing complex to be used for servicing various centers.

The exception was the FCC of TsNIIMash (Central Research Institute of Machine Building) that provided control of unique piloted complexes, including those used under international space programs. The complicated tasks of optimizing control of piloted SC involve the efforts of a large number of developers and testers of piloted orbital stations and spaceships. Also, this implies an access of foreign specialists to FCC. Such activity could not yield the desired results should it proceed in the traditional way at the facilities of the MoD with the use of standard automation technologies.

After completion of flight tests and acceptance for operation all manned SC were handed over to the Space Systems Test and Control Center of the MoD located at the town of Krasnoznamensk, Moscow region. The center was fully responsible for due operation of hardware received. Among such hardware there were GLONASS SC, one of the most sophisticated information systems known to date. Being unique, some SC, for instance manned space complexes and deep SC, were not handed over.

In practice the situation has evolved which is legalized by supporting documents that specify the distribution of functions among FCC and Target Information Processing Center (IPC). The IPC develops a program for using target oriented equipment of SC and transmits it to FCC which develops a program for functioning of onboard service equipment, combines it with the target program and provides for its realization. This practice is characteristic, in particular, of the Resurs-O and Meteor types of ERS SC. The target program for use of those SC is developed by a dedicated organization (NPO Planeta) with involvement of the SC Chief Designer's sector which analyzes the state of target equipment and issues recommendations for controlling SC.

In the 1970s a number of factors emerged that changed the situation in the creation of new space systems and controlling them at various stages of function.

1. The Russian Space Agency had been founded (later renamed the Russian Aviation and Space Agency, or Rosaviakosmos for short), which became the customer of space systems and, as per the law of the Russian Federation "On Space Activity", the entity responsible for their use.

2. The political and economic changes in Russia brought about catastrophic cuts in funds allocated by the Federal Budget to space programs and left Russia's space industry to its own devices. As a result, many SC and their control systems are beyond the guaranteed service life. Some research and development work begun as far back as the 1980s is either suspended or proceeds as a matter of formality. The operation of GBACC and maintaining its systems in the state of operational readiness become increasingly costly. This does not suit Rosaviakosmos and chief designers of space systems who are pressed to pay for services of the MoD in controlling civil SC.

3. Many types of observation, communication, and navigation SC have been declared dual use objects that will be created and operated by Russia's MoD and Rosaviakosmos on shared expenses basis.

4. Various joint stock companies have been established capable of acting as customers or developers of space systems. As a result, the first group of

commercial space communication and TV broadcasting systems has been created since it's these systems that pay off the quickest. Among them are such systems as Express, Gals, Bankir, Gonets, Bonum-1, Yamal. All those systems have specific one-point or few-points (two points) GBCC not integrated with GBACC of the MoD. The creation of such GBCC has become possible due to the use of the many year experience in controlling similar SC and employment of modern computing hardware (mostly of foreign manufacture) in flight control centers as well as in consequence of attempts to gain independence from the MoD in management of commercial SC and to reduce the operating cost of GBCC.

Other trends, too, become evident in the ERS space systems and others being built under the programs of Rosaviakosmos. Those systems are not fully self-repaying economically and are built mainly at the expense of the Federal Budget. The developers of those systems and the leadership of Rosaviakosmos try to reduce the expense for services of the MoD in controlling SC. So they seek to switch to a single point or few-point technology with the use of command and instrumentation system (CIS) and telemetry system from the component package of GBACC and to create a FCC of new generation SC outside the framework of GBACC of the MoD, for example at their own facilities (deep SC FCC at S.A.Lavochkin NPO or on the basis of TsNIIMash FCC (Okean-O SC FCC and Meteor-3M SC FCC). Here at work are objective circumstances that manifest themselves in the fact that the developers can utilize experienced help in controlling SC of their own development which often are not wanted in today's conditions. In addition new FCC can be created directly on the basis of new control technologies with the use of state-of-the-art engineering facilities.

Modern flight control centers (FCC) are created on the basis of personal computers or high-performance work stations integrated into a single local computing network of FCC. They use highly reliable servers with large RAM and long term memory devices for storage of data bases and archiving information. This principle was used in creating FCC of TsNIIMash, FCC of NPO PM, FCC of Bankir system, FCC of Yamal-100 SC and others, some of which are being created on a corporate basis. The computing hardware used in equipping the above listed FCC is of Hewlett Packard origin.

The experience of creating GBCC for SC of various applications shows that as a rule GBCC is created for controlling a specific type of SC. In such a case a new center or SC flight control sector is created and special software (SSW), general system software (GSSW) and information exchange software (IESW) is developed. The interaction of FCC with systems and outside organizations is based on the use of the existing or newly created communication and data transmission equipment (CDTE). Under such circumstances it is possible to additionally develop special elements of CDTE to provide interaction with outside organizations re-involved in controlling SC.

In the conditions of drastic financial cuts and desperate shortage of all kinds of resources it is necessary to reduce to the minimum the cost of creating and operating SC ground-based control complexes. This means rejection of the traditional approaches in creating the SC GBCC. The solution of this problem implies transition to one-point control technologies with integration of FCC into ground-based stations of the command and instrumentation system, combining in the single region the tasks of SC control and information processing. What is meant here by the single-point technology is the one in which all operations of SC control: transmission of CSI on to SC, pick-up of telemetric information (TMI) from SC, measurement of current navigation parameters (MCNP), checking and correction of onboard time scale are conducted by means of one GBCC CIP. The required reservation in order to attain the desired operation reliability is achieved by fitting extra technical systems on CIP.

The experience shows that further development of GBACC of Russia's MoD by traditional means through replacement of obsolete systems by new ones cannot yield the desired effect both on account of physical impossibility to renovate the entire fleet of existing systems and because of emergence of new information technologies based on the use of space communication, navigation and relay service systems.

The above listed circumstances urge the search for new approaches in creating SC GBCCs and revision of the accepted control technologies.

## The existing structure of SC ground-based complexes and control systems

The control of spacecraft for various applications is accomplished in Russia mainly by means of the ground-based automated control system of the MoD. In addition, control over some research and utility national economy SC, execution of flights of manned SC and control of commercial SC are obtained by using the systems and means of other state entities and commercial organizations.

GBACC of the MoD controls all military and dual use SC, as well as most of SC for national economy purposes. In addition, the technical facilities of the MoD's GBACC are used to control the manned vehicles and deep SC whose GBCC have flight control centers insubordinate to the MoD. Overall, the GBACC of the MoD controls 85% to 87% of SC of Russia's orbital constellations (OC).

The GBACC of the MoD is a well organized entity with a centralized management sustained from the Space Systems Test and Control Center (SSTCC). Operating within the framework of GBACC of MoD are the SC control and information processing centers located in SSTCC at facilities known as Shabolovka and Rockot (Moscow) plus 11 command-instrumentation and one instrumentation post located near the towns of Shcholkovo, Krasnoye Selo, Maloyaroslavets, Ulan-Ude, Yeniseisk, Kolpashevo, Vorkuta, Yakutsk, Komsomolsk-on-Amur, Ussuriysk, Yelizovo, Barnaul. Fig. 21 shows one of the command-instrumentation posts of GBACC.

**Fig. 21.** Command-instrumentation post of GBACC

The average daily capacity of the MoD GBACC is around 800 control sessions, the total throughput capacity being up to 215 SC simultaneously operating in orbits.

The SC ground-based complexes and control systems not comprised by the DM GBACC have been and are still being created within the framework of agencies supervised by Rosaviakosmos and other state-owned and commercial organizations. Operating now in this group are the following facilities.

TsUP-M of TsNIIMash (Flight Control Center of the Central Research Institute of Machine Building based at Korolev, Moscow region) performing the functions of FCC of piloted vehicles flown under various programs, including SC of Russia's segment of the ISS as well as FCC of automatic SC Okean-O and Meteor-3M. In addition, TsUP-M of TsNIIMash processes and analyzes measurement data provided by Russia's network of optic-laser stations tracking space objects.

The S.A.Lavochkin NPO has established a FCC for deep SC and high apogee research SC of Interbol and Spektr types.

Based at the facilities of Medvezh'i Ozera (Moscow region) and Kalyazin (Tver region) are radio electronic complexes involved in controlling deep SC, medium- and low-orbital research SC, including foreign ones.

Functioning at the Research Institute of Thermal Engineering (Moscow) is a FCC of the Gonets (messenger) low orbit space communication system.

A number of joint stock companies have created by order of state agencies or commercial organizations few-points (single point) GBCCs for communication and TV broadcasting.

AO Persei (joint stock company) with participation of Rosaviakosmos and the Ministry of Communications (represented by the state-owned enterprise Kosmicheskaya Svyaz (space communications) has created and has been using over a number of years the first phase of the GBCC for communication and TV broadcasting SC like Express and Gals comprised by FCCs and CIPs based at NPO PM in the town of Zheleznogorsk, Krasnoyarsk region,and by CIPs at Space Communication Center in the town of Gus-Khrustalnyy, Vladimir region.

OAO Gazkom (joint stock company) has created GBCC for communications SC of Yamal type. It is comprised by FCC and CIP on the territory of the Energia space rocket corporation in the town of Korolev, Moscow region.

OAO NTV has purchased from Hughes (USA) and put in operation in 1998 NTV SC of Bonum-1 type along with equipment for FCC and CIP, placing it at Skolkovo, Moscow region.

To control deep SC and high apogee research SC the DSC GBCC uses the Ukrainian LDSCC at Yevpatoriya.

The Central Bank of the Russian Federation has created GBCC for the Bankir SC comprised by FCC in Moscow and CIP in Nudol, Moscow region. Fig. 22 shows the antenna and the building of the command and instrumentation system Kashtan at the Nudol facility.

The tendency to create independent GBCC is manifest not only for communications SC but also for some other types.

The organization of independent civil SC GBCC by various departments and agencies outside the currently operating MoD GBACC is urged by the following factors:

1. the desire of customer departments (including Rosaviakosmos) to be independent and obtain the stable control of their SC and space systems since SC control is not among the MoD's top priority jobs and under certain conditions it can ignore the interests of civil departments for the sake of its own primary duties;

2. the high cost of services rendered by the MoD in controlling SC, which in some cases exceeds by far the cost of similar services offered by civil organizations;

3. essential differences in requirements imposed in creation and operation of control systems of military and civil SC: the main requirement applied to military SC GBCC is operability in special conditions, stability and survivability, stealth capability; the main requirement applied to civil SC GBCC is compliance with international standards and recommendations, compatibility with foreign complexes, competitiveness on the foreign market of space services, unclassified nature.

Overall, the current control complexes and systems of SC belonging to state civil departments and commercial organizations do not constitute a single whole. They are being created and developed independently from one another without a general plan or coordination.

**Fig. 22.** Kashtan command and instrumentation system at the Nudol facility

The important tasks now facing the state's major agencies involved in space work (Rosaviakosmos and MoD) are the pursuit of the single engineering policy and coordination of work aimed at creating and operating SC control systems, at bringing together and regulating the construction of new SC GBCC belonging both to state and commercial organizations, at optimizing the development of technical facilities of GBCC for pursuit of federal interests in space.

To resolve the arising problems, to securely control SC for dual (military and civil) application, for tasks of national economy and international cooperation, for subsequent saving of budgetary funds, the management of the MoD and Rosaviakosmos decided in 1999 to establish a single Federal GBACC for SC control and instrumentation.

The Federal GBACC is supposed to provide reliable and effective control of 130-140 SC of various applications being simultaneously in orbit, including 25-30 SC for military, 70-80 for dual (military and civil) and up to 30 SC for scientific, economic and international cooperation purposes.

In this case the facilities of the Federal GBACC are supposed to control and keep up the information exchange with 6-12 modules of ISS and piloted SC, 40-60 low orbital SC and 65-70 SC in high elliptical, high circular and geostationary orbits.

The accepted concept of establishment of a single Federal GBACC proposes to legalize the requirement for management of the Federal GBACC assets by predominantly commercial SC. Should it be impossible, economic conditions must be created that urge commercial organizations to use and develop the Federal GBACC. Rosaviakosmos jointly with Russia's MoD must supervise the creation of commercial GBCC, including the licensing of space activity of commercial organizations and certification of their SC control systems.

The component package of a single Federal GBACC is supposed to contain:

• all structural elements of MoD GBACC, instrumentation complexes of spaceports and test sites of strategic rocket forces;

• FCC of TsNIIMash, FCC of Lavochkin NPO, FCC of Bankir space system with controls placed at the Nudol facility, FCC of NPO PM with SC control systems located at the facilities of Gus-Khrustalnyy and Krasnoyarsk-26;

• the facilities of the Moscow Power Engineering Institute's Experimental Design Bureau, Medvezh'i Ozera and Kalyazin, with radio engineering complexes located at them;

• the federal network of optic-electronic and laser stations.

The interaction of complexes and systems of the single Federal GBACC is ensured by the communications and data transmission system which also supports their interaction and information exchange with outside organizations and complexes.

To address the issues of prospective development and the use of Federal GBACC and its components, an interdepartmental coordination council has been formed consisting of representatives of Russia's MoD, Rosaviakosmos and other interested federal and non-federal entities.

The council is supposed to prepare proposals for Russia's Federal Space Program and the State Armaments Program. The proposals concern SC control complexes, systems and instrumentation.

The distribution of work and responsibilities among Rosaviakosmos and the MoD for controlling various SC within the framework of Federal GBACC is regulated by the law of the Russian Federation "On Space Activity".

The customers of the Federal GBACC complexes and systems are:

• for military systems – Russia's MoD;

• for dual-use application systems – the MoD and Rosaviakosmos on shared expenses basis;

• for systems for civil application and international cooperation – Rosaviakosmos.

The amount of combined funds for maintaining serviceability, operation, modernization and development of the Federal GBACC are distributed among Rosaviakosmos and the MoD in proportion to the load of the Federal GBACC technical assets in interests of the said agencies. Participation in the shared funding is also open to interested federal and commercial organizations and municipal entities.

The structure of the single Federal GBACC is shown in Fig. 23.

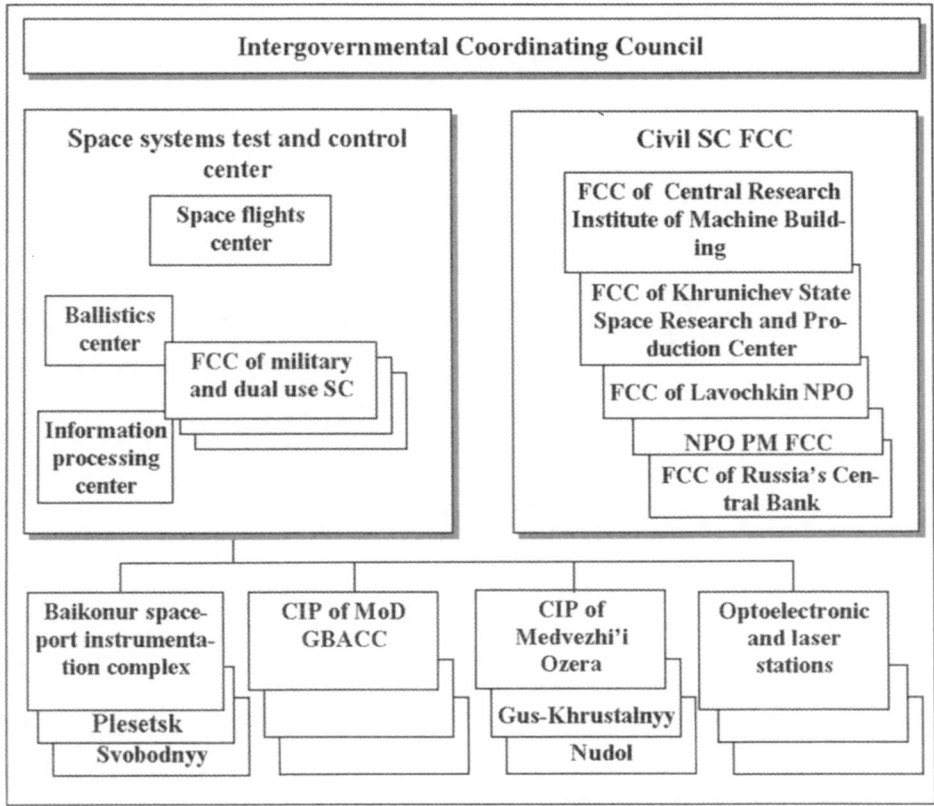

**Fig. 23.** Structure of the single Federal GBACC

## The state and prospective development of complexes and systems of the single Federal SC GBACC

The single Federal GBACC will comprise both modern systems and a large quantity of obsolescent and worn hardware. Fig. 24. presents summarized data showing exhaustion of service life of MoD GBACC which is the core of the single Federal GBACC.

Fig. 24 shows that more than 70% of MoD GBACC assets have exceeded their guaranteed life cycle (are in category 2 or 3).

More than 40% of assets have been in operation for more than 15 years. The GBACC assets having a long service life had been manufactured with the use of old components and have, as a rule, low performance, inferior maintainability and do not meet the requirements of the day.

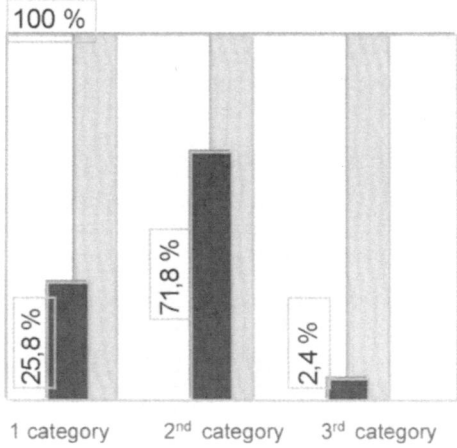

Fig. 24.
Exhaustion of service life of GBACC
systems:
    1st category – within guaranteed life
                   cycle;
    2nd category – beyond guaranteed life
                   cycle;
    3rd category – requires mid-life repair

The adverse factor is the steady tendency of the yearly aging (on average) of GBACC assets and deterioration of their performance. For example, the value of the mean-time-between failures slides down yearly by 10%. Table 2 shows that aging is the main cause of failures in GBACC assets in course of their operation. More than 79% of hardware fails on account of aging.

The progressive aging of technical assets and systems of GBACC results from drastic cuts in 1991 of funding for acquisition, modernization and replacement of obsolete and worn-out equipment, which resulted in reduced procurements and insufficient introduction into operation of new hardware (Fig. 25).

**Table 2.** Causes of failure of GBACC assets

| Causes of failure | Structural | Workmanship | Aging | In-service failures | Others |
|---|---|---|---|---|---|
| Relative number of failures, % | 1.04 | 2.63 | 79.45 | 0.6 | 16.28 |

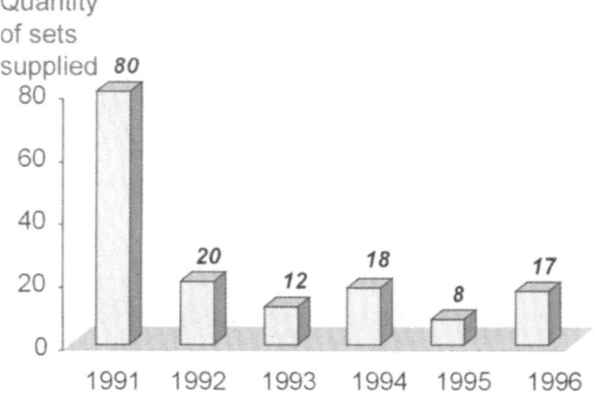

Fig. 25.
Supplies of GBCC systems
by years

The component package of the existing GBACC contains a large number of various types of equipment. Specifically, it has seven types of the command and instrumentation system in sixteen modifications, three types of telemetric stations in nine modifications. Figures 26–27 show some technical assets used in GBACC.

The core of the MoD GBACC are modern PC operating along with obsolescent and worn computers like M-222, BESM (large electronic computer), EC-1033, EC-1045.

The presence of various generation computers in the control configuration necessitates the development, introduction and servicing of software separately for each type of computer.

Under such conditions theprovision of a reliable and effective control of the existing or prospective orbital constellation of SC requires a step-by-step re-equipment of GBACC with a new generation of control and automation equipment with simultaneous keeping the existing systems in workable condition.

The accrued ex-perience in operation of SC, the improvement of on-board control systems (OBCS) of SC and the new requirements emerging in new economic conditions, requirements to minimize the cost of creation and operation of SC GBACC, enabled the revision of both SC control technology and the principles of building GBCC.

**Fig. 26.**
Antenna-feeder device of Mars small size aggregated radio elektronik system

**Fig. 27.**
Zhemchug antenna-feeder
device of PRA-MK

The standard technology is determined primarily by the requirement for the number of SC control sessions effected by GBCC assets for each control function. The analysis of technological charts of SC control shows that the performance of basic functions of information exchange of GBCC and SC in the phase of standard operation necessitates the conduct of the following number of control sessions:

• introduction of operating program (OP) and timing programs (TP) in SC – not more than once in 24 hours in two control sessions (main and stand-by) for most SC and not more than twice in 24 hours respectively in four sessions for ERS SC with refinement of OP (TP) as per weather conditions;

• conduct of SC MCNP by means of GBCC – at two "visible" circuits by two to three CIPs at each circuit for SC with orbit altitude up to 400 km and much more seldom for SC in higher orbits;

• pick-up of information of summarized control (ISC) from SC over CIS line – in all communication sessions scheduled for the given 24 hour period for transmission of CSI and conduct of MCNP;

• checking of onboard time scale (OBTS) and single time scale (STS) – not more than once in 24 hours during CSI transmission sessions or execution of MCNP;

• pick-up of telemetric information (TMI) – not more than once in 24 hours.

To maintain communication with SC and to receive telemetric information beyond radio visibility zones of ground-based systems, ship- and aircraft-borne posts are used (Fig. 28 and 29).

The enumerated requirements should be accepted as maximum in case of standard operation. On many occasions, in view of the coming intellectualization of OBCS they can be essentially slackened. The significant (in the future) increase of SC autonomous functioning will simplify the functional control circuit and, hence, the structure of GBCC. Rising as top priority is the development and realization of the single-point control technology enabling the control of SC from one CIP. The use extra CIPs may appear necessary only for stand-by roles in case of emergency on board SC and the necessity to do the repair.

Considering the analysis of SC control technologies, the minimal time required for informational contact of GBACC systems with all SC of the orbital constellation in course of normal operation could be obtained in the future on the territory of Russia by means of three CIPs spaced over the geographical longitude, one in the central, western, and eastern regions.

The introduction of advanced technologies of SC control is one of the main trends in perfection of GBACC, in improvement of its systems' characteristics, in structure simplification and load reduction which cuts back the cost of control systems' operation. The main elements of SC advanced control technology whose introduction is due to begin early in the 21st century are as follows:

• the use of GLONASS/GPS space navigation system for determination of SC orbit and autonomous fulfillment aboard SC of tasks relating to navigation, ballistics and time evaluation;

• transmission of SC position vector and results of solution of NBS assignments conducted at OBCS via telemetric radio channel to GBCC;

• the use on SC of highly intellectual OBCS with sophisticated software support for automatic diagnosis of condition, for restoration of workability of utility and target onboard systems and their controls;

• the use of space relay system for large scale servicing of low orbit SC;

• the introduction for controlling SC from FCC of more effective command and coordination technique instead of the time-and-software method;

• the use on board SC of equipment that forms "GBCC call" signal and its transfer to GBCC via Farman warning system's radio channel in case of emergency on SC not envisioned by OBCS software support;

• realization at GBACC data transmission equipment of digital network of integral servicing based on prospective telecommunication technology;

• the use in creating and operating GBCC systems of international standards, recommendations provided by the International Consultative Committee on Space Data Transmission (CCSDS), of protocols of information exchange and frequency ranges assigned by radio communication regulations on a primary basis in order to obtain compatibility and assure integration capability of Russian-made SC control systems with foreign ones in implementation of international space projects.

**Fig. 28**. Ship-borne instrumentation complex

**Fig. 29**. Aircraft-borne instrumentation complex

The effect achieved in realization of the new control technologies when applied to the existing and prospective SC is shown in Table 3.

**Table 3.** Advanced technologies of SC control

| The aim of new technology | Schematic solutions and technological operations | Expected effect of introducing the new technology |
| --- | --- | --- |
| Autonomous support of SC serviceability at prescribed level | Automatic diagnosis of condition and restoration of SC systems' serviceability by means of OBCS | Increase of autonomous functioning of SC to 30 days. Reduction of the number of remote tests sessions and tasks of information and telemetric serviceport (ITS) in GBCC by 5 to 10 times3. |
| Autonomous (without participation of GBCC) parameters of SC orbit and checking and correcting of onboard time scale | The use aboard SC of users' navigation equipment (UNE) of GLONASS/GPS space navigation system. Integration of UNE with OBCS. Autonomous performance of NBS duties by OBCS | The increase of duration of autonomous functioning of SC to 30 days. Improvement of NBS accuracy by 3 to 6 times. Reduction of the number of SC MCNP sessions, checks and corrections of OBTS by 10 to 30 times |
| Provision of continuity and global coverage of SC control and supervision of launch services with a limited number of CIPs | Use of relay information exchange modes between GBCC and SC via RS included in control circuit. Transmission of TMI from carrier and booster via RS | Reduction of the number of CIPs in GBCC, and of instrumentation posts in instrumentation complex of spaceport to two (including stand-by). Increase of global coverage of SC to a unity. Provision of virtually real time scale in SC control and in receipt of TMI from carriers and boosters |
| Enhancement of autonomy in fulfilling ERS duties by SC | Employment of coordinates method in controlling target use of SC for ERS. Introduction from Earth of operating programs in "coordinates – operations" format. Autonomous formation in OBCS of programs for operation of SC onboard systems during survey | Reduction of frequency of sessions of communication with SC, and of amount of transmitted CSI by 5 to 10 times. Reduction of the number of CIPs in GBCC to 1-2. Reduction of amount and time of work at FCC devoted to formation of OP of target use of SC for ERS duties. |
| Provision of operative control of condition of SC at FCC without involvement of GBCC assets in course Enhancement autonomy of SC | Combining in one radio channel of transmission of target and control information from SC Formation on board SC and transmission to GBCC of "GBCC call" in case of emergency on board | Reduction of the number of sessions of SC remote control conducted by GBCC CIS in course of normal operation.Increase of autonomous function of SC. Reduction of the number of sessions of remote tests and of tasks tackled by ITS by 5 to 15 times |

The introduction of new technologies allows to radically change the distribution of SC control among ground-based and onboard control complexes in a way that enhances the role of OBCS, to essentially reduce the intensity of information exchange between OBCS and GBCC, to decrease by 3 to 5 times the load of GBCC systems, to increase by 2 to 5 times the effectiveness of fulfilling target missions, to enhance by 5 to 10 times (to one – two months) the autonomous mode of SC operation.

The realization of new SC control technologies hinges upon using the navigation equipment of GLONASS/GPS customers for NBS of SC flight. GLONASS space navigation system is intended to continuously supply navigation and time related data to an unlimited number of users located on Earth or in near Earth space at altitudes up to 2,000 km. The precision of establishing the exact location by such users is (3s) which is within 30 to 50 m in coordinates and 3 to 5 cm/s in components of the speed vector. The GLONASS system also provides for determination by the user of a correction to his time scale and its fixation to the national standard with an error within 1 ms.

The use of space navigation system to provide navigation for SC will increase by an order of magnitude its precision and effectiveness and will reduce the technological cycle of SC control by means of GBCC.

The availability of highly precise information about the locality of SC at all times will enable the switch from time-aided methods of planning the operation of SC and its onboard systems to coordinates-aided methods. This will not only relieve the ground-based control complex thanks to tackling most planning tasks on board SC, but will also cut the cost of fulfilling special tasks owing to activation of target equipment precisely over prescribed areas.

The employment of users' navigation equipment (UNE) of space navigation system (SNS) on board SC will make redundant periodic checks against the GBCC time scale and corrections of onboard time scale. Also, this will slacken requirements imposed on the stability of the onboard frequency and time standard.

Integration of UNE with sophisticated SC onboard digital computing system (OBDCS) will enable in the short term to transfer to OBCS the assignment of practically all NBS jobs, including calculation of basic data for orbit correction in order to maintain it and to perform maneuvers to other orbits.

In the late 1990s specimens of navigation equipment had been built in Russia that operate on signals provided by the national GLONASS space navigation system and the American GPS and are intended for installation on space rockets and other types of hardware. They are installed on board flying vehicles, on ship-borne navigation complexes and on ground mobile systems for various applications.

The SC flight control and coordination center incorporated by the Research Institute of Space Systems (the branch of Khrunichev State Space Research and Production Center) (Fig. 30), plans to develop the technique of processing information received from users' navigation equipment.

Its facilities are used to refine the coordinates methods of controlling ERS SC which, in turn, will enhance the efficiency of formulating at FCC the operating programs used to control the operation of SC target equipment.

The detailed program of time-related operations of preparation and execution of survey must be developed on SC OBCS with the aid of the navigation related information available on board about the orbit's parameters and with the help of software for automatic formulation of time-based programs.

**Fig. 30.** Small SC flight control and coordination center at the Research Institute of Space Systems, the branch of Khrunichev Center

## The principles of building multi-purpose information relay space system

The aim and functions of space relay system (SRS). One of the effective means to enhance the efficiency and global coverage in controlling low orbit SC with simultaneous reduction of the number of CIPs to one or two is the use of the space information relay system based on relay satellites (RS) in geostationary orbit. Before the creation of the SRS the issues of duration increase and of global coverage in information exchange with the objects of space rocketry were tackled by development of GBACC, spaceports' instrumentation complexes (SIC), target information (TI) exchange complexes as systems distributed in space, including tens of CIPs, IPs (instrumentation posts) and special information receipt posts. This is caused by the fact that information exchange of SC, carriers and boosters

with ground-based control and instrumentation systems is accomplished mainly via direct radio channels in the direct mode. The continued use of the spatially separated infrastructure of the said complexes with a large number of telemetric, command-instrumentation and other systems requires considerable funding on the one hand, while on the other it does not allow to essentially enhance the effectiveness of using space hardware, to improve SC control, the collection from launch vehicles of target related, telemetric and trajectory information. In addition, there are a number of SC control tasks and duties in information support of tests and standard operation of space rocketry which in principle cannot be carried out by the existing infrastructure of GBACC and SIC. Those are, specifically, the following:
   • launches of SC to solar-synchronous and polar orbits;
   • launches of ballistic missiles into trajectories not equipped with instrumentation systems;
   • control of piloted SC, orbital stations, ISS modules in a remotely operated mode during long periods of time.

The compliance with prospective requirements imposed on information support of tests and standard operation of space rocketry as well as the considerable assets reduction in the relevant ground infrastructure and, hence, the reduction of operating costs can be obtained only by using space information relay channels between space rocketry and centers for flight control, for target information collection and processing.

The developed in the 1980s and created in Russia SRS based on Luch (ray), Potok (flow) and Luch-2 relay satellites could not due to the technological restrictions of the then operating space hardware fulfill all the tasks pertaining to space rockets flights, the reason why they performed only limited missions in servicing separate SC.

The above described circumstances require creation of a multi-purpose SRS providing the effective solution of all prospective tasks set before them.

The aim of creating the prospective SRS is the securing of control in real time mode of an orbital constellation of SC, the transmission of target information from any portion of an orbit, the relay of telemetric and trajec-tory information transmitted from carrier rockets, boosters and ballistic missiles at an active portion of their launch trajectory, and the reduction of cost of maintaining the serviceability, operation and development of GBACC, SIC and the complexes engaged in target information collection and processing.

The systems engineering requirements imposed on the prospective space relay systems formulated on the basis of analysis of requirements applied to control of SC for military, business and research applications are shown in Table 4.

**Table 4.** Systems engineering requirements imposed on prospective SRS in relation to SC

| Performance indicator | Standard requirement |
|---|---|
| Number of simultaneously controlled SC | Up to 7 |
| SC control rapidity | In time scale close to real (<1 min) |
| Jamming resistance of relay radio lines (threshold ratio noise/signal at receiver input), dB/W | 15-20 (for control radio lines) |
| Mass of subscriber SC CIS onboard equipment, kg | 20-30 |
| Measurement conditions of current navigation parameters of subscriber-SC in relay mode | For single-point SC GBCC, including cases for checking the function of UNE |
| Commonality of prospective SRS with existing SC control systems | Use of S-, Ku (Ka)-ranges (for controlling civil SC and transmission of target information); UHF-range – for transmissionof "GBCC call" signal |

The prospective SRS is created as a national multi-purpose space relay system (MPSRS) to carry out research, economic, military, international and commercial space programs. In fulfilling its tasks the MPSRS must provide for:
• relay exchange with piloted objects of a space program, with SC of Earth natural resources investigation, hydro-meteorological support, military and dual application, scientific research;
• information relay from active portions of the trajectory during trials of ballistic missiles, launches of carriers and boosters at any inclinations;
• information flow exchange for controlling SC, exchange of target and instrumentation related information between control and information processing centers as well as with customers of rocketry and users of target information;
• receipt and relay of "GBCC call" signals in case of emergency on board SC.
The multi-purpose SRS must provide adequate commonality in the use of methods and standardized means of control, collection of instrumentation and target related information.

MPSRS must be compatible with foreign SRS (TDRSS of US, DRTSS of Japan, DRS of ESA) which will enable foreign RS to pick up information from the Russian SC in emergencies and its subsequent transfer to Russia's information collection centers. The realization of compatibility of MPSRS enables rejection from the orbital reserve of RS, providing for substitution in case of failure as agreed upon with other countries. In addition, MPSRS in the event of compatibility with other relay systems can be used commercially for servicing foreign SC, carriers and boosters.

The compatibility of MPSRS with foreign systems is achieved by using common radio frequency ranges, signals structure, modulation types, channels energy patterns, information exchange protocols with regard for recommendation of the International Consultative Committee on Space Data Transmission (CCSDS).

The capabilities, level of engineering and economic performance of MPSRS must not be inferior to the TDRSS of the USA and foreign prospective systems now in development. MPSRS must be competitive in terms of cost of services rendered in relay of various types of information as compared to foreign SRS. The cost of communication sessions with space rockets, which sessions are equivalent in information supply and duration, must be lower than in case of using ground-based instrumentation and command-instrumentation posts.

**The main characteristics of prospective MPSRS.** The space segment of MPSRS must consist of two or three RS in geostationary orbit. The active service life of RS is 10 to 12 years. The servicing zone of space rockets must be global and, if necessary, continuous up to the altitude of 2,000 m. At subscribers' altitude above 2,000 km the MPSRS must provide communication with single objects (for example, with boosters) up to the altitudes of geostationary SC.

In course of development the question must be resolved about the possibility of creating RS – RS channels preferably in the optic range, which channels enable transfer of high speed information flows at the reduction by several times of mass and size characteristics and power consumption of onboard equipment as compared to radio electronic channels.

MPSRS must use the following frequency ranges for relay of information to and from space rockets:

• - range – for low speed information flows of piloted vehicles, automatic SC of research and economic applications, TMI flows from active portions of trajectories of carriers, boosters and ballistic missiles;

• Ku- and Ka-ranges – in the main channel and for transfer of high speed flows of TI (tens and hundreds of megabits per second);

• optic range – for transfer of high speed information flows (hundreds of megabits per second) from observation SC;

• C-range – for commonality and compatibility with the existing SC control systems;

• UHF-range – for transfer of "GBCC call" signals, information concerning collection of data from automatic stations, warning signals about ground and mobile objects in distress. The MPSRS must provide both for multi-station access (MSA) and individual access (IA) to space rockets.

Throughput capacity of the system as per number of objects being serviced at a time:

• *at IA:* S-range – 2 objects; Ku- and Ka-ranges – 2 objects in each; C-range – 1-2 objects;

• *at MSA:* S-range – 1-2 objects per transmission, 4-8 objects per receipt; UHF-range – 1-2 objects.

RS must have in addition to the main channel a re-aiming antenna in *Ku-* or *Ka* range for direct delivery of TI to ground-based information processing centers.

The maximum speeds of information transmission are as follows:

• *at MSA:* S-range – Kb/s for receipt from a subscriber and 1-5 Kb/s for transmission;

• *at IA:* S-range – 0.256-5Mb/s; *Ku*-range – 64-180Mb/s; *Ka*-range – hundreds of megabits per second; *C*-range – 2Kb/s; in UHF-range – up to 1Kb/s;

• in optic range – no more than 300 Mb/s, up to 1 Gb/s in the longer term.

The mass of subscriber equipment for low speed information flows must not exceed 20-30 kg, for medium- and high speed flows should vary between 100 and 200 kg. The diameter of onboard antennas of object-subscriber must not exceed 1-1.5 m.

Promising for transmission of high speed flows of research information (300-600 Mb/s) is the use of the optic range. The mass of subscriber equipment and its power consumption can be reduced by 2 to 3 times as compared to equipment operating in the radio range for transmission of similar information

**Engineering policy in creating MPSRS.** The use of MPSRS is economically justified only when it is used for servicing on a mass scale of many low orbit SC of an orbital constellation and practically all launched carriers, boosters and ballistic missiles. Therefore decisions must be taken at a federal level that provide for installation of MPSRS subscriber equipment on all newly developed SC with a flight altitude up to 2,000 km as well as on carriers and boosters. After creation of MPSRS it is necessary to carry out research and development work on modernization of the already accepted SC, carriers, boosters and ballistic missiles for installing (testing) on them MPSRS subscriber equipment.

Simultaneously with work on creating MPSRS work must proceed on the equipping of space rockets with onboard UNE of GLONASS and GPS space navigation systems which will enable to essentially reduce the demand for ground-based systems of trajectory measurements. The transmission of UNE navigation information from space rockets must also proceed via RS as part of telemetric information or as independent batches of information.

Work must be done on introduction of adaptive-addressing onboard telemetric systems which will allow to compress information by several times and to switch radio channels to S-range. This will decrease the flow of telemetric information and slacken demands imposed on energy performance of MPSRS radio channels. This is particularly important for space rockets operating in emergencies when their antennas are not directed adequately.

The deadline of creation and introduction into service of MPSRS depends on the amount of funding allotted to above said work and will probably be established at a far-off point in the future. The expected completion date of research and development work and creation of the first examples of RS, subscriber equipment of space rockets and the system's ground-based segment is 2005 – 1007. The final introduction of MPSRS into operation and switching the bulk of low orbit SC and other objects of space rocketry to servicing MPSRS are possible after 2010.

Promising for the global control of low orbit SC are also space-based network structures built as part of ground- and space-based information control network on the basis of inter-satellite radio channels. In such systems onboard CIS on each SC of the orbital constellation must be used not only for communication with stations on the ground but also for re-relay of control information between SC. Under certain conditions the network structures have an advantage over SRS based on geostationary artificial earth satellites and can do more effectively some jobs related to SC control (e.g. controlling military SC in special conditions).

Considering that SRS built on principles of inter-satellite network structures essentially differs from its analogs and implies solution of quite a few rather complex engineering problems during its creation, its introduction into service will proceed on a step-by-step basis. The complete deployment of the system is expected after 2010.

## The state and prospects of development of automation systems

The core automation systems (AS) of SC control centers and information processing centers operated by GBACC in the 1990s were low performance computing systems of the second and third generations, more than 50% of which had exhausted their service lives many times over, were obsolescent and worn (SM, M-222, VK-2M45/46, Elbrus-1 series computers and the like). The SC control automation level was 70 to 80%. The inadequate state and insufficient level of automation, the considerable physical wear and heterogeneous types of software and information support systems resulted from their extremely slow renovation on account of insufficient funding. The procurements made by GBACC in 1997-1998 amounted to less than 20% of the required volume. As a result, in terms of development even the newly supplied AS were 10 to 15 years behind their analogs used abroad.

In the late 1990s the necessity to resolve "The 2000 problem" changed radically the situation at GBACC. According to the program approved by Russia's MoD, by the early 2000 around 60% of AS computing systems had been renovated at the GBACC priority facilities.

Compliance with technical specifications and tactical requirements imposed on SC prospective automatic control systems generates a number of serious systems engineering problems that determine the general technical and tactical requirements applied to AS of the Federal GBACC.

The necessity to equip the various elements of the Federal GBACC with prospective automation systems raises the problem of AS unification which in turn will essentially reduce the cost of creating and operating those systems. Of special importance is the problem of creating computing networks of the "client – server" architecture, which networks must become the base of automation systems of prospective SC FCC and the core of the systems that collect and process TM information.

An acute problem in creating prospective AS is the enhancement of efficiency of operators of SC control sectors. The study of modern control methods shows that the essential improvement in speed and quality (reliability) of solution of problems in controlling SC can be achieved by introducing new information technologies aimed to enhance the automation of operators' work. Among those technologies are the introduction of new forms of information presentation (color dynamic graphics combined with digital cartography), "man – machine" natural language interface; realization of expert systems of decision-making support; modeling of situations and other systems based on principles of artificial intellect; standardization of the communication procedure between operators and computers.

Particularly important is becoming here the issue of information protection against unsanctioned access and special information security measures since all types of computers used by GBACC are subject to requirements of guiding directives that specify the standards of information security (SIS), unsanctioned access prevention and special protection. The envisioned measures for information protection include:
- introduction of cryptographic techniques of information protection;
- protection of software and information support against virus programs;
- introduction of a special work station for a SIS officer into computing complexes.

It should be noted that for the single Federal GBACC the issues of SIS are particularly important since here along with military SC planned for servicing also are civil SC that are created and operate under international projects that envision cooperation with foreign partners in information and engineering technologies.

As per application and the natures of tasks tackled, all computing systems comprised by AS of GBACC facilities can be by convention categorized in two main groups.

**The first group** consists of CS that are used as a basis to form operator work stations, graphic stations, collective use information representation systems, terminal concentrators (computers) that are the central element of multi-terminal complexes and other systems comprised by the clients' part of computer networks. CS of the given class must primarily service all possible areas of automation of active duty operators' activities and address various computing problems, information processing assignments for estimating the condition of objects being controlled and the technical means of control systems, formation of information models, technical systems management, etc.

The characteristics of this class of CS must be within the following range: performance – from 5 to 30 million operations per second, RAM capacity – from 8 to 32 MB, HDD – from 0.5 to 3 GB.

The first group of CS must be represented by personal computers (PC). The analysis of the existing practice and proposals made by the industry to build prospective automation complexes shows that most proposals in this area favor the use of Intel PC.

According to operating conditions, all PC comprised by AS objects of GBACC must be manufactured as per state standard GOST V20.39.304-76, execution class 1.1 and 1.3 for use on stationary facilities; execution class 1.7 for use in the package of automatic systems of mobile GBACC.

The second group of CS used in building automation complexes of GBACC control centers must contain PC-servers and multi-processor work stations. Their employment is necessitated by organization of peer-to-peer computer nets and "client – server" nets. PC-servers are intended for use as file-servers of peer-to-peer nets with a number of users not exceeding 10. More rigorous demands are made of them compared to PC in terms of performance and failure-free operation. PC-servers must have Pentium II (III) type of processor, not less than 64 Mb of RAM, HDD – not less than 6 Gb, video memory – 2 Mb, VESA and PCI system bus.

Automation systems used by the Federal GBACC must be built using the principles of "open" systems, modularity and unification. This will provide the desired configuration and efficiency of AS in each element of GBACC (CIP, FCC, ballistics center and others) by including in their component package the required number of functionally oriented modules of various types integrated in local computing networks. The base of such modules must be the prospective high-performance systems manufactured in Russia which will replace the obsolete computers. At the initial stage such systems could be represented by Pentium and Pentium Pro-based computers. In the longer term those could be multi-processor computing systems based on RISC and MIPS whose manufacture will be organized at the Russian facilities under the program "Integration-CBT" (computers of the Baget family and Elbrus-90-micro computing complexes). This will make it possible to meet the requirements applied to information security and technical independence of AS employed and to standardize AS used in objects operating in various conditions and performing various functions.

The practicality of above described requirements imposed on AS as well as on basic CS is confirmed by the results of analysis of specifications of their foreign counterparts.

The following three interconnected stages can be singled out in the trends of development of AS within a single Federal GBACC.

**The first stage** (2000-2002) will be generally influenced by restrictions of budgetary funding, reduced content of orbital constellation of SC and the forced partial use of obsolete CS.

The main tasks in the development of AS at the first stage will be:

1. the completion of re-equipment of GBACC by new examples of computing hardware which began under the program "Problem 2000", with simultaneous introduction into service of a highly effective system of its warranty and post-warranty service;

2. scheduled continuation and completion of modernization of the Federal GBACC technical assets carried out within the framework of research and development effort aimed to create automated systems and objects of GBACC;

3. gradual withdrawal from the forced employment of obsolete hardware, specifically, AS using EC-1045 and EC-1046 types of computers and others;

4. development, coordination and validation of the required documents that regulate the issues of information security, production and technological independence of the country in building new AS of the single Federal GBACC.

The second stage (2003-2010) must be characterized by gradual restoration of the optimum component package of Russia's orbital constellation of SC, by transition to the use of advanced technologies of SC control (autonomous navigation based on GLONASS SNS, reduction of TMI flows to the minimally required level. etc.)

**The second stage** of development of AS is characterized by transition from the centralized processing of information at GBACC to the distributive system with the use of heterogeneous computing systems based on local computing network (LCN), "client – server" technologies, Baget type of PC, Elbrus-3M computing complexes as well as by designing AS based exclusively on CS that meet the requirements of information security and assure the country's independence in production and technology research. All automation systems of the Federal GBACC are supposed to be united into a territorially distributed computer network before 2005. The stages of development the Federal GBACC AS and the expected effect of introduction of SC advanced automatic control systems and new information technologies are shown in Fig. 31 and in Table 5.

**Table 5.** The effect of introduction of advanced AS and new information technologies

| Characteristics of GBCC AS | Improvement in quantity |
| --- | --- |
| Speed of solution of control tasks | Increased by 1.3 – 1.8 times |
| Error probability in decision making | Reduction by 2 – 5 times |
| Automation level | Increase from 70-80 to 90-95% |
| Operator load in fulfilling SC control tasks | Reduction by 1.2 – 1.4 times |
| Manufacturing dates of AS | Reduction by 1.2 – 1.5 times |
| Total power consumption | Reduction by more than 100 times |

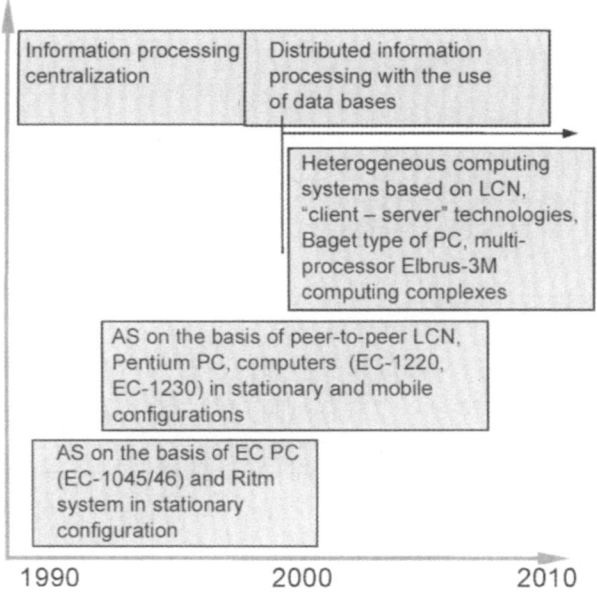

**Fig. 31.** Characteristics of a standard automation module

The main task in development of AS at the third stage (after 2010) is the radical restructuring of GBACC and OBCS computing systems based on large scale introduction of means and methods of artificial intellect, expert systems and the latest achievements in their software support.

The expected integration of expert systems with neuron nets paves the way to creating fully automatic systems of SC control. Neuron nets feature the highest speed of information processing, identification and classification of situations and images even in cases of incomplete or indistinct information.

In the coming 30 years the computing capability of FCC AS and SC OBCS is expected to increase by 100 times (Fig. 32).

The performance of computers built on the basis of neuron technology may reach by the end of the first quarter of the 21st century 1011-1012 neuron switches per second and come close to the performance of the human brain.

The transfer onto SC of functions of NBS, checking and diagnosis of the systems' condition, planning their operation as per FCC assignments will make it possible to create by 2020 practically autonomous SC.

With GBCC operating in standard mode there will be no need to continuously track and control SC.

There will only remain the functions of periodic checking and planning of target use of SC. The security will be assured by improving failure-free operation of computing and expert systems and of neuron nets. All operations pertaining to SC control from FCC will be reduced to reprogramming the software support of OBCS.

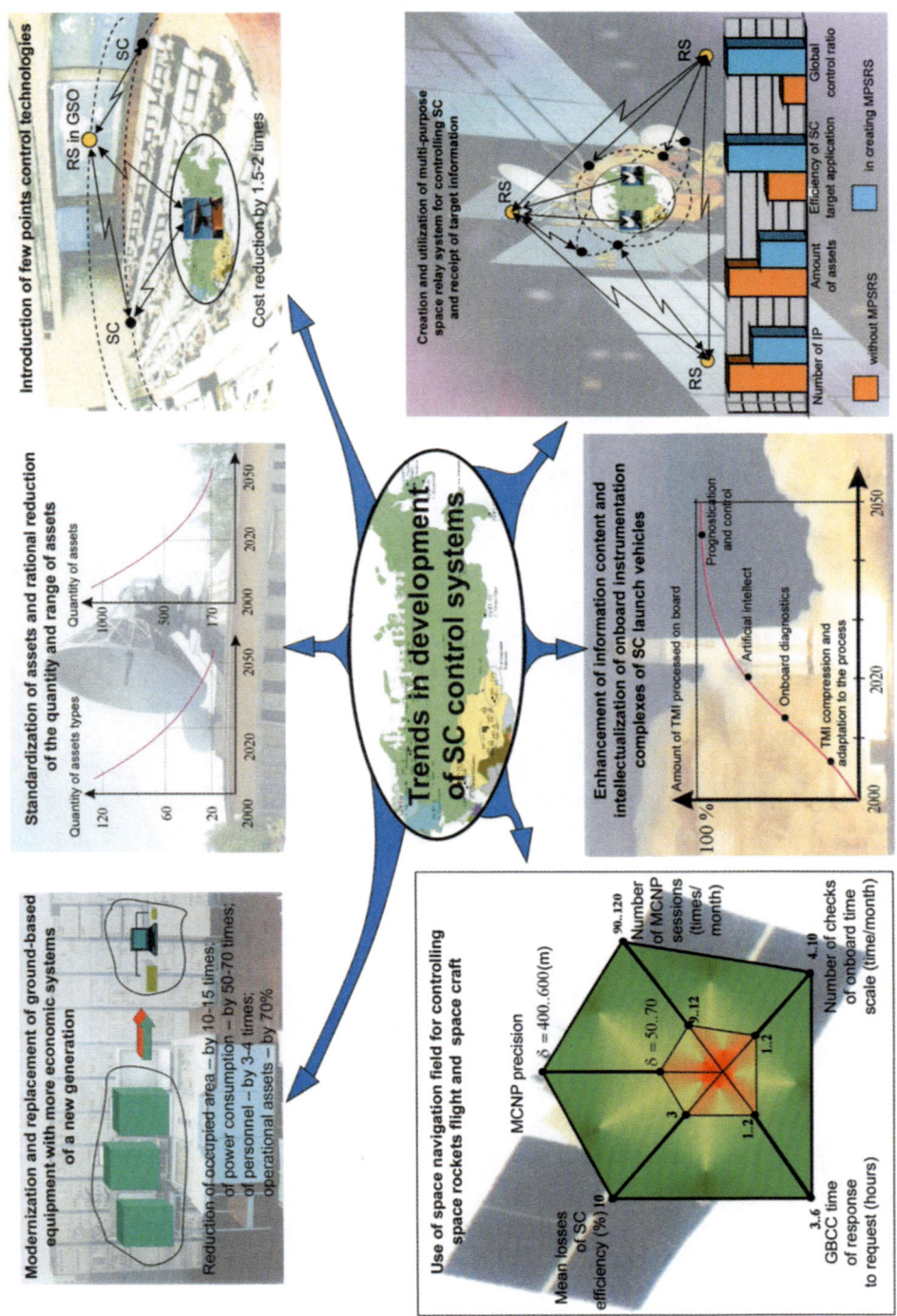

**Fig. 32.** Stages in development of GBACC AS

## 3.3 Operation systems. The state and prospective development

In building space hardware (at the stages of technical specifications definition, construction of space infrastructure, flight tests) a relevant operation system must be created. At the pioneering stages of creation and operation of space hardware the operation system as such was not elaborated and each chief designer applied his own standards of operation to the hardware he devised. This was the cause of erratic functioning of operation systems in space hardware which generated additional loads for personnel and above-the-norm consumption of materials and technical reserves. Further work with space systems called for search of ways to achieve maximum efficiency of the entire package of spaceport equipment. This necessitated the development of methods for creation and optimization of the system for operating space hardware and servicing it in standard conditions.

What is meant by space hardware operation system (SH OS) is a purposefully functioning aggregate of objects, systems and operating controls that ensure the creation and functioning of orbital constellations of SC aimed at tackling the tasks of the entire spectrum of space systems (space complexes). The aim of SH OS is to provide for creation and maintenance of orbital constellations of SC in the prescribed component package and condition. This aim is achieved by tackling the following two groups of tasks:

• keeping ground-based assets in readiness for launching SC and controlling them in orbits;

• maintaining serviceability of orbital constellation SC and ensuring their standard functioning as per target application.

A retrospective analysis shows that the existing SH OS emerged in the mid-1960s on the basis of the operation system of combat missile complexes (which, in turn, used the experience of operating aviation hardware and artillery materiel). That was when engineering and launch complexes of the R-7A ICBM withdrawn from combat duty started being re-equipped and re-configured for military space application and the Raduga (rainbow) and Voskhod (sunrise) types of space rocket complexes began to be deployed. SH OS has gone a long way towards its current state-of-the art (Fig. 3.3.1).

The development and creation of space hardware in the USSR during confrontation of the world's two social and political systems was dictated by considerations of national defense. That is why emphasis was laid on the development of military space equipment, whose order and operation was the domain of the MoD. Therefore SH OS was elaborated with regard to combat use of space hardware and focused on its military application.

The combat use of the operation system determines its component package and structure. The anticipated application and expected performance of space rockets imply such characteristics as operational readiness, fast response, efficiency, stealth capability, survivability and reparability, resistance to hostile counter measures, protection against unsanctioned actions, etc.

Until the mid-1970s there was no clear concept of the single model of operation system for SH OS supported by technical documentation and regulatory instructions. In the late 1970s the working group 50 at the MoD's Central Research Institute along with experts of spaceports of Baikonur and Plesetsk started developing scientific foundations for building and optimization of SH OS operation system, elaborating management technology of SH OS operation in compliance with the achieved level of space technology, consistent with requirements of the military and political situation and with the changes in the organizational structure and staffing policy of operating divisions. As a result of research, requirements were introduced into technical documentation and regulatory instructions that cover the SH OS operation, the principles of its formation and function, organization and maintenance. This made it possible in 1980s to establish the system of SH OS operation in the form of an engineering entity capable of ensuring the use of SH OS both for military and dual purposes (Fig. 34).

The space hardware operation system is a complex system dealing with organization and engineering matters and consisting of three subdivisions (sections): technical, organizational and functional (Fig. 35) whose well-coordinated efforts contribute to the successful fulfillment of the tasks allotted to them. The creation and function of SH OS (Fig. 36) are based on the following three principles:

• the principle of priority of operational capability, i.e. the dependence of the organizational and functional subsystems of SH OS on the achieved level of operational capability of spacecraft, carriers, boosters, space rockets' ground-based equipment and GBACC;

• centralized management and control of SH OS;

• rigorous regulation of rights, commitments and responsibilities at all levels of SH OS management;

• conformity of organizational and functional sections to the complexity and volume of tasks tackled by SH OS;

• balanced development of the assets of SC orbital constellation and engineering facilities of the ground-based operation and maintenance bases (spaceports, control complexes for supervision of orbiting SC and for receipt of special information);

• the unity of principles of planning, management and organization of SH OS operation;

• assured fulfillment of tasks facing the SH OS;

• the uniformity of the information support system used in SH OS operation;

• suitability of the personnel's skill and competence to the technical level of SH OS;

• comprehensive and fully adequate provision of operation with all required types of materials and technical assets and their rational utilization;

• guaranteed (technical) supervision of the industry.

**Fig. 33**. Super-heavy carrier rocket for lunar program, N1-L3

The base of the SH OS technical structure are infrastructure facilities united as per their assigned tasks and consisting of the SC orbital constellation, space rocket complexes, ground-based SC control complexes, SH OS storage facilities, means of production, delivery and storage of space rocketry, electric power supply systems (EPSS), engineering systems, metrological support systems, state technical certification systems, special purpose structures, buildings, etc.

The rapid development of SH OS proceeds in the late 1970s and 1980s, the time when a wide variety of orbital hardware for various applications was created and introduced into service, when space rocket complexes like Zenit, Cyclone-3, Energia-Buran were deployed and the second launch complex with two take-off installations was commissioned for launching the Proton carrier rocket. Also at that time the Soyuz space rocket complex was being modernized and space infrastructure facilities were being overhauled (Fig. 37).The work load on the operation system reached then 100-odd launches of various carriers per year and involved supporting more than 200 various application SC in orbital constellations, keeping up in various stages of serviceability for combat use or in storage more than 300 units of carriers and spacecraft.

The organizational section is a hierarchical system of organs that control, manage, organize and support the operation of SH OS within their assigned duties. The primary task of the SH OS organizational section is to plan and carry out work aimed at maintaining technical readiness of space systems and using them as per prescribed purpose.

**Fig. 34.** Soyuz carrier rocket with Soyuz transport ship

The organizational section of SH OS includes control and executive organs. The control organs are divided into four control levels: operation control at the Center, armament services of associations (spaceports, Space Systems Test and Control Center), formations (CIP), units (independent test engineer units, independent ground-based test posts).

Operation control is the function of control organs aimed at achieving the timely and efficient performance of space hardware operation duties with rational distribution of resources while doing so.

The executive organs include the officers and other personnel of units and divisions engaged in scheduled day-to day operation of space systems.

The work directly associated with preparation and performance of launches and with control of SC orbital constellations is supposed to be done by temporary substitute formations created in units and divisions. Those should be space rockets launch preparation crews, orbiting SC control duty operators, rescue teams and other formations intended for elimination of leakages of rocket fuel components, for search, evacuation and recycle of space rockets, for space hardware disaster recovery, etc.

The organizational section of SH OS formed within the framework of Military Space Forces mainly coped with the tasks of maintaining the operational capability of ground-based assets and orbital constellations. One of the principal manifestations of SH OS organizational section's efficiency is the trouble-free operation of space systems since the early 1980s without failures through the fault of operating personnel and without disruption of SC launching schedules on account of inadequate organization.

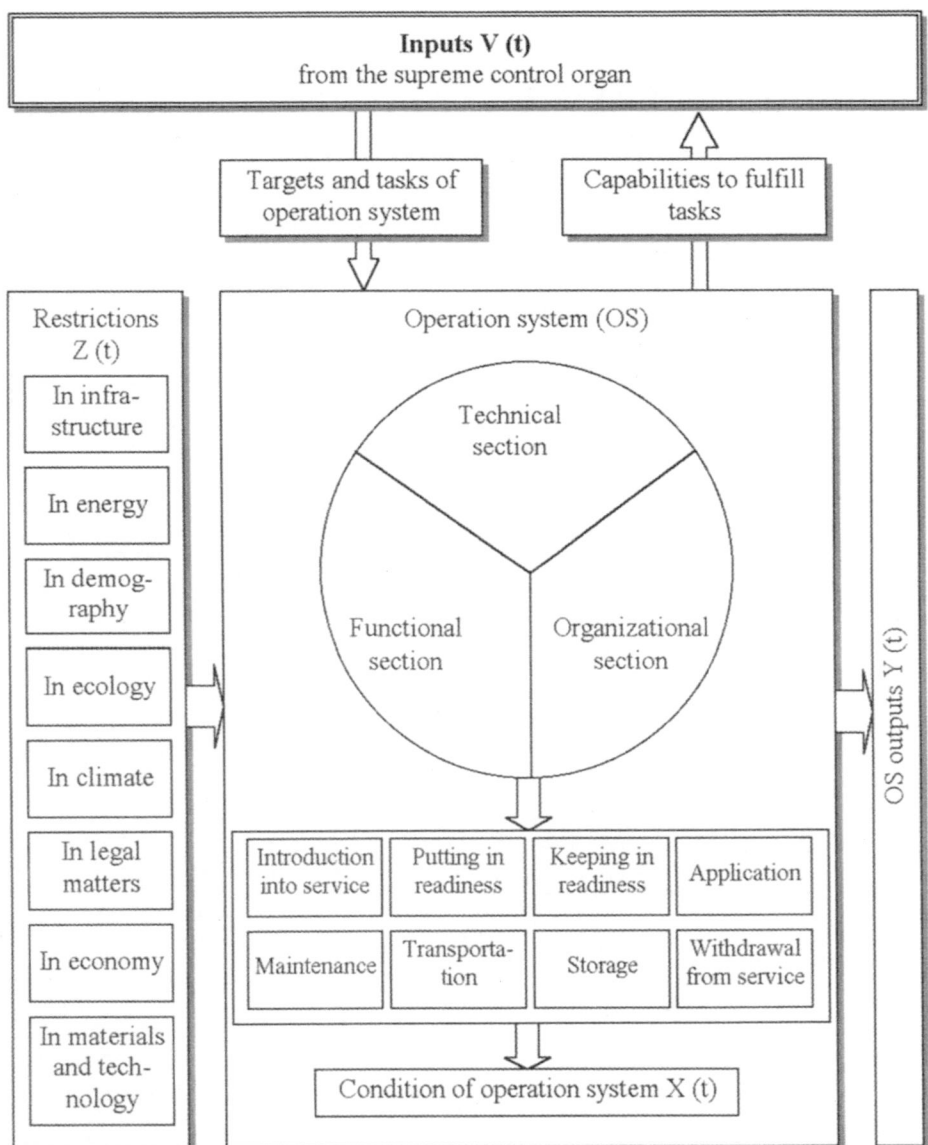

**Fig. 35.** Structural diagram of SH OS functions

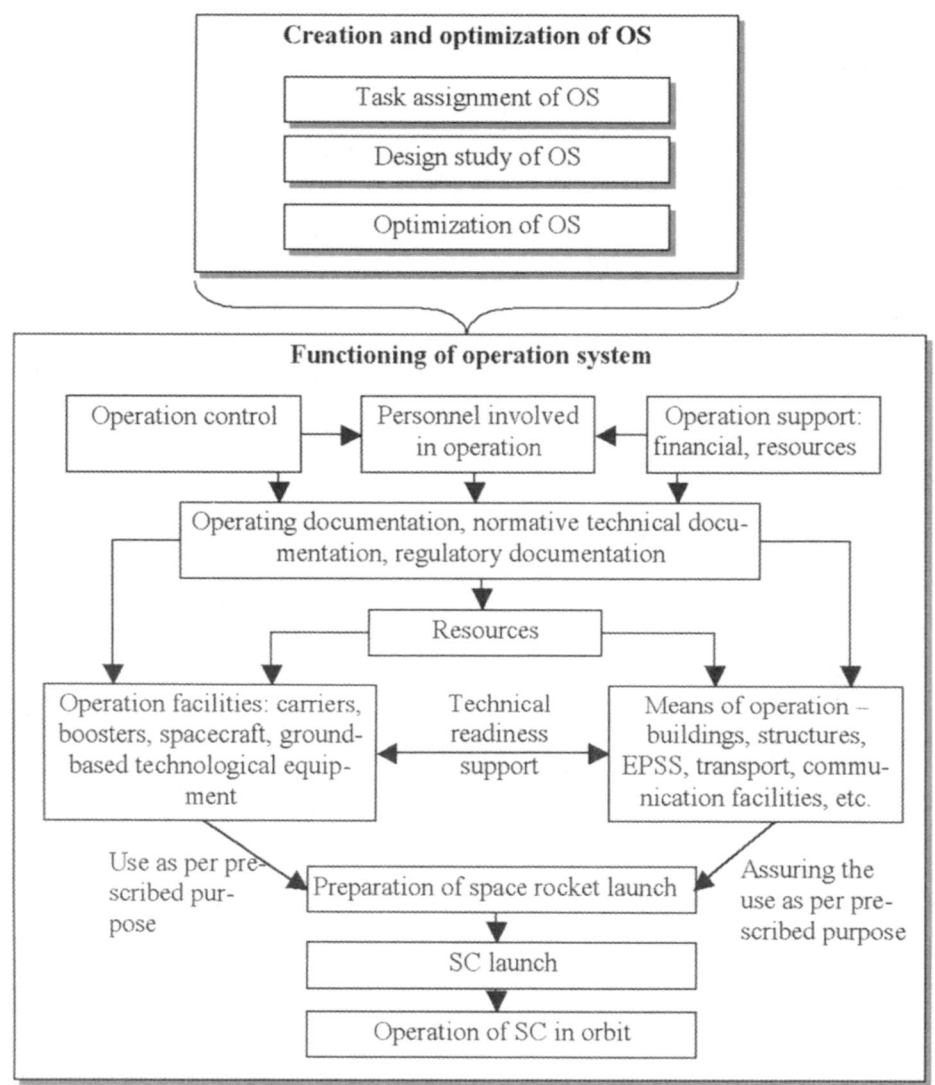

**Fig. 36.** Structure and tasks of SH OS

Functionally, the section is a package of tasks and functions assigned to SH OS whose realization is regulated by requirements of normative technical, regulatory and operating documentation.

The main tasks of the space hardware operation systems are:

• maintaining the operational capability and combat readiness of ground-based equipment of spaceports and GBACC as per prescribed purpose;

• creating and sustaining the prescribed component package of an orbital constellation;

**Fig. 37.** Proton carrier rocket at the launch complex

- maintaining the required technical condition of orbiting SC;
- creating and maintaining the prescribed component package and condition of the resources of space hardware, space rocketry and other necessary materiel;
- maintaining at the required level the operating skills and technical expertise of the attending staff.

The main functions stemming from the space hardware operation system tasks are:

- provision of operating units with all types of space hardware in accordance with a specified product range and quantity;
- special training of operating personnel; putting in operation the space hardware, its modernization and reconstruction; preparation of space hardware for operation and maintenance of operational capability; preparation of space rockets, SC launch, orbital constellation control;
- post-launch repair and restoration; search and evacuation of carriers' discarded components and descended craft;
- transportation, receipt and storage of space hardware; maintenance and repair of space hardware; categorization and service life extension;
- optimization and amendments following the claims made;

checking and evaluation of operation conditions and the state of space hardware;

- provision of safety in space hardware operation;
- logistic support;
- metrological support;
- national technical supervision, power consumption supervision;
- registration of the available stock and traffic of space hardware;
- registration, storage, amendments to operating documentation;
- investigation of causes and space hardware disaster elimination;
- organization of military units' cooperation with industrial organizations in operating space hardware.

The necessity to enhance effectiveness of SH OS called for creation of a single coordinated legal, normative-technical, methodological and informational supporting base. A methodological foundation had been elaborated aiming to create a package of normative-technical and regulatory documents that cover general organizational requirements and a single approach to planning, organization and control of operation related measures applied to space hardware, its registration and accounting.

Based on this, all the required regulatory documents have been elaborated and validated.

They regulate the procedure of putting space hardware (SH) in operation, its registration, storage, transportation, maintenance and repair as well as organizational and methodological documentation for planning, operational control and information support of SH OS.

The military and political situation in the world, the existing plans to create launch vehicles of a new generation have posed before SH OS the task of not only retaining the existing infrastructure and providing for launches and control of SC by the available means, but also of further developing of the technical, functional and organizational sections of SH OS in order to enhance the national defense capability. Demands made of the advanced space rocket complexes' operation system become ever more stringent.

The increasing technological perfection of the space rocket equipment toughens requirements applied to optimization of not only hardware as such, but also of its operation system. The rationale of optimizing space rocket complexes at a spaceport is based on the following factors:

• improvement of output characteristics of the operation system at the test stage which allows to avoid excessive expenditures after the complex's commissioning;

• detection of design errors during optimization which precludes their rapid build-up at the subsequent stages;

• analysis of statistics data obtained prior to standard operation which allows to refine the operation system's elements for improvement of its performance.

During refinement of the space rocket complex's operation system (Fig. 38) subject to optimization are the following assets:

• space rocket;

• launch and engineering complexes, refilling and neutralization station (RNS), information measurement, collection and processing complex (IMCPC);

• storage facilities for preservation of supplies and training aids, for repair work, metrological support, state engineering supervision;

• elements of technical infrastructure of a spaceport and position area (communication and control equipment, power saving and heat supply systems, industrial effluent systems, buildings and structures, road network) which ensure the serviceability of space rocket complexes (SRC);

• operating documentation for SRC and their components;

• organizational structure of operation subdivisions;

• system of normative documents and documents for organizing the operation of SRC.

The optimization of SRC operation system starts with the beginning of SRC construction, continues during installation, balancing and commissioning, systems and aggregates autonomous tests, integrity tests, flight tests, and ends as a rule in pilot operation of the complex.

The critical pre-requisite to taking correct decisions in analysis of results of SRC operation system optimization is the sound estimation of its efficiency. The efficiency performance indicators are obtained using the analysis of targets, tasks, properties and characteristics of SRC operation system with formation of its output parameters Y(t) (see Fig. 35).

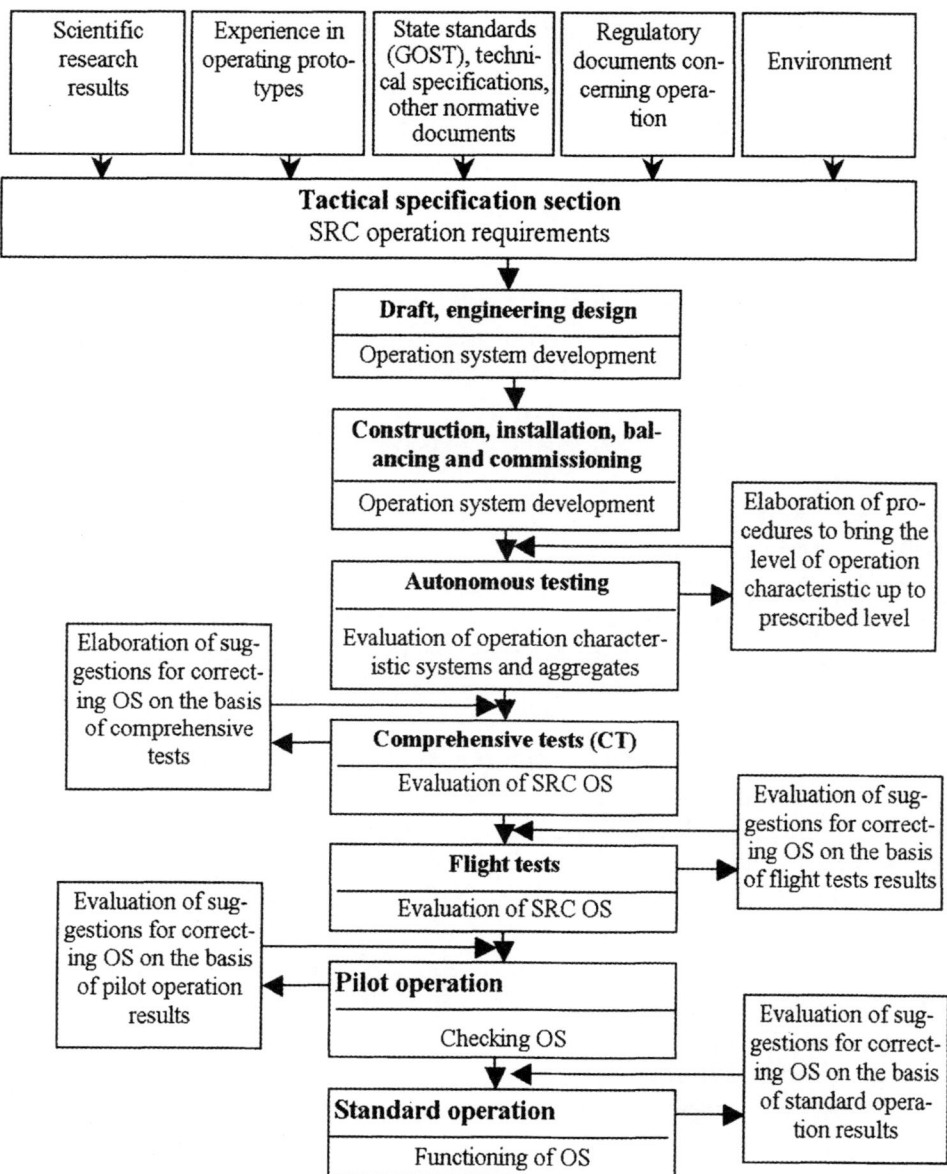

**Fig. 38.** Stages of optimizing SRC operation system

In choosing the SRC specifications it is good practice to use such operating properties as efficiency, reliability, maintainability, safety, operational readiness, and resource effectiveness. In this case the vector of SRC operation system output performance characteristics will look as follows:

$$Y = Y\{W_O, \ni_{TO}, P, P_{\delta.\ni}, K_{T.\Gamma}, T_{O\cdot K}, C_{c.\ni}, ...\},$$

• where $W_O$ - performance characteristic of the complex operation system efficiency; $Э_{TO}$ - summarized characteristic of the effectiveness of the complex's maintenance system; $P$ - summarized characteristic of the complex's reliability;
• $P_{δ.э}$ - probability of the complex's safe operation; $K_{T.Г}$ - the complex's technical readiness ratio; $T_{O·K}$ - the complex's operation system's optimization time; $C_{c.э}$ – the cost of optimizing the complex's operation system.

As a performance characteristic of W¿ it is possible to use:
• time characteristics of preparing a space rocket for launch (mathematical predictions, quantiles of time distribution in preparing a space rocket for launch);
• output capacity of SRC and its operation system (Fig. 39, 40).

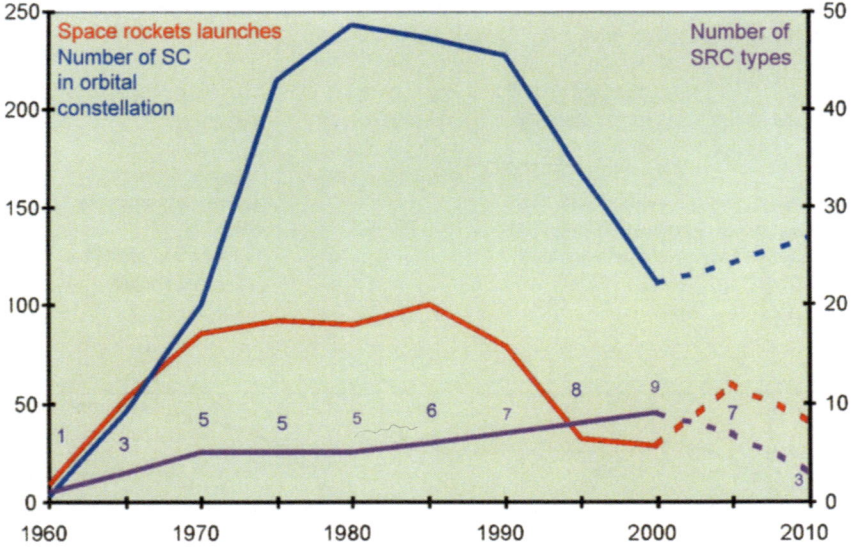

**Fig. 39.** Dynamics of load on space hardware operation system

In estimation and optimization of space hardware operation system it is advisable to use for the system (aggregate) as output parameters the following operating properties: efficiency, reliability, maintainability, safety, transportability, ergonomic factors, technical readiness, standardization capability, resource effectiveness. In this case the vector of output indicators of effectiveness for the operation system (system, aggregate) will look as follows:

$$Y_j = Y_j \{W_{Oj}, Э_{TOj}, P_j, P_{δ.эj}, ΔP_{Tpj}, K_{эрηj}, K_{T.Гj}, Y_{Hj}, C_j, T_{O.Cj}...\},$$

where $W_{Oj}$- efficiency characteristic (time characteristic showing conduct of work on the system in realizing theμ operation processes); $Э_{TOj}$- summarized characteristic of the effectiveness of the system's maintenance; $P_j$- summarized characteristic of the system's reliability; $P_{\delta.\mathfrak{z}j}$ - probability of the system's safe operation; $\Delta P_{\text{T}pj}$ - relative deterioration of reliability during transportation; $K_{\mathfrak{z}prj}$ - summarized characteristic of ergonomic conveniences; $K_{\text{T.}\Gamma j}$ - system's technical readiness ratio; $Y_{Hj}$ - standardization level; $Cj$ - cost of system's optimization; $T_{O.Cj}$ - aggregate (system) optimization time.

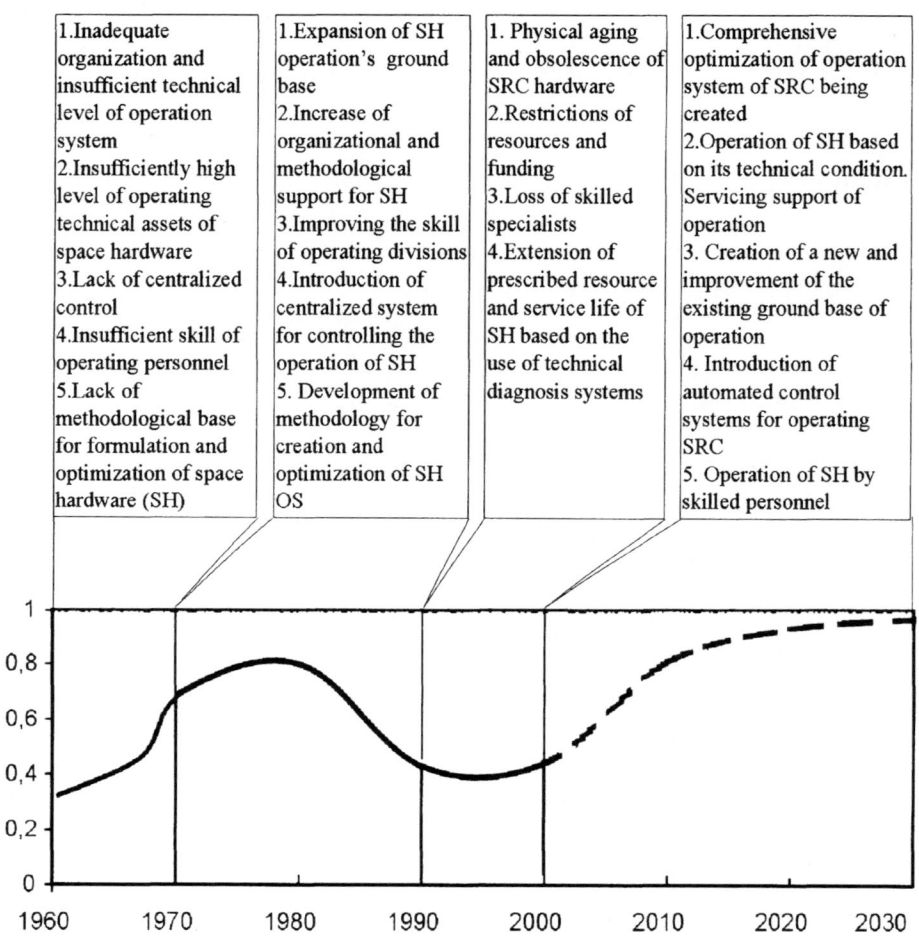

**Fig.40**. Effectiveness of functioning of space hardware operation system

In optimization of SRC operation system it is quite important to use modern diagnostics techniques that enable to evaluate correctly the technical condition of SRC being tested and its components. The processes and systems of control, diagnostics and recovery (CDR) are a sort of a core in the process of testing SRC. This is so because in the course of tests it is necessary to control at all times the serviceability of on board and ground-based facilities and systems of SRC, detect and diagnose their malfunctions and fix up their repair.

However, the general research background against which the systems and methods of CDR of complex objects, SRC in particular, are developing is rather controversial. On the one hand, the number of various objects subject to control and their variety are huge. In a situation like this it's impossible to find the ways of building a standardized CDR system. On the other hand, however, theoretical investigations show that the principal tasks facing the CDR with reference to various objects have much in common. This drives theorists to seek ever greater standardization and abstraction of CDR theoretical algorithms, which, in turn, hampers practical engineers to use them in practice.

These contradictions have generated a situation in which to date there isn't a single qualitatively independent CDR with hard- and software of its own. As regards the existing SRC, applicable to them at the high level of technological development are only serviceability control systems. But they, too, are confined exclusively to allowable control techniques and realize only their basic advantage – fast monitoring of a large number of automatically controlled parameters. The remaining diagnostic capabilities, should any test fail during control, are very scarce and invariably call for involvement of experts in a semi-automatic mode. Till now, the post-disaster diagnostics is not yet formalized in terms of algorithms and is not studied theoretically.

Anyway, even now while designing advanced SRC a realistic solution could be recommended how to overcome the outstanding difficulty – instead of an attempt to create all-embracing systems of their technical diagnostics it would make sense to use the hierarchical network of such systems in which each of them is responsible for control and diagnostics of its own level. This would enable a construction of a net that unites the systems of technical diagnostics from the level of built-in on-board diagnostics systems to the control system at the Flight Control Center. It's noteworthy that each level offers a capability to restore the serviceability of a lower level. Technically, however, these tasks cannot be fulfilled in a fully automatic mode. But this can be done on the basis of expert systems of technical diagnostics, control and planning at the relevant levels.

For a flying vehicle (FV) this approach is presented in Fig. 41 which shows FV on-board subsystems; CDR built-in systems; sensor elements (SE) and executive organs (EO) in the control circuit of the subsystems operation; the central system of FV CDR; the control system for controlling FV flight and accomplishing the flight's target; flight (test) control center.

The lack of knowledge until recently as to how subsystems interact precludes creation of a CDR central system. In their turn, delays in creation of the subsystems affect at a later date the economy of the CDR subsystems.

The possible means of overcoming this difficulty is the employment of a two-level hierarchy system of CDR. In doing so a subcontractor must from the very outset assume the complete responsibility for the problems of the internal CDR of its subsystem. As regards the CDR central system, it will only takkle the problems of CDR, particularly, those of the interface and the subsys-

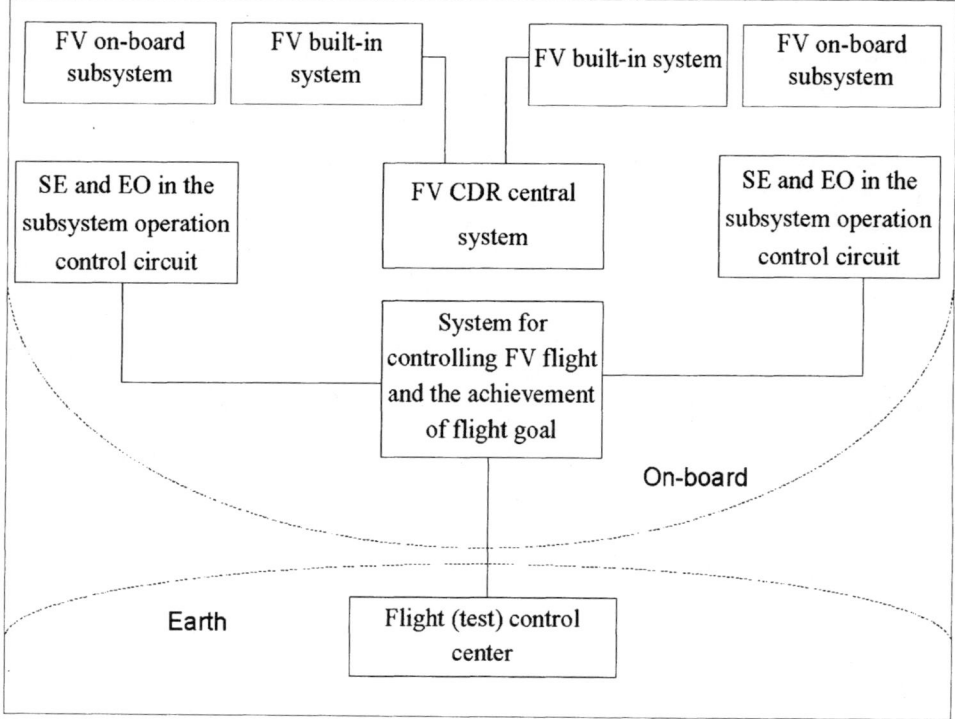

**Fig. 41.** FV technical diagnostics and control system

The CDR subsystem makes its evaluations based only on the "inner" information obtained inside the subsystem. It is only when the CDR subsystem is unable to establish the system fault with the aid of information available to it or to compensate for it using its own reserves, does it turn "for help" to the FV CDR central system.

The CDR central system attempts to resolve the outstanding problem with the help of analysis of interaction of subsystems and its change. Once the system exhausts its resources, it addresses itself to one level above it, i.e. the system that controls the flight and the tasks of the entire FV. This system can change the flight (test) mode or change the level of task fulfillment, adapting itself to the arising technical condition. Once its resources are exhausted, the FV control system addresses itself to the flight (test) control center. The latter can change the flight (test) program in an automatic or automated mode, adjusting it to the newly arisen situation.

As this brief description shows, at issue is a hierarchical structure (network) of expert systems specializing in CDR assignments and operations planning.

In addition, it is necessary to have one more expert system which must receive after each flight (test) complete information about the functioning of all individual subsystems of SRC and its behavior as a whole. This expert system must carry out an in-depth post-flight (post-test) analysis and is supposed to supply global current estimations of the SRC optimization level.

The construction within a reasonably short period of time of a single system for CDR and for planning SRC operations is practically impossible. However, breaking it down into a hierarchal net makes it possible to solve the problem quickly.

Such a network of expert systems (on board and ground-based) was created by Boeing Aerospace during tests of individual B-1B aircraft and then adapted to tackle the problem of CDR carrier rockets and communications spacecraft. It's noteworthy that its high effectiveness in post-disaster diagnostics was confirmed right after it went into service.

Special emphasis in the course of SRC ground and flight tests and its operation system's optimization is laid on testing and verification of software comprised by on-board and ground-based SRC systems. The significance of such measures is likely to grow in the future. The software reliability issue is all-inclusive and concerns such matters as:

elaboration of optimum technology of software development since it is at this stage of development that it "gets" many errors and malfunctions that will occur in the future;

• evolution of the theory and practice of autonomous and comprehensive tests (optimization) of software;

• organization of software servicing which is currently confined mainly to its incessant rework.

Among these top priority jobs quite a few minor problems crop up which sometimes look isolated and unrelated to each other, but all of which must be tacked in a systematic fashion and get their share of attention. Among them is the task of estimating the reliability of software support.

The models and techniques of evaluating the reliability of software support essentially affect the stability and correctness of functioning of control systems. The expenditure on verification and optimization of the software support exceeds by far the cost of developing control systems hardware. For example, the control systems hardware costs a third of that of the relevant software support. This trend has become of late so pronounced that the hardware can now be regarded as a sort of "container" for software. Therefore it is expressly software that determines the reliability of on-board and ground-based control systems.

Optimization of software is becoming an ever more labor-intensive and responsible job in testing SRC on the ground and in flight. The accrued experience of testing SRC allows to assert that failures (errors) in software can often bring about graver effects than the usual faults in hardware.

It should be noted that the growing amount of software leads to growth in error probability of software support. The software support of Space Shuttle system, including on board software and ground-based software support of automated system for controlling launch preparation contains over 3,000,000 codes. All this distributed software must function as a single whole, being integrated into various on board and ground-based computer systems. With to-day's level of technology, creating a reliable software support in such a volume is extremely difficult. According to Russia's statistics of SRC tests, the reliability of their on-board software support is estimated at only 0.990. The standard reliability requirement applied to the carrier rocket control system equals 0.999. Thus the mean statistic reliability level of the software established in course of tests is below that defined by the performance specification. Hence the necessity to test the software in most various modes for it is only through detection of software errors that the reliability level stipulated by specifications can be attained.

The software reliability problem can't be resolved without tackling two critical issues: reliability support and reliability evaluation.

To solve this problem, analysis has been carried out of the principal mathematical models of reliability evaluation in order to establish their weak and strong points and suggestions have been elaborated as to how to improve them.

Analysis showed that all individual mathematical models stem mostly from suggestions to make a progressive parametric evaluation of a certain unknown and unobserved parameter. Generally, this parameter could be associated with a number of errors in the software that remain undetected up to the current moment.

The observed parameter contains statistical data on software error detection during testing. For example, those could be working periods between times of error detection.

Most models make evaluation using the software maximum reliability likelihood technique as a probability of its setting the prescribed time without errors.

The individual nature of a mathematical model is normally linked to the function of distributing the observed parameter on the condition that the observed parameter has a certain value.

Thus the job of evaluating the software reliability is limited to stochastic filtration of an unobserved component. This is precisely the technique that should be used to obtain further generalizations concerning all specific models in evaluating the reliability of software comprised by on-board and ground-based SRC systems.

At the same time, a technique has been proposed that aptly combines measures of both local optimization of one chosen model and global optimization of software tests with a well-timed switch from one model to another depending on results observed. The use of concept of uniting the local and global optimizations allows to single out a sort of a complex "by-passing" model that monitors quite adequately the real state of software being optimized.

Working most successfully on this problem are the Khrunichev State Space Research and Production Space Center, Design Bureau of Transport Engineering, Energia Rocket Space Corporation.

Based on the needs of national defense, plans to evolve space systems, develop economic capability and scientific and industrial potential of the country, the authors assume that the main trends in the progress of SH OS will be as follows:

1. Reduction of space rocket hardware both in variety (product types) and quantity owing to further improvement of its performance characteristics, maintainability, standardization, tasks integration and introduction of the latest technologies.

2. Improvement of infrastructure of spaceports, ground-based orbital constellations control posts, support bases (storage of space hardware, production and storage of space rocketry, state technical supervision, etc.), adapting their structure and technical capabilities to the requirements of assured launches of all types of military SC from Russia's territory.

3. Development and introduction of advanced technologies of preparing space rockets for launches with the use of standardized work stations based on testing procedure and with the aid of principles of flexible technology using the achievements of robotics.

4. Extension of service life of ground-based engineering complexes, launch complexes, GBACC and orbital hardware. The creation at spaceports of laboratory facilities that offer superior capabilities for diagnostics and prognostication of SH technical condition.

5. Continued study of the possibility and feasibility of transition from scheduled preventive maintenance of SH to its operation based on technical condition determined by diagnostics.

6. Development of research and methodology base employed to improve SH operation with the aid of computerized scientific research and greater practical utilization of expected results of scientific research.

7. The creation of SH automated control system, development and introduction of automated information supply system of efficient registration and utilization of stocks of carrier rockets, boosters, spacecraft, their means of launch and flight control, spare parts tools and accessories (SPTA) kits.

8. Improvement of technology of accumulating and storing the stocks of carrier rockets, boosters, spacecraft, rocket fuel components, spare parts tools and accessories kits, expendable materials, equipping ammunition storage facilities with up-to-date maintenance and diagnostics gear.

9. Elaboration of normative technical and regulatory documents that specify the issues of control, organization, support and supervision of SH operation in the light of new tasks and restructuring.

10. Optimization of SH OS organizational structure, improvement of SH operation control and refinement of functions it performs. Transition from

predominantly administrative to predominantly economical relationship of SH OS with interested agencies and organizations.

11. Improvement of the system of professional and psychology-oriented selection of personnel and their special training with the aid of automated testing equipment and simulation facilities.

**Fig. 42**. Before launching the first space rocket Rockot at the Plesetsk spaceport (left to right: V.Ivanov, G.Stamerjohanns, V.Inden, G.Kovalenko)

Realization of the above said trends in development of SH OS can be accomplished through solution of the following problems.

1. **In respect of development of SH OS technical structure:**
• step-by-step development of infrastructure of the Plesetsk spaceport which ensures storage, preparation and launch of military SC performed from the spaceport of Baikonur; reconstruction of engineering complex and launch complex for Rockot and Rus SRC; creation of multi-purpose engineering and launch complexes for preparation of the Angara carrier rocket; building a multi-purpose refueling and neutralization station; extra equipping of the spaceport's instrumentation complex; reconstruction of the spaceport's power supply system (Fig. 42);
• creation of a technical infrastructure at the Svobodnyy spaceport required to prepare and launch Start, Rockot, Strela carrier rockets;

• maintaining the serviceability of Baikonur spaceport's infrastructure providing primarily for operation of the Proton heavy class carrier and its modifications (Fig. 43–44);

**Fig. 43.** Docking the Proton carrier rocket with the Briz booster unit

• reconstruction and major overhaul of the existing carriers and SC assembly and test buildings, reconstruction and construction of launch complexes that ensure the preparation and launch of advanced SC;
• provision of the required amount of production and storage of rocket fuel and compressed gases supplies on reconstructed and newly built facilities;
• re-equipment of the existing and creation of new storage facilities for stocks of SH, optimization and introduction of advanced technologies of checking and maintaining the technical condition of space hardware during its storage;
• reduction of deleterious impact of operating SH on environment, improvement of ways and means of neutralizing toxic rocket fuels and recycling of SH;
• improvement of orbiting spacecraft automated control complex through integration of spaceports' instrumentation complexes with automated control complex and through replacement in its component package of physically worn and obsolescent control systems and computing hardware;
• creation of a network of GBACC repair facilities outfitted with modern diagnostics equipment and instrumentation apparatuses.

2. **In respect of improvement of SH OS functional structure:**
• creation of a single interconnected normative-legal and normative-technical base for the functioning of SH OS with due regard for the emerging economic conditions in the country, the expansion of the range of missions tackled by

space hardware and the restructuring evoked by reformation of the Armed Forces;

**Fig. 44.** Putting the Proton carrier rocket with Zvezda service module onto launch ramp

• optimization of orders and enhancement of timeliness and completeness of provision of spaceports with space hardware, SPTA kits, expendable materials;
• extension of active service life of SC and life cycle of ground-based equipment of the engineering complex, launching complex and GBACC beyond warranty periods by developing and introducing the technical condition control system with creation of a relevant base for diagnostics and prognostication of the technical condition of space hardware being monitored;
• improvement of the contents and organization of SH operation through introduction of advanced cost-effective and resource saving technologies used to prepare the hardware for action;
• reduction of labor input and duration of preparing SC at an engineering complex by bringing the craft to ready-to-launch condition, leaving to spaceports such operations as filling with fuel and gasses and a minimum set of tests for checking workability of SC at engineering and launch complexes;
• enhancing the security (including ecological) of SH operation through complete automation of work on the launch complex and rejection of toxic rocke fuels on advanced carriers, ensuring rigid compliance with technological requirements while working with hardware;
• development and utilization of new high-performance ways and means of search, evacuation and recycle of booster components landing in a normal or

abnormal fashion, neutralization of leakages of rocket fuels in debris impact areas;

**Fig. 45.** Directors of the launch of Zvezda module: Yu.P.Semenov, A.I.Kiselev, L.T.Baranov

- reduction of expenditure of materials and technical reserves during operation of space systems and complexes through their rational distribution and consumption in conditions of financial and material constraints, attraction of non-budgetary sources of funding;
- revision of regulatory documents that specify the procedure of organizing the functioning of SH operation system and their adaptation to new conditions.

3. **In respect of organizational structure of SH OS:**
- improvement of control on the basis of a newly evolved rational structure of controlling and executive organs, perfected professional training of operating personnel and optimization of their number at all levels;
- provision of rapid response, continuity, stability and stealth capability of control through automation of control and introduction of automated information system for registration of availability, traffic and technical condition of SH;
- perfection of the system of vocational selection of psychologically suitable operating personnel for launching and controlling SC;

- improvement of university students' training in operation related subjects;
- development of training aids, all-purpose simulators and automatic equipment resorted to in decision making and eliminating distress situations; provision of such equipment.

The realization of the above said trends and tendencies in evolution of SH OS will call for significant funding; however, its necessity and feasibility is determined by a number of factors:

- enhancement of the role of SH OS in combat effectiveness of armed services and arms of service of the Russian Federation in performance of strategic and tactical operations;
- the presence of military threats from the outside coming in the wake of NATO's eastward expansion and its intervention in recent years without UN sanctions with domestic affairs of sovereign states;
- intention of the US to embark on deployment of a modern full-scale ABM defense system;
- obsolescence and physical wear of up to 80% of ground-based equipment of spaceports and GBACC which jeopardizes the assured fulfillment of military tasks by space hardware;
- the necessity to develop and switch to resources-friendly, safe and environmentally clean technologies of SH operation;
- enforced concentration of practically all types and classes of space rocketry at the spaceport of Plesetsk and the necessity to deploy there the appropriate infrastructure to ensure the fulfillment of defense tasks.

Having considered in general terms the state of and the main trends in development of operation systems of military space hardware, it should be pointed out that over recent years a system of operation of civil space hardware starts to materialize. This is evoked by significant changes that are taking place in the country in the field of space exploration.

The drastic cuts in funding of the MoD led in the mid-1990s to a considerable recession in space activity and, at the same time, to a relative expansion of space activity in international cooperation and commerce. This activity proceeds under the aegis of the Rosaviakosmos created as a federal organ responsible for realization of the state's policy in the area of space research for civilian purposes.

The expansion of the range of tasks fulfilled under international cooperation programs and provision of services to foreign and national customers in space work urged the government to take a decision according to which the MoD handed over to Rosaviakosmos a number of facilities of Baikonur spaceport leased from Kazakhstan which enable storage, maintenance, preparation and launch of Soyuz, Proton, Zenit, and Cyclone-2 carrier rockets. Currently, a number of the spaceport's facilities are used by civil departments. Functioning now is a Ground-Based Space Infrastructure Operation Center assigned to Rosaviakosmos. It coordinates their interaction with industrial enterprises, military units and spaceport's services. A state unitary enterprise, Federal Space Center Baikonur has been created. Its assigned mission is to perform the functions now being in charge of the spaceport's management and supporting services. Along with the technical and organizational entities a functional entity is being now created for operating space hardware that is in charge of Rosaviakosmos. A number of regulatory documents are being prepared for enactment, among them "The direction for organization, preparation and execution of SC launches by civilian crews", "The direction for safe operation of space hardware" and others.

It should be noted that the formation of civil space hardware operation system proceeds now using the experience of functioning of military SH operation system. The organizational and functional entities being created within this system are in some way or another a semblance of similar entities in the military hardware operation system. It's obvious that such continuity is justifiable during some transitional period so long as the creation of the operation system is based on the existing space hardware accepted from the MoD and on the existing normative-technical base while space activity of Rosaviakosmos proceeds in a close cooperation with the MoD. Further on, as space work under the aegis of Rosaviakosmos will escalate accompanied by creation of advanced orbital equipment and launch vehicles, the formation of the civil space hardware operation system must go on independently with recourse to the required scientific, methodological, technical and economical substantiation of its establishment and development.

The trends, methods and rate of building advanced space hardware for exclusively civil use will depend on how actively the state will be utilizing space for scientific research and socio-economic progress. Also, this will be determined by the country's economic potential, by demand and competition on the foreign market of space services. It has already become evident that the tendency for commercialization of space work has led to a greater activity of the space industry in building advanced SRC for civil applications.

Further activity in this area will necessitate the creation of a full-scale and effective operation system based on the use of advanced space hardware for civil applications. That is why as early as today the formulation of requirements applied to such an operation system, the definition of its targets, tasks and principles of functioning together with the development and substantiation of its structural characteristics, legal and engineering standards of its functions are regarded as top priorities.

The difference of targets, tasks and organization in the use of civil and military hardware calls for revision and change in the methods of setting requirements (previously traditionally oriented to the military use of SH), of developing advanced orbital and launching hardware, its specification and

performance, organization of its trials and optimization at spaceports, introduction into operation and operation of the system on the whole. Evidently,
some of the requirements applied now to such features of SRC as performance,
service-ability, survivability, efficiency, stealth capability, resistance to adverse
exposure and others will be essentially changed.

More rigorous requirements will be applied to the operational performance
and technical specifications of the ground-based equipment, to its standardization, to introduction of advanced economical technologies involving low
operations input in preparation and maintenance of carriers, spacecraft, engineering and launching complexes. Also, more stringent demands will be made
of the equipment safety, of automation of space rockets launch preparation,
of improvement of operational procedures and their cost reduction.

Currently, in development and creation of advanced SRC a necessity has
arisen to move from the scheduled preventive maintenance of the ground-
based technological equipment on to servicing as per technical condition. This
method is realizable thanks to the available technical assets and fairly effective methodological support enabling the detection of a pre-failure condition
and prevention of the failure proper.

The investigations of the maximum realization of safety of ground-based
aggregates and systems of engineering and launching complexes based on
the development of optimum system of maintenance with regard for the
specific operating conditions carried out at the Baikonur spaceport as far back
as the 1980s have shown that the transition to maintenance as per technical
condition of the equipment improves performance characteristics of maintenance, specifically, the readiness ratio by 7 to 12%, resources consumption
ratio by 2.1 to 3.7 times, controllability ratio by 3.3-5 times. The specific labor
input in this case is reduced by 3 times, the costs are reduced by 1.5-2.9 times
which eventually compensates for the increased cost of development and
creation of diagnostics equipment.

This makes it necessary to:
• organize investigations aimed at choosing and substantiating the technical,
operational and economic characteristics whose requirements must be established while creating advanced space rockets depending on their target application;
• develop the methods of setting requirements for such characteristics;
• validate the type requirements applied to operational and technical characteristics of SRC and their components specified in the customer's task assignment for carrying out research and development of civil SRC.

The creation of advanced orbital hardware and launch equipment of aerospace complexes will necessitate research into the problem of creating and
rational placing of ground-based space infrastructure facilities for optimization and operation of such hardware and equipment. In particular, the use of

modular principles of assembling carrier rockets and spacecraft with creation of reusable elements may necessitate research into economic and technical feasibility and the possibility of removing part of manufacturing works' functions (assembly, repair, checks and trials, etc.) to spaceports as such which was previously considered inadmissible.

It is known that until recently the development of the ground base of SH operation proceeded on the whole without planning as to the future. There was no single technical policy and no centralized management. This resulted in protracted delays in construction and unjustifiable loss of finance and materials.

In conditions of constrained resources and financial shortages this is unacceptable. Further development and improvement of the ground base for operating SH must proceed under the guidance of centralized federal agencies that ensure a coordinated comprehensive evolution of all elements of space infrastructure in accordance with a single general plan that takes account of the prospects of space hardware as well as the interests of all entities involved in space research.

The steady improvement of space systems will make ever more stringent demands of the operating personnel and their professional skills.

The quality of fulfilling the tasks that will be set before the civil operation system will largely depend on its functional subsystem whose foundation is constituted by organizational and methodological assets ensuring SH operation and by provision of normative, regulatory and methodological documentation. This documentation must meet the requirements of the day, the real economic and social conditions and secure the unity of understanding and action on the part of managing organizations and executive entities in organizing the operation of SH for its maximum utilization as per prescribed purpose.

## 3.4 The prospects of development of software for space

The complexity of space rocketry hardware (SRH) stems from the variety of tasks it is supposed to tackle in scientific, economical and military fields. In terms of capabilities SRH will approach in the future automatic flying robots while their constellations and control complexes will acquire the features of large intellectual systems set out in space. Topologically such systems can be presented in the form of ground- and space-based intellectual information network. The level of intellect as well as the efficiency of the system is largely determined by the progress of software. The objects within such network that receive the software and urge its further development are:
 • information systems of engineering and launch complexes, command posts, information supply and analysis centers and spaceports' communication stations;
 • spaceports' automated control systems;

- ground-based automated control complex (GBACC) for controlling space-craft as part of independent command and instrumentation complexes (ICIC), flight control and flight monitoring centers (FCC and FMC), ballistics centers (BC), communication and data transmission equipment;
- information system of government agencies engaged in space work;
- information systems of corporate management of space centers, engineering organizations and space industry companies;
- on-board computing systems of space transport vehicles (carrier rockets, booster units, aerospace flying vehicles, inter-orbital towing vehicles, etc.);
- on-board control complexes of piloted SC and orbital stations;
- on-board control complexes of automatic SC and interplanetary stations;
- technical assets of telecommunication systems.

The software being created for and introduced into such systems and objects is divided into general, general for systems and special.

**General software (GSW)** is intended for organizing the computing process in local and global computing nets (fragments of ground- and space-based information net), for supporting distributed and users' data bases, for protecting information against unsanctioned access. The GSW is comprised of net and users' operating systems (OS) along with distributed software for information protection.

The principal purpose of the **general system software** (GSSW) is to organize:

- comprehensive functioning of interrelated computing systems of fragments of ground- and space-based information network;
- telecommunication data exchange;
- support of a "user-friendly" interface for work with intellectual software complexes;
- functioning of information presentation system.

The component package of GSSW contains software for:

- supporting network technologies and telecommunications;
- e-mail;
- geo-information systems;
- processing graphic and textual information, audio and video information;
- problem tackling expert systems;
- automatic design of rocketry components;
- systems that realize the formal model of a neuron network (neuron computers).

Special software (SSW) is a combination of software packages that perform specific functions in the course of controlling objects and systems. SSW functions in GSW environment and is as a rule conjugated through information and software with GSSW. SSW is created to automate those jobs that can't be done by virtue of GSW and GSSW alone.

In more general terms a number of SSW groups can be singled out intended for fulfilling such critical tasks as:

navigational and ballistic support;

information-telemetric support;

planning and command-and-software controlling of space rockets and their groups;

processing, exchange and presentation of telemetric and other information.

Without recourse to detailed description and enumeration of the entire software package, including hundreds of above mentioned SSW, it will suffice to point out that taken together they constitute an intellectual core of technical assets and provide for automated operation of interrelated complexes of space rocket equipment, their groups and control centers.

The requirements applied to the capabilities of the advanced software (SW) proceed primarily from the nature of the tasks being tackled as well as from economic factors taken into account at the stage of its creation and operation. Among the most important requirements applied to the advanced SW are:

mobility, i.e. the capability to transfer SW from one computing platform to another as well as from one operating environment to another;

capability to operate in open computing networks;

"user-friendly" interface, including the one in a natural language;

compatibility, i.e. capability to operate in conjunction with a wide range of universally recognized software for general and special purpose applications;

meeting the requirements of national and international standards of software engineering.

Compliance with those requirements will assure not only operability of SW with a wide spectrum of related software products, but also the possibility of using during its development the advanced instrumentation that has proven itself highly efficient. It should be pointed out that in the longer term more attention must be focused on the technology of development of SW, which technology must be based on standards. It is the standards whose number in the area of software is now nearing a thousand that are the quintessence of the huge Russian and foreign experience in the development of SW the use of which ensures satisfaction of the above said general requirements.

To develop SW for space work instrumental linguistic and technology supporting types of software (TSSW) are used. TSSW contains:

• algorithmic languages of high and low levels, algorithmic languages oriented to problems and objects;

• translators and compilers of algorithmic languages;

• means of automated designing of special software supplements (means of CASE technology).

Because of wide use of multi-purpose space rockets and their active control it is becoming ever more important to use software enabling realization of verbal command control of their state and functions based on recognition of voiced commands. This essentially facilitates the control and increases its efficiency. Under such circumstances, in the context of the intense internatio-nal cooperation in space, the language barrier between cooperating parties can be cleared on the basis of automatic recognition of voiced commands and their translation from one language into another. The inherent problem in creating such interface is the development of appropriate physical sensors, algorithms and software ensuring in aggregate the recognition of commands conveyed either by text or by voice.

One of the biggest problems is security. Security assets use a number of mechanisms to protect the transmitted control signals, speech and data. They include the means of authentication, confidentiality provision, key controls (including the transmission of the latter by broadcasting) and blockin (de-blocking) of terminals. In addition, an end-to-end coding is obtained which utilizes the technology of synchronous coding of an information flow, a tech-nique that reliably protects the user's information traffic. Cryptographic methods and means of information protection are now economically more profitable than other technical and organizational measures. On a number of occasions it's only they that can be effective. The producers already propose a variety of cryptographic means to give a stealth capability to documentary, voiced or any other type of information. Such means operate in networks at the information transmission speed varying between tens of bits and hundreds of megabits per second.

In the longer term an industrial technology must be elaborated for deve-lopment of software using the above mentioned TSSW. The economic data testify to the fact that as early as today the earnings obtained from the deve-lopment of software contributes in a big way to gross national product of the world's leading powers. With informatization growing wide and deep, this trend will become still more pronounced. That is why the creation and development of software, including that for space rocketry, are assured to receive a great deal of attention.

In space rocketry the software is used and developed for research, enginee-ring and manufacture, for flight tests and control, i.e. in all phases of space rockets' life cycle. Today and in the future, no effective scientific research is possible without software that supports data and intelligence bases. Modern designing of elements of space rockets and space rocket complexes involves wide use of automated design systems (ADS). The base of ADS is constituted by developed hard- and software comprised by local computer networks (LCN). In the course of production of space rockets the software is used as an intellectual core of the product quality control and diagnostics systems. At manufacturing works, at test stations and spaceports the effective trials are only possible if highly intellectual hard- and software complexes (HSWC) are available. Such HSWC are used to create advanced ground-based test and launch complexes (GBTLC). One such complex developed by Motor Building Design Bureau Mars is shown in Fig. 46. This is a multi-processor computing system with reserved functions that operates in a synchronous mode with an on-board digital computer (OBDC) of the on-board flying vehicle (carrier rocket, booster unit, space craft) control system. GBTLC offers such features as a high rate of information supply and multi-purpose capability. Also, it assures fast and efficient amendments to its own software and to that of OBDC.

**Fig. 46.** A.S.Syrov, chief designer of Motor Building Design Bureau Mars presiding over soft- and hardware test and launch complex

Program subsystems integrated into the so called corporate control information system are now being manufactured and put in operation at the plants of space rocket industry. Such a system is supposed to monitor and control a flow of documents, materiel and personnel. It is linked to the technological automated control system (ACS).

The creation of high-performance information control systems is envisioned as a means of developing spaceports. The information system of spaceport control is a complex structure consisting of the following elements:

• spaceport ACS as per types of activity on the basis of LCN of control organs and subordinate divisions;

• ACS for preparing launches of transport injection vehicles (carrier rockets, booster units, etc.) and launches of SC at an engineering complex;

• ACS for preparing and launching transport injection vehicles at the launch complex;

• information supply and analysis center as part of the spaceport's instrumentation complex;

• computerized communications centers and telecommunications-aided facilities of data collection and transmission.

The software of above listed pieces of equipment ensures automatic performance of functions, information exchange and interaction with pieces of equipment as well as with the flight control center, with information systems of corporate control in charge of plants and departments of space rocket industry, engineering organizations and research establishments.

Controlling and monitoring of transport launch vehicles and spacecraft at GBACC, FCC and ballistics centers is impossible without software complexes.

Further progress of space research will enhance the role, raise intellectualization and expand the application of software.

# PART 4
# Space exploration and ecology

Space exploration does not exert a critical impact on the environment, but as it adds to ecological pressure as well, the impact should be analyzed. As ecology has a major importance for the further development of civilization, the authors devoted a part of the book to the topic and tried to describe both the contribution of spacecraft into the noble cause of environmental protection and the possibility to minimize in the 21st century the already existing negative impact of space technology on environment.

## 4.1 Space contribution to ecology

Space technology has been used to resolve ecological problems since the very beginning of the space age. The major method was the monitoring of the Earth's surface with spacecraft equipment. If at the initial stage the equipment worked only in a visible range and had low-resolution levels, its capabilities increased with the development of space technology. At present the equipment used in ecological interests allows to monitor the Earth's surface with a high-resolution level (up to 1 meter) and in various sections of the spectrum both of the optical and radar bands. Spectral images containing much data are of specific value for ecological tasks. (Fig. 1). Earth survey data arrival has considerably accelerated – from weeks at the dawn of the space age to real time at present.

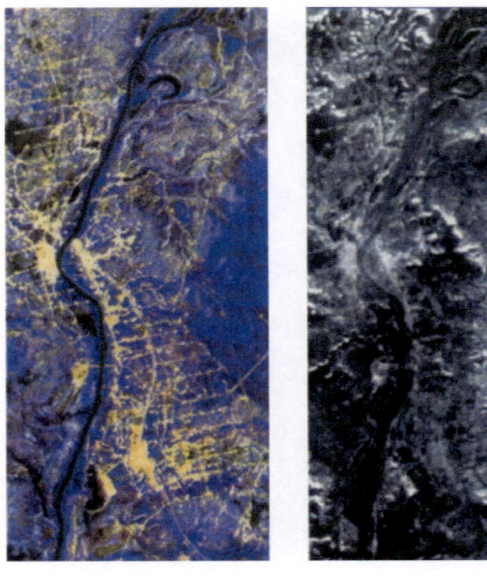

a) resolution 30 m          b) resolution 10 m

**Fig. 1.** Multi-spectral survey (left) provides more information even at poor resolution compared to panchromatic survey

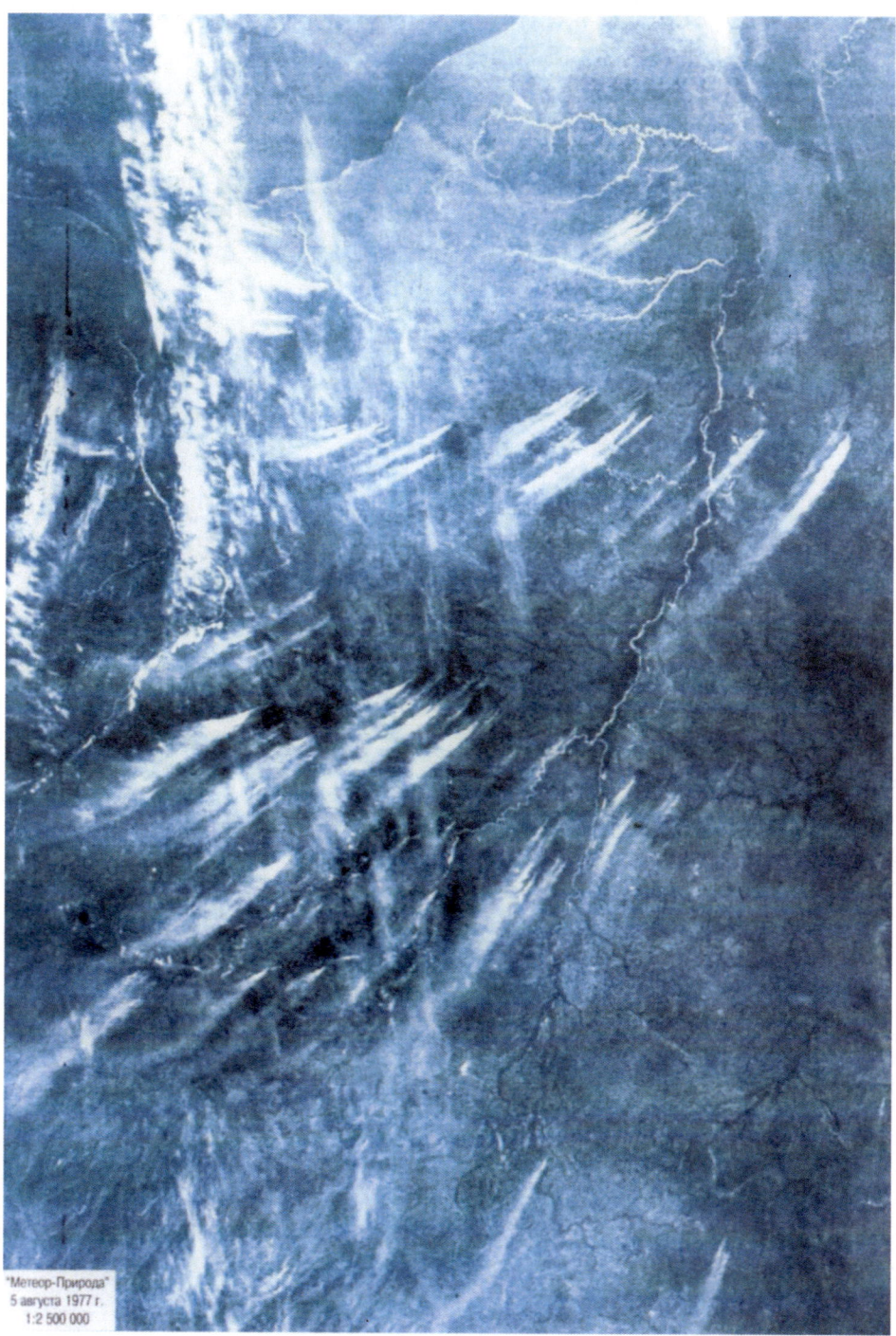

**Fig. 2.** Forest fires in Eastern Siberia. Picture from "Meteor-Priroda" satellite

The wide range of ecological problems resulted in a long list of spacecraft solving the problems. They include spacecraft for the research of natural resources ("Resourse-F", "Kometa", Landsat, Radarsat, Spot and others), meteorological spacecraft ("Meteor-Priroda", "Meteor-3", Meteosat and others), and research spacecraft (Uosat, Posat, Rocsat, and others). The role of small spacecraft for remote Earth monitoring has been constantly growing. They include existing US Ikonos, Orbview satellites and future spacecraft, in particular, the "Monitors" that were designed on the basis of the multi-purpose platform by the Khrunichev Scientific and Production Center and that allow to monitor the Earth's surface both in the optical and radar bands.

Space monitoring data allow to study the anthropogenic influence on the environment and take timely measures to decrease the impact. Monitoring of natural processes allows to issue timely warnings about natural calamities and thus decrease their catastrophic consequences.

The dynamics of hydrogenous ecosystems is one of the major objects of space monitoring.

The analysis of space photos taken at various time of the Aral Sea area, in particular, allows to design proper measures to avert the development of the territory into a desert and promote its rational use.

One of the most damaging types of anthropogenic impact is the ravine erosion. Space pictures allow to study the process and correctly plan the use of land.

Such a global problem as the depletion of the ozone layer cannot be resolved without spacecraft. The network of ground stations does not provide for a global assessment of the phenomenon. Information from satellites gives a global view and the necessary timeliness, cycle and precision of measurements of ozone layer changes.

Major damage to forests is inflicted by fires that in most cases are caused by anthropogenic impact. Timely information from satellites allows to increase fire protection and concentrate the fire-fighting effort (Fig. 2).

One of the burning global problems facing humanity is the forecast of upcoming catastrophes and earthquakes, in particular.

Space technology is already contributing to the solution of the problem. In particular, the GPS navigation system can make high-precision (up to several centimeters) measurement of the movement of tectonic platforms and determine the force of the processes taking place in the Earth crust that are the main cause of earthquakes and volcanic eruptions.

It is also possible to create specialized spacecraft for a timely detection of the signs of an upcoming earthquake. It is known that earthquakes are triggered by processes in the lithosphere when the system tends to decrease the potential energy to the minimum.

Changes in the potential energy of the Earth are caused, in their turn, by external reasons – displacement of gravitation centers of the Solar system and of the Earth-Moon system. In this case the nucleus begins to "rotate" in magma and hit the coating, gravitation forces in faults are changing. Deep fissures and displacements of ground create seismic waves that are divided into volumetric and surface. The waves in the ground spread faster than surface waves. Coming out of the depth the waves can spread to the atmosphere as sound waves (close to 15 Hz).

Changes in space (solar bursts, the position of the Earth against the Sun, the Moon, the planets) affect the living habitat on the Earth, while all global Earth anomalies influence the lithosphere, the atmosphere, the exosphere, and the near-Earth space and can be detected and measured from space. Gravitational and electromagnetic anomalies herald catastrophes. The possibility to use space-craft for measuring anomalies allows to timely warn about upcoming earth-quakes, volcano eruptions, emergence of powerful cyclones (anticyclones), tsunami, droughts, solar storms, depletion of atmospheric ozone.

The signs of upcoming earthquakes include:
• low-force tremors,
• low-force tremors following major quakes,
• gradual elevation of the ground,
• radon emission,
• abnormal behavior of animals,
• electromagnetic fluctuations.

Despite the existence of the mentioned signs most attempts to predict an earthquake failed. However, there are processes preceding an earthquake that can be detected by technical means from space.

Mechanical processes of potential energy accumulation in the lithosphere are connected with the movement of rock and, consequently, with the creation of electric charges. The quazi-steady fields of the Earth are overlaid by regio-nal fields that are conditioned by the dynamics of geophysical processes. The places where tectonic formations meet electric currents are the sources of local magnetic fields that carry information on deep underground processes. The same processes change the magnetic field in the biosphere, atmosphere, ionos-phere and magnetosphere.

It is known that an upcoming earthquake is accompanied by electromag-netic emission ranging from decimal Hertz to dozens mega Hertz for several hours or several days.

Some very powerful earthquakes can generate in the epicenter considerable gravitational-acoustic fluctuations capable of reaching the  ionosphere and casing considerable changes there.

In particular, the "Spektr sound-c" device installed on the "Interkosmos-19" spacecraft registered 61 minutes before an earthquake on July 9, 1979 a burst of electric and magnetic components of the emission field.

Calculations show that an increased negative charge of up to $10^5$ V/m appears on the surface of the region in a radius of 500 kilometers from the center of an earthquake compared to 130 V/m in an unagitated state. Maximum current of the polarized charge comprises $10^5$-$10^6$ A and generates impulse magnetic fields of up to 10 A/m. That impulse decreases the external magnetic field of the Earth by 20 percent which can be detected by a magnetometric method.

The space can help improve the ecological situation not only with the help of monitoring, which is passive contemplation, but also through active measures. In particular, the radioactive waste burial in space can considerably contribute to environmental protection. The stockpiles of waste produced by nuclear reactors is enormous and not only inhibits further development of atomic power engineering, but also poses a potential hazard for humanity.

The existing radioactive waste processing methods cannot cope with the growing amounts of waste fuel. Therefore, the space burial option is a possible guideline for the solution of the problem. It shall not be an alternative method, but an additional one that is used when other methods become inefficient.

Depending on characteristics the radioactive waste can be stored for a long time (up to several hundred thousand or million years) in space or be sprayed into small (close to 1 mcm) particles.

Where should be radioactive waste containers sent to in space? Safety reasons prompt they should be outside the sphere of Earth activities. That can be the remotest planet from the Earth, the research of which is hardly possible in future. Among these planets Jupiter is of most interest due to delivery costs.

However the safest ecological option, but a costly one would be to burn the waste in the plasma of the Sun or transfer it outside the Solar system.

Radioactive waste containers can be delivered two the Sun by two ways. The first is a direct flight to the Sun by the heliocentric trajectory with the pericenter less or equal to the radius of the Sun. It takes relatively short time (some 140 days) and needs mission control efforts for possible flight correction only in the first 10-20 days. However considerable disadvantages of such trajectories is that they demand a major accelerating burn (23 km/s).

The Sun-destined flight speed can be decreased to 7.5-7.8 km/s if radioactive waste containers enter the heliocentric trajectory by a gravity-assisted maneuver in the gravitation field of the Jupiter (flyby under the influence of the gravitation field of the planet). However in this case the total flight time to the Sun will comprise close to three years and an active mission control effort should last at least 18 months until the fall of containers on the Sun can be reliable predicted after the flyby of the Jupiter.

The launch of radioactive waste outside the Solar system can be accomplished by a direct entry to the hyperbolic trajectory of the Sun (the necessary impulse of 8.7 km/s) or by an additional boost in the gravitational field of the Jupiter that will allow to decrease the accelerating burn to 6.5 km/s, but simultaneously increase the flight time compared with the direct entry.

The most efficient transportation of radioactive waste outside the Solar system can be accomplished with the help of the solar wind, which is an uninterrupted flow of plasma that radially spreads from the Sun at a speed of 400 km/s. The option demands to place the waste into the heliocentric orbit that guarantees it would leave the near-Earth space and would then turn onboard the spacecraft into dispersed or gaseous state. The ionized particles that joined the flow of the Sun wind will not return to the internal part of the heliosphere as no ions flying against the Sun wind have ever been spotted in the interplanetary environment. Experts estimate that the minimum safe distance from the Earth for the dispersion of radioactive waste comprises from one to several million kilometers.

The orbits with an average radius of 0.8 astronomic units (between the orbits of the Earth and Venice) and 1.2 astronomic units (between the orbits of the Earth and Mars) can serve as heliocentric orbits of artificial planets for the launch of radioactive waste containers. The necessary boost impulse comprises 4.5 and 5.1 km/s correspondingly.

In case the waste efficiently and quickly disperses onboard the spacecraft (from several days to dozens of days) it is possible to use less energy-consuming orbits for the burial – with the spacecraft travelling by elliptical trajectory to the boundary of Earth activities or to the point of Sun-Earth libration with the orbital apogee of up to 1-1.5 million km. The boost speed necessary to move the craft from the low orbit to the burial orbit decreases to 3.25 km/s. However in this case an additional operation is necessary to return the used booster jointly with the spacecraft structure from the high-elliptic orbit and guarantee its landing in a target point to keep the near-Earth space free from junk.

The key issue of radioactive waste space burial is the provision of ecological safety on all stages of the project that involves the launch of hazardous waste to space. The use of existing and future launchers will demand a substantiated change of the flight scheme, a proper set of standard and safety features, a general-arrangement layout and spacecraft structure, as well as special protection facilities that taken together must rule out the possibility of a direct contact between the radioactive waste and the biosphere of the Earth in any regular or emergency situation.

Naturally, special safety measures must be taken during a launch of spacecraft with radioactive waste, including:
• booster structure of increased reliability and launch safety,
• emergency launches that exclude a contact of radioactive nuclides with the biosphere of the Earth,

- a launch path controlled by ground facilities, which in case of emergency can definitely determine the landing point of the capsule with waste along the whole launch path,
- the launch azimuth shall be chosen on conditions of safety guarantees along the whole launch path, while base line orbits shall be chosen on condition of ensuring radioactive safety of the start of the orbiter with the waste and exclude a possible collision with space junk,
- massive launches shall not pollute the environment, deplete the ozone layer of the Earth and fill the near-Earth space with separating booster fragments.

It is necessary to properly develop the safety concept for the design of the craft that will take the waste to space. Such a craft shall at least meet the radioactive safety requirements for nuclear and isotope space energy sources.

In case radioactive waste is buried in space through dispersion of radioactive nuclides in orbit, the spacecraft shall have additional equipment for the dispersion of radioactive substances or their transfer into gaseous state. The structure of the craft will be even more complicated and will have to be equipped with onboard heat regulation system if is necessary to bury in space radioactive nuclides with a high level of heat generation.

Both traditional spacecraft and future shuttles can be viewed as a means of delivery of radioactive waste to the burial orbit.

According to estimates, the annual number of launches to the most economical space burial areas of extremely hazardous waste in the volumes that correspond to the current capacity of the world atomic power engineering will be comparable to the payload of the currently implemented world program of space launches.

Massive launches will affect the atmosphere of the Earth, including the ozone layer, by considerable exhaust products. That means the solution of one problem will aggravate the other. Therefore, non-traditional launch services of a new quality are necessary.

Promising non-traditional services may include mass driver reaction engines, laser engine systems, light gas guns. Such systems have relatively small (mass) accelerated payloads and a high productivity due to the launch rate. Their use excludes the influence of exhaust products on the atmosphere as it happens with thermo-chemical engines.

In general the problem of space burial of hazardous radioactive waste needs concrete research and development and a complex feasibility study. The present development level of space technology is insufficient for a complete implementation of the task. It is necessary to design special systems for the transportation of radioactive waste to space with guaranteed safety and acceptable ecological and economic terms.

Some scientists propose to use space to fight tornado – atmospheric gale winds that blow at a speed of up to 500 km/s and sow damage and death on their path (Fig. 3).

The scientists say that tornado has one weak point despite the enormous energy of up to 70 GW. Tornado receives the initial energy due to the difference of physical parameters of air layers at various altitudes. If we interfere into the process and make it unstable, the formation of a hurricane can be stopped. To achieve that it is necessary to heat up some parts of the tornado column with microwave emitters installed on satellites. Such satellites can serve as space electric power generators, the creation of which was actively discussed in the '70s.

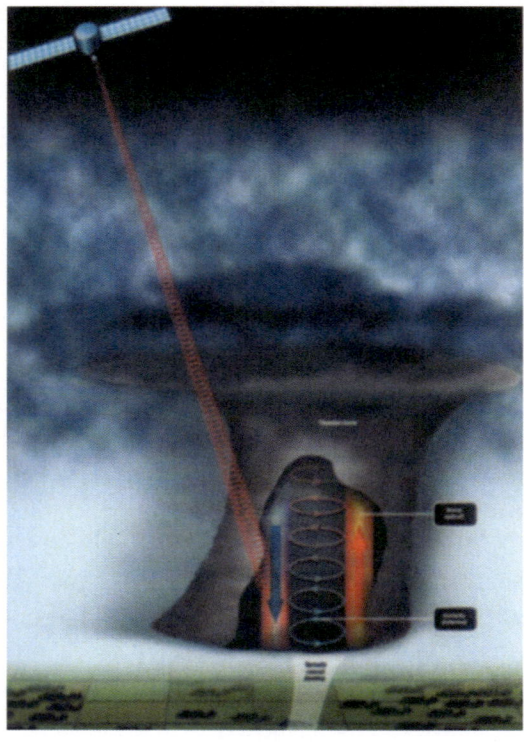

**Fig. 3.**
The use of spacecraft to fight tornado

Some scientists proposed a similar scenario to fight catastrophic cyclones, however with the use of high-altitude nuclear explosions.

One of the guidelines for the use of spacecraft to ensure ecological safety of humanity comprises the fight against the danger of asteroids. Initially, it is possible to use the service that ensure the search for and monitoring of objects aproaching the earth. In a more distant future it is possible to create space vehicles for an activ protection, which allow to destroy dangerous objects or change their trajectory.

In general it is to be noted that yet in the '70s (the pictures that) the data from spacecraft were actively useed in ecological interests. The accumulated experience shows there is no alternative to space technologies in resolving many ecological problems.

## 4.2 The impact of space and missile technology on the environment

Any human activity to master the natural habitat results in considerable ecological burden. The 20th century showed that if no measures are taken to decrease the burden, the result may be a global catastrophe. It includes the "hothouse effect", the ozone layer depletion, and the emergence of new deserts across the globe. If these factors are ignored, the consequences may be irreversible.

As any other type of human activity, space exploration makes its own contribution to environmental pollution.

The impact of missile technology and of space exploration on the environment is shown as a schedule on figure 4.

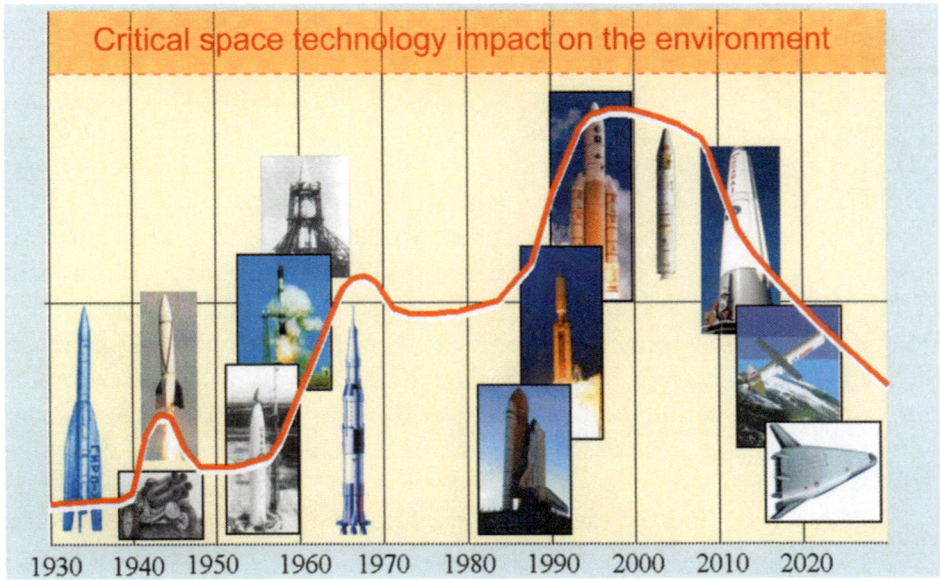

**Fig. 4.** Space technology impact on the environment

Missile technology began to exert considerable impact on the environment already in the '30s of the 20th century when first testing stands appeared and experimental rockets were launched. World War Two triggered the first burst in the impact of missile technology on the environment as rockets were massively used for military purposes (jet mortars, cruise missiles) and future weapons were tested (e.g. V-2).

The second wave emerged in the middle of the '50s as missile test ranges were constructed, ballistic missiles were tested and deployed, and space exploration began. It peaked in the middle of the '60s when over a hundred spacecraft were annually launched.

Then the activities somewhat subsided as regular operation of the already created space boosters began.

The next wave of pressure of space missile technology on the environment came in the middle of the '80s when the launch services became more powerful, space exploration turned commercial and the number of involved countries increased. That burst brought the impact of space and missile technology on the environment to a critical level after which its contribution to general pollution can become "the last drop" that may trigger catastrophic consequences.

In the beginning of the new century the ecological hazard of extensive space exploration development was comprehended and results followed. However, we shall speak of them later. Now let's get to the roots of the problem.

As it has been already mentioned, the negative impact of missile technology on the environment began much earlier than the first space launches. The reason was that a testing base was created in the '30s with stands for real-life firing tests of engines and their parts.

The experimental base threatened the environment with hazards that emerged in case of unstable work of the first liquid-fuel engines, their explosions during tests. Such dangers appeared in the Gas-dynamic laboratory, in particular, in 1931 when an ORM engine was tested. One of the first most hazardous situations took place in Germany on December 24, 1932. The combustion chamber of an experimental liquid fuel engine exploded on a stand of the West testing laboratory at the Kummersdorff range near Berlin. The stand was completely ruined. However the dangerous beginning of the history of test beds did not stop the creation of the experimental base in European countries, which continued to actively develop it. The creation of efficient missile engines was possible at the time only thanks to real-life tests of the created designs.

Already then some experience was accumulated in designing measures to decrease the impact of the experimental base on the environment that mostly focused on the safety of personnel that ran the tests. Thus, R. Goddard began to build concrete launch shelters in the '30s in the United States, to design remote engine starts, and carry out engine tests and missile launches in remote areas.

That first experience was taken into consideration in the design of major experimental bases in post-war years. However, it is another question how fully the experience was used.

It was hard not to notice that the experimental base of missile technologies affected the environment. Thus, missile engines roared so loudly during tests at the stands in Penemunde (Germany) that the noise and flares were heard and seen on the coast of Sweden some 50 kilometers away. However, the test beds posed a major threat for the personnel that mastered new technology. Several accidents occurred in war years during missile engine tests in Germany and took a toll of over 40 lives.

The Americans treasure human life and took enhanced safety measures to protect the personnel while constructing similar objects, and in particular, the test laboratory of the Convair Company that was built in Florida in the '50s. Enhanced shelters were built to protect the personnel from accidents. In particular, the engine start center was located in a concrete bunker with walls and ceiling some 7.5-meter thick.

If the testing personnel were more or less protected, no efficient measures were designed to exclude the negative impact of firing tests on nature. Thus, the test bed of the North American (St. Susanne, California) Company used enormous amounts of water to cool the operating engine (1850 tons per minute). It was so badly polluted after tests that no further use was possible. The blast waves that occurred during engine tests at such stands spread dozens of kilometers and hit the local fauna with a sharp and unpleasant noise.

In the '50s major firing stands were built by several US companies and the Aerojet, in particular. They all created ecological problems. The Soviet Union built a missile testing ground of the Research Institute of Chemical Machinery in the same years in the settlement of Novostroika, near Sergiyev Posad in the Moscow region (Fig. 5).

**Fig. 5.** Testing complex in Novostroika

Testing experts often risked a lot in those years by accelerating the work. Thus, in 1957 a whole set of first-stage engines of the "Sputnik" booster (400 tons thrust) was tested at a stand in a flight regime. Such tests are extremely dangerous from the ecological and technical points of view as they can trigger a major accident or a catastrophe. However Sergei Korolev and other experts, who had to be sure that the booster was ready to orbit the first artificial satellite, decided to test the first-stage power unit by launching all the engines simultaneously. However, such tests were rare.

The creation of missile fuel was and remains a major problem in missile technology design. Various chemical substances and their compounds, even those poorly studied, were considered as missile fuel in the '50s. They included flammables – liquid ammonia, ethyl alcohol, aniline mixtures, asymmetric dimethylhydrazine, various hydrocarbons: oxidizers – liquid oxygen, concentrated nitric acid, fluorine, trifluoride chlorine, ozone and its mixtures with oxygen, concentrated hydrogen peroxide, nitromethane, etc.

One of the most efficient, but very toxic missile fuel components was developed in the '50s – asymmetric dimethylhydrazine NH2-N(CH3)2, which represents a derivative of hydrazine – compound of nitrogen and hydrogen N2H2. In the Soviet Union the asymmetric dimethylhydrazine was known as heptyl, while in the United States as dymazine. Hydrazine derivatives result from replacement of one or several hydrogen atoms in hydrazine by hydrocarbon radicals. Depending on the number of replaced hydrogen atoms there are mono, di, tri, and tetra derivatives. Asymmetric dimethylhydrazine was widely used in missile technologies because boosters and spacecraft could be fueled with it and kept for a long time.

In the United States the asymmetric dimethylhydrazine was used in the middle of the '50s in the second-stage liquid-fuel engines of the Vanguard carrier that orbited the first US satellites. In 1957-1959 the Vanguard made 11 launches from Cape Canaveral and only three were successful. This was the beginning of contamination with heptyl, a toxic rocket propellant component.

Due to its good power and operational characteristics the asymmetric dimethylhydrazine was widely used in missile and space technology which contributed to environmental pollution. In the United States it was later used in the second stages of the Thor-Delta, Thor-Able, Thor-AbleStar, Atlas-Agena boosters that were widely used in the '60s, and in the powerful boosters of the Titan family. The Soviet Union used it in the "Kosmos" carriers designed by Yanghel, the first launches of which took place in early '60s, in the "Proton" booster of Chelomei, in intercontinental ballistic missiles and some air defense missiles.

In the '50s a compound of boron and hydrogen – borohydrogen (boranes) was actively researched as rocket propellant. There are several borohydrogen compounds – diborane liquid ($B_2H_6$), pentaborane, decaborane ($B_{10}H_{14}$), tetraborane ($B_4H_{10}$), dihydropentaborane ($B_5H_{11}$), and others. They produced much more combustion heat than other compounds. Borohydrogens were proposed as propellant for liquid-fuel missile engines by Kondratyuk in 1929.

In the '50s borohydrogens were considered as a promising fuel. Intensive research confirmed both their high power characteristics, as well as high toxic content, which was unpleasant for ecological and technical safety. The toxic characteristics even exceeded those of fluorine, which is a record holder in this aspect. Besides, research showed that production of borohydrogen was difficult, expensive and energy consuming. As a result, borohydrogen was not used in missile technology. This is one of the cases when toxic characteristics of fuel inhibited its practical use.

The possibility to use metals as missile fuel was considered in those years. Research showed that considerable amount of energy produced by metal fuel went to evaporate the resulting metal oxides. Thus the final efficiency dropped and the fuel was rejected. Ecological characteristics of metal fuels were not researched, but can hardly be consoling because of the test results of the use of metal components in solid missile fuels. The aluminum oxides that are exhausted together with exhaust gases of solid-fuel engines create a several centimeter thick layer on the ground of the launching pad and make the landscape Moon-like.

The research to select the oxidizers for missile fuel showed that high-production oxidizers are either very toxic or extremely corrosive, like the fuel itself. In particular, it was established that pure ozone is specifically sensitive to strikes and easily explodes, while long-distance transportation of liquid fluorine is hazardous for environment as the fuel is very toxic and should be produced at the launch place. Therefore, the oxidizers were not used in missile technology which slowed down environmental pollution.

The research carried out in the '50s to find efficient components of missile fuel allowed, despite certain failures, to get not only a better idea of its power, physical and chemical characteristics, but also to assess the ecological parameters. The upper concentration limits for the vapor of many missile fuel components were determined, as well as their physiological impact on a human being – smell, color, skin and eyesight, etc. (Table 1)

**Table 1.** Top allowable concentrations of missile fuel components and their physiological impact on a human being.

| Components | Allowable limit, $cm^3/m^3$ | Smell, color | Poisoning effect |
|---|---|---|---|
| Ethylene oxide | 25-100 | – | Irritates eyes, affects kidneys |
| Diborane | 0.1 | – | Very toxic, affects respiration system |
| Pentaborane | 0.01 | Garlic | Affects central nervous system. Loss of coordination, weakness, convulsions |
| Decaborane | 0.05 | Unpleasant, sharp | Same effect. Affects kidneys, liver |

| Components | Allowable limit, cm³/m³ | Smell, color | Poisoning effect |
|---|---|---|---|
| Fluorine | 1-3 mg/m3 | Sharp | Very toxic, Affects skin, throat, irritates eyes |
| Nitric acid | 10 | Sharp, no color or from yellow to orange | Irritates eyes, skin. At 200-700 cm3/m3 considerably ruins lungs |
| Hydrogen peroxide | – | No smell or color | Irritates eyes and skin, may explode |
| Liquid oxygen | Over 21% in air | Light blue | Non-toxic, ruins skin because of heavy is cooling effect, Inflammable. |
| Nitroglycerine | 0.5 | Ester | Decreases blood pressure, causes headache. Explosive. |
| Nitromethane | 50-200 | Ester | Affects the nervous system, lungs |
| Ozone | 0.1-1.0 | Caustic | Irritates eyes and respiratory ways at a concentration of 300 cm³/m³. Death occurs two hours after concentration exceeds 400 cm³/m³. Smell can expose it at concentration of 0.002 cm³/m³. |
| Tetranitro-methane | 0.1 – 0.5 | Ester | Very toxic. Affects eyes, liver, lungs. Death occurs in 10 minutes at concentration of 10 cm³/m³ |
| Aniline | 5.0 | Weak ammonia | Absorbs through skin, affects eyes. |
| Ammonia (water-free) | 100 | Caustic | Irritates eyes, respiratory ways, skin. Can be detected by smell at concentration of 53 cm3/m3. |
| Ethyl alcohol | 1000 | Spirits | Non-toxic. Irritates eyes and lungs |
| Furfuryl alcohol | – | Almond | Affects central nervous system |
| Gasoline | 500-1.000 | Petroleum | Suffocating effect |
| Hydrazine | 5.0 | Amine | Can accumulate in human body, Ruins lungs, kidneys, liver. |
| Hydrogen | – | – | Non-toxic. Inflammable. |
| Methanol | 200 | Alcohol | Affects eyesight nerves. Causes blindness. |
| Kerosene | 1.000 | Petroleum | Causes headache, nausea. Weakly toxic. |

The table shows that all missile fuel components were characterized in the '50s as hazardous for human health and consequently for the environment. At the dawn of space exploration there were no devices capable of detecting a small concentration in the air which complicated the work with toxic substances. At that time they had to smell concentrations, which were hazardous for health. The smell sensitivity threshold of fuel components for a human being is sometimes higher than the allowable concentration. Thus, borohydrogen diborane can be smelled by a person at a concentration of 3 - 4 $cm^3/m^3$, while the maximum allowable concentration is 0.1 $cm^3/m^3$. Therefore, the creation of devices for the detection of small doses of substances in the air was very important. As far as the toxic impact on the flora and fauna was concerned, they did not pay much attention to the problem in the '50s.

Space technology that was created by the principle of major effect at low costs was unready, like many other industries, to meet ecological requirements. Otherwise it would be necessary to introduce considerable constructive and structural changes, which demanded material expenses and time comparable to the total costs of space exploration, and to create a fleet of boosters anew.

On the other hand, the absence of reliable estimates of pollution scope and forms from the space industry promoted the spread of extreme opinions: from claims that space technology does not affect the environment altogether to outcries about global catastrophic consequences.

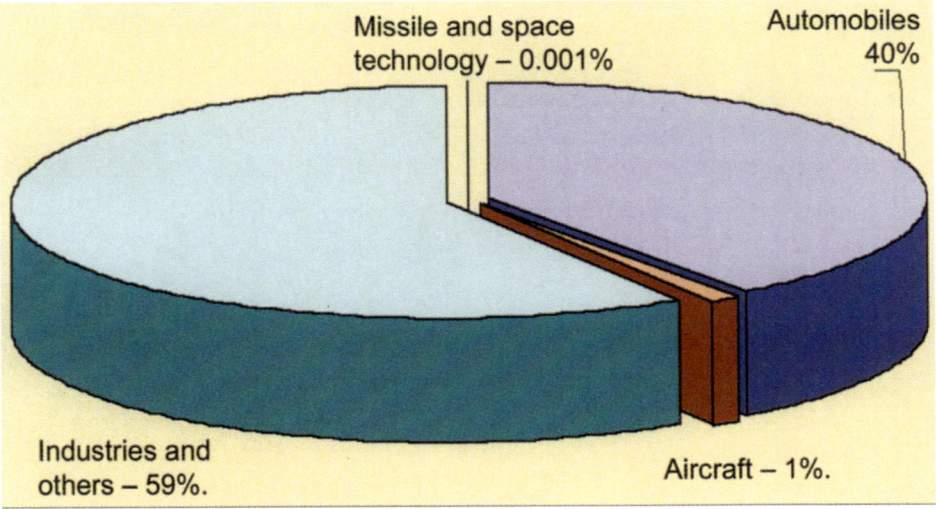

**Fig. 6.** Annual pollution breakdown of various human activities

In such conditions it is important to rate the industries by their impact on the environment. That would promote more efficient and oriented effor to preserve nature. So far they only state that the most contaminating activities include industrial production (up to 60% of total pollution) and automobiles (some 40%). The pollution share of the space industry comprises only 0.1 percent of aircraft exhaust or 0.001 percent of total exhaust (Fig. 6).The diagram data are of a general nature and do not reflect the hazard degree of each industry.

Surely, it is possible to ignore the pollution share of space exploration. But according to modern ecological outlooks, the environment has reached the limit of protective capabilities. Thus, an increase of each polluting factor by even one thousand decimal point of a percent will result in such a total impact that could ruin the natural balance. Therefore, the main task of space industry experts, like all other experts, is to limit environmental pollution. Besides, the public is attentively following space exploration and that adds to the necessity to decrease the ecological impact.

Space technology has certain unique polluting characteristics (Fig. 7), which means that only space industry experts are capable of dealing with emerging ecological problems. Major specific environmental impacts of the space industry are listed below.

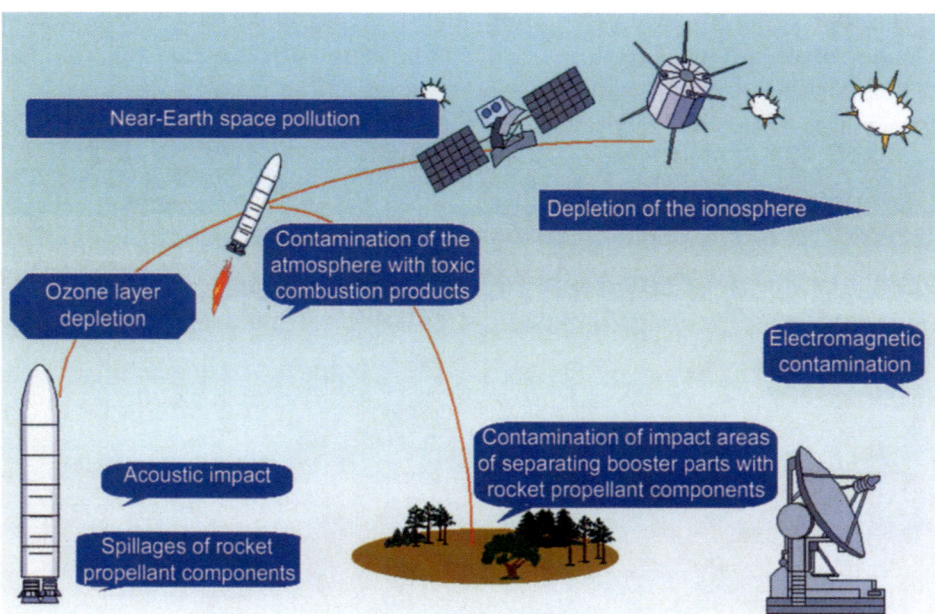

**Fig. 7**. Space technology impact on environment

Space technology has some polluting characteristics that do not exist anywhere else. Pollution continues along the whole flight path of the craft up to the upper layers of the atmosphere and space. That means that the surface of the planet, its atmosphere and space are polluted.

Some experts believe that missile and space pollution is growing rapidly. They say that it took the automobiles some 100 years to pollute the air of the cities, but it will take the space industry some 40 years to pollute the near-Earth space and some regions of the Earth's surface. Although atomic power engineering is not far ahead (45 years), the space technology is considered to lead environmental pollution. Still there can be more emotions than substantiated data. A question arises whether the respected experts know how much time it took to eliminate 3,000 km² of forests by one mining enterprise? Only 10 years. That is also a problem to think about.

Highly toxic missile fuel components are used for certain types of boosters. Such fuel sprinkled or spilled to the surface of the planet may get into a human body, into animals and cause heavy diseases and in some cases a lethal outcome.

Potential threat is posed by orbiting nuclear-powered spacecraft that in case of an accidental fall on the surface of the Earth can spread the radioactive substance on a wide area (10 · 20 km). The launches of such spacecraft from the Earth are also hazardous.

The space industry occupies wide territories, although they are not the best ones. They include launching sites, the impact areas for separating stages, testing stands, plant territories, storage bases, etc.

The falling spacecraft and their fragments pose a danger. Besides planned (sanctioned) deorbiting cases which experts estimate at several dozens, catastrophic and emergency falls also occur. The total number of annually falling objects of various weight and size is estimated at least at 400.

There are other forms of space activities that contribute to the general environmental pollution. Thus, pollution by missile and space technology producers has not been sufficiently researched. And it can be considerable. Thus, the aircraft industry, which has similar technological processes, annually exhausts to the atmosphere 153 thousand tons of hazardous substances and consumes 485 mln. m³ of water. It is to be noted that only the Norilsk Nickel metallurgical giant exhausts twice as much.

Thus, the specifics of space exploration show that it poses a certain environmental hazard and the absence of measures to minimize the negative impact can considerably affect the ecology of the planet.

To design measures for decreasing environmental pollution by space technology it is first necessary to determine the pollution impact of its various components.

Pollution of the Earth's surface, the character and intensity of accumulations are of a major importance. The major polluters are the ground equipment, construction and production facilities that are followed by engines, rocket units and other falling fragments of the launch services, and finally the falling deorbited spacecraft and fragments that sometimes reach the surface of the Earth causing pollution and posing a danger for people, animals and buildings.

## 4.3 Pollution of the Earth's surface

### 4.3.1. Launching sites and ecology

Nobody doubts the fact that launching sites are ecologically hazardous objects. Experts admit that "a cosmodrome is the zone of enhanced danger" and in some cases is a "gun powder barrel", as it has explosives and electric current sources, flammable materials and inflammable components, high-pressure pipelines and toxic liquids.

Although various measures are taken to ensure a safe operation of launching sites, the safety problem has not been completely resolved. Most attention was paid to the safety of personnel and to the aversion of accidents and catastrophes, however environmental protection was left aside.

When such complicated objects as launching sites were constructed in the '50s, no attention was paid to the problems of flora and fauna preservation at the place of their location. Several years later that resulted in major ecological problems.

Practically all the launching sites operating in the end of the 20th century and the beginning of the 21st century exert a negative impact on the environment – flora and fauna, landscapes, the atmosphere and water, existing microclimate, etc.

The construction of launching sites considerably interferes with the existing natural conditions: the average annual temperature grows, ground winds change direction and force, ground water level changes. The landscape after the construction of a major launching site (Fig. 8) comprises dozens of scattered buildings – assembly and test facilities, service towers, launching pads that are often over 100 meters high, and they are all located at a distance of 2.5 kilometers from each other. The erections usually have mighty concrete foundations that go 50 and more meters into the ground, specifically in case of construction on marshy lands. Thus, the underground part of the landscape is also affected.

Special attention is paid to the reliability of launching pads that have big hollows in the underground part to divert exhaust gases from booster engines. Thus, the foundation pit of the "Gagarin start" – the launching pad from where Yuri Gagarin blasted off to space for the first time in history – is 45 meters deep, over a hundred meters wide and 250 meters long. Over a million cubic meters of the ground were dug out during the construction, which considerably affected the landscape.

Besides launching pads, various infrastructure constructions are build at launching sites – storages for fuel, oxidizers, compressed gases, water, and command posts, launch control centers, etc. Thus, the first years of the Baikonur launching site construction included 115 kilometers of railways, over 200 kilometers of concrete roads, over 200 kilometers of water pipelines, over 100 concrete reservoirs, over 300 kilometers of communication lines, over 150 kilometers of power transmission lines, a settlement for 15,000 people, a network of measuring centers.

**Fig 8.** Overview of the Kourou space center

Intensive construction of launching sites (from mid '50s to late '60s) undoubtedly contributed to the demolition of the landscape of the planet, although the impact was not the biggest one. The Soviet Union constructed the Baikonur, Plesetsk sites, expanded the Kapustin Yar range. The United States built the Eastern and Western test ranges.

The construction of the Cape Canaveral site resulted in major changes of the landscape, with the biggest one caused by a special dirt road for the movement of a unique transportation unit with a weight of over 2000 tons that carried cargoes of a thousand tons, including the Saturn-5 boosters with a docked Apollo spacecraft and the service tower. The fully assembled booster with the docked craft was transported in a vertical position to the launching pad. It was again checked there, fuelled with various components and compressed gases and launched.

The necessity to strictly observe the allowable inclination angles of the road, the radius of the turn, and the elevation above the marshy land demanded considerable melioration work. It made the territory even marshier forcing animals, birds, and fish to abandon the area. The dirt road was over 40 meters wide. It was 5.8 km long to one launching pad and 7.5 km long to the other. The road went along a two-meter high dam.

Major landscape violations – foundation pits, water ways and canals, railways and hard-cover roads, dug-in reservoirs, concrete sites, etc – were carried out at all launching sites and pads.

One of the specifics of launching site construction that triggers environmental pollution is a high concentration of launching pads and unpredictability in the construction of new ones as it is often unclear when new boosters will be put into operation. Launching pads were built for boosters designed on the basis of existing ballistic missiles. As each ballistic missile has its own constructive specifics, it demands a special launching pad (Table 2). Thus, considerable territories were used, but as boosters became outdated, they were abandoned and turned into an environmental pollution source.

Soviet spacecraft designer Sergei Korolev created a long-range R-7 missile and envisaged a possibility of its peaceful use. Thus, the R-7 was easily modified into the "Vostok" carrier that blasted off from the R-7 launching pad. As it was unnecessary to build a new launching pad and as the existing one survived over 200 launches, the impact on the environment decreased.

**Table 2.** The number of launching pads for US boosters in the '50s – '60s of the 20th century

| Booster. New launching pad construction time | | Booster all-up weight, t | Test range, launch site | Cost' US$ mln. | Number of pads | Annual planned launch number | Distance between launching pads, m |
|---|---|---|---|---|---|---|---|
| Atlas, | 1954–1958 | 116 | 1 | 10-14 | 1 (1) | 3 | — |
| Atlas-D, | 1954–1958 | 115,6 | 2 | 10-14 | 1 (2) | 4 | — |
| Thor-Agena, | 1955–1960 | 51 | 1 | 10 | 1(2) | 17 | 250 |
| Thor-Delta, | 1955–1960 | 50 | 1 | 8 | 2 (2) | 3 | 250 |
| Thor-Burner, | 1955–1960 | 50 | 1 | 6-8 | 2 (2) | 4 | 250 |
| Thorad-Agena, | 1955–1960 | 70 | 1 | 8-10 | 1 (2) | 5-10 | 250 |
| Titan-2, | 1958–1960 | 150 | 2 | 20 | 1 (1) | 6 | — |
| Saturn-1, 1B, | 1958–1962 | 490 | 3 | 43-60 | 2 (3) | 2-3 | 360 |
| Scout, | 1959–1960 | 18 | 4 | 6 | 2 | 7-8 | — |
| Scout, | 1960 | 18 | 4 | 1.7 | 1 | 4 | — |
| Atlas-Agena, | 1960–1966 | 120 | 1 | 30 | 7 (27) | 5-7 | 300 |
| Titan-III, | 1960–1968 | 140 | 1 | 15-18 | 1 (3) | 3 | 300 |
| Titan-IIIA | 1960–1968 | 150 | 2 | 15-18 | 1 | 2 | — |
| Titan IIIC, | 1960–1965 | 560 | 2 | 185 | 1 | 3 | 360 |
| Atlas-Centaur | 1961–1965 | 132 | 3 | 20 | 1 (2) | 4 | 300 |
| Saturn-5, | 1961–1967 | 2950 | 3 | 400 | 1 (3) | 4 | 2650 |
| Scout, | 1963 | 18 | 4 | 1.7 | 1 | 4 | — |
| Titan-IIIM, IIID | 1968 | 805 | 1 | 65-68 | 2 | — | 350 |

Note:

- Figures in column "Test range, Launching site" signify:
1 – Western test range
2 – Eastern test range
3 – Kennedy Space Center
4 – Wallopse Island range
   - In brackets – number of pads at the site

A launching site comprises buildings, constructions and technological equipment for the mounting of the booster-spacecraft system on the launching pad, its pre-start preparation, testing, fuelling with fuel and compressed gas, and launching. The sites can have one or more launching pads located at a safe distance from each other (from 250-350 meters to 2.5 km). An explosion of a fully fuelled carrier on one of the pads should not damage other constructions and affect the personnel.

The safe distance between launching pads and other constructions considerably increases the territory of the launching site in general. The site with one pad occupies a territory of 5 km². In case of two pads the territory increases 1.5-fold, and in case of three pads – 2-fold. Thus, the launching site for the "Energiya" booster occupies a territory of over 10 km² on Baikonur. In the United States there were 48 pads at the Eastern test range in the '80s, and 53 pads at the Western range. All of them naturally changed the landscape of the territories and the microclimate.

A special commission that studied the Eastern test range and the Kennedy Space Center concluded that 30 years of intensive operation turned the territory and a part of launching pads into an ecological plight. Some pads serviced a small number of launches as corresponding boosters were replaced by new ones and new pads were constructed for them. Some pads were not used altogether (Fig. 9). The commission concluded that the reason for such a situation was the unique structure of the pads intended for a concrete booster, on the one hand, and a fast development of space technology that accelerated the moral and physical ageing of the equipment, on the other.

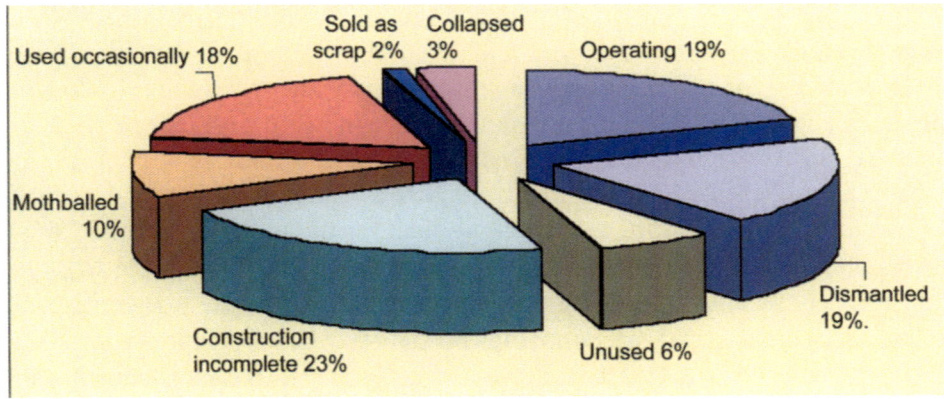

**Fig. 9.** The state of launching pads at Cape Canaveral at the end of the '70s

Flora and fauna in the area of a launching site are subject to heavy ecological pressure. The sites located in tropical and subtropical areas drain out marshlands and change the hydro regime of the locality. Thus, the landscape changes at Cape Canaveral partially eliminated marshes, polluted them with various technogenic waste, and changed the habitat of living organisms there.

Launching sites in steppe and semi-desert areas considerably spoil the vegetative cover that keeps the upper soil from splitting into various fractions. Plants are ruined by vehicles running off the roads, by movements of big crowds of people, the construction of objects and industrial and other waste dumps. As a result, deserts begin to develop and it will take decades for the nature to restore the vegetative cover at such launching sites.

The forest-located sites affect the nature by wood felling. Remaining trees become sick because of toxic pollution in the atmosphere and the soil and acid rains, in particular.

Wildlife in the area of the sites either gradually dies out or has to abandon the place. Animals, birds and insects are constantly affected by such negative factors, as noise, major concentrations of people and operating mechanisms, bright light. The change in the chemical composition of water due to missile and industrial dirt also affects the living organisms, and often kills them.

Flora and fauna in the areas of the sites are also affected by burning steppes and forests after accidents and falls of flying boosters, the pollution of the territory by debris of metal constructions, armed concrete, garbage, leaks of fuel and lubricants, and so on.

Launching sites interfere with the normal life not only of animals and smaller living organisms, but of local inhabitants as well.

One of the types of such interference is the resettlement of locals, the elimination of cultural and spiritual objects and symbols of the indigenous population. In some cases the withdrawal of land is very painstaking for the locals. Thus, a part of the withdrawn territory at Cape Canaveral houses a cemetery of the indigenous population of America, the Indians who voiced dissatisfaction over the withdrawal and demanded the territory back.

Another negative factor is that usual activities of the locals are inhibited in the areas adjacent to the sites and along the flight path. Resolute protests of Japanese fishermen against the use of the ocean as a burial place for falling booster parts decreased the number of launched to three in a year.

Space equipment dumps are also a problem for the local population. Thus, during 40 years of the operation of the Baikonur site huge dumps of rocket, industrial, construction and other waste appeared. Unfortunately, nobody controlled those unsanctioned dumps at the dawn of space exploration. As a result, the end of the 20th century faced an urgent need to clear the territories and revive them.

The Baikonur site accumulated some 5,000 tons of metal scrap (metal constructions of spacecraft and boosters) on an area of 2.5 million hectares. The junk falling on arable land impedes fieldwork and cattle breeding. The local population is specifically concerned by the pollution with the remains of toxic rocket fuel.

What are the future pollution prospects of launching sites in the 21st century?

There are a number of factors that give ground both for a pessimistic and optimistic assessment of the ecological impact of the sites.

The **negative factors** include:

- increased number of space launches,
- continued operation of polluting boosters,
- increased number of launching sites.

Various countries and organizations want to have their own launching sites or create them in cooperation with other participants (Table 3). The reasons that deter the construction of new sites are mostly economic. The cost of a site construction for a light booster in Australia is estimated at US$ 1.1 billion.

Numerous projects to construct new and expand existing sites and launching pads cause concern over a possible increase in environmental pollution in the 21st century. The construction and operation of any new launching site will undoubtedly add ecological tension to its territory and adjacent areas.

The quantitative growth is accompanied by expanding geography of launching sites. Their new location areas may include international equatorial waters and Pacific islands, the territory of South Africa, Indonesia, central areas of South America, the Far East, and others.

**Table 3.** Plans to expand the launching site network voiced in the end of the 20th century and their implementation

| | Geographic location | Participants (countries or corporations) | Time of proposal, essence and implementation by the end of the 20th century |
|---|---|---|---|
| 1 | Indonesia, Kiribati | France and other countries | 1985. Plan to create an international site |
| 2 | USSR, Kazakhstan | USSR | Mid '80s. Construction of a launching site for "Energiya-Buran" complex |
| 3 | Australia (Cape York) | International project | 1987. Under construction. In 1995 the first launch was scheduled |
| 4 | Equatorial islands | Japan | 1987. Launching site |
| 5 | International waters | England | 1987. Marine floating launching platform |
| 6 | International airspace | USSR | 1988. Air-based rocket complex (MAKS). Under implementation |
| 7 | Hainan Islands, South China Sea | China | 1988. Details unclear. |
| 8 | Hawaii | USA | 1989. Area selected. Launches planned for 1995–1996 |
| 9 | Pacific island | International project | 1991. Research planned for 1987–2007. Construction planned for 1995–2008. |

| | Geographic location | Participants (countries or corporations) | Time of proposal, essence and implementation by the end of the 20th century |
|---|---|---|---|
| 9 | Pacific island | "Pacific space center" | Full capacity in 2019. First launch in 2000. Cost US$ 44.5 billion. |
| 10 | Plesetsk | Russia | 1992. Increase in the number of launching pads |
| 11 | South Africa | South Africa | 1992. Intention to build a site near Cape Town. |
| 12 | International waters | Joint project | 1995. Sea launch. Cost US$ 2 bln. (Russia-Ukraine-Norway–the USA) Test launch carried out in March, spacecraft launched in October 1999. |
| 13 | Ukraine | Ukraine | 1995. Idea to build a site on the national territory. First stages are to fall into the Black Sea, second stages into the Mediterranean. |
| 14 | International airspace | Ukraine | 1995. Idea to create an air-based site on the basis of the AN-225 plane. |
| 15 | South America (Kourou) | France and others | 1995. The third launching pad is constructed (for Ariane-5 booster). |
| 16 | Russia, Amur region | Russia | 1996. Combat missile base rebuilt into Svobodny launching site. Spacecraft launched in 2000. |
| 17 | New Guinea Island | Australia, Papua New Guinea, Russia | 1996. Areas for possible launching site inspected. |
| 18 | Australia (near Dublin) | Australia, Russia | 1996. Intention to build a launching site for "Proton" boosters |
| 19 | Brazil (Alcontara) | Brazil, Russia | 1996. Intention to build a launching site for "Rokot" booster |
| 20 | Canada ("Churchill" range) | Canada, Russia | Mid '90s. Idea to build launching site for "Proton" booster |
| 21 | International waters | Russia | Mid '90s. Idea to build a fishing trawler-based sea launch "Riksha". |
| 22 | Ecuador | United States | 1998. Intention to build a mountain site in the Andes. |

On the other hand, the growth in the number of launching sites and their spread across the world can decrease the environmental impact in the area of existing sites, which is a positive factor.

Other positive factors that can decrease ecological pressure of launching sites include:

- growing trend to use boosters operating non-toxic fuel components;
- design of boosters with reusable stages ("Baikal"-type designed by the Khrunichev center);
- use of uniform launching equipment allowing to use one pad for various-class boosters (e.g. light, medium and heavy options of the "Angara" booster);
- enhanced work to design boosters with engines based on new physical principles (e.g. gravitational engines).

It is possible that the ecological pressure on the surface of the Earth can be decreased by a wider use of sea and air-based launching pads. Anyway, that can free (or leave in peace) much land. However it is hard to say whether the damage to marine environment and the atmosphere would be inconsiderable. These issues have not been properly researched yet.

### 4.3.2 Pollution from operating launching sites

Most pollution at launching sites occurs during pre-launch preparations of the booster, the blast-off and the powered flight, as well as in case of accidents and catastrophes. Some experts say pre-launch preparations account for 10% of total pollution, while the figure for the launch is 20%, powered flight – 40%, passive flight – 10%, in orbit – 20%. During accidents and catastrophes the bulk of pollution appears at the launch stage at which it took place, without considering previous pollution.

In pre-launch preparations most environmental pollution occurs during the fuelling of the booster and spacecraft. What are the main ecological hazards during that complicated process? It is necessary in a short time to fill in the fuel tanks of the carrier with a big amount of propellant components that have various toxic, flammable and corrosion degrees. The chance of pollution increases when the components leak, pipelines burst, in case of drainage, discharge and neutralization of fuel tanks, pipelines, and equipment that occur when the launch is cancelled. Modern space technologies use various aggressive components of liquid rocket propellant that include:

oxygen, nitrogen tetraoxide mixed with tetramethane, hydrogen peroxide, chlorine acid, chlorine pentafluoride, fluorine perchlorile, as well as promising oxidizers – liquid fluorine, liquid hydrogen difluoride, etc., and propellants – asymmetric dimethylhydrazine, ammonia, hydrogen, aerosene–50, pentaborane, hydrazine.

Solid fuel is also aggressive as it has ptastisizers, stabilizers, hardeners, combustion catalysts and other additives. Besides, solid fuel has less oxidizing elements than it is necessary for complete combustion. As a result, unburned substances appear as soot. Thus, solid fuel is ecologically hazardous.

Despite accepted tough measures, vapors of liquid fuel components continue to leak to the atmosphere and propellants still spill on the ground during fuelling. The most hazardous are hydrogen leaks during pre-launch preparations of boosters with oxygen-hydrogen driven engines, as the oxygen-hydrogen mixture can trigger an explosion. Leaks of toxic components pollute the ground, water and atmosphere of the launching site and adjacent areas. Resulting toxic vapor of acids is carried by the wind causing burns to people and animals and suppressing plants and living organisms in water.

Another ecological hazard is the launch of a booster that is accompanied by active chemical and physical processes, which affect the environment.

The launch pollutes the environment with unburned propellant or its oxides. This mostly concerns solid fuel engines. Thus, the launch of a Space Shuttle results in a several centimeter thick layer of silver-color soot around the pad. The soot is the oxide of aluminum that appears during the com-bustion of 1,017 tons of solid fuel in the launch accelerators. Although the dust is believed to be ecologically and chemically inactive, it considerably pollutes the ground and the atmosphere. Some experts fear the aluminum environment may trigger cancer tumors. Soot and other combustion products have a negative impact on the atmospheric ozone and result in acid rains.

Mankind is lucky that the number of Space Shuttle flights is less than the initially planned figure (10% according to some reports). Otherwise, experts estimate that the soot layer could be 1.5-meter thick in some places.

Some US experts see a way out in the creation of liquid boosters for the Space Shuttle. But that demands considerable replacement expenses and does not trigger enthusiasm in solving the problem.

The launch of a booster burns out oxygen in the atmosphere and pollutes it with combustion products. A considerable amount of nitric oxide, carbon dioxide and other poisonous substances is exhausted into the atmosphere. In certain cases water is supplied to the exhaust flame to decrease temperature of the exhaust stream and consequently its impact on the launching pad, as well as noise (Fig. 10). High temperature of several thousand degrees turns water vapor and exhaust gases into complex aggressive compounds that are

**Fig. 10.** A vapor cloud appears as water is supplied to the flame

spread by the wind. The blast-off of a booster burns out oxygen, creates chloride and fluorine compounds and, as a result, may create ozone holes in the atmosphere of the planet and trigger acid and other chemically aggresive rains.

The launch of heavy boosters creates a powerful acoustic field up to 150 decibel that considerably surpasses the top level acceptable for human ear (some 70-80 decibel), for animals, birds and other living organisms. It is a very complicated technical problem to decrease the noise of a booster.

The most negative impact on the environment is produced by accidents and catastrophes of boosters on the launching pad or during the first seconds of the flight. The gravity of consequences grows because of explosions and fires of fully fuelled tanks of boosters with docked payloads that may be both manned and unmanned spacecraft. The power of the explosion, according to certain estimates, can reach 70-90% of the blast power of trotyl equivalent of the amount of the booster fuel (Fig. 11). A blast-off explosion is not a rare case in the history of space technology.

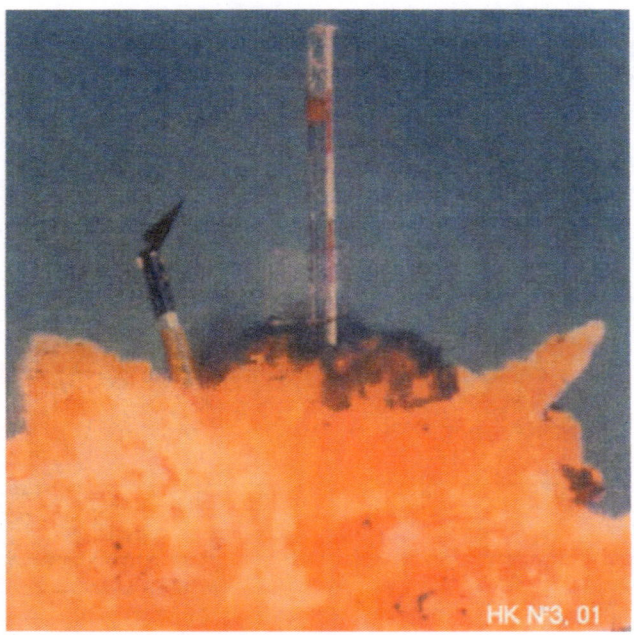

Fig. 11.
Explosion of the Vanguard
booster on December 6, 1957

Thus, on July 3, 1969 the explosion of an H-1 carrier ruined the launching pad completely. It took two years to rebuild it.

On September 23, 1989 a fire occurred and an explosion followed on the first stage of the booster during preparations for the start of the "Soyuz-T" spacecraft with cosmonauts Titov and Strekalov. The emergency safety system (SAS) saved the lives of the cosmonauts. This solid-fuel engine mounted on top of the landing capsule operates for 2-6 seconds, which is enough to fire off the capsule in case of emergency to a height of 1-1.5 km that is sufficient for safe landing. SAS switched on one second after the fire broke out, while the explosion followed seven seconds later. The cosmonauts were fired off in the capsule 1-2 km away from the launching pad and safely landed.

On October 4, 1990 the Zenith booster exploded shortly after blast-off when it fell back on the launching pad. The explosion was so powerful that booster engines weighing 2-3 tons were scattered 2-3 kilometers away.

In case of H-1 some 2600 tons of rocket fuel exploded, while Zenith had some 450 tons. In the first 20 years of operation of the Baikonur launching site three pads were ruined by explosions.

The environmental impact of an explosion of a fully fuelled booster on the launching pad is similar to the sanctioned blasts of solid-fuel missile engines that were eliminated in the framework of the international treaty for the reduction of short- and medium range missiles. The explosion of three medium-range missiles in one unit with a total weight of 90 tons created a damage envelope of 1.5 kilometers. A cloud of toxic substances rose to 1-3 kilometers. The ground gas and smoke outburst of the explosion was spread by the wind for dozens of kilometers and posed the main toxic pollution hazard for the environment.

Heavy boosters contain much more fuel compared to medium-range missiles. Thus, their impact on the environment will be several times bigger.

Luckily, explosions at launching pads took no toll. However accidents and catastrophes during blast-off and on the first seconds of the flight resulted in loss of life. Accidents at the early stage of the flight are dangerous both for the crew of the spacecraft and residents of nearby settlements and for people staying hundreds of thousand kilometers away from the launching site. Damage is inflicted to buildings, cattle, vehicles, etc.

The list of such catastrophes and accidents is provided in Table 4.

**Table 4.** Some accidents and catastrophes of boosters in 1995–1999

| Date | Type | Booster, country | Booster explosion Height Flight Time m./s. | | Technical reasons and environmental consequences |
|------|------|------------------|--------|--------|------------------|
| January 26, 1995 | Catastrophe | CZ-2E, China | – | 51 | 6 civilians killed and 23 wounded, environmental pollution |
| March 28, 1995 | Accident | Start, Russia | – | 49 | Environmental pollution. Engine stopped making debris fall in the mouth of the Yana River on the Laptev Sea coast. |
| February 11, 1996 | Catastrophe | CZ-3E, China | 2000 | 20 | Six civilians killed and some 100 wounded or injured |
| May 14, 1996 | Accident | Soyuz, Russia | 1500 | 49 | Nose cap ruined. Environmental pollution |

| Date | Type | Booster, country | Booster explosion Height Flight Time m./s. | | Technical reasons and environmental consequences |
|------|------|--------|--------|--------|--------|
| June 20, 1996 | Accident | Soyuz, Russia | 1500 | 49 | Nose cap ruined. Environmental pollutionage. |
| June 4, 1996 | Accident | Ariane-5, France | 3660 | 45 | Control system failure. Environmental pollution |
| January 17, 1997 | Accident | Delta-2, USA | 600 | 13 | Solid-fuel accelerator cracked. Debris damaged cars and buildings. |
| May 21, 1997 | Accident | Zenith-2, Russia | – | 49 | Explosion during stage separation |
| November 2, 1997 | Accident | VSL, Brazil | 3230 | 65 | One engine did not start. Environmental pollution |
| July 6, 1999 | Accident | Proton, Russia | 120 km | Some 200 | Second stage engines explode. Debris fell on a village. No casualties or damage. |
| October 27, 1999 | Accident | Proton, Russia | 140 km | Some 210 | Second stage engines explode. Debris fell 25 km from a village. |

The data in the table show that an average of two-three grave accidents and catastrophes annually take place at the flight stage.

One of the major catastrophes involved the US Challenger Space Shuttle on January 28, 1986, which exploded 74 seconds after blast-off, and thousands of spectators witnessed it. Nearly a thousand ton of liquid oxygen and liquid hydrogen exploded. Seven crewmembers were killed. The atmosphere, the ground and the sea surface were polluted. The live TV broadcast of the catastrophe shocked the people around the globe.

Another grave accident took place on August 12, 1998 with the US Titan-4A booster (Fig. 12). It exploded upon a mission control command on the 41st second of the flight at an altitude of 5 kilometers. Car alarm systems switched on the ground in a range of 20 kilometers. However nobody was injured. A poisonous orange cloud moved to the ocean and dispersed in half an hour. The material damage was estimated at US$ 1 billion and the figure was second only to the Challenger catastrophe.

Explosion-resulting debris can fly to big distances. Thus, the Challenger explosion debris fell into the sea on an area of 50 km² at a distance of 1-5 km from the launching pad. The explosion of the Chinese CZ-3E booster on February 14, 1996 sent most of the debris two kilometers away from the pad. There were cases when debris were scattered at bigger distances. Thus, traces of debris were reported 13 kilometers away from the pad after the Delta-2 explosion on January 7, 1997.

Space technology explosions in the air can kill or injure a person, who has nothing to do with space research. Thus, the explosion of the Chinese CZ-2E in 1995 and the already mentioned explosion of the CZ-3E in 1996 killed 12 and injured over 120 civilians.

In the history of manned flights 11 cosmonauts were killed in catastrophes onboard, while over 230 people were killed on the ground by booster explosions on the launching pad and during the first seconds of the flight.

**Fig. 12.** Accident with the US Titan-4A booster on August 12, 1998

**Table 5.** Ground casualties of space technology in 1960-1999

| Date | Booster, event | Casualties |
|---|---|---|
| | Casualties among rocket engineers and staff | |
| October 24, 1960 | Fire of Yanghel-designed long-range ballistic missile at Baikonur | 92 |
| October 24, 1963 | Fire in the silo of Korolev-designed intercontinental ballistic missile R-9 due tohigh oxygen content (32%) in the surrounding atmosphere was caused by a spark during a change of the bulb. Baikonur | 8 |
| August 9, 1965 | Fire and explosion of Titan-2 in silo during repairs | 53 |

| Date | Booster, event | Casualties |
|---|---|---|
| January 27, 1967 | Fire in Apollo cockpit during crew training | 3 |
| October 1970 | Fire near Vostok booster during pre-launch preparations caused by personnel negligence. Baikonur | 4 |
| June 26, 1973 | Fire during preparations for fuel discharge from Kosmos-3M booster. Plesetsk site. | 9 |
| March 18, 1980 | Explosion (fire) of Vostok booster during pre-launch preparations. Plesetsk site. | 49 |
| April 2, 1994 | Explosion of a Chinese booster on the launching pad. Sichan site. | |
| Civilian casualties | | |
| January 26, 1995 | Chinese CZ-2E explosion. Sichan site | 6 |
| February 14, 1996 | Chinese CZ-3E explosion. Sichan site. | 6 |

The relatively small number of casualties among rocket engineers and space crews should in no way be consoling. It is to be noted that manned space flights are currently at the initial stage of development and have not become massive, like aviation, which annually kills some 700 people. So far the ratio is one space flight casualty to 20 deaths on the ground. However, the use of boosters with fully automatic pre-launch preparation allows to hope that the situation will change. Thus, the fuelling of the R-7A booster needs some 100 people on the pad, while there will be nobody there during the fuelling of the Angara booster.

### 4.3.3 Pollution by spillage of rocket fuel components

Pollution of the ground and the lower layers of the atmosphere with rocket propellant components is very hazardous. There are heptyl, nitrogen, fluorine, and kerosene fuel contaminations.

The most dangerous rocket fuel component due its toxic characteristics and a wide use is the asymmetric demithylhydrazine. Its has the $(CH_3)_2N_2H_2$ chemical formula and is also known as heptyl. This chemical compound is an colorless hygroscopic liquid with ammonia smell. From the ecological point of view heptyl is a very toxic fuel component.

Heptyl was synthesized in the United States in the '50s of the 20th century. Its wide use in space technology is conditioned by high power-generating capacity and the possibility of long storage in booster and spacecraft fuel tanks.

Besides high toxic qualities, heptyl is also hazardous as it decomposes at temperatures over 3500C emitting heat and inflammable gaseous products. It explodes if overheated in a closed space (in orbit) or if the rocket hits the ground. As a rule, heptyl is used together with toxic oxidizers, e.g. nitric acid substances. It is also used with liquid oxygen as a component of US rocket propellant – aerosene, hydene, etc.

Heptyl gets on the ground during production, transportation, fuelling of boosters and spacecraft before the launch, and as fuel residue in tanks of separated stages, as well as during accidents and catastrophes.

Most heptyl contamination of the ground comes from falling fuel tanks. After a stage engine stops working its tanks still contain some residual fuel (Fig. 13). The propellant also remains in pipes, pumps, and valves. The residual fuel in tanks may be conditionally divided into the safety and unburned reserve. Tanks are loaded with a small excess so that the engines do not stop during the flight because of fuel shortage. The safety reserve usually comprises 2-3% of the total amount. The ratio may seem small at first sight, however in heavy boosters it may comprise dozens of tons in absolute figures (Table 6).

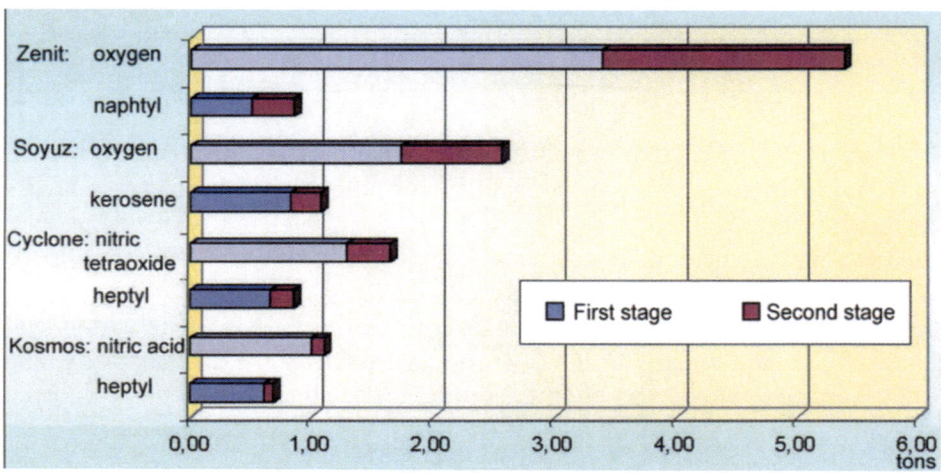

**Fig. 13.** Residual weight of rocket propellant components

**Table 6.** Approximate mass of safety fuel reserve of some boosters

| Booster, year of commissioning | Launch mass, t | Rocket propellant (oxidizer and fuel) | Safety reserve, t |
|---|---|---|---|
| Atlas-Agena, '60s | 130 | Liquid oxygen and hydrogen. Nitric acid and asymmetric hydrazine (second stage) | 2.4-3.5 |
| Saturn-5, 1967 | 2950 | Liquid oxygen and kerosene. Liquid oxygen and liquid hydrogen | 54-80 |
| Ariane, 1979 | 260 | Nitric tetraoxide and asymmetric dimethylhydrazine | 4.1-5.9 |
| Space Shuttle, '80s | 2000 | Solid fuel (in accelerator), liquid oxygen and liquid hydrogen | 36-54 |

Ecological experts have some interesting estimates of the amount of heptyl that returns to the ground in falling stages.

Heptyl production in Russia by mid '90s was estimated at 200-300 thousand tons. To estimate the world production ecologists use the 2.5 ratio (1.0 for Russia, 1.0 for the USA, 0.5 for other countries). Thus, they estimate the world heptyl output at 500-750 thousand tons in the same period. If 2-3% of the safety reserve is not burned out, then the amount of heptyl falling on the ground was 10-20 thousand tons. It remains to wonder how the nature has survived it.

Depending on the design, the separating stage either falls with residual fuel on the ground or discharges the fuel in the atmosphere and falls without it.

In the first case the stage hits the ground and falls apart and up to 90% of heptyl burns out due to the contact with oxygen in the air. The remaining fuel is spread as droplets and vapor by the wind to long distances and then gets into reservoirs with atmospheric precipitation. Some stages fall into seas and oceans and contaminate their waters.

The second case rules out ground contamination. Thus, the accident with the Proton-5 on July 5, 1999, when it collapsed in the upper layers of the atmosphere, yielded no contamination with unburned fuel in the fragment impact area. That means the fragments, including the fuel tanks of the Breeze-M booster, were dry when they fell on the ground.

What is the ecological and biological hazard of heptyl?

**Firstly,** it is very toxic. It is six times as toxic as the hydrocyanic acid and belongs to first-class hazards as the most poisoning chemical compound among non-traditional poisonous substances.

**Secondly,** heptyl may get into human body both through respiration tracks and through skin, alimentary canal, and mucous membrane which makes it specifically hazardous.

**Thirdly,** heptyl is highly volatile, fully dissolves in water, preserves its composition in the ground and in living organisms. But if disintegrated, it produces even more toxic agents.

**Fourthly,** heptyl impact on living organisms was insufficiently studied and it is difficult to fight it therefore. There are no methods to detect its small con-cen-trations and no ways to decontaminate it. There are no efficient prophy-lac-tic medicines and treatments of heptyl-caused pathologies.

**Fifthly,** heptyl affected considerable territories and waters not only in Russia, but also in the whole world.

Heptyl and its metabolites were found in plants, in soil, bottom sediments, ground and surface waters. Its concentration often surpasses the maximum allowable norm (Table 7). Heptyl may be present in vegetables, in meat and fish. Heptyl compounds may get into human body with food and water and cause grave diseases (blood, cardio-vascular diseases, hemorrhagic attacks, cancer, and liver cirrhosis).

Contamination hazard of rocket propellant components has increased of late. Various-purpose boosters are launched not only by leading space powers, but by other countries as well. The ecological pressure on the World Ocean grows specifically. Thus, Russia and Norway discharge their booster parts into the Arctic Ocean, Israel – into the Mediterranean. The United States and the ESA pollute the Atlantic Ocean and will soon be joined by Brazil, Argentina and Canada. The Pacific Ocean is the burial site for space technologies of the USA, Japan, the ESA, the international Sea Launch Consortium, as well as for space objects deorbited by command from mission control centers.

**Table 7**. Maximum allowable concentrations of heptyl and its oxidizing products

| Substance | Working air, mg/m3 | Atmospheric air, mg/m3 | Water, mg/l | Soil, mg/kg |
|---|---|---|---|---|
| Heptyl | 0.1 | 0.001 | 0.02 | 0.1 |
| Heptyl oxidizing products | | | | |
| Tetramethyl-tetrazone | 3.0 | 0.005 | 0.1 | – |
| Nitrosodimethylamine | 0.01 | – | 0.01 | – |
| Formaldehyde | 0.05 | 0.035 | 0.05 | 7.0 |

Which of the two evils of space junk pollution is less hazardous – to contaminate land or ocean? So far there are no substantiated scientific recommendations on that account. However, one can suppose that land pollution has less negative consequences. Many toxic propellant components and heptyl, in particular, unlimitedly dissolve in water. Heptyl contaminations spread in water faster than on land. They cannot be localized like on land by sealing off the area and decontaminating it. In water heptyl easily gets into sea products and consequently into human organism.

How can such contaminations be avoided? The only way out is to create nature-preserving space technology and engines burning the so-called "pure" fuel.

It is believed that the amount of fuel falling on the ground can be consideably decreased by dispersion in the atmosphere. The toxic impact on the ground diminishes, however contamination of the atmosphere grows. Fuel components are carried by wind to major distances. As a result, the total impact on the environment does not decrease.

A serious step to decrease environmental pollution by residual fuel would be a complete elimination of the safety reserve and a full combustion of the fuel in flight. Such an approach became possible thanks to high precision fuelling, on the one hand, and the creation of optimal engine control systems, on the other. This approach has been already implemented in the new modification of the heavy Proton-M booster of the Khrunichev center.

The major method to fight contamination with toxic fuel components is to reject them altogether and switch to ecologically safe rocket propellants also for heavy boosters. An example of such a booster is the Angara carrier, as its combustion product is water.

### 4.3.4 The fall of rocket parts

The fall of various space technology elements on the Earth poses a serious environmental hazard. This impact is conditioned by constructive specifics of present-day boosters.

Since the first launches the space carriers had and will have for a long time one specific constructive characteristic – they are multi-staged. That characteristic, which was proposed yet by Konstantin Tsiolkovsky and which allowed to launch space research, is gradually developing into an environmental hazard as space flights intensify.

When a space carrier dashes though the atmosphere, its used stages fall on the ground from an altitude of several dozen kilometers. The time of the fall, as well as the weight and size of the objects are well known. Thus, there is a good chance to avert a negative ecological impact.

Booster parts fall on scarcely populated localities along the flight path, which are called "the impact area of separating units", and break into fragments (Fig. 14). The areas can be located at various distances from the launching site: some 800 km for two-stage boosters, and some 2500 km for three-stage boosters.

**Fig. 14.** Impact area of the second stage of Proton-K booster.
Employees of the Moscow State University study the impact of fallen rocket parts on nature

The impact area on the ground has an ellipse form that stretches along the flight path. The size of the impact area may be considerable, up to 1500 - 5000 km$^2$.

Historically, the space industry has been using a somewhat irrational practice when new impact areas had to be designed for each new booster and, with some exceptions, they did not coincide with the existing impact areas. As a result, impact areas considerably increased, as well as the size of the territory that was withdrawn from economic use and polluted with debris (Fig. 15).

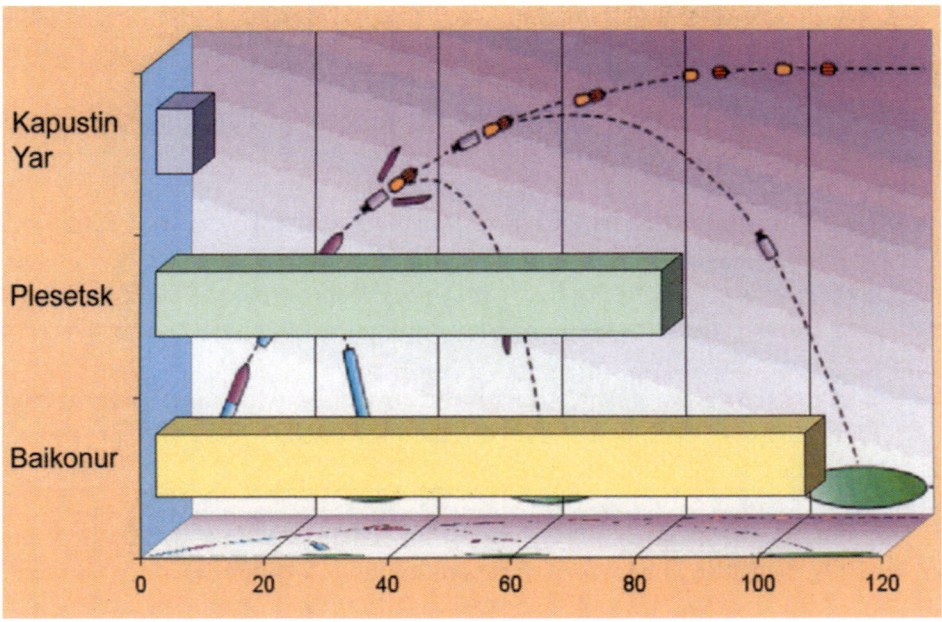

**Fig. 15.** The size of impact areas of Russian cosmodromes in 1995, thousand km²

There is another, national peculiarity in choosing the impact area. Russia and China prefer land, although some parts of Russian boosters fall on the coast of the Arctic Ocean and into its waters. The United States, France and some other countries prefer the oceans (Fig. 16). However, it is a complicated issue, both ecologically and legally, to say what is more hazardous for humanity and nature – to pollute international waters or own land.

Major contamination occurs in the impact area when booster parts hit the ground, fall apart and spill the remaining fuel. As the impact areas are located in uninhibited places, the booster parts do not cause damage to people if the flight goes as planned. However, deviations from the norm happen from time to time and it is therefore impossible to guarantee full security to people living close to the impact areas.

Nobody has yet dared calculate how much metal has fallen on the ground since the beginning of space exploration. Its amount differs in each booster. It is estimated that some 15 tons of metal constructions are falling on the ground from each Russian heavy and medium boosters. The figure may rise to several dozen tons in case of super-heavy boosters. Thus, some 180 tons of metal constructions fell on the ground after the launch of Saturn-5 booster with an Apollo spacecraft. Experts estimate (Table 8) that an average of up to 70 tons of metal constructions or 10% of the all-up booster weight fall on the ground after each blast-off (Fig. 17).

**Fig. 16.** Ariane-5 blasted off to space, but a trace remained in the ocean (splash-down of ARD-200 demonstrator).

**Table 8.** Weight of some returnable booster constructions (first and second stages)

| Booster | Weight, t | | Booster | Weight, t | |
|---|---|---|---|---|---|
| | All-up[1] | returnable | | All-up[1] | returnable |
| Thor-Delta | 55 | 5 | Saturn-1B | 570 | 53 |
| Thor-Agena | 125 | 7 | Titan-3C[2] | 645 | 170 |
| Vostok | 290 | 21 | Saturn-5 | 2,930 | 166 |
| Total average | | – | Total average | | 70 |

Note:
1 – without payload
2 – weight of three stages

Pollution of the impact area with booster debris and fuel residue turned them into the zones of hazardous space impact on the environment, including human beings.

What is the impact like? There are different extreme opinions on that account which include statements of local population and authorities who want to get all imaginable compensations and are backed by scandal-seeking media, on the one hand, and estimates of experts who deal with space technology design and operation, on the other.

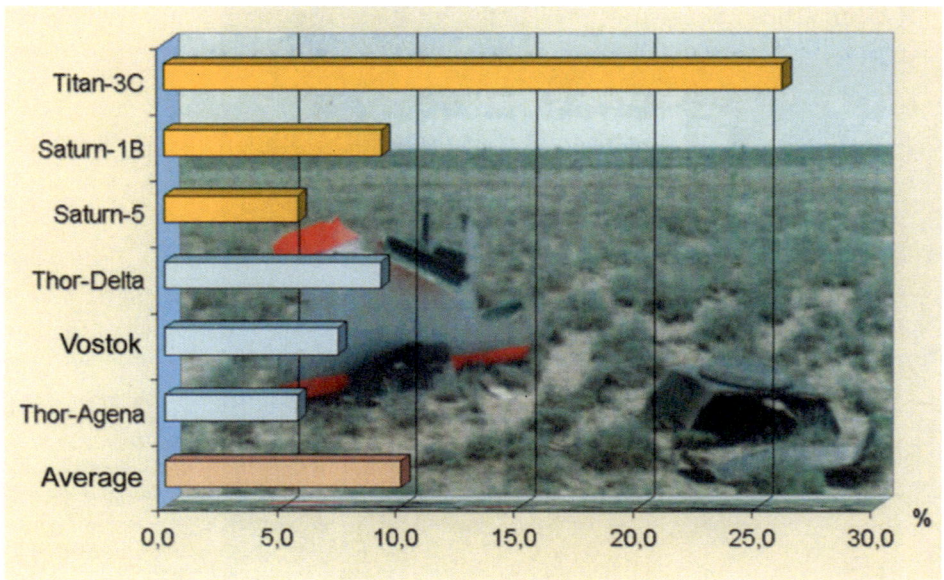

**Fig. 17.** The share of all-up booster weight that falls back on Earth

The opinion of the media and local authorities is often emotional and rarely objective. Here are some quotes from the press.

"The leadership of Dzhezkazgan region issued an ultimatum to the military and space departments: the best pastures have been sown by various objects of their research. 890 complete and fragmented stages of space rockets have been scattered on the territory of two districts and contaminate the area with toxic materials. For a long time the used stages simply exploded. The debris were scattered on fields and turned them into a desert. As a result, state farms annually sustained irreparable damage of four million rubles. Attempts to clear the pastures were rare. Last year (1989) only forty objects were gathered. There were negotiations on the elimination of spacecraft launch aftermath. It was decided to limit the metal-from-the-sky fall areas, decrease the toxic characteristics of fuel, change the stage explosion method, clear pastures from debris".

Another report about a different launching site said: "The used stages of boosters launched from the Plesetsk site fall on the Nenets tundra... The stages were exploded at the impact place thus sowing sharp debris into reindeer moss. The deer injured legs in the "space dumps" and perished. Hundreds of hectares of reindeer moss have been actually withdrawn from pastures. Reindeer breeders sustain losses... This year (1990) the Soviet Glavkosmos was forced by public pressure to create a special unit to clear territories of space junk... Two helicopters in three months... cleared 215 places, as Glavkosmos has no money for bigger areas".

"As no efficient measures to clear the territory from space junk were taken, residents of Dzhezkazgan region of Kazakhstan themselves collected over a thousand tons of space metal scrap. The local administration proposed to various enterprises to buy it. It contains chrome, nickel, silver, platinum and other precious metals. Hard currency buyers appeared".

"Businesslike northerners have been engaged in space conversion for a long time. Nose caps have been adjusted as framehouses. The fuel tank of a booster is used in the "Rossiya" reindeer farm as a fodder storage facility".

One of the first ecological agreements was concluded in 1992. It was signed by a representative of the Plesetsk cosmodrome and authorities of Uvat district of Tyumen region and envisaged collection by the polluter of booster stages that fell on the territory of the district. A total of 100 stages fell there. They were collected by a Mi-8 helicopter and kept at a local airfield.

Since then space industry enterprises stopped ignoring ecological problems
of the local population. Although they did not always have the necessary funds (in the '90s specifically), they nevertheless took the possible cleaning effort.

Thus, a special 500-men strong battalion was set up on Baikonur to clear the impact areas (Fig. 18). Ballistic experts worked to decrease the impact areas to ease debris collection. Besides, an ecologically progressive method of breaking the fallen blocks apart was designed to replace ordinary explosions. Now the so-called explosion cable is placed on the unit, which brakes it into easily transportable fragments. Besides easing the space junk-collecting task, the method demands 100 times less explosives and produced no debris.

A storage facility for used booster parts was built at the Plesetsk cosmodrome close to the settlement of Koifa in Arkhangelsk region. It stockpiled strap-on boosters, first and second-stage engines, other falling constructions of Soyuz, Cyclone boosters, and strategic missiles. In 1990-1994 1,858 tons of space junk were collected and a territory of 318 thousand hectares cleaned.

Separating booster parts falling on the territory of the Komi Republic are not only collected in a centralized way, but also reprocessed as scrap at enterprises of the "Treasures of Urals" financial and industrial group in Yekaterinburg.

The falling booster parts pollute not only land, but also the World Ocean. What can be done to decrease pollution? First of all, it is necessary to strictly fix the time and place of the impact. The United States, for instance, conducts telemetric and trajectory measurements of test flights of space objects to locate the impact area of the fore parts, and uses stationary hydrophones that ease the location task. The data on the place of the fall, the depth and water temperature, local currents and wind directions, the amount of residue fuel, the state of the ocean allow to assess the ecological hazard by determining pollution in the impact area and its spread.

**Fig. 18.**
Clean-up of impact areas of booster parts

The choice of impact areas used to depend on the necessity to optimize booster characteristics, including those of separating parts, on the one hand, and on the desire to make the parts fall in scarcely populated localities, on the other. Therefore, a change in impact areas results in such changes in booster characteristics that it actually means the creation of new generation boosters. Such research is being conducted by the Khrunichev center, in particular. For example, constructive solutions implemented in the modernized Proton-M booster allow to decrease the necessary impact area by 3.000 km². The use of the promising Angara booster allows to have no impact area for the first stage at all.

### 4.3.5 Impact of spacecraft and their fragments

Impact, or as specialists put it deorbit, of spacecraft and their fragments is another kind of environment pollution. It is necessary to distinguish controlled (authorized) deorbit and abortive, i.e. accidental or emergency impact of spacecraft and their fragments.

Controllable descent with normal functioning of all systems as a rule poses no danger to population or ground structures. At a preplanned time, brake engines switch on and work during a certain period of time directing the unburned fragments to an ocean area forbidden for navigation. The area should be large enough as trajectory of fragments that vary in size and mass is practically unpredictable. It is not a difficult task to estimate roughly borders of the impact area and time of entry into the atmosphere provided the object's velocity and aerodynamic resistance are known. It is possible to forecast ground touching time with the precision of 15 percent of the period remaining before the touchdown. In the majority of cases it is quite enough to take necessary safety measures.

Descent of spacecraft and their parts to the atmosphere or ground level is a rather complicated operation from technical and management viewpoints. A descending aircraft shall meet certain requirements including primarily controllability. Then specialists determine the current trajectory of a space-craft, location of a point where it will enter the atmosphere (usually 110-130 kilometers above the earth) and time of approaching this point. When the spacecraft approaches the point of "digging" into the atmosphere, onboard descent engines are ignited to slow down the spacecraft motion.

The threat of falling spacecraft fragments increases substantially in case of uncontrollable descent of spacecraft or its major fragment as nobody knows whether these fragment will burn down completely in the atmosphere or reach ground level and in what form they will fall. Emergency impact is even more dangerous. It might be caused by a variety of reasons including the so-called "swelling" of the atmosphere due to solar activity, mission center's inability to control a spacecraft, sudden breakdown of a spacecraft into fragments, propellant run-out or spacecraft inability to maneuver.

Spacecraft and their debris impact threaten human life physically and psychologically, initiating fear and anxiety ruinous to health. Economic consequences are also important. Evacuation of population to safe areas and suspension of normal economic activities might be rather costly.

This chapter deals with most important cases of spacecraft and their debris fall to the earth reported by the end of the 20th century. But apparently in the 21st century potential impact of spacecraft or their debris will cause major concern of specialists and public opinion.

The first registered impact took place in 1962. Its was a spacecraft tested in the framework of preparations to the first manned flight. This space vehicle was launched on May 15, 1960. It was lifted to a wrong orbit due to malfunction of the guidance system and ceased to exist in September 1962. After a command from the mission center, instead of brining the spacecraft down its engines boosted it to a higher orbit adding about 12 months to its life in outer space. After that it entered dense layers and partially burned down. Its fragment fell on a street in an U.S. town and burned a whole in the pavement. Analysis showed that the fragment contained a gray cast iron bolt presum-ably used for fastening. Most likely the fragment served to define the center of gravity aboard the spacecraft. A part of it was returned to the USSR.

The 1970s saw two major incidents that caused creation of space junk and impact of spacecraft fragments. On March 29, 1973 the orbiting Salyut station disintegrated into major units and other heavy fragments. They failed to burn down in the atmosphere and reached the ground.

An incident with Kosmos-954 satellite was even more serious because the spacecraft carried a nuclear power plant. Its debris fell in Canada on January 24, 1978 causing environmental as well as political and diplomatic tensions.

The accident occurred due to poor design of the spacecraft. Instead of planned burning down and dispersing in the high atmosphere its power source fell to the earth.

The impact area stretched along about 800 kilometers in the artic zone of Canada. The USSR had to spend 10 million dollars to eliminate consequences of the incident that were so serious that designers had to make serious amendments to the spacecraft nuclear power unit design. Under the new design, in case of deorbit of spacecraft with nuclear sources aboard it was to eject radioactive fuel in portions (nor all at once as before) when it reaches the order between outer space and the upper layers of the atmosphere. Engineers believed they would thus reduce consequences of such incidents.

The downfall of the U.S. Skylab space station was another serious incident than marred the environment situation in the 1970s. The 71.5-tone station was launched to the orbit in 1973. In late 1978, orbiting 421 kilometers above the earth it started rapid descent. Experts predicted that about 25 tons of its fragments weighing from kilograms to hundreds of kilograms could hit the ground and cause serious damage if they fall in densely populated areas.

Estimates said the debris would fall within a band 4,800-6,500 kilometers long and 80-160 kilometers wide. NASA hoped that 75 percent of them would fall in the open sea and forecasted that the splashdown would occur somewhere from mid-1979 to mid-1980. Despite all efforts from NASA, the station became uncontrollable. In July 1979, it entered the atmosphere, disintegrated and partially burned. Unburned fragments made a relatively lucky fall – onto the ocean and deserted areas of western Australia, where they alarmed only remote farm dwellers.

Australia government and population demonstrated a rather peculiar response to this violation of the country's sovereignty. After a short period of uncertainty the incident was turned into a joke making happy U.S. officials and NASA leadership. The Australian prime minister informed the U.S. president about the incident in a special message and jokingly said that his country will return to the United Stated the debris of the station because of their complete uselessness. He also asked to plus its weight to lamb deliveries that his country was to dispatch under a contract. Thus the conflict was exhausted. But for several weeks media reports about the fall of the station's debris added tension to Australians and NASA staff.

While the first half of the 1980s saw no serious accidents with spacecraft impact (Kosmos-1402 alone fell in 1983), in the late 80s and the early 90s a string of incidents occurred. Perhaps this was associated with the peak of solar activity which is normally registered each 11 years. The solar activity reached its maximum and caused the "swelling" of the atmosphere. Upper layers of the atmosphere swept over low orbiting space junk. Four major accidents were registered involving three spacecraft and an upper stage of a launcher. One of the falling spacecraft, orbiting platform LDEF, was removed from outer space by a Space Shuttle. This prevented damage of its debris possible impact. Three other spacecraft – Kosmos-1900, Solar Max and a third stage of a Proton launcher disintegrated and their debris fell onto the earth.

Kosmos–1990 carried a nuclear power unit and a considerable amount (up to 50 kg) of nuclear fuel Uranium-235. In April 1988, radio contact with the satellite was lost. For several months it continued uncontrolled orbiting factually turning into space junk. In 1988, it entered dense layers, disintegrated and fell close to the U.S. - Canadian border. Aftermath of the fall and efforts to liquidate it resembled those taken in 1978 after the aforementioned similar crash.

Kosmos-1900 had a better design, and a more efficient system of self-liquida-tion. At least, it was designed to prevent concentrated contamination of the impact area. It had been planned that at the end of its service life, an automatic system will be activated and lift the satellite to a higher orbit (up to 800 kilometers) where radioactive decay of reactor fuel will take place some time. In case this system fails to activate, another sub-system was designed to disintegrate the satellite into fragments that would comfortably burn down in the atmosphere. The former system was to activate automatically at the altitude of 115-120 kilometers when the satellite starts heating up from atmospheric friction. Designers believed that satellite disintegration and burning of its fragments in dense layers during a rather long path will take a long period of time and thus prevent underneath areas from radioactive contamination.

Unfortunately both the main system and its sub-system failed to work as it had been planned. And a Soviet satellite with nuclear fuel aboard fell again on the territory of two sovereign states. The incident caused tension in political and diplomatic circles as well as in public. Media of practically all countries reported about it.

The authorized disintegration of 17-ton American AFP-731 (KH-12) satellite (as big as an ordinary street bus) caused a far lower public response. The satellite was designed for optical and electronic reconnaissance. It was lifted to the orbit by a Space Shuttle from Cape Canaveral in March 1990. A few days after it was exploded in the orbit upon a command from the missioncontrol center for unknown reasons, presumably because of irreparable technical malfunction. Apparently, controllers wanted to destroy secret equipment aboard the satellite. But the explosion produced a mass of debris. Soviet monitors detected four major fragments. It was presumed that the impact path will be above northern territories of the USSR, but the first fragment entered dense layers 1,500 kilometers north from the Midway Island in the Pacific Ocean. Other three fragments splashed down somewhere unregistered. Specialists say they posed no danger to population.

But the deorbit of a similar in size astronomic GRO satellite managed to make headlines. The 15-ton observatory was lifted to the orbit by the same Atlantis Space Shuttle 12 months after AFP-731 and worked in outer space successfully for nine years. But then it started registering malfunctions. It had been planned that Space Shuttles would service the orbiting GRO. But this promising concept remained unrealized due to poor planning. And it became easier to splash GRO down into the ocean than to repair it in the orbit. In July 2000, it was deorbited under control from the mission center. Experts say about six tons of its debris reached the earth with six fragments measuring about 800 kilograms each. The splashdown took place in a remote area of the Pacific Ocean along the band 4,000 kilometers long and 25 kilometers wide.

This was an example of NASA specialists ecological thinking. They preferred a guaranteed safe splashdown of a quite repairable satellite (with estimated up to ten years of would-be service life) to a possibility of gaining important research data. But if they decided to leave it on the orbit they risked an uncontrollable fall of the station as they could well lose control over it.

And, of course, one of the most remarkable events of space exploration last century was the splashdown to the ocean of the Russian veteran Mir orbiting station on March 23, 2001. The weight of the complex (about 130 tonnes) and its size (about 30x30x30 meters) caused concerns because of certain unpredictability of such unprecedented operations. But, nevertheless, media paid too much attention to the event, while, for example, the deorbiting of the Skylab back in 1979 was fraught with a far bigger menace. The Mir splashdown was performed in strict accordance with preliminary estimates and its debris fell in the southern Pacific. Only the third, and the last, braking impulse was more powerful than it had been planned. But this only increased steepness of the descent trajectory and reduced the debris impact area.

Mankind preoccupied with thoughts about the menace which literally hangs over the head, unfortunately ignores another hazardous aspect of spacecraft deorbiting. Even in case of controllable descent, is it wise to be happy of another successful splashdown. How many fragments have fallen into the ocean and how many will fall in the future? What is the effect of these "aliens" to the ocean ecology? These questions are still waiting for an answer. Meanwhile, the government of Chile expresses dissatisfaction over the use of the Pacific as a dust bin for space junk.

However, apparently, this problem arouses concern of not only environmentalists and potential victims (God knows who will be a new one) but also of producers and operators of spacecraft. This alone is an important step towards its solution. Indeed, in the mid-1960s very few people thought about it. Today, there are not only plans but real samples of environment-friendly spacecraft, such as Angara launcher with the reusable Baikal booster, that will help considerably reduce, if not bring to zero, the hazardous effect of space technologies on the environment.

## 4.4 Near-earth space and space exploration

Near-earth space (NES) comprises upper atmosphere, ionosphere and magnetosphere (< 1,000 km). Its specific features are linked with direct effect from ultraviolet solar radiation. Main absorbers of ultraviolet radiation in the upper atmosphere are molecular oxygen, nitrogen and ozone. The lower border of the upper atmosphere is lower than the thickest ozone layer, which immensely depends on processes in the NES.

The NES is the place where basic space activities occur such as purposeful use of spacecraft. Hence, its considerable influence on the environment.

Factors affecting the NES include launchers' engine emission, space junk and radioactive pollution.

### 4.4.1 Launchers' engine emission

Engine performance creates emission of huge masses of various chemicals mainly in the form of gas, which derives from the very nature of jet propulsion.

These chemicals form in the upper atmosphere a gas cloud with complex chemical composition along the entire trajectory of stages flight. These gases react with components of the upper atmosphere and ionosphere during the slow process of the cloud's diffusion.

Any gas in the upper atmosphere has three stages of spread in the environment. The first stage is collision-free, as gas expands to a distance responding to the length of free run of molecules in the upper atmosphere. The expansion range in this case varies from several meters at the altitude of 100 kilometers to several kilometers at the altitude of 300 kilometers. Respectively, this stage lasts from tiny fractions of second to seconds.

The second stage is hydrodynamic. It is characterized by the expansion of a cloud of emitted gas which is more dense than the environment and thus replaces normal gases in the upper atmosphere. Gas density in the expanding cloud declines until it equals pressure of the surrounding atmosphere. The duration of the hydrodynamic stage and the final size of the cloud depend on the mass of emitted gas and density of the surrounding atmosphere at a given altitude. Bigger emission and higher altitude result in a greater range of the cloud expansion.

The third, diffusion, stage of the cloud expansion consists in the diffusion of the emitted gas molecules into the surrounding gas until they are uniformly spaced in the environment. This stage lasts much longer than the first two and the emitted gas has enough time for chemical reactions.

During the diffusion stage the emitted gas spreads over hundreds, thousands and even scores of thousands kilometers depending on its mass and altitude of emission.

The main substance originating in the upper atmosphere as a result of booster engine performance is hydrogen. Initially, a hydrogen cloud, which is a result of hydrodynamic expansion, has the form of a mushroom because atmospheric density decreases at higher altitudes. For instance, for Proton boosters the roof of the cloud is about 200 kilometers above ground level. Then, the cloud diffuses upwardly and horizontally. In addition to the diffusion spread at high altitudes, most quick atoms start to escape from the upper ayers of the atmosphere. So, in case of a single emission all excessive hydrogen will escape the upper atmosphere and original gas composition will restore, i.e. the balance between natural hydrogen inflow from lower air and its outflow to the geo-corona (Fig. 19).

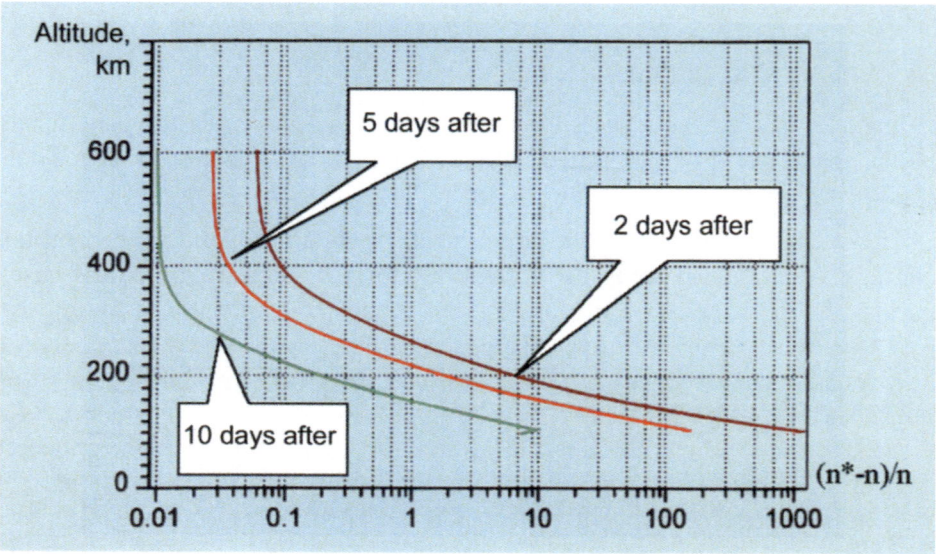

**Fig. 19.** Distribution of anthropogenic hydrogen excess along the Proton path

$(n^* - n)$    hydrogen excess;
$n -$    background hydrogen concentration

Diagrams on Fig. 19 show that an excess of emitted hydrogen against natural background at the altitudes between 100 kilometers and 150 kilometers remains considerable even 10 days after the emission, while at the altitudes over 300 kilometers the excess is insignificant. Serious excess occurs only close to lower parts of the launcher trajectory. At the altitude of 150 kilometers excessive hydrogen is registered scores-fold less already 100 kilometers from the trajectory, while about 600 kilometers from the trajectory there is practically no emitted hydrogen at all. At the altitude of 300 kilometers the hydrogen excess falls ten times about 500 kilometers from the trajectory.

Space Shuttle launches give a generally similar picture of gas diffusion, but as they produce a considerably bigger amount of emitted hydrogen (about

40 tonnes) its excess over the background is five times bigger than during a Proton launch. Such a considerable excess and its long duration could be explained not only by a bigger amount of emitted hydrogen but also by the fact that Space Shuttle engines stop working at a relatively low altitude of 110 kilometers compared with 200 kilometers of a Proton engine. This prevents hydrogen atoms from reaching the region of escape.

Proton engine performance at altitudes higher than 100 kilometers produces a considerable amount (about 40 tonnes) of carbon dioxide. Molecules of carbon dioxide are much heavier and bigger than those of hydrogen. Hence, a far slower diffusion spread of their cloud. Even ten days after the launch, the carbon dioxide excess in the vicinity of the trajectory exceeds background figures at least three-fold. Meanwhile, 300 kilometers from the trajectory it is no more than fractions of one percent. Heavy molecules are practically unable to escape, but $CO_2$ molecules can dissociate under influence of solar ultraviolet radiation. Normally, carbon dioxide dissociates in the upper atmosphere within 60 days.

The aforementioned data correspond to a bigger part of NES from 100 kilometers to 500 kilometers above ground level. As concentration of all components of the upper atmosphere is much higher closer to the earth booster engine performance in case of single launches is insignificant. High gas density simply prevents emitted gases from spreading. But emissions of chlorine and its compounds typical for Space Shuttle launches make an exception. The mass of their emission is comparable to the mass of natural chlorine compounds even hundreds kilometers away.

Table 9 shows composition and amount of products of rocket fuel combustion for some types of boosters.

**Table 9.** Composition of rocket fuel combustion (tonnes)

|  | $H_2$ | $H_2O$ | CO | $CO_2$ | $N_2$ | HCl | Cl | NO | $AL_2O_3$ |
|---|---|---|---|---|---|---|---|---|---|
| Kosmos-3M | 0.69 | 34.77 | 2.15 | 37.17 | 30.52 | – | – | 0.34 | – |
| Tsiklon-2 | 0.35 | 51.62 | 4.55 | 55.23 | 58.0 | – | – | 0.45 | – |
| Soyuz-2 | 0.95 | 68.56 | 29.0 | 149.0 | – | – | – | – | – |
| Proton-K | 1.4 | 178.22 | 14.2 | 222.4 | 212.9 | – | – | 0.74 | – |
| Space Shuttle | 2.12 | 826.2 | 181.4 | 55.2 | – | 193.1 | 4.5 | 1.7 | 279.6 |

Thus, water fumes are one of the main products of launcher engines' operation emitted to the near-earth space. At altitudes higher than 100 kilometers water molecules dissociate producing hydrogen in amounts comparable with its natural background even if only one heavy rocket goes off. Hydrogen spreads to tens of thousands of kilometers forming a mushroom-shaped cloud in the NES containing much more hydrogen atoms than natural background.

Violation of hydrogen's natural balance causes changes that affect all parts of the NES. Ionosphere develops so-called "ozone holes," i.e. vast areas where concentration of electrons falls several times.

Through connections between various parts of the NES these changes affect the entire NES and deform it.

As it has been mentioned earlier, carbon dioxide is another principal product of rocket fuel combustion emitted in great quantities. It spreads much slower than hydrogen because its molecules are bigger and heavier. That is why excessive concentration of carbon dioxide is registered mainly near the trajectory of a booster.

Carbon dioxide plays an extremely important role in the thermal balance of the thermosphere established as a result of heating by ultraviolet radiation and cooling by infrared radiation, a considerable part of which depends on carbon dioxide molecules.

Excessive carbon dioxide resulting from human activities can cause heat loss of the thermosphere of up to tens percent. Moreover, changes occur in the entire vertical thermal profile of the atmosphere including exosphere. Changes in temperature are accompanied by changes in density and chemical composition.

Solid-fuel booster engines emit considerable amounts of chlorine and its compounds, up to 200 tonnes in case of a single Space Shuttle launch. They are extremely hazardous to the environment as they kill ozone, one of the main components of the strato-mesosphere. Ozone is in fact the shield that protects life from deadly ultra-violet radiation. According to some estimates, monthly launches of space shuttles during four years decrease ozone total contents by 0.3 percent in medium latitudes and by 0.4 percent to 0.6 percent in high latitudes.

## 4.4.2 Space junk

Another human activities' effect on the NES consists in the creation of space junk which is sometimes called orbital junk. Space junk originates from a variety of sources. First, it consists of parts and fragments of boosters that remain on the orbit, such as final stages of launchers and inter-orbit tugs. Second, spacecraft that effectively terminated their active performance and their fragments after deliberate explosions. For instance, a command to explode a spacecraft may be given when a secret satellite exhausts its service life. Third, fragments of space vehicles after accidental explosions that occur primarily with fuel remaining in engines or boosters. Fourth, fragments of spacecraft after collisions in outer space.

The rapid development of space activities has resulted in a growth of the space junk mass in spite of the fact that spacecraft and their fragments tend to fall from their orbits, as the upper atmosphere gradually slows down their speed, and burn in thicker layers of the lower atmosphere.

So there is a certain menace of a space collisions of spacecraft (especially manned) with space junk.

Fig. 20 shows percentage of various types of space junk.

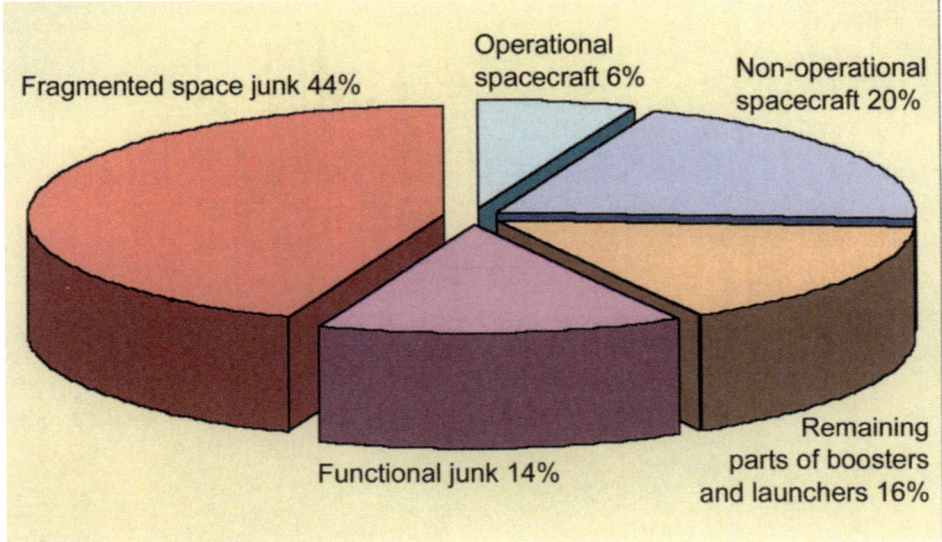

**Fig. 20.** Space junk percentage in the orbit zone

Relative speed of collision between space junk and operative space vehicles might exceed 10 km/sec considerably. This makes potentially dangerous collisions even with particles that measure more than 1 cm. It is estimated that orbiting today are up to 150,000 such particles, which cannot be detected visually or by radar. A number of particles measuring less than 1 cm has surpassed one million.

Today, space junk includes about 8,000 fragments measuring 10 cm with the aggregate mass exceeding 3,000 tonnes. All these fragments have been registered in special catalogues and are permanently monitored. NORAD in the most complete catalogue of such fragments.

Apart from the aforementioned particles, space junk includes a great number of smaller particles, the existence of which is confirmed by numerous caverns on solar batteries of spacecraft. Moreover, the rain of artificial objects measuring more than 1 cm exceeds thousands times the flow of similar natural meteorites.

Any further increase of space junk might cause an avalanche growth in the number of orbiting particles that will multiply their inter-collisions.

There is a bright example of danger arousing from a spacecraft collision even with the smallest particles of space junk. A cavern 0.63 mm deep an 2.4 mm in diameter appeared on a Space Shuttle window glass after its collision in the orbit with a small particle of paint that was no more than a salt grain (0.2 mm) at the relative speed of 6 km/sec.

As it is still impossible to make a catalogue of space junk objects measuring less than 10 cm, the only way to describe them quantitatively and determine their location in space is to make a statistical model of space junk.

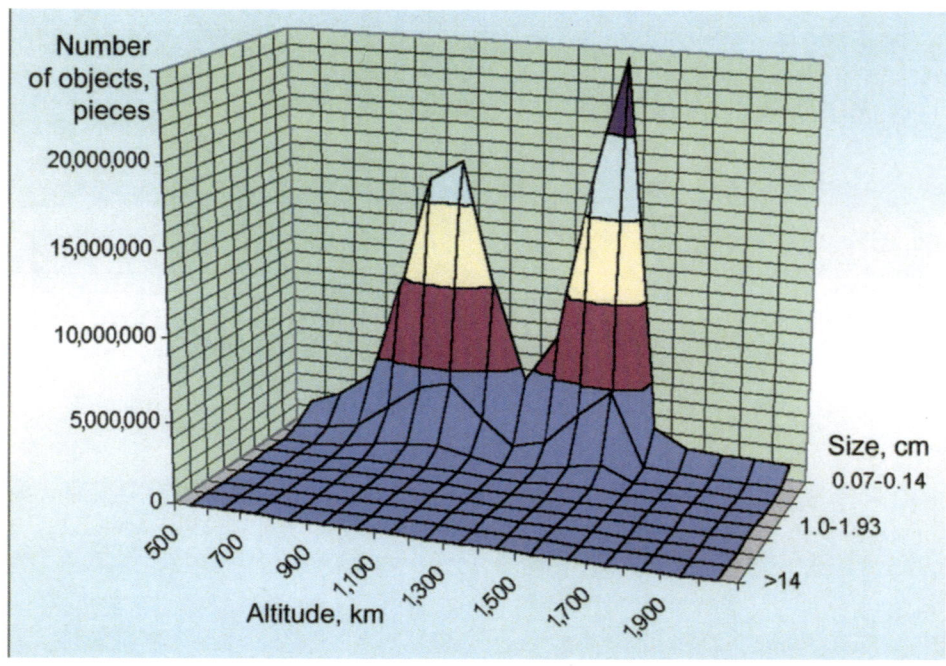

**Fig. 21.** Distribution of space junk particles to various altitudes

Several such models have been built. One of them was developed at the Center of Program Research of the Russian Academy of Sciences. It deals with solid particles measuring more than 0.07 mm that have not been included in any catalogue. The whole range of all possible sizes was divided into eight sub-ranges. Intensity of fragment formation was presumed to be in reverse proportion with their size. It was also presumed that 14,600,000 particles measuring more than 0.7 cm add annually to space junk. A diagram on Fig. 21 shows distribution of particles at various altitudes depending on their size. The diagram shows two areas of maximal space junk density – at 900 kilometers and at about 1,500 kilometers above ground level. The bulk of space junk concentrates below the 1,000-kilometer mark.

Fig. 22 shows distribution of small space junk particles in various altitudes calculated from the model. Obviously, particle density grows considerably around 90-kilometer and 1,450-kilometer marks.

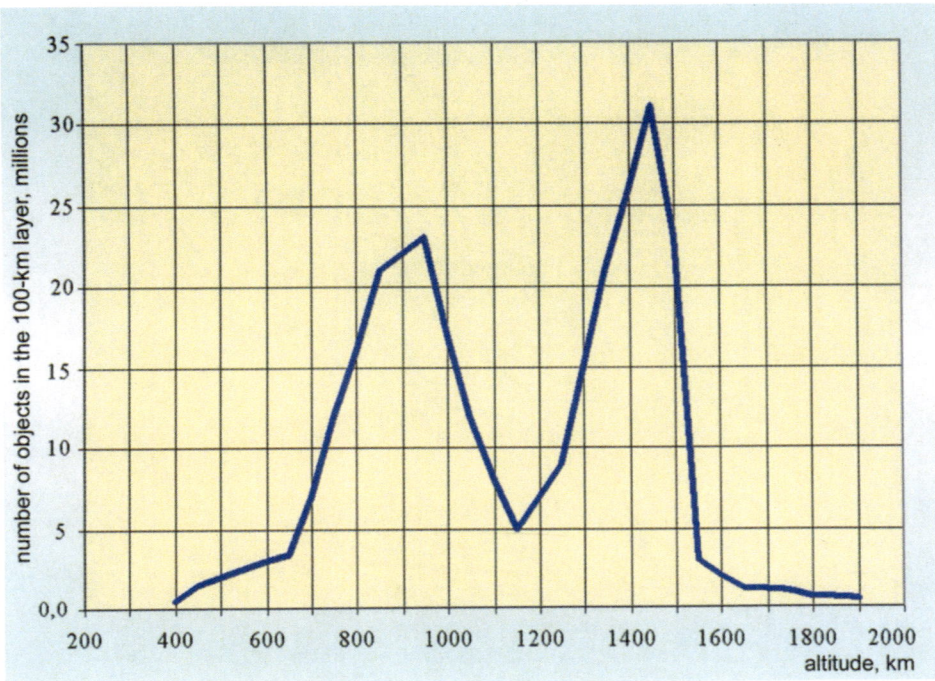

**Fig. 22.** Distribution of small space junk particles to various altitudes. Estimates of 1995

Space junk density varies periodically depending on upper atmosphere density, which in turn depends on the 11-year cycle of solar activity. When solar activity reaches its maximum, upper atmosphere effectively slows down the motion of space junk particles and thus cleans itself. Vice versa, in the years of low solar activity atmosphere density drops weakening the process of self-cleaning.

But the fact that the aggregate mass of space junk is steadily growing proves that this self-cleaning ability is not enough to overcome pollution caused by human activities (Fig. 23).

Experts differ in their forecasts of the situation with space junk till the year 2100. There is at least four scenarios of the development of the space junk problem:

1) the intensity of fragments and particles origination preserves at the quantitative level of 1991-1995;

2) the number of explosions drops five-fold while the intensity of launches and fragment formation remains on the previous level according to specific features of space technologies;

3) explosions never occur but all other factors remain; and

4) the rate of artificial fragment formation drops two-fold, explosions do not occur, which the intensity of launches remains on the same level.

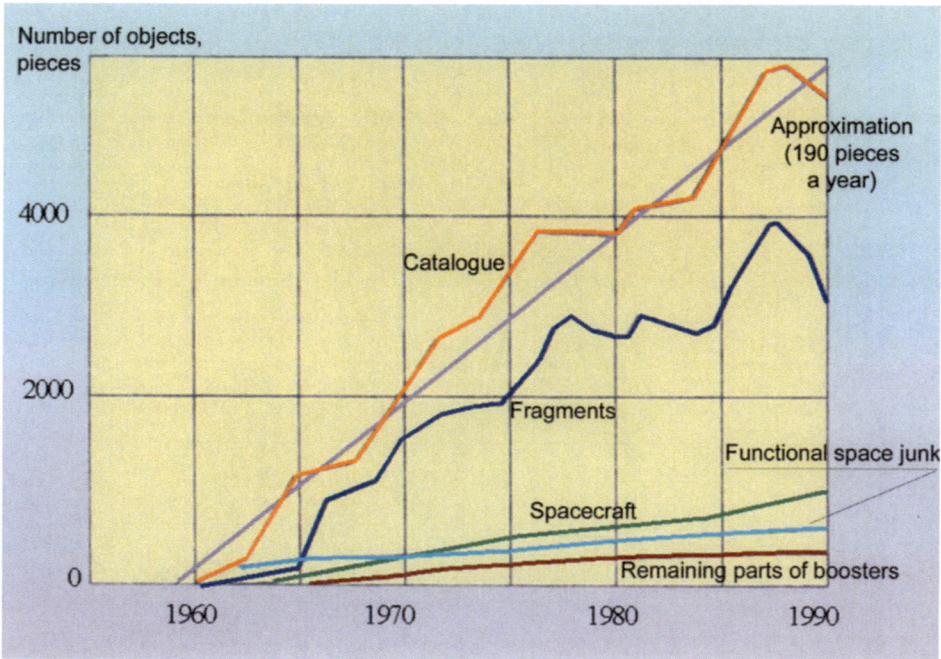

**Fig. 23.** Dynamics of low orbits pollution

Estimates show that if the situation develops in conformity with first three scenarios, the aggregate amount of space junk will grow. If today's trends preserve, fragments measuring more than 1 centimeter will double during the next hundred years. The only way to overcome the trend and improve the situation considerably is to exclude explosions completely as a reason of fragment creation. If the number of explosions drops five-fold the rate of fragment formation will decrease by 70 percent. In order to keep the space junk mass on today's level it would be necessary to decrease the rate of fragment formation by 70 percent.

But if we take into account a possible rise in space activities that is not envisaged by any scenario, it would be difficult to count on any drop in the rate of space junk formation. So, the first scenario looks more realistic and by the end of the century the NES will have about 650,000 artificial objects measuring more than 1 centimeter. The mass of space junk will grow up to 7,000 tonnes to 8,000 tonnes or practically up to 30 percent of the total mass of the upper atmosphere where space junk may exist (>400 kilometers). Moreover, the zone of the upper atmosphere which normally hosts spacecraft orbits will consist primarily of solid particles. The kinetic energy of just 3,000 tons of speedy space junk exceeds more then ten-fold thermal energy of all molecules in the upper atmosphere. Hence, an evident menace from space junk not only to practical cosmonautics.

Thus, space junk accumulated in the NES in a wide range of altitudes (from 400 to 2,000 kilometers) is comparable in mass with the entire material of the NES higher than 400 kilometers above ground level. Space junk consists of numerous fragments. The NES has more than 8,000 registered objects measuring more then 10 centimeters, about 300,000 fragments measuring more then 1 centimeter and hundreds of millions smaller particles.

### 4.4.3 Radioactive contamination of NES

The main source of NES radioactive contamination are nuclear power units of some spacecraft, operational with the activated nuclear reactor or idle with the shut down reactor.

The active zone of a normally operating reactor is a powerful source of gamma and neutron radiation. According to expert estimates, its significant affect could be registered up to 1 kilometer from the reactor. In the NES zone close to a working reactor, it changes the natural gamma and neutron background depending on the orbit altitude. If a spacecraft with working reactors orbits 200-300 kilometers above ground, the density of gamma quanta comparable with natural background could be registered up to 700-1,000 kilometers above ground in low latitudes and up to 500-700 kilometers – in high latitudes. Neutron flows are comparable with natural background up to 100 kilometers on low orbits and up to 300 kilometers on geo-stationary orbits.

When service life of such a spacecraft is over it is normally boosted to an orbit higher than 700 kilometers where it remains for many years till its radioactive material decays to the safe level. As for today, about 50 objects with radioactive materials are orbiting in the NES at altitudes between 800 kilometers and 1,000 kilometers.

These altitudes were chosen as safe in the late 1960s when space activities were rather low. But today space junk concentrated exactly between 800 and 1,000 kilometers high and there is a serious danger that their collision with fragments of space junk might ruin nuclear reactors long before their radioactive materials decay down to a safe level. According to some estimates, a spacecraft with a nuclear reactor due to remain on a "safe" orbit for 200 years risks about 20 collisions with space junk fragments that might cause an accident. On the average, it faces one collision in six years with a fragment measuring 0.5 centimeters and one in 26 years with a particle measuring 1 centimeter. Such a collision might result in a breakdown of the nuclear power unit and expansion of radioactive materials in the NES with their eventual descent to the lower atmosphere. The main threat to environment is associated with a possibility of an impact of nuclear power units to earth and a descent of radioactive matters to the lower atmosphere and earth.

So the growth of the space junk mass causes greater probability of nuclear power units breakdown and eventually of greater nuclear contamination of the

NES. Although nuclear contamination does not affect the environmental characteristics of the NES directly, it threatens the performance of navigational and meteorological space systems as well as natural reserves monitoring systems that use almost similar orbits.

## 4.5. Environment-friendly space technology has no alternatives

Of course, the afore-presented situation hardly adds optimism and looks somewhat threatening. But is it so bad indeed? Previous sections of this book dealing with space technologies' effect on the environment have mentioned measures designed to reduce this effect. In this chapter the authors will try to present the sum and substance of such measures and outline guidelines for minimization of hazardous effect of space technologies on the environment.

It is hardly possible to turn launching pads into blossoming gardens. But quite realistic and, moreover, necessary is to reduce the menace and minimize the hazardous effect of space technologies on nature.

It is possible to increase environmental safety of launching pads through lessening their size by brining launching equipment to commonality and make them capable of launching various types of boosters.

First steps in this direction have been already made. For example, partially modernized launching equipment of decommissioned Saturn-5 booster is being successfully used for Space Shuttle launches. Another example is the modernization of the LC-41 Titan launching complex at Cape Canaveral for launching promising Atlas-4 and Atlas-5 rockets (Fig. 24).

A similar approach was demonstrated in this country where the foundation of an N-1 launching ramp was used for building a launching complex for the Energia booster. Although economically this made little sense, hazardous effect on the environment was considerably reduced.

**Fig. 24.** Before you build a new house, you must ground the old one.
Refurbishing of the LC-41 launching complex for start-up new modifications of Atlas launchers

Today, at the Plisetsk cosmodrome a launching complex that had been designed for Zenit boosters is being refurbished to launch promising environment-friendly Angara boosters designed by the Khrunichev State Space Research and Production Center (Fig. 25).

**Fig. 25.** A. A. Medevedev and A. A. Kalinin before the dispatch of an Angara booster to Le Bourget

For instance, it will be able to launch various types of Angara – heavy, medium and light.

Leakages of propellant toxic components on the territory of space launch bases could be terminated if only environment-friendly boosters (of Angara class) are used. But not all methods of fighting this evil have been tried yet. Promising results have been achieved from experiments with recultivation of heptyl-contaminated areas at Plisetsk.

But the main achievement is the environment-friendly thinking developed among specialists in space technologies. Already in the mid-1990s Plisetsk specialists drafted plans to build a national park and ecological reserve Cosmodrome Plisetsk.

The creation of ecologically clean space technologies is another important way to preserve environment, including the design of ecologically clean boosters and spacecraft.

The initial stage of design works envisages reduction and then exclusion of propellant spillage during the downfall of separating sections. For this purpose, a variety of methods seem possible:

• propellant components depletion (even full depletion) with the use of "flexible" control system and a system of propellant flow;

• chemical methods using special reagents added into propellant tanks;

• afterburning of remaining propellant for instance through adding unsymmetrical dimethylhydrazine (UDMH) to nitrogen tetroxide (NT) tanks and NT to (UDMH) tanks;

• ejection of unburned components from the tanks promptly after the separation of a burnout stage.

Today Russian specialists broadly use the method of individual tanking that helps reduce by half unburned components in spent stages. In combination with specially developed programs determining booster's flight, it helps reduce at least by 30 percent the area where spent stages might fall.

Another measure that could be implemented in the near future consists in the performance of the first stage engine until propellant full depletion and subsequent neutralization of unburned fuel before the stage touches the ground.

Proton and Proton-M boosters eject unburned components from tanks immediately after a spent stage is separated. This occurs at altitudes between 43 and 70 kilometers. Propellant components ejected so high partially interact, boil up, split into drops measuring from 1 mcm, freeze, sublime in the process of descent, thaw entering dense layers from gasdynamic heat and solar radiation, fully evaporate before they reach troposphere. Then their fumes are indefinitely diluted by air because of diffusion and turbulent (primarily horizontal) atmospheric flows. So, they never reach earth either in the impact area or elsewhere. The constantly spreading cloud in 24 hours will horizontally measure about 100 kilometers and concentration of fumes will drop down to less than 0.01 mkg/m$^3$ which is $10^3$ times lower than the maximum allowable concentration (MAC) which is 0.1/m$^3$ for an eight hour stay in the zone.

Second stages of three-stage rockets may reach speed of 5,000 m/sec at the moment of separation, while their third stages (or second stages of two-stage rockets) reach circular or semi-circular velocity. Entering dense layers of atmosphere with such speed, constructions of a stage with a liquid-propellant engine normally disintegrate. For instance, a spent second stage of Proton separating at an altitude of 130 kilometers enters dense layers with the speed of about 4,500 m/sec (Fig. 26) and partially (primarily its tank) breaks down at an altitude of 45-30 kilometers depending on its orientation. Tank fragments slow down because of the "sail effect" and have enough time to heat up to 300-350°C which eliminates completely even traces of unsymmetrical dimethylhydrazine (UDMH). They fall close to the nearer (along the flight line) border of the impact area. Engines have higher density and decelerate slower so, that their fragments heated up to 500-600°C fall near the opposite border of the impact zone and may cause wildfires. Propellant components emitted during the tanks ruining partially vanish through interaction, vaporize and are dispersed by turbulent atmospheric flows and diffusion processes down to indefinable concentrations and never reach troposphere and soil.

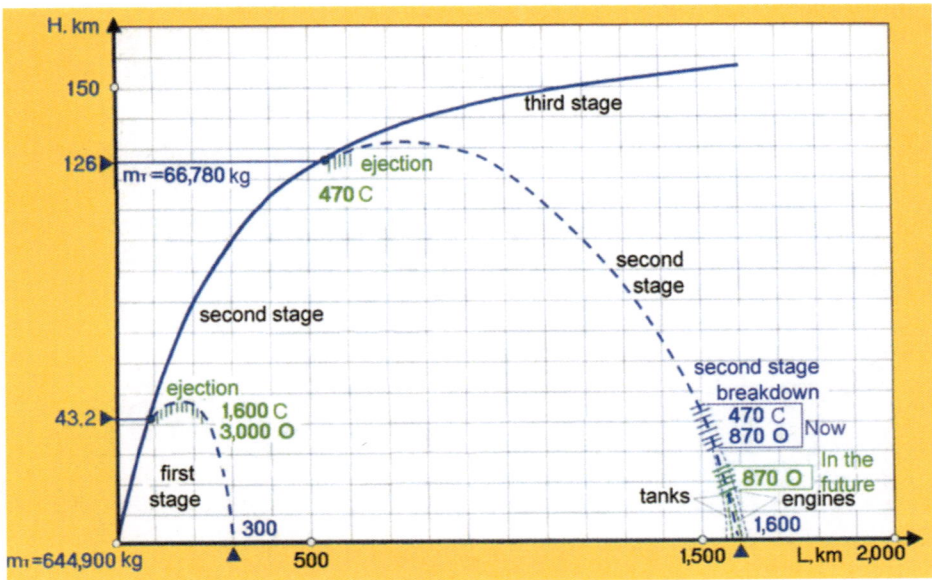

**Fig. 26.** Proton launcher flight diagram

Proton-M booster has a flexibly control system with a discrete on-board computer that ensures full combustion of guaranteed reserves of both components. This reduces their unburned propellant by 50 percent. Together with a bigger payload this makes this booster more environment-friendly.

For many years already specialists have been trying to reduce the amount of residual propellant in spent stages. Kosmos-3M was the first booster where they used the method of incomplete fuelling of the first stage. Without any changes in the rocket design this helped reduce unburned fuel in the spent stages by almost 20 percent and remaining oxidizer – by almost 10 percent. Similar technique applied to Tsiklon-3 booster helped reduce residual propellant by almost 35 percent, and with meager changes in design (individual tanking) – by 50 percent. Perhaps, someone might say these figures are inconsistent. But the tendency shaping out is extremely important.

Another promising method of reducing harmful effect on the environment is the use of separating stages with controllable descent.

Dispersion could be significantly reduced through aerodynamic stabilization as well as through stage orientation with a zero angle of attack using gas control devices before they enter dense layers. A practically zero dispersion could be reached through controllable descent (for example on a radio beacon) with the use of aerodynamic bearing capacity of the stage body and aerodynamic or gas control devices (flat rudder, flow fence, etc). This method was designed for the first stage of Proton booster.

Plus to the aforesaid spread, fragments of second and third stages disperse after their explosive breakdown upon entering dense layers. In order to reduce this kind of dispersion, Proton-M drops pressure in tanks (simultaneously ejecting unburned components) at the altitude of about 130 kilometers (Fig. 26). This eliminates explosive dispersion and makes the stage disintegrate at a lower altitude. Hence, its fragments disperse to a smaller territory. Already today specialists operating Proton-M managed to minimize the breakdown of the first stage and make tanks touch the ground practically side by side (Fig. 27).

**Fig. 27.** Proton-M first stages impact area. Recultivation efficiency control

Stage breakdown could be prevented by strengthening its body and reinforcing its thermal insulation. Together with stabilized and controlled descent, this will add to a more comprehensive solution of the problem. But in this case, payload will be inevitably reduced. Economic consequences of this decision should be correlated with economy from the reduction of the impact area from the viewpoint of efficiency-cost criterion.

Guided descent may involve any aerodynamic constructions enabled during the passive section of the separated stage trajectory that ensure its gliding to a distance of no less than 30 kilometers or hanging up on a planned altitude.

Reusable stages are another important step in this direction, such as reusable cruising module Baikal of the Angara booster (Fig. 28).

**Fig 28.** Reusable launcher Baikal in the assembly shop of the Khrunichev Center

Baikal has the all-up weight of 150 tons and is about 27 meters long. It carries non-toxic propellant components – oxygen and kerosene. Its pivoting wing is fixed along the body axis during the take-off and thus does not cause any aerodynamic resistance. After the separation, the module's wing turns to normal position and its air-breathing engine is switched on to ensure the landing of the stage-turned-aircraft on an ordinary airfield.

Fig. 29 shows the diagram of Baikal's performance. Such reusable stages will allow not only to avoid spillage of propellant components but will also make unnecessary impact areas for first stages.

A cut in the number and square of impact areas for separating fragments is an important aspect of works aimed at the reduction of hazardous effect of space technologies on the environment. Albeit a final solution of the problem is still far away, some serious steps in this direction have been already made. For instance, the impact area No 25 for Proton first stages is a 30x18-kilometer ellipse of 1,400 square kilometers. However, Proton-M's control device provides the cross-range possibility and hence, allows to reduce the impact area by 7 percent. Moreover, ejection of unburned propellant in second stages will delay their breakdown and will allow to reduce the impact area No 326 (one of the largest) by almost 8 percent. These measures alone will reduce Proton debris impact areas by almost 3,000 square kilometers. Proton second stage impact areas could be reduced several-fold in case of some changes in booster design (Fig. 30).

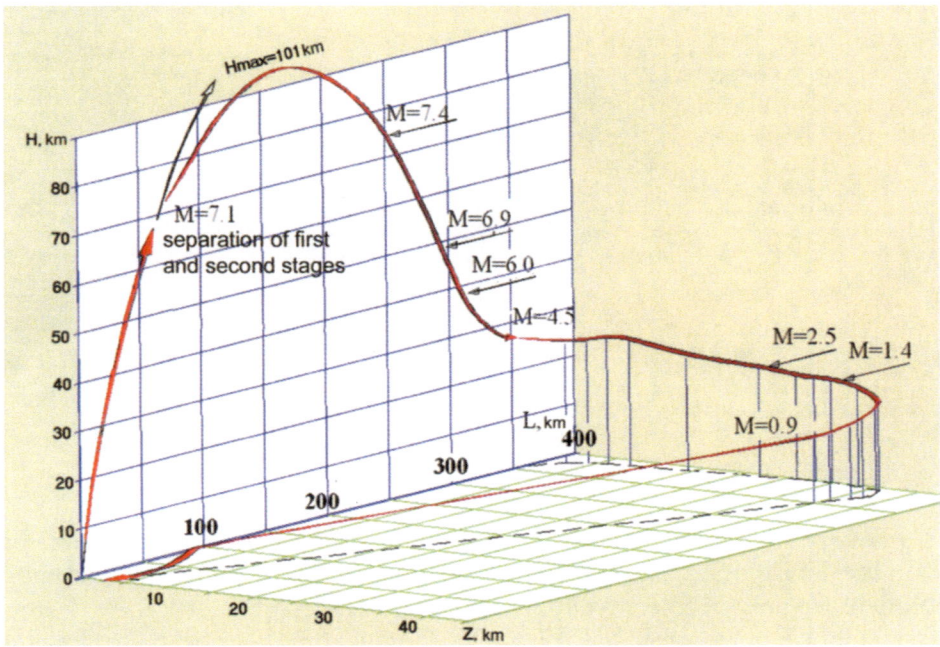

**Fig. 29.** Baikal flight diagram

The current efforts shall result in a considerable overall reduction of impact areas. In Russia, out of 160,000 square kilometers of impact ground areas almost 1,500 square kilometers were not used already by mid-90s. By the beginning of the 2,000 s, separating parts stopped falling on another 10,000 square kilometers. By 2010, the use of Baikal-class reusable stages will make unnecessary impact areas for first stages and the total impact area will thus reduce by almost 50 percent (Fig. 31).

In the more distant future (2020-2050) space technologies will shift to the use of single-stage-to-orbit space vehicles (of Venture Star class) and impact areas will be eliminated and replaced by runways returning vast lands to national economies.

Later on (2050-2100) engines using other physical principles (such as gravitation) might appear. They will probably allow to eliminate environment pollution with spilled propellant and combustion products and will substantially reduce areas used for take-off and landing of spacecraft.

These measures should be taken against the background of steady abandoning of toxic propellants. A radical solution of the problem is the use of nontoxic propellant components such as kerosene-oxygen or hydrogen-oxygen which is typical, for example, in the family of light and heavy launchers of Angara-class designed by the Khrunichev Center.

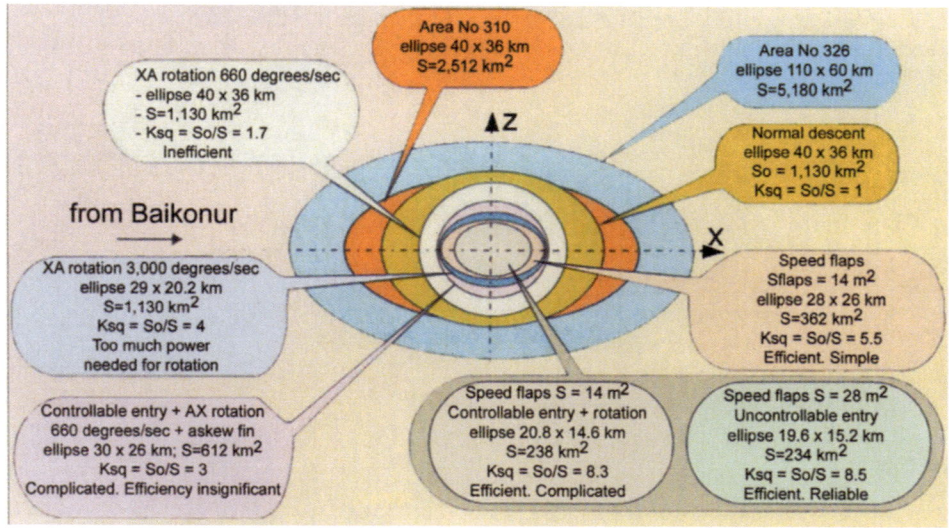

**Fig. 30.** Changes in ellipses of touching points of Proton second stages

However, the problem of space junk is becoming more and more serious with the development of space technologies. Creation of fund-saving and environment-friendly boosters will only aggravate this problem.

A solution of the problem of space junk prevention boils down to the problem of the se of residual power of spacecraft remaining on the orbit. Energy reserves could be exhausted by the end of their service life. In order to minimize reserves of kinetic energy, it is necessary either to destroy the mass or to bring to zero the relative velocity. Since in the majority of cases it is impossible to make spacecraft orbit with reducing relative velocity, it is necessary to ensure the absence of their mass in the areas of space where it might have undesirable effect on operation of other spacecraft directly or by its fragments. In some cases the mass leaves the orbit due to natural reasons, but generally its removal should be envisaged by a flight program.

In order to slow down the intensity of space junk creation, resources of chemical and mechanical energy aboard spacecraft should be regulated. Reliable constructions should prevent spacecraft from explosions. In order to prevent explosions after the end of a spacecraft service life, it is necessary to exhaust or withdraw residual pressure, fuel or mechanic energy. If spacecraft remain orbiting during a rather long period of time they will be destroyed into fragments by collisions on the orbit. So, it is necessary to remove spacecraft from their orbits instantly after their mission is over. It is also necessary to ensure adequate protection of spacecraft from space junk and other ruinous factors in order to have the possibility to remove them from the orbit before they collide and break down.

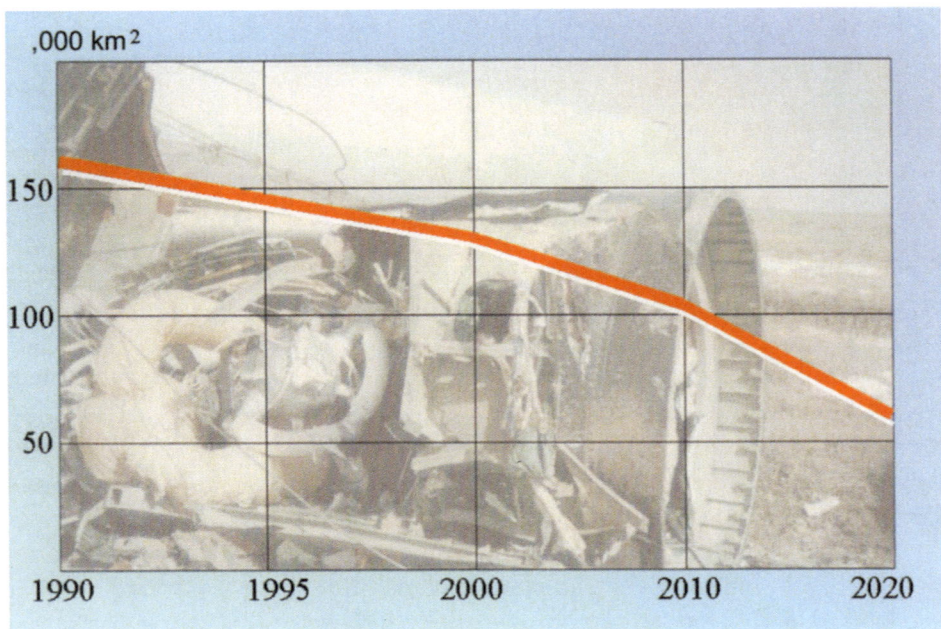

**Fig 31.** Diminishing of separating fragments impact areas

Today nations and organizations leading in space technologies take efforts aimed at the minimization of space junk creation. For instance, NASA already in 1993 issues an instruction limiting space junk creation. Under this instruction, any space program should be certified on its compliance with certain requirements as a potential source of space junk. Later on, NASA enriched the instruction by a special standard setting up requirements of a procedure and technique of space programs certification. Measures aimed at minimizing space junk life in the orbit include junk separation on low orbits, ensuring of a high surface to mass ratio, timing of junk separation to lunar or solar disturbance, design changes aimed at limiting the number of separating fragments, etc.

Both in Russia and the United States organizations engaged in space technologies primarily express concern over the growth of space junk. For example the Khrunichev Center worked out a normative document "Guidelines for the organization of works aimed at space junk reduction." In particular, the document envisages the following measures at the end of spacecraft service life:

• to burn out propellant components and remove as completely as possible other working fluids or gases such as used for tank pressurization through aftercombustion and venting in order to prevent accidental breakup because of high pressure of chemical reactions;

• to design and produce batteries properly so that their electric circuits and construction be breakup-proof. At the and of spacecraft mission batteries should be discharged and charging line should be switched off;

• to vent high-pressure bottles down to the level preventing their breakup;

• auto-destructing systems should be designed so that they could not be accidentally disintegrated because of a false command, excessive heat or radio-frequency effect;

• flywheels and gyroscopes should be switched off.

The documents envisages other measures aimed at reduction of space junk in the zone of low orbits and geo-stationary orbits. Particularly, is suggests avoiding deliberate breakup of space objects, using non-separating engines for launching spacecraft to geo-stationary orbits, etc.

The Khrunichev Center does not limit its desire to fight space junk by mere declarations, but realizes it through practical samples of space technologies. In order to prevent accidental breakup, Briz-KM, Briz-KS and Briz-M launchers can release pressure from their tank pressurizing systems. Angara boosters will be capable of similar operation.

The growing threat of an operational aircraft collision with space junk, on the one hand, and involvement into space operations of a growing number of countries and organization, on the other hand, make this problem international. Pretty soon the industrialized nations sick and tired of the constant menace to lose costly aircraft will forward a common initiative of international measures to combat the growth of space junk. One of these measures could become obligatory certification of space technologies, including environmental certification with all relevant consequences.

## 4.6 International law aspects of space ecology

The basic document dealing with the problem of pre-serving stable ecology of outer space is the 1967 Outer Space Treaty. Article 1 of the Treaty reads that, "the exploration and use of outer space ... shall be carried out for the benefit and in the interests of all countries." Article 4 stresses that nations "shall bear international re-sponsibility for national activities in outer space, including the Moon and other celestial bodies, whether such activities are carried on by governmental agencies or by non-governmental entities." Article 7 stipulates that "each State Party to the Treaty that launches or procures the launching of an object into outer space ... and each State Party from whose territory or facility an object is launched, is internationally liable for damage to another State Party to the Treaty or to its natural or juridical per-sons by such object or its component parts on the Earth, in air space or in outer space..." According to Article 9, nations should conduct appropriate inter-national consul-tations if its activities or planned experiment "would cause potentially harmful interference with activities of other States Parties in the peaceful exploration and use of outer space." The Treaty however gives no clear criteria what interference shall be regarded as "harmful" and offers no procedure or technique of such consultations. The same Article 9 calls on the nations "to inform the Secretary- General of the United Nations as well as the public and the international scientific community, to the greatest extent feasible and practicable, of the nature, conduct, locations and results" of space activities, including those associated with Earth and outer space pollution.

A special role in the legal regulation of space ecology matters is played by the 1972 Convention on Inter-national Liability for Damage Caused by Space Objects. The Convention reads that, "a launching State shall be absolutely liable to pay compensation for damage caused by its space object on the sur-face of the earth or aircraft in flight" (Article 2). In the event of damage being caused to a space object of one launching state or to persons or property on board such a space object by a space object of another launching state, the latter shall be liable only if the damage is due to its fault or the fault of persons for whom it is responsible. Defining a "space objectl" as including "component parts of a space object as well as its launch vehicle and parts" irrespective operational or not, the Convention presumes liability for outer space pollution.

Determining liability for damage caused by a space object to outer space environment makes a serious problem. Liability appears only with damage, which is defined by Article 1 as "loss of life, personal injury or other im-pair-ment of health; or loss of or damage to property of States or of persons, natu-ral or juridical, or property of international intergovernmental organizations.". appar-ently the list deals strictly with physical damage and does not include damage to environment.

The 1968 Agreement on the Rescue of Astronauts, the Return of Astronauts and the Return of Objects Launched into Outer Space also carries some provisions indirectly relevant to space environment protection. Under this agreement contracting parties which discover that a space object or its component parts has returned to Earth in territory under its jurisdiction shall notify the launching authority and the Secretary-General of the United Nations. Nations "which has reason to believe that a space object or its component parts discovered in territory under its jurisdiction, or recovered by it elsewhere, is of a hazardous or deleterious nature may so notify the launching authority, which shall immediately take effective steps, under the direction and control of the said Contracting Party, to eliminate possible danger of harm."

The 1975 Convention on Registration of Objects Launched into Outer Space also deals with legal regulation of space ecology matters. It obliges nations to keep jurisdiction and control of their space objects during all stages of their flight.

The 1977 Convention on the Prohibition of Military or Any Other Hostile use of Environmental Modification Techniques directly, albeit declaratively, bans space pollution and thus is essential for the matter in question. Under this convention each state party "undertakes not to engage in military or any other hostile use of environmental modification techniques." "The term "environmental modification techniques" refers to any technique for changing - through the deliberate manipulation of natural processes – the dynamics, composition or structure of the Earth, including its biota, lithosphere, hydrosphere and atmosphere, or of outer space." However due to the universal character of the Convention its contribution to the solution of space ecology problems is rather declarative. The Convention lacks concrete provisions about space pollution by artificial objects and defines no machinery of practical measures.

There is an evident and publicly acceptable understanding of the problem of space ecology – the necessity to reduce the scale of space activities (its regulation concerning environmental perfection of space technologies i.e. the reduction in the number of inactive space object remaining on their orbits). But so far, nobody has assessed the efficiency of these measures and their compliance to the costs-efficiency criteria. Thus, prospects of the perfection of an environment-friendly international legal regime in outer space seem rather vague.

Starting from late 1980s, the issue of outer space pollu-tion has been discussed by scientific and legal sub-committees of the U.N. Committee on the Peaceful Uses of Outer Space (COPUOS). A report on the situation with space junk was prepared by COSPAR and IAF and submitted to the committee upon its initiative in 1988. From 1989 to 1991, a number of countries including Germany and the Netherlands, submitted to the Committee working papers concerning spacecraft and space stations collisions with other space objects and Space pollution in general.

In December 1990, the U.N. General Assembly in its Resolution 45/72 noted for the first time that the problem of space junk causes concern of all nations and called on the international community to pay more attention to the appraisal of the threat of collisions with space junk and other aspects of this problem and to enhance national research in this sphere.

As it has been mentioned, environmental problems of today's space exploration include the problem of on-board nuclear reactor safety which arouses when NPP carrying spacecraft enter atmosphere or touch the earth. As to the menace of nuclear contamination of celestial bodies the aforementioned 1967 Outer Space Treaty seems sufficient for today's level of space technologies development.

Prevention of radioactive pollution of outer space caused by objects of human activities is regulated by three international accords:

the 1963 Treaty Banning Tests of Nuclear Weapons in the Atmosphere, Outer Space and Under Water that prohibits explosions of nuclear weapons and other nuclear explosions in the atmosphere and outer space; the 1986 Convention on Early Notification of Nuclear Accidents that obliges nations to announce a possible threat of an accident with a nuclear reactor or caused by the use of radioisotopes in outer space;the 1986 Convention on Assistance in the Case of Nuclear Accident or Radiological Emergency that established frameworks within which one nation may render assistance to the other in case of a nuclear accident or radiological emergency caused by radioactive space debris.

On December 14, 1992, the U.N. general assembly ap-proved Principles Relevant to the Use of Nuclear Power Sources In Outer Space. A special resolution stressed applicability of international laws to activities associated with the use of nuclear power sources.

This document obliges nations launching spacecraft with nuclear power sources aboard to take efforts to protect individuals, population and biosphere from radiological threats. The design and use of space objects with nuclear power source aboard shall ensure with confidence the degree of safety where under all predictable normal and emergency situations it would be lower the allowable levels. The guiding principle of the safe use of nuclear power sources shall be the minimum amount of radioactive matter in outer space.

Normal operation of space objects with nuclear power sources aboard shall meet the population safety require-ment of the international Commission for Radiological Protection. Specially mentioned were accidents involving space objects having nuclear power sources as elements of their construction.

The approved Principles formulated requirements to spacecraft carrying nuclear materials, including:

to safety systems of devices with nuclear power sources;

to nuclear reactors and radioisotope generators used aboard spacecraft in the capacity of nuclear power sources;

to the nuclear power source operation safety.

The Principles pay special attention to the issue of noti-fication about spacecraft return to earth.

The analysis of treaties, agreement and principles of activities associated with the problem of space ecology shows that these documents create a solid basis for future international procedures applicable to environmental safety in outer space.

Thus space technology and ecology are closely interrelated.

On the one hand, sizable contribution is already made to the solution of many environmental problems and it will be even greater in the future thanks to unlimited poten-tialities of space technologies. First and foremost, environmental monitoring helps discover anthropogenic impact on the environment and take prompt measures to reduce it. Space technologies are useful in early warning about natural calamities such as earthquakes, tidal waves and typhoons, and their prevention in future. Radioactive and toxic waste storage in outer space might also contribute to the solution of environmental problems.

On the other hand, space activities have an increasingly negative effect on the environment causing pollution of land, oceans, lower atmosphere and NES. The main source of the pollution are launchers that are still far from being perfect. This is the target for concentrated efforts of de-signers and producers. Steps taken today give hope that the problem will be solved. Serious measures have been taken to reduce propellant spillage. There is a trend towards the rejection of toxic components. Impact areas will be considerably reduced and cease to exist some day. The 21st century will open the era of environment-friendly reusable launchers.

But space junk makes the problem that will only enhance with the development of space technologies. While today its growth impedes mainly spacecraft, time may come when it affects environment making it harmful to the very existence of humanity. This is another sphere of fight for lesser pollution of the environment by space activities.

However laws and normative acts regulating space ecology matters are normally strictly declarative and fail to encompass the entire scope of space activities. On this legal basis international community under the U.N. auspices should develop necessary documents boosting the environmental security of space activities.

## Conclusion

The simultaneous change of centuries and millenniums is a unique temporal occurrence characterized by events of historic significance in the history of mankind. Such is the magic of numbers. This is the time for summing up the results in activities that determine the progress of mankind and for assessing the prospects of future developments. This is true of astronautics as of any other science.

The rapid development of astronautics in the second half of the 20th century gave a great impetus to the progress of science and technology. True, it had been provoked primarily by the needs of "the cold war".

The periods of global confrontations in the history of mankind had always brought forth revolutionary achievements in science and technology. The First World War boosted an unprecedented development of aviation, motor building, materials study, radio engineering, chemistry, biology and other sciences. A mere thirty years later the anti-Hitler allied powers achieved still more impressive results in these areas and scored a convincing victory in the Second World War. The cold war that came in its wake stimulated such branches of science and technology that were only being speculated upon in researchers' works while their translation into practical life was projected for the not-too-near future. But the dire necessity stemming from the confrontation of two political systems gave birth to nuclear power engineering, rocket engineering, cybernetics and electronics. Astronautics holds here by right one of the leading positions. It has revolutionized such branches of science and technology as communications, television, navigation, meteorology, geodesy, cartography, Earth remote sounding, astronomy, materials study and many others.

To ensure space activity, unique space rocket complexes and manufacturing infrastructure had been built whose operation at the turn of the millennium is analyzed in this book.

In all fairness it must be said that astronautics today is having difficult times awaiting the next outbreak in its development based on radically new technology of the 21st century. The prospects of the outbreak will be primarily determined by military factors. This is, above all, the possibility of using the space component for waging both local and major wars, including world conflicts. The key role in the multi-polar world of today belongs to informational neutralization of the enemy. This action implies the domineering position of the space segment of information systems.

In the longer term, space hardware can be used as an informational, naviga-tional, meteorological and cartographic means of supporting a conventional or nuclear war. In addition, the threat of using strike space systems for attacks in or from space is becoming very real indeed. As early as the start of the 21st century it will become possible to carry out large scale non-lethal strikes against individual regions or entire states.

The terrorist attacks against the American people on September 11, 2001, showed to mankind that the world has entered a new epoch when economic might and military superiority do not fully ensure national security. Terrorism, which until some point in the past manifested itself as sporadic acts of violence, has grown to really global proportions and become an international hostile force with which the progressive humanity cannot coexist. Thus a necessity arose to seek new methods to oppose that phenomenon and eliminate it by all means available, including military. That's where space can play a crucial role. Space information support systems with their global coverage capability allow to collect data regarding the situation practically in any place on the globe that might be controlled by terrorists or where their training camps and bases might be situated. Space hardware is indispensable in anti-terrorist operations, especially major ones like Noble Eagle in Afghanistan in the aftermath of the September 11 attack.

Another significant event was the statement made by President G. Bush concerning the US unilateral withdrawal from the 1972 antiballistic missile (ABM) defense agreement which confirmed the declared by certain western experts determination to militarize space. The withdrawal from the 1972 agreement clears all barriers before the US for building weapon systems (anti-missile defense systems) for which space will play an important – and in the longer term critical – role. Space hardware that ensures functioning and combat use of the new ABM system will call for protection against possible attack from those forces whose protection the ABM system will be supposed to provide. This will entail the creation of "space vs. space" strike assets. As for creation of systems capable of striking from space it will be just a matter of manufacturing technology.

Apart from purely military tasks, it will become necessary to address ecological issues with the help of space hardware. Those include earthquakes and volcano eruptions forecast; prognostication of hurricanes, anticyclones and typhoons and development of techniques to fight them; ozone layer restoration; clearing space of "space litter" etc.

The energy crisis on the planet will urge the search for alternative sources of energy, including the search aided by space systems.

Throughout its history the Earth had always been and is still being jeopardized by asteroid threat. There was not a century without a space catastrophe. Therefore the use of space hardware for detection and destruction of asteroids and comets, calculation of probability of their fall onto Earth or flying past, are the missions whose successful fulfillment increases the mankind's chance to survive.

Demographic factors, too, call for development of astronautics. If not in this century, then in the next, the removal of mankind to other planets may become quite an important practical job. The problems of the fundamental sciences, astronomy and cosmogony as well as the study of yet unknown physical phenomena in the Universe will also require to increase the number of space flights.

In other words astronautics faces still lots of problems whose solution will be the concern of our descendants in the 21st century.

# Bibliography

1. **Kiselev A.I., Menshikov V.A., Medvedev A.A.** Prospects of Astronautics Development in the 21st Century // Kosmos na Strazhe Rodiny (Space in defense of the nation). Military sciences conference in memory of M.K.Tikhonravov,: NITs KOSMO (Scientific Research Center) 1999, V.1.

2. **Kiselev A.I., Medvedev A.A., Kuzin A.I.** Space Launch Vehicles Developed by Khrunichev Research and Production Space Center for Implementation of Federal Space Programs // Dvoinyye Tekhnologii (Dual Use Technologies). 1999, No 4.

3. **Kiselev A.I., Medvedev A.A.** Rockot Steady on Course to the World Market // Aerospace courier. 1999. No 6.

4. **Kiselev A.I., Medvedev A.A.** Angara as a Highly Competitive Family of Advanced Carrier Rockets // Aviakosmicheskii Kuryer (Aerospace courier). 1999. No 4.

5. **Kiselev A.I., Medvedev A.A., Karrask V.K., Motornyi Ye.I., Yur'yev V.Yu. et al.** Advanced Light Class Carriers of Khrunichev Center // Proceedings of the first international show conference Small Satellites. New Technologies, Achievements, Problems and Prospects of International Cooperation in New Millennium. Korolev: 1998.

6. **Kiselev A.I., Medvedev A.A.** The Effective Way to Create Competitive Carrier Rockets of the Angara Family: Report // Le Bourget-9 air show conducted at Paris. Paris: 1999.

7. **Kiselev A.I.** Engineering and Design Support in Creation and Protracted Operation of Space Rocketry // Dissertation in the form of thesis for a degree of a doctor of engineering. M.: 1999.

8. **Kiselev A.I., Medvedev A.A.** Angara as a Competitive Family of Carrier Rockets // International Aerospace Journal Aviapanorama. 1999.

9. **Kiselev A.I., Medvedev A.A., Nagavkin V.F.** The Modular Principle in Creation of Carrier Rockets as the Strategy of Ground-Based Processing in New Economic Conditions // Vesti MAI (New of the Moscow Aviation Institute). 1999. V.1. No 1.

10. **Medvedev A.A., Karrask V.K., Lyagin V.G., Nesterenko A.A., Petukhov V.Yu., Yur'yev V.Yu, et al.** Preliminary Investigation of the Possibility to Use the SS-19 Convertible Missile for Space Research Missions. // Second Symposium of IAA on small AES for Remote Sounding. Berlin: 1999.

11. **Medvedev A.A.** Analysis of Advanced Trends in Development of Transport Launch Vehicles for Ejections into Near-Earth Space // Proceedings of 15th Science and Technology Conference Design and Structure. M.: 1982.

12. **Medvedev A.A., Menshikov V.A., Silant'yev A.Yu.** Stochastic Differential Modeling of Complex Engineering Systems. M.: Nauka, 1999.

13. **Medvedev A.A.** The Techniques of Evaluating Design Decisions for Carrier Rockets Based on Probability Models // Dissertation in the form of thesis for a degree of a candidate of engineering. M.: 1985.

14. **Medvedev A.A.** Reusable Boosters of the Angara Carrier as an Indispensable Phase on the Way to Fully Reusable Launch Vehicles: Report summary // International conference of UNISPACE-III. Vienna: 1999.

15. **Ivanov V.L., Menshikov V.A., Pchelintsev L.A., Lebedev V.V.** Space Litter. – M.: Patriot, 1996.

16. **Menshikov V.A., Kalinin A.A., Mirosh Yu.M., Medushevsky L.S.** Quality Quadrants. Reduction of Systems' Defectiveness // Elektronika. 2001, No 4.

17. **Menshikov V.A., Kalinin A.A., Mirosh Yu.M., Medushevsky L.S.** Quality Control Methods in Low-Rate Production of Complex Items // Polet. 2002, No 4.

18. **Menshikov V.A.** Firing Range Trials. M.: KOSMO, 1997, Book 1.

19. **Menshikov V.A., Bogdanov Yu.V.** Optimization of Space Rocket Complexes' Operation System. M.: KOSMO, 1997.

20. **Menshikov V.A.** Stochastic Optimization of Space Rocket Complexes' Testing Procedure // Dissertation in the form of thesis for a degree of a doctor of engineering. M.: 1993.

21. **Menshikov V.A.** Firing Range Trials. M.: KOSMO, 1999. Book 2.

22. **Menshikov V.A., Klimenko Yu.L., Lysyi S.R., Medushevsky L.S.** The Main Results of Operating the Sputnik Family of Rockets in 40 Years // 34th scientific conference on the development of K.E.Tsiolkonsky's scientific heritage. Kaluga, 14-16 September 1999. Report summaries. M., IIYeT RAN (Institute of Natural History and Engineering at Russia's Academy of Sciences), 1999.

23. **Menshikov V.A.** Orbital Constellation of Space Assets in Restructuring of Russia's Armed Forces // Voyennaya Mysl (Military Thought). 1999. No 6.

24. **Menshikov V.A., Kuzin A.I., Lebedev V.V.** Background and Target Situation in Littered Space and Improvement of Objects Observation // Kosmos na Strazhe Rodiny (Space for Defense of the Nation). Second scientific conference on military astronautics in memory of M.K.Tikhonravov. M.: NITs KOSMO, 1999. Vol.2.

25. **Menshikov V.A., Medushevsky L.S.** The Russian Carrier Rockets Outperform Others in Reliability // Military Parade. 2000 No 3 (39).

26. **AIAA-90-5055.** Proceedings of the 2nd international symposium on space information systems, September 1990.

27. **AIAA-90-5056.** Proceedings of the 2nd international symposium on space information systems, September 1990.

28. **Bobkov B.D., Sevost'yanov V.V., Filimonov V.M., Yakovlev I.P.** Firing Ranges and Launch Complexes of the USA. B.m. GONTI-1, 1970.

29. **Vlasov M.N., Krichevsky S.V.** Ecological Dangers of Space Activity. – M.: Nauka, 1999.

30. **Vlasov M.N., Pokhunkov A.A.** Impact of Emissions of Some Chemically Active Gases on the Upper Atmosphere // Geomagnetism and Aeronomy, 1986, v.26, No 5.

31. **Vlasov M.N., Pokhunkov A.A.** Impact of Emissions of Nitrogen Dioxide on the Thermo-atmosphere // Geomagnetism and Aeronomy, 1987, v.27, No 2.

32. **Galushkin A.I.** Neurocomputers in the Developments of US Military Technology (review of declassified information). M.: International Informatization Academy, 1995.

33. **GOST 25866-83** (State Standards) Operation of Equipment. Terminology and Definitions.

34. **GOST 25883-83.** (State Standards) Operation and Repair of Defense Hardware. Terminology and Definitions.

35. **DARPA** Director's Report. Washington: 1996

36. **Foreign** Military Review. 1996. 1996. No 1.

37. **Foreign** Military Review. 1997. No 12.

38. **Foreign** carrier rockets of medium and heavy classes and power plants for them (review). M.: NITs (scientific research center) Kompas, 1998.

39. **European Space Operations Center** in 2023 // Rocketry and Space Hardware. News bulletin, ser.1. GONTI TsNIIMash, 1994, No 31 (1811).

40. **Analysis of Experience of Restructuring** Major Aerospace Companies of the US. M.: NITs Kompas. 1998.

41. **Klimenko Yu.L., Lysyi S.R., Medushevsky L.S., Fastovets I.I.** The Cost and Reliability of Advanced Types of Carriers as the Determining Factors in Effectiveness and Competitiveness; Reports summary // Public scientific conference in memory of Yu.A.Gagarin. Gagarin: 2000.

42. **Klimenko Yu.L., Lysyi S.R., Medushevsky L.S.** Provision, Control and Support of the Required Reliability and Safety on the International Space Station Alpha during its Creation and Operation // Dvoinyye Tekhnologii (Dual Use Technologies). 1999. No 1.

43. **Klimenko Yu.L., Lysyi S.R., Medushevsky L.S.** Optimization of Strategy of Construction and Replenishment of Space Systems Based on Small SC by Criteria of Reliability and Cost // Dvoinyye Tekhnologii (Dual Use Technologies). 1999. No 1.

44. **Computers of the future** // Izvestiya (News). 12.02.2000.

45. **Foreign** Space Hardware. Awareness news. M.: NITs Kompas, 1966. No 11, p.12,13.

46. **Cosmodrome.** Edited by A.P.Volsky. M.: 1997.

47. **Cosmonautics.** Encyclopedia – M.: Sovetskaya Entsiklopediya. 1985.

48. **International Space Rocketry Engineering Standards** // (Astronautics and Rocket Production). 1996. No 7.

49. **Medushevsky L.S.** The Phenomenon of the Rocket Tribe // Russkiy Vestnik (Russian Courier). 1999. No14, 15.

50. **The Methods** of assessing Proton-M Space Complex's Impact on the Environment. – M.: Khrunichev Center, 2000.

51. **Mikhailov V.P.** Pollutions by Rockets and Space Litter: the Root of the Problem. – M.: Publication of the Institute of Natural History and Technology at the Russian Academy of Sciences, 1999.

52. **Mikhailov V.P.** Pollution of Earth by Rockets and Space: the Incipient Trends. – M.: Publication of the Institute of Natural History and Technology at the Russian Academy of Sciences, 1999.

53. **Mukhamedzhanov M.Zh., Chekalin S.V.** The Prospects of Isolation of Extra-Dangerous Waste of Nuclear Power Engineering // Kosmos I Ekologiya (Space and Ecology). – M.: Znaniye, 1991.

54. **Pavutnitsky Yu.V., Mazarchenkov V.A., Shilenkov M.V., Geresimov A.B.** Domestic launch vehicles. – S-Peterburg, 1996.

55. Novosti Kosmonavtiki (Astronautics News). 2001. No 1.

56. Novosti Kosmonavtiki (Astronautics News). 2001. No 2.

57. Novosti Kosmonavtiki (Astronautics News). 2001. No 6.

58. Novosti Kosmonavtiki (Astronautics News). 2001. No 8.

59. Novosti Kosmonavtiki (Astronautics News). 1998. No 21/22.
    P. 40 - 41, 44 – 45.

60. Novosti Kosmonavtiki (Astronautics News). 1998. No 11. p. 13.

61. Novosti Kosmonavtiki (Astronautics News). 1999. No 5. p. 44, 45.

62. Novosti Kosmonavtiki (Astronautics News). 1998. No 3. p. 41.

63. Novosti Kosmonavtiki (Astronautics News). 1998. No 6. p. 29 – 32.

64. Novosti Kosmonavtiki (Astronautics News). 1998. No 7.

65. Novosti Kosmonavtiki (Astronautics News). 1998. No 14.

66. New Approach to European Standards in Aerospace Technology // Kosmonavtika I Raketostroyeniye (Astronautics and and Rocket Production). 1996. No 7.

67. Forecast of the World AES Market for 1998 – 2007. M.: NTTs Kompas, 1998.

68. Ecological Security Program for Proton-M Space Rocket Complexes. – M.: Khrunichev Center, 2000.

69. Ecological Security Program for Rockot Space Rocket Complexes. – M.: Khrunichev Center, 1999.

70. Raketnaya I Kosmicheskaya Tekhnika (Rocketry and Space Technology): Express information. 1999. No 9, P.5 – 8.

71. Raketnaya I Kosmicheskaya Tekhnika (Rocketry and Space Technology): Express information. 1996. No 3, P.5 – 7.

72. Raketnaya I Kosmicheskaya Tekhnika (Rocketry and Space Technology): Express information. 1997. No 5, P.9, 10.

73. Raketnaya I Kosmicheskaya Tekhnika (Rocketry and Space Technology): Express information. 1999. No 15, P.11.

74. Raketnaya I Kosmicheskaya Tekhnika (Rocketry and Space Technology): Express information. GONTI, 1997, No 43.

75. Light Class Carrier Rockets and Propulsion Units for them Produced by European countries, China, Japan, India and other countries (Review). NTTs Kompas, 1998.

76. Ground-Based Monitoring and Control of Several AES's Flights Including Elements of Artificial Intellect // Astronavtika I Raketodinamika (Astronautics and Rocket Dynamics). Express information. 1996. No 39. P.3 – 21.

77. Stromsky I.V. Spaceports of the World. – M.: Mashinostroyeniye, 1996.

78. Slepov N. Soliton Networks // Seti (Networks). 1999. No 3. P. 90 – 100.

79. USA: Economics, Politics, Ideology. 1996. No 3. P. 4, 5, 9.

80. Technologies of the Future // Computer Today. 1999. No 9. P. 34 – 64.

81. **Umansky S.P.** Carrier Rockets. Cosmodromes. – M.: Restart + Publishing House, 2001.

82. Ecological Problems and Risks of Rocketry Impact on the Environment. Reference book. – M.: Ankil Publishing House, 2000.

83. **Aviation** Week and Technology. 20.02.1989. P. 34

84. **Aviation** Week and Space Technology. 26.10.1998. V. 149, N 17. P. 71.

85. **Aviation** Week & Space Technology. 16.09.1996. V. 145. N 12. P. 51.

86. **Air Force** Magazine. 1990. N 5.

87. **Air Force** Magazine. 1991. N 5.

88. **Computer Weekly.** 1998. N 8. P. 1, 18 – 20.

89. **DARPA** Neural Network Study / AFCEA Press. 1998.

90. **Flight International.** 1996, 23 – 29/X. V. 150. N 4546, P. 23.

91. **Flight International.** 1997, 23 – 29/VII. V. 153. N 4584, P. 26.

92. **Flight International.** 1998, 9 – 15/IX. V. 154. N 4642, P. 84.

93. **Menshikov V.A., Aleshenkov M.S., Radionov B.N., Griniaev S.N.** Prospects of Using Space Assets for Development of Transportation in the 21st Century. International Conference on the Use of Photonics for Transportation, PhT 99 Prague, March, 1999.

94. **Menshikov V., Kuzin A., et al.** Advanced Development and Use of Space Nuclear Power Systems as Part of Power Supply Modules for General Purpose Spacecraft. International Nuclear Power Conference, Albuguergue NM, USA, 7 – 11 January, 1996.

95. **Military Space.** 15.02.1986.

96. **Military Space.** 25.07.1994. V. 11. N 15. P. 1 – 3.

97. **Space News.** 1996. 7 – 13/X. V. 7. N 39. P. 1, 19.

98. **Space News.** 1998. 14 – 20/XII. V. 9. N 48. P. 8.

99. **Space News.** 1999. 8/III. V. 10. N 9. P. 13.

# Abbreviations

| | | |
|---|---|---|
| ABM | – | antiballistic missile |
| ACLV | – | Ariane Complementary Launch Vehicle |
| ACS | – | automated control system |
| ACV | – | assembly and command vessel |
| ADS | – | automated design system |
| AES | – | artificial Earth satellite |
| AFSCF | – | Air Force satellite control facility |
| AFSCN | – | Air Force Satellite Control Network |
| AI | – | artificial intellect |
| AIRP | – | autonomous information reception post |
| ALF | – | analysis of likely failures |
| ALV | – | air–launched vehicle |
| AMD | – | anti–missile defense |
| ANS | – | autonomous navigation system |
| APAA | – | active phased antenna array |
| APSC | – | Asia Pacific Space Center |
| ARC | – | Atlantic Research Corporation |
| AS | – | automation system |
| ASFV | – | aerospace flying vehicle |
| ASI | – | Italian Space Agency (Italy) |
| ASL | – | active service life |
| ATV | – | advanced technology vehicle |
| AVATAR | – | Aerobic Vehicle for Advanced TransAtmospheric Research |
| AWS | – | automatic work station |
| BB | – | basic block |
| BC | – | ballistics center |
| CBC | – | common booster core |
| CC | – | control center |
| CCB | – | common central block |
| CCC | – | carbon-carbon composite |
| CCS | – | California Commercial Spaceport |
| CCSDS | – | Consultative Committee on Space Data Systems |
| CDR | – | control, diagnostics and recovery |
| CDTE | – | communication and data transmission equipment |
| CEO | – | chief executive officer |
| CIC | – | command and instrumentation complex |
| CIP | – | command and instrumentation point |
| CIP | – | command and instrumentation post |
| CIS | – | command and instrumentation station |
| CIS | – | Commonwealth of Independent States |
| CPU | – | combined propulsion unit |
| CRMP | – | consolidated risk management program |
| CSI | – | command software information |
| CSTS | – | Commercial Space Transportation Study |
| CSTS | – | Commercial Space Transportation System |
| DARPA | – | Defense Advanced Research Projects Agency |
| DG | – | directive gain |
| DoD | – | Department of Defense |
| DPAF | – | dual payload attach fitting |
| DRS | – | data relay system |
| DSC | – | deep [distant] spacecraft |
| DSC | – | durable spacecraft |
| DSCS | – | Defense Satellite Communications System |
| DSN | – | deep space network |

| | | |
|---|---|---|
| DSS | – | deep space station |
| EC | – | engineering complex |
| ECS | – | energy conversion system |
| EELV | – | evolved expendable launch vehicle |
| EL | – | east longitude |
| ELV | – | expendable launch vehicle |
| EMS | – | European Missile Systems |
| EO | – | executive organ |
| EPSS | – | electric power supply system |
| ERE | – | electromagnetic rocket engine |
| ERE | – | electric rocket engine |
| ERPP | – | electric rocket power plant |
| ERPU | – | electric rocket propulsion unit |
| ERS | – | Earth remote sounding |
| ES | – | expert system |
| ESA | – | European Space Agency |
| ESI | – | European Satellite Industries |
| ESOC | – | European Space Operations Center |
| ESRIN | – | European Space Research Institute |
| ESTEC | – | European Space Research and Technology Center |
| ETS | – | Eastern Test Site |
| ETSC | – | energy transportation spacecraft |
| FCB | – | functional cargo block |
| FCC | – | flight control center |
| FCU | – | functional cargo unit |
| FMC | – | flight monitoring center |
| FMS | – | Federal Monitoring System |
| FSP | – | Federal Space Program |
| FV | – | flying vehicle |
| FY | – | fiscal year |
| GBACC | – | ground-based automated control complex |
| GBCC | – | ground-based control complex |
| GBCS | – | ground-based control systems |
| GBS | – | ground-based station |
| GBTLC | – | ground-based test and launch complex |
| GDP | – | gross domestic product |
| GEO | – | geostationary orbit |
| GES | – | Global Environmental Service |
| GLONASS | – | global navigational satellite system |
| GN | – | ground network |
| GNSS | – | Global Navigation Satellite System |
| GPS | – | global positioning satellite [system] |
| GRO | – | gamma-ray observatory |
| GSLV | – | geosynchronous space launch vehicle |
| GSO | – | geostationary orbit |
| GSSW | – | general system software |
| GSW | – | general software |
| GTO | – | geostationary transfer orbit |
| HEO | – | high-elliptical orbit |
| HIF | – | horizontal integration facility |
| HLV | – | heavy lift vehicle |
| HSWC | – | hard- and software complex |
| IA | – | individual access |

| | | |
|---|---|---|
| IAEB | – | Institute of Aeronautics and Space |
| ICBM | – | intercontinental ballistic missile |
| ICC | – | instrumentation and command complex |
| ICIC | – | independent command and instrumentation complex |
| ICS | – | instrumentation and control station |
| IESW | – | information exchange software |
| IMCPC | – | information measurement, collection and processing complex |
| IMEWS | – | integrated missile early warning satellite |
| IMI | – | Israel Military Industries |
| INTA | – | Spanish National Institute of Aerospace Technologies |
| IP | – | instrumentation post |
| IPC | – | Information Processing Center |
| IR | – | infrared |
| IRIG | – | Interange Instrumentation Group |
| ISAS | – | Institute of Space and Aeronautical Science |
| ISC | – | information of summarized control |
| ISO | – | International Standardization Organization |
| ISP | – | international space program |
| ISRO | – | Indian Space Research Organization |
| ISS | – | international space station |
| ITS | – | information and telemetric support |
| ITU | – | International Telecommunications Union |
| JPL | – | Jet Propulsion Laboratory |
| LACE | – | liquid air cycle engine |
| LARE | – | liquid-air rocket engine |
| LC | – | launch complex |
| LCN | – | local computing network |
| LDSCC | – | long distance space communications center |
| LLOS | – | long life orbital station |
| LMA | – | Lockheed Martin Astronautics |
| LMLV | – | Lockheed Martin Launch Vehicle |
| LNG | – | liquefied natural gas |
| LPE | – | liquid propellant engine |
| LPM | – | liquid propellant motor |
| LPR | – | liquid propellant rocket |
| LPRE | – | liquid propellant rocket engine |
| LPRM | – | liquid propellant rocket motor |
| LTR | – | low thrust rocket |
| MCNP | – | measurement of current navigation parameters |
| MHD | – | magnetohydrodynamic |
| MLV | – | medium lift vehicle |
| MMS | – | Matra Marconi Space |
| MoD | – | Ministry of Defense |
| MPAS | – | multi-purpose aerospace system |
| MPCE | – | magnetic plasma chemical engine |
| MPD | – | magnetoplasmadynamic |
| MPL4C | – | multi-purpose launch complex |
| MPRM | – | multi-purpose rocket module |
| MPSRS | – | multi-purpose space relay system |
| MSA | – | multi-station access |
| MSLS | – | multi-service launch system |
| MSOCC | – | Multi-satellite Operations Control Center |
| NAL | – | National Aeronautical Laboratory |

| | | |
|---|---|---|
| NASA | – | National Aeronautics and Space Administration |
| NASDA | – | National Aerospace Development Agency (Japan) |
| NASP | – | national aerospace plane |
| NBS | – | navigational and ballistic support |
| NEO | – | near Earth orbit |
| NES | – | near Earth space |
| NEU | – | navigation equipment of users |
| NL | – | north latitude |
| NOAA | – | National Oceanic and Atmospheric Agency |
| NOSS | – | National Orbital Space Station |
| NPI | – | nuclear power installation |
| NPP | – | nuclear power plant |
| NPRE | – | nuclear power rocket engine |
| NPS | – | nuclear power station |
| NRE | – | nuclear rocket engine |
| NT | – | nitrogen tetraoxide |
| OBCS | – | onboard control system |
| OBDC | – | onboard digital computer |
| OBDCS | – | onboard digital computing system |
| OBTS | – | onboard time scale |
| OC | – | orbital constellation |
| OHBU | – | oxygen/hydrogen booster unit |
| OP | – | operating program |
| OS | – | orbital stage |
| OS | – | orbital space ship |
| OS | – | operating system |
| OSC | – | Orbital Sciences Corporation |
| PAA | – | phased antenna array |
| PDE | – | pulse detonation engine |
| PEC | – | photoelectric converter |
| POW | – | prisoner of war |
| PP | – | power plant |
| PU | – | propulsion unit |
| RAM | – | random access memory |
| RAP | – | reliability assurance program |
| RCS | – | reaction control system |
| R&D | – | research and development |
| RLV | – | reusable launch vehicle |
| RNS | – | refilling and neutralization station |
| RPSS | – | remote power supply system |
| RR | – | refrigerator-radiator |
| RRB | – | reusable rocket booster |
| RS | – | relay satellite |
| RSP | – | reliability support program |
| RSRG | – | reusable space rocket glider |
| RSS | – | reusable space system |
| RSSC | – | reusable single–stage carrier |
| RSTS | – | reusable space transportation system |
| SAC | – | Space Activities Commission |
| SB | – | solar battery |
| SC | – | spacecraft |
| SCDE | – | systems of communication and data exchange |
| SCRAMJET | – | supersonic ramjet |

| | | |
|---|---|---|
| SCTS | – | State Central Test Site |
| SDI | – | strategic defense initiative |
| SDS | – | satellite data system |
| SE | – | sensor element |
| SFA | – | Spaceport Florida Authority |
| SGLS | – | Space – Ground Links System |
| SGTI | – | solar gas turbine installation |
| SH | – | space hardware |
| SH OS | – | space hardware operation system |
| SIC | – | spaceports' instrumentation complex |
| SIS | – | standards of information security |
| SL | – | south latitude |
| SM | – | service module |
| SMP | – | Soldier Modernization Plan |
| SNS | – | space navigation system |
| SPB | – | solid propellant booster |
| SPE | – | short–pulse electrodynamics |
| SPE | – | solid propellant engine |
| SPE | – | stationary plasma engine |
| SPOT | – | satellite positioning and tracking |
| SPRE | – | solid propellant rocket engine |
| SPTA | – | spare parts tools and accessories |
| SRB | – | solid rocket booster |
| SRC | – | space rocket complex |
| SRF | – | Strategic Rocket Forces |
| SRH | – | space rocketry hardware |
| SRS | – | space relay system |
| SS | – | space system |
| SSC | – | small spacecraft |
| SSE | – | super–wide band short–pulse electrodynamics |
| SSG | – | space support group |
| SST | – | single-stage launcher |
| SSRMS | – | Space Shuttle remote manipulator system |
| SSTCC | – | Space Systems Test and Control Center |
| SSTS | – | Space Shuttle Transportation System |
| SSW | – | special software |
| STDN | – | Spaceflight Tracking and Data Network |
| STS | – | space transportation system |
| STS | – | single time scale [system] |
| SW | – | software |
| TC | – | thermotaxic coating |
| TDRSS | – | tracking and data relay satellite system |
| tf | – | ton-force |
| TFT | – | tree-of-failures technique |
| TI | – | target information |
| TM | – | telemetric |
| TMEC | – | turbo-machine energy conversion |
| TMI | – | telemetric information |
| TP | – | temporary program |
| TP | – | timing program |
| TPS | – | thermal protection system |
| TSSW | – | technology supporting types of software |
| UCIS | – | unified command and instrumentation system |

| | | |
|---|---|---|
| UDMH | – | unsymmetrical dimethylhydrazine |
| UFO | – | ultra-high frequency follow–on |
| UHF | – | ultra-high frequency |
| UNE | – | users' navigation equipment |
| USC | – | United Space Command |
| USOC | – | united space operations center |
| USTC | – | united space tests center |
| VSAT | – | very small aperture antenna |
| WITC | – | Wallops Island Test Center |
| WTS | – | Western Test Site |

# Index

**Booster units**
- Breeze-KM 308, 311, 314, 316
- Breeze-M 303, 306–308, 319, 320
- DM 300–306
- Fregat 296, 305
- Korvet 322
- 12KRB 70, 308–310
- KVRB 318

**Ecology**
- catastrophes and emergencies of launchers 535–539
- radioactive waste 561

**Insurance of space engineering**
- insurance rates 205
- losses 207
- market capacity 205
- Megaruss 198, 199
- premiums 207
- types of space insurance 198

**Launchers**
- ACLV 81
- ALV 86
- Angara 110, 288, 308, 317–325, 350, 361, 370, 410, 415, 416, 418, 419
- Ariane-4 65–67, 79
- Ariane-5 50–52, 308
- ASLV 83, 84
- Athena-1 70, 71
- Athena-2 71
- Atlas-2 53–55, 319
- Atlas-3 55
- Atlas-5 63
- Avrora 111, 322
- Black Arrow 111
- Capricornio 79
- Conestoga 42, 71, 72
- Cosmos 284, 286, 287, 290, 291, 312, 313, 408, 418
- Cyclone 284, 286, 287, 291–293, 312, 316, 405, 406, 408, 409, 418
- CZ-1D 87
- CZ-2C 87, 106
- CZ-3 67, 68
- Delta-2 55
- Delta-3 56
- Delta-4 59, 60
- DLA 81, 82
- Dnepr 287, 310, 315, 316
- Eclipse Astrolinear 94, 95
- EELV 43, 44, 64, 65
- Energia 287, 298, 317, 350, 375, 405
- ESL 82
- GSLV 70, 308
- H-2 52, 53, 69, 86
- HD-1L 87
- ILV-1 42
- J-1 86, 87
- K-1 93
- Lance-Proteus 81

  – LLV 70
  – LM EELV 61–65
  – LMLV 70, 71
  – M-5 85
  – Minuteman 73, 74
  – Molniya 252, 286, 287, 295–297, 323, 325, 406, 408, 418
  – Mu-3 85
  – Next 82
  – Pegasus 74, 75
  – Proton 286, 288, 300–304, 306, 308, 349, 363, 370, 405–407, 418, 483, 495, 496
  – PSLV 84, 85
  – R-7 284, 293, 295, 296, 401–404, 418
  – Rockot 288, 308, 311–314, 405, 406, 408, 418, 419, 494
  – Saturn-5 47, 48, 334, 350
  – Scorpios 72
  – Scout 79
  – Shavit 82
  – Shtil 310, 316, 425
  – SLV 83
  – Soyuz 286, 293, 294, 296, 297, 305, 320, 333, 405, 406, 408, 418, 419, 479
  – SpaceRay 73
  – Sputnik 284, 519
  – Start–1 286, 287, 310, 311, 418, 419
  – Strela 287, 314, 418, 419
  – Taurus 75–78, 80
  – Titan-2 57, 58, 78
  – Titan-2SLV 78, 79
  – Titan-3 42
  – Titan-4 47–50
  – Vega 79, 80
  – VLS 83
  – Vostok 284, 335
  – Yamal 284
  – Zenit 298–301, 304, 308, 317, 320, 322, 405, 406, 415, 418, 419

**Operation of space engineering**
  – monitoring, diagnostics and recovery system 490
  – parameters, effectiveness of a system of operation 486, 488

**Propulsion systems**
  – AUS–51 82
  – Castor-4 42, 53, 58, 72, 79
  – Castor-120 77, 78, 80
  – EAP 81
  – electromagnetic rocket engines (ERE) 355, 365, 371, 374, 382, 383
  – KVD1 309
  – L-9 50
  – LE-5B 52
  – LE-7A 52
  – Linear Aerospike 91
  – liquid-air cycle engine (LACE) 379
  – liquid propellant engines (LPRE) 290, 292, 295–297, 299–302, 304–306, 309, 311, 322, 349, 354, 378
  – magnetic plasma dynamic (MPD) engine 371
  – NK-32 94
  – NK-33 94

– NK-43 94
– nuclear power station (NPS) 367, 370
– nuclear rocket engine (NRE) 354
– P–230 50
– pulse detonation engines (PDE) 375, 376
– RD-58M 304, 305
– RD-107 294
– RD-108 294
– RD-120 96, 299, 300
– RD-0146 308, 309
– RD-170 299, 319, 379
– RD-174 317
– RD-180 43, 55, 61, 319
– RD-191 319, 323, 350
– RD-0210 302, 349
– RD-0211 302
– RD-0212 302
– RD-0214 302
– RD-216 290
– RD-218 292
– RD-219 292
– RD-253 302, 349
– RL-10 56, 60, 63, 64, 309
– RS-56SA 94
– RS-68 43, 59
– S1-5400 297
– Star-48 42, 78, 94
– Star-63F 94
– three-component engine 351, 379
– Vicas 70
– Vulcan–HM60 50
– Zefiro 79–81

**Reliability of space engineering**
– principles of maintenance of reliability of launchers 172
– principles of maintenance of reliability of orbital stations 184
– reliability of an International Space Station 180
– reliability of launchers 170
– specific cost index 191

**Reusable transport space systems**
– Ajax 331, 332
– ATV 92, 93
– AVATAR 98
– Baikal 323, 324, 566, 567
– Burlak 327
– Energia-Buran 326, 327, 336, 350, 404, 406, 407
– Hermes 98
– HOPE 97
– Hotol 98, 330, 380
– MPAS 329
– Pathfinder 92, 95
– Roton 96, 97
– Space Shuttle 87–90, 336, 337, 339
– Tu-2000 331
– VentureStar 91, 92

– X–33 90–92
– X–34 90–92

**Rocket fuel**
– unsymmetrical dimethylhydrazine (UDMH) 520

**Rocket units**
– Common Booster Core 43
– universal rocket module 319

**Solar batteries** 362, 364

**Spacecraft**
– AFP-731 551
– Bars 389
– Brillant Pebbles 216
– Dialog 266
– EarlyBird 38
– Ekran 389
– Electro 391
– Envisat 52
– Express 389
– Gals 389
– GLONASS 389, 464
– Gonets 390
– Ikonos 38
– Kometa 389
– Kosmos-1900 551
– Landsat 38, 39
– Luch 388
– Meteor 390
– METEOR-1 71
– MightySat-2 74
– Monitor 252
– Nadezhda-M 389
– OrbView 38
– Oreol 257
– Orion-3 57
– Phobos 258
– Resurs 254, 388, 390
– SPOT 38, 387
– Yakhta 266
– Yamal 390

**Spacecraft control systems**
– AFSCN 114, 116–122
– ASGLS 127
– DSN 115, 124, 127–129
– ESOC 130–132, 144
– ESRIN 130, 131
– ESTEC 130, 131
– ESTRACK 132
– Goddard Space Flights Center 125, 126
– Johnson Manned Space Flights Center 125, 126
– Marshall Space Flight Center 126
– MCC 125
– MSOCC 125
– Remote Agent 143
– STDN 115, 123–127

– TDRSS 117, 123–127, 438
– USBS 126, 127

**Spaceports**
– Al Akbar 99, 112
– Alcantara 83, 99, 112
– Baikonur 290, 291, 293, 303, 313, 316, 338, 342, 405–407
– Canaveral 49, 57, 61, 64, 78, 529
– Christmas Island 99, 110, 111
– Eastern Test Site 100, 101
– Florida 104
– Hummock Hill 110, 111
– Kapustin Yar 290, 399, 402, 412, 413
– Kennedy Space Center 99, 100, 101, 104
– Kodiak Island 104
– Kourou 81, 99, 105, 106, 527
– Palmahim 82, 99, 112
– Plesetsk 266, 289–293, 311, 314, 319–321, 324, 325, 404, 407, 408, 494
– San Marco 99, 112
– Shriharikota 70, 84, 99, 108
– Shuang Cheng-tze 106
– Xi Chang 106, 107
– Svobodny 314, 404, 410, 411, 417
– Tai Yuan 99, 106
– Tanegashima 53, 99, 107, 108
– Uchinoura 99, 107, 108
– Vandenberg 49, 54, 55, 78, 102, 103
– Wallops Island Flight Center 99, 100, 103
– Western Test Site 99, 100, 102
– Woomera 99, 109, 111
**Space stations**
– International Space Station 340
– Mir 336–340
– Salyute 335, 337
– Skylab 550
**Space systems**
– ARGOS 278, 279
– FSM 280
– GLOBALSTAR 278
– ORBCOMM 279
– ROSTELESAT 279
– TDRSS 467, 468

# About the authors

**Anatoly Ivanovich Kiselev** –
General Director of Khrunichev State Research and Production Space Center (till 6 February 2001), Hero of Socialist Labor, doctor of engineering sciences, professor, USSR Lenin prize winner, laureate of the Russian Federation Government's prize, author of more than 200 research works. The fields of scientific and practical interests are the prospects of space technology, utilization of space expertise in various areas of human activity, the theory of competition in the market of space services. Worked in aerospace industry, was deputy head of the Directorate of USSR Ministry of General Engineering, director of Khrunichev Machine Building Plant (from 1975 to 1993), general director of Khrunichev State Research and Production Space Center (from 1993 to 2001).

**Alexander Alexeevich Medvedev** –
general director of Khrunichev State Research and Production Space Center (since 6 February 2001), doctor of engineering sciences, merited machinist of Russia, laureate of the Russian Federation Government's prize, author of more than 200 research works, including 4 monographs. The fields of scientific and practical interests are optimization of space hardware configuration and manufacturing technologies, organization of market research of space technologies and services. Worked in aerospace industry, making the career from an engineer to the general director.

**Valery Alexandrovich Menshikov** –
assistant director of Khrunichev State Research and Production Space Center, research director of Space Systems Research Institute, doctor of engineering sciences, professor, merited research worker of Russia, laureate of the Russian Federation Government's prize. Author of more than 400 research works, including 23 monographs. The fields of scientific interests are space applied mechanics, development, trial and application of space technology, forecast of space equipment development. Worked as deputy director and chief engineer of Baikonur spaceport, as head of Tikhonravov Central Research Institute of the Russian Defense Ministry.

# Wherever you go take Heath Lambert with you...

Why? Because last year Heath Lambert Aerospace, together with the Megaruss Insurance Company, was involved in more than 70% of all C.I.S. launches. **Which means that** today, the Heath Lambert and Megaruss partnership is the foremost space insurance broking force in the C.I.S.

Our specialists arrange programmes that take away some of the uncertainty of a space launch.

They can negotiate coverage for all phases of both commercial and military launches - including unique coverages such as Third Party Liability within an impact zone.

Visit our website: www.heathlambert.com

  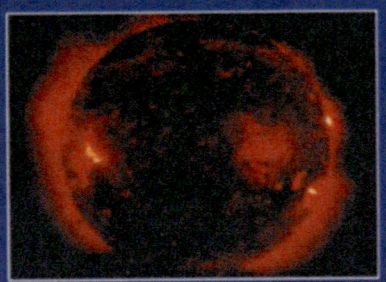

Quite simply, in what will always be a high risk business, we have the extensive experience and professional resources to keep your mission looking good.

To find out more contact:

Gary Bryant
Heath Lambert Group - Aerospace
133 Houndsditch
London
EC3A 7AH

t: + 44 20 7234 4511
f: + 44 20 7234 4177
e: gbryant@heathlambert.com

Kevin Eley
Heath Lambert Group - Aerospace
133 Houndsditch
London
EC3A 7AH

t: + 44 20 7234 4786
f: + 44 20 7234 4177
e: keley@heathlambert.com

Vladimir Shcherbakov
Heath Lambert Group
Moscow Representative Office
16 Marksistskaya Str.
109147 Moscow Russia

t: +7 (095) 234 0388 / 956 0434
f: +7 (095) 234 0387
e: vladsch@heath.df.ru

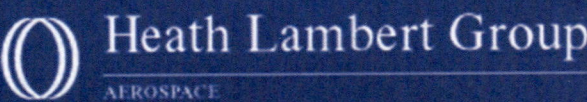

Heath Lambert Group
AEROSPACE

International
Moscow Bank

**International Moscow Bank**

SHAREHOLDERS:

| | |
|---|---|
| Bayerische Hypo-und Vereinsbank AG (Germany) | 43% |
| Nordea Bank Finland Plc | 22% |
| Eurobank (France) | 20% |
| European Bank for Reconstruction and Development | 10% |
| The Industrial Bank of Japan, Ltd (Japan) | 3% |
| Sberbank of Russia | 2% |

GENERAL LICENSE No.1 OF THE BANK OF RUSSIA 1989

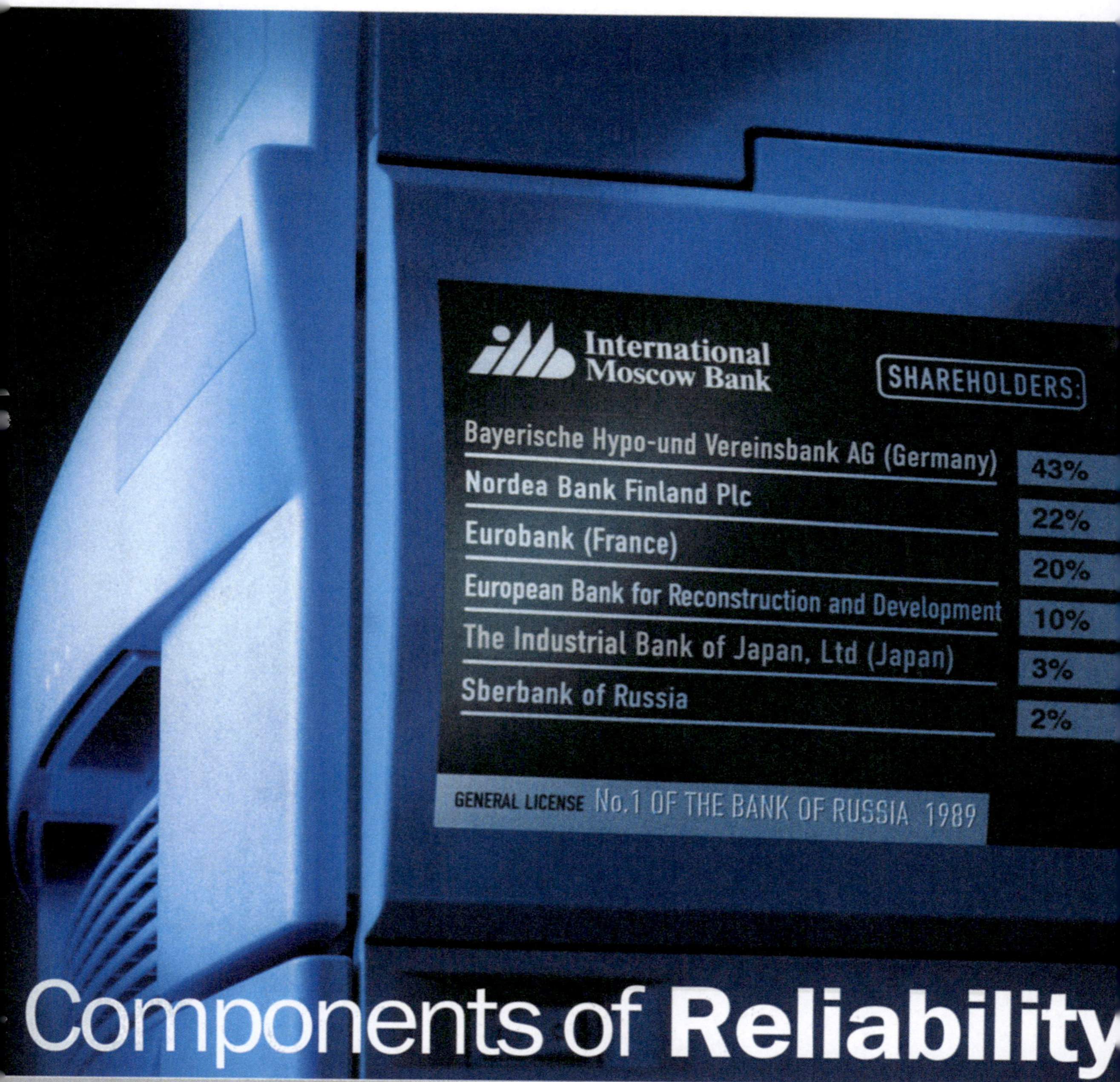

# Components of **Reliability**

International Moscow Bank's shareholders are leading international banks. Since its inception it has always been guided by sound international banking principles. Driven by a highly dedicated and professional staff, IMB is one of the most reliable and dynamically developing Russian banks. These are the components of reliability.

IMB Head office: 9 Prechistenskaya Emb., Moscow, 119034. Tel.: +7 095 258-7258
E-mail: imbank@imbank.ru

**Springer**Engineering

Martin Tajmar

# Advanced Space Propulsion Systems

2003. VIII, 130 pages. 121 figures, partly in colour.
Softcover **EUR 29,–**
(Recommended retail price)
Net-price subject to local VAT.
ISBN 3-211-83862-7

Space propulsion systems have a great influence on our ability to travel to other planets or how cheap a satellite can provide TV programs. This book provides an up-to-date overview of all kinds of propulsion systems ranging from classical rocket technology, nuclear propulsion to electric propulsion systems, and further to micro-, propellantless and even breakthrough propulsion, which is a new program under development at NASA. The author shows the limitations of the present concepts and how they could look like in the future. Starting from historical developments, the reader is taken to a journey showing the amazing technology that has been put on hold for decades to be rediscovered in the near future to questions like how we can even reach other stars within a human lifetime. The author is actively involved in advanced propulsion research and contributes with his own experience to many of the presented topics. The book is written for anyone who is interested in how space travel can be revolutionized.

Springer Wien New York

A-1201 Wien, Sachsenplatz 4–6, P.O. Box 89, Fax +43.1.330 24 26, e-mail: books@springer.at, Internet: **www.springer.at**
D-69126 Heidelberg, Haberstraße 7, Fax +49.6221.345-229, e-mail: orders@springer.de
USA, Secaucus, NJ 07096-2485, P.O. Box 2485, Fax +1.201.348-4505, e-mail: orders@springer-ny.com
Eastern Book Service, Japan, Tokyo 113, 3–13, Hongo 3-chome, Bunkyo-ku, Fax +81.3.38 18 08 64, e-mail: orders@svt-ebs.co.jp

**Springer**Geosciences

Bernhard Hofmann-Wellenhof,
Herbert Lichtenegger,
James Collins

# Global Positioning System

Theory and Practice

**Fifth, revised edition.**
2001. XXIII, 382 pages. 45 figures.
Softcover **EUR 51,–**
(Recommended retail price)
Net-price subject to local VAT.
ISBN 3-211-83534-2

This new edition accommodates the most recent advances in GPS technology. Updated or new information has been included although the overall structure essentially conforms to the former editions. The textbook explains in comprehensive manner the concepts of GPS as well as the latest applications in surveying and navigation. Description of project planning, observation, and data processing is provided for novice GPS users. Special emphasis is put on the modernization of GPS covering the new signal structure and improvements in the space and the control segment. Furthermore, the augmentation of GPS by satellite-based and ground-based systems leading to future Global Navigation Satellite Systems (GNSS) is discussed.

Springer Wien New York

A-1201 Wien, Sachsenplatz 4–6, P.O. Box 89, Fax +43.1.330 24 26, e-mail: books@springer.at, Internet: **www.springer.at**
D-69126 Heidelberg, Haberstraße 7, Fax +49.6221.345-229, e-mail: orders@springer.de
USA, Secaucus, NJ 07096-2485, P.O. Box 2485, Fax +1.201.348-4505, e-mail: orders@springer-ny.com
Eastern Book Service, Japan, Tokyo 113, 3–13, Hongo 3-chome, Bunkyo-ku, Fax +81.3.38 18 08 64, e-mail: orders@svt-ebs.co.jp

*Springer-Verlag*
*and the Environment*

WE AT SPRINGER-VERLAG FIRMLY BELIEVE THAT AN international science publisher has a special obligation to the environment, and our corporate policies consistently reflect this conviction.

WE ALSO EXPECT OUR BUSINESS PARTNERS – PRINTERS, paper mills, packaging manufacturers, etc. – to commit themselves to using environmentally friendly materials and production processes.

THE PAPER IN THIS BOOK IS MADE FROM NO-CHLORINE pulp and is acid free, in conformance with international standards for paper permanency.